John Phillips and the Business of Victorian Science

First published in 2005, this book represents the first full length biography of John Phillips, one of the most remarkable and important scientists of the Victorian period. Adopting a broad chronological approach, this book not only traces the development of Phillips' career but clarifies and highlights his role within Victorian culture, shedding light on many wider themes. It explores how Phillips' love of science was inseparable from his need to earn a living and develop a career which could sustain him. Hence questions of power, authority, reputation and patronage were central to Phillips' career and scientific work. Drawing on a wealth of primary sources and a rich body of recent writings on Victorian science, this biography brings together his personal story with the scientific theories and developments of the day, and fixes them firmly within the context of wider society.

John Phillips and the Business of Victorian Science

Jack Morrell

Routledge
Taylor & Francis Group

First published in 2005
by Ashgate

This edition first published in 2017 by Routledge
2 Park Square, Milton Park, Abingdon, Oxon, OX14 4RN
and by Routledge
711 Third Avenue, New York, NY 10017

Routledge is an imprint of the Taylor & Francis Group, an informa business

Publisher's Note
The publisher has gone to great lengths to ensure the quality of this reprint but points
out that some imperfections in the original copies may be apparent.

Disclaimer
The publisher has made every effort to trace copyright holders and welcomes
correspondence from those they have been unable to contact.

A Library of Congress record exists under LC control number: 2004006876

ISBN 13: 978-1-138-21478-1 (hbk)
ISBN 13: 978-1-315-44508-3 (ebk)
ISBN 13: 978-1-138-21483-5 (pbk)

John Phillips, a lithograph by T.H. Maguire of 1851, an Ipswich Museum portrait issued in celebration of the visit of the British Association. National Portrait Gallery.

John Phillips and the Business of Victorian Science

JACK MORRELL
University of Leeds, UK

ASHGATE

Published by
Ashgate Publishing Limited
Gower House
Croft Road
Aldershot
Hants GU11 3HR
England

Ashgate Publishing Company
Suite 420
101 Cherry Street
Burlington, VT 05401-4405
USA

Ashgate website: http://www.ashgate.com

British Library Cataloguing in Publication Data
Morrell, Jack, 1933-
    John Phillips and the business of Victorian science. -
    (Science, technology and culture, 1700-1945)
    1. Phillips, John, 1800-1874 2. Geologists - England -
    Biography
    I. Title
    551'.092

Library of Congress Cataloging-in-Publication Data
Morrell, Jack.
    John Phillips and the business of Victorian science / Jack Morell.
    p. cm. - (Science, technology, and culture, 1700-1945)
    Includes bibliographical references and index.
    ISBN 1-84014-239-1 (alk. paper)
    1. Phillips, John, 1800-1874. 2. Geologists - Great Britain - Biography.
    I. Title. II. Series.

QE22.P477M67 2005
551'.092-dc22   2004006876

ISBN 1 84014 239 1

Printed and bound in Great Britain by MPG Books, Bodmin

Contents

Preface

In writing this book I have profited from the support of many people and institutions. For general encouragement and particular help I am indebted to David Allen, Simon Bailey, Anne Barrett, Douglas Bassett, Bill Brock, John Brooke, Neil Brown, Janet Browne, Bill Bynum, Gordon Herries Davies, Brian Harrison, Tony Heywood, Roger Hutchins, Ian Inkster, the late Bobby Jenkins, David Knight, Frank James, David Levene, David Miller, Dorinda Outram, John Pickstone, the late Roy Porter, Munro Price, Anne Secord, Tom Sharpe, Tony Simcock, and the late John Thackray. At the Oxford University Museum of Natural History I have been welcomed by Jim Kennedy, Philip Powell, and above all by Stella Brecknell, its ever-helpful librarian who looks after the Phillips papers to which she has produced an excellent hand-list. I began sustained research on Phillips in 1985 when I enjoyed a visiting fellowship at Brasenose College, Oxford, and received the first of several grants from the Royal Society of London for this particular project. I have benefited from the help given by many librarians and archivists, and especially from the first-rate services provided by the university libraries of Leeds and Bradford. Since 1992 I have had the privilege of being an honorary visiting lecturer in history of science at the University of Leeds, where Sam Alberti, Geoffrey Cantor, John Christie, Steven French, Graeme Gooday, Jonathan Hodge, Chris Kenny, Richard Noakes, Suzanne Paylor, Gregory Radick, Jonathan Topham, and Adrian Wilson have given friendship, stimulus, and useful references. I owe much to three distinguished historians of geology, Martin Rudwick, Jim Secord, and Hugh Torrens for advice, encouragement, and information freely given over many years. With characteristic generosity Jim Secord has read meticulously the entire manuscript of this book. It has been greatly improved by his incisive comments.

For expert typing I thank Christine Lawlor. At Ashgate, Thomas Gray, Celia Hoare and their colleagues have aided me beyond the call of duty. For enabling me to finish the book I am grateful to the staff of the Heaton Medical Practice, Bradford. For forbearance I am beholden to my family, to whom this book is dedicated.

Acknowledgements

For permission to cite manuscripts in their care or ownership I am grateful to the American Philosophical Society (W. Hutton and Darwin papers, miscellaneous correspondence); the Bodleian Library, Oxford (BAAS archives, including General committee minutes, Council minutes, Foundation volume, York reception committee proceedings, documents pertaining to mines; H.W. Acland papers; Ashmolean Society papers); Bradford District archives (W.Danby papers); Bristol University Library (Eyles collection); British Geological Survey, Keyworth (Geological Survey archives); British Library (Peel papers); British Museum (Natural History) (Owen Papers; miscellaneous correspondence); Buxton Museum (Dawkins collection); Cambridge University Library (Greenough, Henslow, Kelvin, Sedgwick, and Stokes papers); Cambridge University Museum of Zoology (Strickland papers); Devon Record Office (Buckland papers); Edinburgh University Library (Lyell, Murchison Gen 523, and Geikie Gen 525 papers; miscellaneous correspondence, including letters to Phillips, Gen 784/1); Fitzwilliam Museum, Cambridge (miscellaneous correspondence); the Honourable Mrs C Gascoigne (Harcourt papers); Geological Society of London (Murchison papers); Humberside County archives (H Robinson papers); Imperial College, London, archives (Huxley, Playfair, and Ramsay papers); King's College, London, archives (KCL Council minutes, incoming correspondence, letter book 1834-43); Leeds Public Libraries archives, Sheepscar, now known as West Yorkshire archive service, Leeds (T. Wilson papers, Yorkshire Naturalists' Club archives); Leeds University Library (Leeds Philosophical and Literary Society archives, Yorkshire Geological Society archives); London University Library (Phillips autobiographical notebook, MS 517); Magdalen College, Oxford, Library (Daubeny papers; Phillips' foreign travel journals, 1829, 1830); Manchester Public Library (Royal Manchester Institution archives, including incoming letters M6/1/50-2, letter books M6/1/49/1-7, and syllabuses M6/1/70; Massachusetts Institute of Technology Library (W.B. Rogers papers); Mitchell Library, Glasgow (Phillips' journal of tour of Scotland, 1826); John Murray (Murray papers); Museum of History of Science, Oxford (Phillips' notebook c. 1819, Gunther MSS 64, 65); National Library of Ireland (Larcom and Monteagle papers); National Library of Wales, Aberystwyth (Ramsay papers, Dolaucothi MSS); National Museum of Wales, Department of Geology (De la Beche papers); Northamptonshire Record Office (Fitzwilliam papers); Oxford University archives (Hebdomadal Board minutes, Hebdomadal Council reports, Burdett-Coutts letters, Ashmolean keepership letters, report on Hope collection, papers concerning University Museum 1849-63, N.W.2.1, Oxford University Museum archives cited as OUA/UM, including minutes of Museum delegacy M/1.1-3, correspondence C/3/2, and capitals documents F/7/7); Oxford University Museum of Natural History, Library (Buckland, Phillips, and W. Smith papers, historical archive cited as OUM/HA); Public Record office (Sabine papers, BJ3); Pusey House, Oxford (Pusey papers); Royal Institution, London (Managers' minutes, RI archives, Grove papers, M. Reid's diary); Royal Society of London (Buckland, R.W. Fox,

Herschel, J.W. Lubbock, Sabine, and Terrestrial magnetism papers, Council minutes, Miscellaneous correspondence MC, Miscellaneous manuscripts MM, Referees' reports RR); St Andrews University Library (J.D. Forbes papers); St Bride's Printing Library, London (R. Taylor collection); Sheffield Public Library (Sorby papers; Wentworth Woodhouse muniments); Torquay Museum (Pengelly collection); Trinity College, Cambridge, Library (Whewell papers); Trinity College, Dublin, Library (TCD Board minutes, Hamilton papers); Turnbull Library, Wellington, New Zealand (Mantell papers); University College, London, Library (Brougham, Chadwick, F. Galton, and SDUK papers; UCL archives including College correspondence and applications for chairs); Wellcome Historical Medical Library, London (autograph letters); Yale University Library (J.B. Murray collection); York City archives (York Mechanics' Institute archives including Committee minute book, monthly meetings minute books, book listing lectures, classes and members 1827-51); Yorkshire Philosophical Society (Building committee minutes 1825-9, Council minutes, evening meetings minutes 1836-48, General meetings minutes, letter book 1822-8, miscellaneous correspondence, scientific communications 1822-38, and Phillips' day-book 1826-8).

Abbreviations, Conventions, and a Note on Monetary Values

AAAS	American Association for the Advancement of Science
APS	American Philosophical Society
BAAS	British Association for the Advancement of Science
BAAS *18xy report*	*Report of the meeting of the BAAS held in 18xy*, published the year after
BL	British Library
BP	Brougham papers
Bu P	Buckland papers
CP	Chadwick papers
DAB	*Dictionary of American biography*
Devonshire Commision	Royal Commission on scientific instruction and the advancement of science, first and second reports, *Parliamentary papers*, 1872 vol.25
DLB P	De La Beche papers
DNB	*Dictionary of national biography*
DP	Daubeny papers
DSB	*Dictionary of scientific biography*
EUL	Edinburgh University Library
FGS	Fellow of the Geological Society
FP	J.D. Forbes papers
FRS	Fellow of the Royal Society
GP	Greenough papers
Gr P	Grove papers
Gu P	Gunther papers
HCR	Hebdomadal Council reports
HP	Harcourt papers
Ham P	Hamilton papers
Hen P	Henslow papers
Her P	Herschel papers
Hu P	Hutton papers
KCL	King's College, London
KP	Kelvin papers
LP	Lyell papers
La P	Larcom papers
Lit-and-phil	Literary and philosophical (society)
Lub P	J.W. Lubbock papers
LPLS	Leeds Philosophical and Literary Society
LPS	Literary and Philosophical Society
MP	Murchison papers

Ma P	Mantell papers
Mu P	Murray papers
OP	Owen papers
OUA	Oxford University archives
OUA/UM	Oxford University Museum archives
OUM/HA	Oxford University Museum, historical archive
PGS	*Proceedings of the Geological Society of London*
Pl P	Playfair papers
PP	Phillips papers
PRO	Public Record Office
PYGS	*Proceedings of the Yorkshire Geological Society*
q	quotation
QJGS	*Quarterly journal of the Geological Society of London*
RCO (1850)	Royal commission appointed to inquire into the state, discipline, studies, and revenues of the University and colleges of Oxford, *Parliamentary papers*, 1852, vol. 22
RCOC (1872)	Royal commission appointed to inquire into the property and income of the Universities of Oxford and Cambridge, and of the colleges and halls therein, *Parliamentary papers*, 1873, vol. 37
RDS	Royal Dublin Society
RI	Royal Institution
RMI	Royal Manchester Institution
RP	Ramsay papers
RSL	Royal Society of London
SP	W. Smith papers
Sa P	Sabine papers
SCOC (1867)	Select committee on the Oxford and Cambridge Universities Education Bill, *Parliamentary papers*, 1867, vol.13, pp. 183-560
SDUK	Society for the Diffusion of Useful Knowledge
Se P	Sedgwick papers
Sorby P	Sorby papers
St P	Stokes papers
TCD	Trinity College, Dublin
Terr Mag P	Terrestrial magnetism papers
TGSL	*Transactions of the Geological Society of London*
UCL	University College, London
UM/M	University Museum, Oxford: minutes of delegates
WP	Whewell papers
YAS	Yorkshire Agricultural Society
YGS	Yorkshire Geological Society
YMI	York Mechanics' Institute
YPS	Yorkshire Philosophical Society
YPS 18xy report	*Report of Council of the YPS for 18xy*, published the year after

A conversion factor of about 50-60 in the value of sterling gives a rough comparison of the monetary values of 1820-70 with those of today. In Phillips' time there were 12 pence (d) to a shilling (/-), 20 shillings to a pound (£), and a guinea was worth 21 shillings.

List of Illustrations

Introduction

The way of death and burial of John Phillips, professor of geology in the University of Oxford and keeper of the University Museum, showed that he had achieved the *ne plus ultra* of distinction. On 23 April 1874 after dinner in All Souls College, Oxford, with his host Mountague Bernard (professor of international law), and his friends the reverend Francis Leighton (warden of All Souls), and the reverend Charles Williams (principal of Jesus College, Oxford), Phillips stumbled over a mat at the top of a flight of fifteen stone steps and fell headlong and backwards to the bottom. He suffered instant paralysis, never recovered consciousness, and died the next day. Knowing that in his will Phillips had stipulated that his funeral be held at York, his home for almost thirty years, the vice-chancellor of the University, the reverend Henry George Liddell, nominated three distinguished Oxonians and close associates of Phillips to go to York to represent the University. They were the reverend John Griffiths (warden of Wadham College), Henry Acland (Regius professor of medicine), and Henry Smith (professor of geometry).

On 29 April a procession of 150 people, with the University members in academic dress, accompanied the hearse from Museum House, Phillips' residence, to the Oxford railway station from where it was carried to York to lay overnight in the vestibule of the Yorkshire Museum of which he had been the first keeper. The funeral of 30 April was arranged by William Gray, junior, an old friend of Phillips and former Lord Mayor of York, who ensured that it was a civic occasion. After a service at St Olave's, the Anglican church where Phillips had worshipped regularly, the hearse was followed by over thirty carriages of mourners through the city to the York Cemetery where the bachelor was buried next to his unmarried sister, Anne, who had been his housekeeper from 1829 until her death in 1862. Many shops were closed, the Big Peter bell of York Minister tolled for ninety minutes, and even the hard-headed directors of the York Gas Company met early in order to attend the funeral. The local press played its part by eulogising Phillips: he was a 'great and good man', 'a great geologist' who was 'widely loved and esteemed'; by 'sheer hard work' he had secured 'world wide celebrity' and an 'imperishable reputation'.[1]

What had Phillips done to elicit such tributes? One of the main aims of this book is to answer this question. Let us make a start by giving here an overview of Phillips' remarkable career. He was born in 1800 in Wiltshire, the second of four children of an excise officer. Orphaned at the age of seven, he was brought up by his uncle, William Smith, a surveyor who was later dubbed the father of English geology because he was the first in England to discover, to exemplify, and to diffuse the notion that different rock strata could be distinguished by their characteristic fossils. After several years at a local school Phillips was sent by Smith to live and study with a local Anglican clergyman, Benjamin Richardson, who expanded the boy's interest in science. At the tender age of fourteen, Phillips joined his uncle in London where he helped in geological work and trained under him as a mineral surveyor. In 1819

Smith was imprisoned for debt and relinquished his London home. After this disaster Smith and Phillips wandered for four years from 1820 in northern England where they took surveying jobs and produced Smith's county geological maps.

In 1824 the Yorkshire Philosophical Society (henceforth YPS) invited Smith to lecture in York. Having assisted his uncle, Phillips shared subsequent courses with him elsewhere and then launched himself as a solo lecturer. He was so adept at arranging the fossils in the museum of the YPS that he became in 1826 its first keeper and made York his home. Now separated from Smith, Phillips established himself as a successful lecturer and the leading expert on the geology of the Yorkshire coast.[2] As the right-hand man of the reverend William Vernon Harcourt, the leading figure in the YPS, Phillips took a leading part in 1831 in organising at York the first meeting of the British Association for the Advancement of Science. Next year he was appointed its assistant secretary, an office he held for thirty years. Through the peripatetic Association, which quickly becme Britain's parliament of science, Phillips made many contact usually denied to the isolated provincial. He spread his ambitions to the capital: in 1834 he succeeded Charles Lyell as professor of geology at King's College, London, began to lecture extensively in London, and brought out the first edition of his *Guide*, a successful textbook.[3] In the 1830s he popularised geology by writing influential articles and treatises for no fewer than four encyclopaedias. In 1836 he added to his reputation with his enduring monograph on the carboniferous limestone of Yorkshire in which he took preliminary steps towards a statistical palaeontology based on the number and distribution of fossil species.[4]

While his multifarious career in York and London was prospering, Phillips worked voluntarily for the youthful Geological Survey from 1836 to 1838 when he began to be paid, as its first palaeontologist, to draw and describe the fossils of parts of southwest England. The resulting monograph, *Palaeozoic fossils*, revealed a more developed statistical palaeontology and proposed that there were three great periods of past life on earth called the palaeozoic, the mesozoic, and the cainozoic, terms which are still used today.[5] Meanwhile he resigned his London chair in 1839 and his York post in order to spend more time with the Survey in which he taught young colleagues new methods of field work. In 1844 Phillips assumed the new professorship of geology and mineralogy at Trinity College, Dublin, which he hoped to combine with the directorship of a new Irish branch of the Survey. He was forced to leave the Survey in autumn 1844 because it was deemed incompatible with his Dublin chair. The Irish Survey scheme having come to nought, Phillips resigned in 1845 from his Trinity post and rejoined the Survey. Though in 1845 he received the Wollaston medal of the Geological Society of London, its highest accolade, and showed himself to be the Survey's leading publisher with his memoir on the Malvern Hills (1848), he resigned from it in 1849.[6]

After four years based in York as a freelance writer, lecturer, government commissioner, and promoter of field clubs devoted to natural history and archaeology, in 1853 he replaced Hugh Strickland as deputy reader in geology in the University of Oxford. In 1856 he was made reader and in 1860 professor, even though he had never attended any university as an undergraduate. His interest in antiquities led to the keepership of the Ashmolean Museum, Oxford, from 1854 to 1870. More importantly, from 1857 to his death he was the first keeper of the new University Museum, which he helped to plan and arrange. As an experienced keeper, a fluent lecturer, a good organiser, an accomplished textbook writer, and a distinguished researcher, he soon became the unifying president of science at Oxford and the leading spirit in the Ashmolean Society there. A keen astronomer, particularly from 1852, Phillips secured eponymous fame in the 1860s when a crater on the moon and two features of Mars were named after him. He also catalysed the first stage of the founding of the University Observatory completed in 1875.

Nationally he was prominent as president of the Geological Society (1858-60), a post he had previously declined twice, and as president of the British Association in 1865. Ever active in pursuing geology in an exceptionally wide way, he published when seventy his *Geology of Oxford*.[7] As a devout but non-sectarian, irenic, and ecumenical Anglican, Phillips played a leading part in the Darwinian debates. In his *Life on the earth* (1860) he reaffirmed his belief in divine design, the reality of species, the relative novelty of humans, and a reverential reading of the book of geological strata.[8] For Phillips the discontinuities in the fossil record were explicable only in terms of separate creations which were transcendental and inscrutable acts of God. Though he was not a dogmatic opponent of evolution or of natural selection, it was characteristic of Phillips that in the controversy of the 1860s about the age of the earth he primed William Thomson, the physicist, with the latest semi-quantitative estimates made by geological means.

From the vignette just given it is clear that Phillips was a scientific Dick Whittington. His career was a variant on the theme of rags to riches: he rose from orphan to Oxford professor worth about £14,000 at his death.[9] The way in which lowly origins were transcended appealed strongly to Victorians such as Samuel Smiles who published not only lives of engineers but also in the 1870s those of two Scottish naturalists, Thomas Edward and Robert Dick, a shoemaker and baker, who exemplified self-help and a host of related virtues. Indeed, in his famous homily, *Self help*, Smiles devoted a chapter to scientists and sang the praises of William Smith for his patient and laborious efforts. [10] Though Phillips was the epitome of the assiduous and triumphant provincial of humble birth, he did not find a Smiles to chronicle his heroic individualism and struggles against adversity in that favourite Victorian genre, a life-and-letters volume. After his death there was no one able and willing to compile it. Phillips' sister Anne could have done so but she predeceased him. As he was a bachelor he had no loving daughter to pay filial tribute to him as the two Mrs Gordons did for their fathers, William Buckland and Sir David Brewster. He had no devoted relatives keen to do for him what Katherine Lyell did for Sir Charles Lyell, her brother-in-law, and what Phillips himself had done for his uncle in his account of William Smith. Phillips never asked any grateful protégé to write his biography as Sir Roderick Murchinson did six months before he died when he commissioned Archibald Geikie to do so. Over in Cambridge, Thomas Hughes, the professor of geology, paid homage to Adam Sedgwick, his teacher and predecessor, in a biography which he found impossible to complete solo. In Oxford Joseph Prestwich, Phillips' successor in the chair of geology, admired his predecessor but never contemplated writing a biography: his mature views about Phillips were confined to his inaugural lecture.[11] Thus Phillips suffered the same fate as William Henry Fitton, Henry De la Beche, William Daniel Conybeare, and George Greenough, all 'heroic' geologists and associates of Phillips, none of whom was 'biographised' at book length in the Victorian period.

Yet the historiography of Phillips goes back almost 150 years. From the 1850s when he became professor of geology at Oxford and president of the Geological Society of London, he was seen as occupying a special position among British geologists. As the nephew and chief pupil of William Smith he was a unique link with the beginnings

in the 1810s of what he called inductive geology, i.e. as a distinctive science with
its own methods of enquiry, authority, and expertise. As a provincial he was not
tainted by the partisanship of London but he was up to date and knew the leading
metropolitan geologists well. His clear and balanced judgement, when exercised on
controversial questions, was widely regarded as the best available, particularly as
he suspended it when he felt the evidence was inadequate.[12] In the 1860s Phillips
became even more widely known, particularly through his presidency of the British
Association and the articles about him which it generated. *The Leisure hour* and
John Timbs' *Year-book of facts* followed the line of *The critic* five years earlier in
characterising him as the most accomplished British geologist of his time while
admitting that others surpassed him in particular specialisms. Timbs emphasised that
Phillips was a designer of instruments for use not only in geology but also in physics
(electricity, magnetism, and meteorology) and astronomy.[13] Robert Hunt, in Edward
Walford's *Portraits of men of eminence* (1866), added three more dimensions: he
stressed that Phillips was a good astronomer who had profitably delved into 'the
arcana of space'; he claimed that the success of the British Association was almost
entirely due to Phillips; and he drew extensively and accurately on a short account of
his early life provided early that year by Phillips.[14]

From that time Phillips enjoyed a reputation as an eminent and veteran geologist
who with Buckland, Conybeare, De la Beche, Sedgwick, Murchison, and Lyell
constituted the pioneering first generation of British geologists. Andrew Ramsay,
who rose to be director of the British Geological Survey, regarded Phillips as an
old master. In 1869 the American Philosophical Society elected him as a member
because with Sedgwick and Murchison he had laid the foundations of geology in
Britain. In private Phillips was told he had a European reputation as an eminent
maître.[15] Shortly after Phillips' sudden death, Robert Hunt printed in full in *The
Athenaeum* the autobiographical account of his early days that Phillips had written
in 1866. Along with the biographical notices of that decade it became the basis of
most of the obituaries of Phillips, which in the main recycled what was already in
print. The exception was the tribute produced for the Ashmolean Society by George
Rolleston, professor of anatomy and physiology, and Henry Smith. They saw Phillips
as 'a centre of union' in science at Oxford, as a Christian philosopher who regarded
nature as sacred and science as sacramental, and as a geologist trained as an engineer
to employ strict and sobering methods and to design and make instruments. They
also confessed that Phillips' avoidance of factious controversy and his conciliatory
personality were seen by some younger scientists as impairing his critical judgement:
generally he condemned or commended only when asked. Rolleston and Smith
inadvertently gave public credence to the view held privately by the Darwin circle
that Phillips was a wishy-washy fence-sitter who was dull and boring and to the
opinion promulgated by the *Quarterly journal of science* that he was an excessively
cautious figure who lacked the nerve to support the new causes of the 1860s such as
science education in schools.[16]

Even so his death was widely lamented. Sir Charles Bunbury epitomised a general
feeling when he alluded in his diary to the sudden and startling death of the famous
geologist. For Bunbury Phillips was 'a geologist of a high order, and connected with
the old *heroes* of geology, and especially interesting as the nephew and pupil of
William Smith; and he was, moreover, a remarkably pleasant and genial man, not
solely a geologist, but intelligent and active minded in various ways'. Years after
his death he was remembered for his '*pleasant* ways of act and thought'.[17] Publicly

Phillips' reputation was kept alive by two posthumous works published under his name but written by others. In 1975 Robert Etheridge brought out the third edition of Phillips' *Geology of the Yorkshire coast*. Ten years later Etheridge and Harry Seeley completely re-wrote Phillips' *Manual of geology* which they admired for its excellent structure and its emphasis on the succession of stratified rocks and their fossils. For them Phillips was a great textbook writer and a genial eloquent teacher 'who if not a brilliant discoverer had verified much of what was known, and was a sound geologist of balanced philosophical habit'.[18]

In the following years Phillips received sporadic but not sustained or extensive attention. Von Zittel paid tribute from a European perspective to Phillips' research on jurassic strata, the carboniferous limestone, palaeozoic fossils and terminology, and to his monograph on belemnites produced for the Palaeontographical Society (1865-70). In his still valuable centennial history of the Geological Society of London, Woodward presented Phillips as an illustrious pioneer geologist but said little new. Phillips was deeply attached to Yorkshire, where some of his happiest years were spent and his most enduring friendships were formed; so it was appropriate that his name was kept alive in his adopted county, especially by local curators and the *Proceedings of the Yorkshire Geological Society*. In his golden jubilee history of the Society Davis devoted on orthodox chapter to Phillips. In 1904 W.H. Thompson, who regarded Phillips as a pioneer of Yorkshire geology, made the wild claim that he was such a famous geologist that he had left his unmistakable impress upon the world's life and thought. In their well-known work on the geology of Yorkshire, Kendall and Wroot claimed that Phillips' own writings still repaid examination: they emphasised that Phillips was a structural geologist with a great interest in faults, a glaciologist who examined erratic blocks and gravel deposits, and an archaeologist with a sustained interest in cave exploration. Also in the inter-war period Collinge and Sheppard painted familiar portraits of Phillips but they offered new information by drawing on the records of the YPS and on ephemera and letters preserved at Hull. Sheppard also gave a full bibliography of Phillips' works. Collinge's approach was subsequently developed by Melmore, who reproduced unpublished correspondence which illuminated the early history of the YPS. At the same time Cox, capitalising on Sheppard's long account of William Smith, threw new light on Smith and on his nephew, pupil, and lieutentant, by examining the Smith manuscripts held in the University Museum, Oxford. In 1961 the Yorkshire Geological Society commemorated Phillips eponymously when it inaugurated the Phillips medal for distinguished research on the stratigraphy or palaeontology of northern England.[19]

From the early 1970s there has been a sustained interest, often based strongly on unpublished sources, in several aspects of Phillips' career. His role as a keeper has been explored directly or obliquely by Orange, Pyrah, Ovenell, Fox, O'Dwyer, and Yanni.[20] His important contributions to the notions of geological systems, particularly in the controversies about those called Devonian, Cambrian, and Silurian, have been illuminated by Rudwick and Secord.[21] Burchfield and Smith and Wise have stressed Phillips' contribution to the controversy which raged in the 1860s about the age of the earth.[22] Herries Davies has demonstrated clearly why Phillips did not become director of the Irish Geological Survey in 1844.[23] In a pioneering study Hutchins has confirmed in detail that Phillips was a ranking astronomer in his Oxford period.[24] Secord has written incisively about the early Geological Survey as a research school.[25] Desmond has shown that in the 1860s Thomas Henry Huxley was keen to secure Phillips' support for his contentious view that certain saurian fossils had ostrich-like features and could therefore be construed as organisms intermediate between land creatures and birds.[26] At a more biographical level Edmonds has published meticulous and valuable articles on Phillips as an apprentice geologist, on his debut as a lecturer, on the failure of the scheme to make him the first professor of geology at University College, London, and with Douglas on the complicated history of his published maps.[27] Recently Knell has illuminated Phillips' approach to fossil collecting and fossil collectors.[28] For over twenty years I myself have had the temerity to publish a few aperçus about Phillips. With Thackray I showed that Phillips was a pivotal figure in the early British Association, while solo I have examined the genesis and features of Phillips' first monograph, his work for the early Geological Survey as its first palaeontologist, his ambivalent relation with the Yorkshire Geological Society, his life as a curator, and his contributions to geochronology.[29]

The time is now ripe for bringing together all the existing materials and insights pertaining to Phillips in a biography which tries to clarify and highlight his roles in Victorian culture, without committing the sin of hagiography. As no book-length study of him exists, I have tried to write a tolerably complete and reliable account of Phillips' highly productive and varied life. In portraying its changing features and flavour, with its numerous simultaneous projects, I have therefore taken chronology seriously. Yet I have not offered an exhaustive day-by-day chronicle because that would have downgraded important analytical themes. This book is therefore an attempt at an analytical biography focussed on important themes within chronological divisions which are not totally arbitrary because they have been suggested by the changing patterns of Phillips' own career and his perceptions of it. In this way I hope to show that his life was fascinating, not just in its own right but also in shedding light on wider themes.

One major and pervasive theme was career-making. It had its tensions, of which Phillips was only too well aware: 'the unfortunate part of my scientific career has been this: that attaching myself to *science* from pure love of it, want of financial power forced me into *scientific business* (if some of my engagements may be so termed) and I am caught in nets of my own forging'.[30] For much of his life Phillips had to make ends meet, so the formation and pursuit of his career are central to this book. My model here is Outram's penetrating analysis of Georges Cuvier's life. She focussed on specific problems, such as power, authority, and patronage, which attended his career. She refused to separate the scientific and the social: though she did not offer an extensive exploration of Cuvier's science, she showed that Cuvier's pursuit of desired kinds of science was both a political and an intellectual project.[31] In this book I have tried to take a similar approach to the historical politics of science while at the same time discussing in some detail the content of Phillips' science. It would be foolish to disparage or ignore his science because it was through science that he rose to eminence as a 'self-made' man.

Throughout this book I assume that science was the socially organised attempt to set and solve problems concerning the understanding and sometimes control of the natural world, the value of the problems and the adequacy of the answers being subject to scrutiny and dispute. Thus scientific knowledge was a social product and scientific theories were socially constructed representations of the natural world. Like Oldroyd I believe that over time some scientific knowledge has become more verisimilitudinous because greater correspondence has been achieved between certain theoretical representations and the natural world, important criteria of correspondence being the coherence of these representations with those from other parts of science, practical efficacy, and specific predictive power. Of course what counted as coherence, practical success, or prediction was itself subject to debate.[32] In the case of geology, the rocks were silent and perceptions of them were mediated by interpretative conventions. Yet the rocks existed and are still open to inspection by the historian. I have found it useful to supplement literary sources by examining on foot some of the terrain and sites where Phillips worked. Without making a fetish of such historical fieldwork, it has paid dividends in clarifying the problem-situations that Phillips and his contemporaries faced and the answers they offered to their problems. I have learned time after time in the field that natural objects were seen as something or other and that phenomena were sometimes not seen at all. This experience has reinforced my belief that what has counted as scientific knowledge was socially shaped from empirical materials concerning the natural world.[33]

Notes

1. *Yorkshire gazette*, 2 May 1874; *Jackson's Oxford journal*, 25 April 1874; Phillips' will, 13 June 1864, copy in Eyles collection, University of Bristol Library; YPS Council minutes, 28 April 1874; Mountague Bernard (1820-82), *DNB*, fellow of All Souls, professor of international law and diplomacy; Francis Knyvett Leighton (1807-81), warden of All Souls; Charles Williams (1807-77), principal of Jesus 1857-77; Henry George Liddell (1811-98), *DNB*, dean of Christ Church 1855-91, vice-chancellor 1870-74; John Griffiths (1806-85), *DNB*, keeper of University archives 1857-85, warden of Wadham 1871-81; Henry Wentworth Acland (1815-1900), *DNB*, Regius professor of medicine 1858-94; Henry John Stephen Smith (1826-83), *DNB*, professor of geometry 1860-83; William Gray jun (1806-80) a York lawyer.
2. Phillips, 1829b.
3. Phillips, 1834a.
4. Phillips, 1836a.
5. Phillips, 1841a.
6. Phillips, 1848a.
7. Phillips, 1871a.
8. Phillips, 1860b.
9. Court of Probate for Phillips.
10. Smiles, 1859, pp. 91-8, 1874, 1878; Thomas Edward (1814-86), *DNB*, Robert Dick (1811-66), *DNB*.
11. Gordon, E.O., 1894; Gordon, M.M, 1869; Lyell, 1881; Phillips, 1844a; Geikie, 1875; Clark and Hughes, 1890; Prestwich, 1875, pp. 3-5.
12. Huxley, 1855.
13. Anon, 1860; Anon, 1865a; Timbs, 1866, pp. 3-9; Timbs to Phillips, 19 Dec 1865, 18 Jan 1866, PP; John Timbs (1801-75), *DNB*, editor.
14. Hunt, 1866; Hunt to Phillips, 24 and 27 Jan 1866, PP; Phillips to Hunt, 7 and 10 Feb 1866, British Museum (Natural History), London; Robert Hunt (1807-87), *DNB*, keeper of the Mining Record Office, London, 1845-83 and writer on science; Edward Walford (1823-97), *DNB*, journalist and compiler.
15. Anon, 1870, Ramsay, 1894, p. 46; Phillips to American Philosophical Society, 12 June 1869, APS, Philadelphia; Bigsby to Phillips, 1 Feb 1869, Gaudry to Phillips, 15 Sept 1872, PP; Andrew Crombie Ramsay (1814-91), *DNB*, director general

of Geological Survey 1871-81; John Jeremiah Bigsby (1792-1881), *DNB*, renowned cataloguer of palaeozoic fossils; on Albert Jean Gaudry (1827-1908), *DSB*, professor of palaeontology, Muséum National d'Histoire Naturelle, Paris, see Rudwick, 1976a, pp. 245-9.

16. Hunt, 1874; Anon, 1874; Evans, 1875; Rolleston and Smith, 1874; Anon, 1865b, pp. 729-30; George Rolleston (1829-81), *DNB*, Linacre professor of anatomy and physiology 1860-81.

17. Bunbury, 1890-3, entry 16 May 1874; Prestwich, 1899, p. 338; Charles James Fox Bunbury (1809-86) popular fossil botanist.

18. Phillips, 1875; Phillips, 1885, vol. 1, p. vi; Robert Etheridge (1819-1903), *DNB*, in 1885 assistant keeper in geology, British Museum (Natural History), president Geological Society 1880-2; Harry Govier Seeley (1839-1909), *DNB*, in 1885 professor of geography, from 1896 professor of geology, King's College, London.

19. Phillips, 1865-70; Zittel, 1901, pp. 401, 436, 451, 497-8; Woodward, 1907, pp. 40, 113, 165-6; Davis, 1889, pp. 119-35; Thompson, 1904; Kendall and Wroot, 1924, pp. 237, 444, 464, 529, 581; Collinge, 1925; Sheppard, 1917, 1934; Melmore, 1942, 1943; Cox, 1942; for Phillips medal, *PYGS*, 1961-2, vol. 33, pp. 256-7, 1963-4, vol. 34, p. 96.

20. Orange, 1973; Pyrah, 1988; Ovenell, 1986; Fox, 1997; O'Dwyer, 1997; Yanni, 1999.

21. Rudwick, 1985; Secord, 1986a.

22. Burchfield, 1974; Smith and Wise, 1989.

23. Davies, 1983, 1995.

24. Hutchins, 1994.

25. Secord, 1986b.

26. Desmond, 1982, 1994.

27. Edmonds, 1975a and b, 1982; Douglas and Edmonds, 1950.

28. Knell, 2000.

29. Morrell, 1983, 1988a and b, 1989, 1994, 2001; Morrell and Thackray, 1981, 1984.

30. Phillips to De la Beche, 26 Jan 1839, DLB P.

31. Outram, 1984; Pickstone, 1988.

32. Oldroyd, 1990.

33. Pickstone, 1995, justifies this belief; Rudwick, 1985, Secord, 1986a, and Oldroyd, 1990, exemplify it.

PART I
THE SCIENTIFIC APPRENTICE
1800-1834

Chapter 1

The Apprentice Mineral Surveyor

1.1 The education of an orphan

John Phillips was born on Christmas Day 1800 at Marden in Wiltshire. What was
in many ways the determining event in his life occurred just seven years later, when
both his parents died within the space of six months. Inevitably 1808 was a year of
affliction: he was old enough to be able to remember the death of his father and his
mother's death-like woe. For years afterwards his recollection of happy childhood
was tinged with deep sorrow: looking backwards was like re-opening half-closed
wounds. Phillips realised that he was dependent on others for shelter, food, clothing,
and education. It seems that he was left financially destitute. In all his tribulation there
was, however, help at hand: William Smith, his uncle, became his kind protector, and
he derived support from the closer attachment to his sister, Anne, after his parents
had died. It may well be the case that Phillips' suffering early in his life made him
yearn for a contented and secure life free from affliction, helplessness, and poverty.
Perhaps, too, the love and charity shown to him as an orphan boy indicated that, cruel
as fate might be, neither the hope of God's blessing nor Christian faith was illusory.[1]

Phillips became associated with William Smith as a result of the mobility and
marriage of his father, also John. He was the younger son of a Welsh family which for
generations had lived at Myddfai in Carmarthenshire. Born in 1769 he was educated
to follow several of his relatives into the church but in 1795 he moved to Bristol
to pursue a career as an assistant excise officer. Though often reviled, such a post
was not humble and required training in mathematics and the use of instruments.
Next year Phillips was promoted to be in charge of the Stow-on-the-Wold area in
Gloucestershire and moved to nearby Oddington where he met Elizabeth Smith
whom he married in her home village of Churchill, Oxfordshire, in 1798. She
was the daughter of John Smith, a blacksmith, and had three brothers, William (an
engineer and mineral surveyor), Daniel, and John. Soon after his marriage Phillips
was transferred to Market and West Lavington in Wiltshire and the couple set up
home in Marden where their first two children, Ann (who died in infancy) and John
were born. In 1801 Phillips resigned his post with the excise and moved to Midford
near Bath where a second daughter, Anne, was born in 1803. The family might then
have lived at Tucking Mill House, between Midford and Monckton Combe, which
William Smith had bought in 1798. In mid-1803 Phillips rejoined the excise, was
put in charge of Deddington, Oxfordshire, and his family moved to nearby Steeple
Aston where a second son, Jenkin, was born in 1806. That year he was promoted
to an excise post in Coventry to which the family moved. He died there in January
1808. His widow and children moved to Churchill, but she died in July 1808 leaving
three orphans. Faced with this double tragedy, her husband's family did nothing but
her three brothers responded nobly: Jenkin stayed at Churchill in the care of Daniel
Smith while John and Anne were looked after by William Smith (Fig. 1.1). Early

Fig. 1.1   Hughes Forau's painting of 1837 of William Smith, then aged sixty-
eight, shows his sturdy strength of mind and body.

in 1809 he took them to his brother, John, who lived at Broadfield Farm, between
Hinton Charterhouse and Midford.[2]

Phillips' early life was nomadic: he lived in at least five different places and
attended four different schools. Clearly William Smith realised that his nephew
needed stability. Having taken the advice of his friend, the reverend Benjamin
Richardson, in 1809 Smith sent Phillips to David Thomson Arnot's school at Holt

in Wiltshire and paid for his education there for five years. John Smith also gave his nephew security by welcoming him and his sister into his household, either at Broadfield Farm or Tucking Mill House, from 1809 to 1815. Phillips never complained about the education he received at Holt. His skill in drawing, so useful to him subsequently, was encouraged. He enjoyed mathematics, via Mole's *Algebra* and Simson's widely used edition of Euclid, an interest he retained for the rest of his life. He became a good Latinist who was able to read and to write the language. He read, spoke, and wrote French in abundance, an acquisition which was particularly useful later when he studied French scientific works and met French savants. He learned enough Greek to be able to master it subsequently. Though there was apparently no formal teaching of science at Holt school, Phillips was given a small microscope which he employed enthusiastically on shells, plants, and insects. This was the origin of Phillips' interest in instruments, which was subsequently expanded to designing and making them himself.[3]

By 1814 William Smith's financial circumstances had deteriorated so much that he was twice arrested in London for debt, being freed only through the intercessions of Sir Joseph Banks, his chief patron and president of the Royal Society. In any event Holt school itself was on the verge of financial collapse: its owner became bankrupt in 1815. In summer 1814 Richardson came directly to Phillips' aid: he invited the thirteen-year-old boy to live and study with him at his rectory at Farleigh Hungerford, six miles south-east of Bath. Presumably he liked what he had seen of Phillips who, when not at Holt school, lived no more than three miles from Farleigh. Smith had good reasons for approving the invitation. He knew that Richardson would 'polish' Phillips and, further his education in science and literature. Presumably Smith welcomed the prospect of his nephew being a private pupil of an Anglican clergyman. No doubt, too, Smith looked forward to saving on school fees. The arrangement with Richardson worked so well that Phillips stayed with him for twelve months from summer 1814. For the rest of his life he revered Richardson's kindness and generosity. In 1844 he took good care to dedicate his life of Smith to Richardson as 'the loved associate of his early studies'. Though Phillips never attended a university he regarded Richardson's home as the equivalent.[4]

Richardson had graduated in 1781 at Christ Church, Oxford, where he studied classics and mathematics. Though he valued polite literature highly, he esteemed natural history even more and became an accomplished naturalist. After ordination and several curacies, he finally settled in Farleigh as its rector in 1796. In that very year Smith, then surveyor to the Somerset Coal Canal Company, joined the exclusive Bath and West of England Agricultural Society through which by 1797 he met Richardson, an active member who was devoted to experimental agriculture and the provision of allotments for the poor. Though he was a botanist of Linnean persuasion and an archaeologist, his favourite pursuit was collecting and naming local fossils, the provenance of each being noted by him. At that time he had no notion of either the laws of stratification of rocks or the connection between the order of the strata and their fossils. When Smith saw Richardson's fossil specimens, he applied to them his general principles that the same strata were always found in the same order of superposition and contained the same peculiar fossils. Without knowing their provenance, he arranged Richardson's specimens in the order of strata from which they had been derived. Richardson was so astonished that Smith knew which strata the fossils exclusively belonged to that he agreed to accompany Smith in a field trip to test his principles. The site chosen was Dundry Hill, south of Bristol, where they were joined by Richardson's close friend, the reverend Joseph Townsend, rector of Pewsey, Wiltshire. To their amazement, Smith's principles and inferences were illustrated and confirmed. In June 1799 when this geological triumvirate was dining at Townsend's house in Bath, Richardson wrote down Smith's dictated list of strata in the vicinity of Bath, in descending order from the chalk to the coal. To this description was added where possible a list of the most remarkable fossils in each stratum. Their names were mainly supplied by Richardson who also helped Smith to decide finally on terminology for the strata. Copies of this pioneering stratigraphic document were circulated in 1799. Two years later Richardson disseminated extensively copies on cards of the list of strata in the Bath area. By these means Richardson ensured that Smith's discoveries became widely if informally known. The original version of 1799 in Richardson's handwriting and Smith's accompanying geological map of the Bath area, completed in 1799, were presented to the Geological Society of London in 1831 when Smith became the first recipient of its Wollaston medal. The table of strata remained unpublished until 1815 when it was inserted in the memoir which accompanied Smith's geological map of England and Wales.[5]

After 1799 Richardson tried to spur Smith into publication by giving him useful local information and by urging him to assert his originality. Even in his early seventies Richardson was still lamenting that Smith had not finished what he had begun: he had not completed his *Strata identified* as the best introduction to Smith's '*own science*'. In addition to prodding Smith, Richardson made an important contribution to Smith's memoir, which accompanied his famous map of 1815, by pointing out omissions and mistakes in a proof of a list of strata. Publicly he supported Smith on two important occasions. In February 1815 he certified to the Society of Arts, apropos Smith's successful application for its £50 prize and medal for a mineralogical map of England and Wales, that it was accurate and the most important work of its kind ever published. One of his last acts before he died was to write in 1831 to Sedgwick, then president of the Geological Society, to testify to Smith's originality and to justify the award of the Wollaston medal to him. Both affidavits, vital for Smith's reputation, were published.[6] Unlike Townsend, a prolific polymath who gave a full exposition of Smith's discoveries in 1813, Richardson published nothing but he did promulgate Smith's views whenever he could. The young William Buckland, reader in mineralogy (1813) and geology (1818) in the University of Oxford, was happy to learn from Richardson about the succession of oolitic rocks as enunciated by Smith. Richardson was long and widely known for promoting geology in general. When the reverend William Daniel Conybeare wanted help in fieldwork on the lower oolite strata south of Bath, he was guided by Richardson and published the results in his well-known book. In the late 1820s William Lonsdale, though then resident in the area, was deeply indebted to Richardson for information about the oolitic rocks around Bath. Richardson made the cabinet of his fossils so freely available and donated so many specimens to museums that on his death it was almost empty.[7]

As a person Richardson's dominating feature was 'very singularly great benevolence'. Phillips was not the only youngster to benefit from it. Harry Jelly, a relative of Mrs Richardson, was orphaned in 1801 and Richardson brought him up. When Phillips went to Farleigh he began a life-long friendship with his exact contemporary. Jelly went on to become, like Richardson, a parson-geologist and fossil collector. Jelly's cabinet of fossils was so rich that the adult Phillips used it for two of his monographs. Unlike Richardson, Jelly did field work, discovering while an undergraduate at Oxford fossils in the ironsand on Shotover Hill and in the

1830s publishing papers on Bath fossils. Though nothing is known in detail about Richardson's religious views, it is clear that for him there was no clash between Christianity and Smithian geology. It is also clear that Richardson was the epitome of generosity and benevolence in his private life. It is not fanciful to suggest that he presented to the impressionable and grateful Phillips an inspiring example of a Christian philosopher. He was a model of practical as opposed to pontificatory Christianity. He showed to the young Phillips that contentment was attainable provided one spurned avarice and used one's financial affluence unselfishly and charitably.[8]

Richardson opened Phillips' mind. The clergyman was au fait with recent geological works, he was an expert on zoological and geological nomenclature, and he talked with Phillips endlessly about plants, shells, and fossils.[9] The surrounding area of south Bath was the boy's geological playground where he avidly collected oolitic fossils. It is inconceivable that Richardson did not expound to him Smith's views about stratigraphy and the identification of strata using characteristic fossils. Phillips enjoyed the free run of Richardson's well-stocked library and used its reference works when sorting, classifying, and naming Richardson's fossil specimens. By March 1815 and aged fourteen, Phillips was capable of making lists of fossils for Smith and furnishing him with mineralogical information derived from Richardson's copy of Werner's account of the formation of metallic veins in rocks. Phillips also improved his skill as a drawer by copying the illustrations of fossils made by Townsend for his *Character of Moses*.[10] At the end of his idyllic interlude with Richardson, Phillips, not yet fifteen years old, was well informed about natural history and geology, especially the systematic classification of fossils. As Phillips put it in 1819 in a draft preface to a projected work on fossil conchology, he learned from Richardson, not Smith, the general principles of geology and 'the proper regulative system of classifying fossil shells'.[11] When Phillips arrived in London in late 1815 to help his uncle, he did not need to serve a long apprenticeship in cleaning, naming, and cataloguing Smith's collection: he knew enough to be an effective assistant immediately and much of his knowledge had been derived from Richardson. Though Phillips presented himself from the late 1820s as the nephew and pupil of Smith, his first teacher and father-figure had been Richardson. Indeed, between 1809 and 1815 Smith's incessant travelling precluded him from teaching his nephew regularly. The boy had no access to his uncle's main collection of maps and fossils which were housed in London. Perhaps when Smith visited the Bath area Phillips may have learned from him something about surveying. Apart from receiving instruction from Smith in 1810 about fossil sea lilies in the cornbrash stratum, Phillips learned no geology directly from his uncle until late 1815.[12]

1.2 Cabinet assistant to Smith

When Phillips' pleasant interlude at Farleigh ended in summer 1815, he spent some time at Tucking Mill before joining his uncle in London in November. The fourteen-year-old orphan lost his beloved oolitic hills of Somerset: they were replaced by a view of the river Thames from his uncle's house at the eastern end of Buckingham Street, off the Strand. Phillips exchanged the rural calm of the rectory at Farleigh for the bustle, noise, and impersonality of the metropolis. He left all his friends and all his relatives back in the west and south-west of England. His sister, Anne, did not come with him and he was not to meet her for over a decade. The only person he knew in London was his uncle and guardian who needed the help of his nephew in order to keep afloat financially. Smith's large geological map of England and Wales had been published in August 1815 and by autumn he was trying to make money by issuing copies to subscribers and purchasers. His financial circumstances were so desperate that he had begun to negotiate with the government about selling his fossil collection to the British Museum. In both these projects Phillips' knowledge, ingenuity, and ability were prospectively useful. He was therefore brought to London for financial as well as altruistic reasons. As Smith's six-year-old marriage was childless, he may well have seen his nephew emotionally as a substitute son. Phillips was grateful to his uncle for his kindness: after all, Smith could have put his nephew in an orphanage or left him destitute. At the same time Phillips was immediately caught up in his uncle's financial crises and personal calamities; and he presumably realised that his own prospects would be very unattractive if his uncle, on whom he was dependent financially, were to become bankrupt. Phillips also apprehended, one presumes, that Smith was living beyond his means. From 1804 he had rented in London a large four-storeyed house, 15 Buckingham Street, at an annual cost of £88: he did not relinquish this tenancy even when his creditors were pressing him hard in 1814 and 1815.[13]

Smith was a canal engineer and land surveyor by profession. These occupations focussed his interest on strata; hence his nickname 'Strata Smith'. As a canal engineer he had the opportunity to study outcrops over long lengths of countryside and became an expert on relative porosities of strata and drainage. As a land surveyor he learnt about the vertical succession of strata and, when advising about the feasibility of any mining operation, had to extrapolate from that knowledge. Not all surveyors turned their attention to geology: most of them had little leisure for reading or writing; and in a highly competitive business they fought each other for work and pay.[14] But some, like Smith and John Farey, tried to pursue and advance geology, often hoping to gain practical advantages from it.

It is well known that Smith, in using his professional income to support his geological mapping, ruined himself financially. Farey also suffered the same sad fate. As a pupil of Smith he urged publicly from 1806 the importance of Smith's discoveries and his priority. Farey made many geological sections but they were not published: they circulated via manuscript copies. From 1806 Farey opposed on Smithian stratigraphic grounds the costly attempts then being made to find coal at Bexhill, east Sussex: true coal deposits, he claimed, lay way beneath the east Sussex rocks, now known as the Hastings beds. For his advice he received nothing, but the Sussex Mining Company lost at least £30,000. In 1811 Farey published in a book on the agriculture and minerals of Derbyshire the first geological map of any English county. On a scale of six miles to the inch it showed the sequence of what are now called carboniferous limestone, millstone grit, coal measures, magnesian limestone, and keuper marl. In drawings Farey revealed in two dimensions the effects of faulting on strata and of denudation on faulted rocks. As Rudwick has stressed, these diagrams were unsurpassed at the time for their theoretical sophistication in illustrating three-dimensional structures. Equally remarkable was Farey's paper on Ashover, Derbyshire, with geological section and map, read to the Geological Society of London in 1813. It was never published, a prohibition which exacerbated Farey's ill-will to the Society of which he was not a member. The map, on a scale of one and a half inches to the mile, showed in sequence twenty-four rocks belonging to the carboniferous limestone and coal measures. Its rich detail confirms the skill in descriptive stratigraphy possessed in England by those whom Banks described as 'practical men well versed in stratification'. Particularly after the death in 1820 of Banks, who had so perceptively patronised both Farey and Smith, Farey did not prosper financially: in 1824 he was reduced to offering himself as a copyist to James de Carle Sowerby, the naturalist and illustrator, at one shilling an hour. The Geological Society, which had ostracised him, neither rehabilitated nor vindicated him either before or after his death.[15]

By late 1815, when the young Phillips joined his uncle in London, Smith had done most of the work which subsequently led to his being called the father of English geology. He had recognised the regular succession of English strata, first in the south-west and then in the rest of the country, in which he travelled extensively. Geological history favoured him: in England all the sedimentary rocks, from the pre-Cambrian to the tertiary, are represented without big gaps in the succession of strata; many of these rocks are neither folded nor compressed, and they are undisturbed by intrusions of igneous rocks. Smith had also recognised that rock strata which are similar in lithology, ie in physical appearance to the naked eye, contain characteristic fossils by which they can be distinguished. Before 1815 these discoveries were not published by Smith in what Fitton was later to call 'the best and least disputable form'. They were, however, diffused by Smith and his supporters in several ways. Smith himself promulgated his views at landowners' meetings and reported his findings to no less than Banks. From 1804 his collection of fossils, arranged on sloping shelves one above the other with each shelf corresponding to a particular stratum, was freely exhibited in his London house and in 1808 was twice inspected by members of the Geological Society, then one year old. His discoveries became 'by oral diffusion the common property of a large body of English geologists'. They were also spread in print by authors such as Richard Warner as early as 1801, Farey from 1806, James Parkinson in 1811, and Townsend in 1813. Smith's list of strata in the Bath area was copied and widely circulated from 1799 by Richardson.[16]

These wayward means of spreading knowledge and securing reputation were supplemented in 1815 when, after years of effort and delay, Smith published his geological map of England, Wales, and part of Scotland, which was dedicated to Banks. On a scale of five miles to the inch, measuring 105 by seventy-four inches, and costing £5-5-0, it was the first large-scale stratigraphical map of any country. Smith recognised and coloured differently twenty-one sedimentary rocks, some of them being only a few metres thick. Each colour indicated the surface area occupied by a particular rock. In a novel move Smith represented the edges of successive strata with more strongly tinted colours, thus highlighting the three-dimensional structure of the rocks and showing their order of superposition. Though Smith was not the first to apprehend that strata could be identified by their fossil content, his solo achievement was remarkable because he virtually created the *stratigraphical* map. It showed how local detail could be incorporated into a form of visual depiction which with great economy revealed large-scale geological structure and instantiated Smith's views in such a way that it could be used interpretatively and predictively. As Smith himself averred, his map gave in one view the locality of thousands of specimens and it therefore generalised the information which he had accumulated for twenty-four years.[17] It became the basis of descriptive stratigraphy in Britain, not least through its terminology of rock formations. Some of his names, based either on places where a particular rock was distinctly visible or on local usage, remain in use today. Witness

Kelloways rock, named after a village in Wiltshire, and cornbrash, a Wiltshire dialect term used to describe limestone rubble on which corn was grown.

Phillips' year with Richardson presumably enabled him in late 1815 to assimilate rapidly from Smith all these aspects of his uncle's discoveries and mapping. He also realised quickly that Smith's finances were parlous. His uncle had spent on geology much of his income from surveying, his employers too often had not paid him for work done, and a big business venture had failed in 1812. In extremis he decided to try to sell to the British Museum his own fossil collection, which had been expressly assembled and laid out to illustrate his doctrines. The sale was protracted, occupying from July 1815 to February 1818, and it realised for him £600. Charles Konig, an influential keeper at the British Museum, was opposed to fossil-labelling of strata and made the transfer of Smith's collection difficult. It lay for decades, unpacked and unused, in its new home.[18] The sale was not only an emotional wrench for Smith but also involved a crucial loss of an intellectual resource. Until June 1816, when his fossil collection was delivered to the British Museum, it was available in his house. Those who inspected it could check for themselves that the shelves, on which his fossils were laid out, did indeed correspond to the rock strata which he had coloured on his 1815 map and had identified by characteristic fossils. As Rudwick has stressed, Smith had determined via fieldwork the structural order of strata, had brought it indoors, and replicated it in a fossil collection. He was thus the first to unite the traditions of field and museum which were concerned with three-dimensional structures and specimens respectively.[19] The prospective loss of Smith's fossil collection spurred him to produce as a substitute, in sellable, portable, and published form, illustrations of characteristic fossils of the various rocks shown on his 1815 map. Furthermore the British Museum required from him a catalogue of the collection. Presumably Smith realised even before his map was published that it would be desirable to supplement it in three ways: by illustrating the specimens in his fossil collection, by describing them, and by publishing a traverse section (a vertical slice on a particular traverse) of the interior of England and Wales to reveal that it conformed with the ranges of strata depicted on his map. Smith made heroic attempts to fulfil these desiderata. In 1817 he published, as an epitome of British geology, his *Geological section from London to Snowdon, showing the varieties of the strata* which were coloured to correspond with his map. Between 1816 and 1819 in the four parts of *Strata identified by organized fossils* he illustrated the characteristic fossils of sixteen strata. The set of fossils from each stratum was printed on coloured paper which matched the colour used on his 1815 map. In 1817 Smith published his *Stratigraphical system*, a descriptive but not illustrated catalogue of specimens from his collection of fossils in which he explained their use in identifying British strata.[20] It seems that Phillips gave no help with the London-Snowdon section. But with the illustrations and description of Smith's fossil collection, the boy's knowledge of zoology and conchology was vital. Smith even admitted publicly that he was 'weak' as a conchologist.[21] For Phillips his work on the illustrations and catalogue of Smith's fossil collection not only made him utterly familiar with his uncle's discoveries but also widened his own geological reading. Crucially it was in the period November 1815 to September 1817, for all of which Phillips was in London, that he first studied the notions of Georges Cuvier of whom he became a life-long admirer.

Phillips came to London, not yet fifteen years old, not primarily to train as an apprentice mineral surveyor-civil engineer under his uncle but chiefly to help him with publications arising from the sale of the fossil collection. Until September 1817

Phillips did no fieldwork, under his uncle or anyone else, in either mineral surveying or geology. His task, in return for food, shelter, and clothing, was to help Smith to illustrate and describe the fossil collection so that its features would become generally available; and to do so expeditiously, thus avoiding the long delay which had dogged the 1815 map. Smith was a sturdily independent man who had not previously co-operated in geological research with others. But he trusted his nephew and saw the advantages of employing him as an assistant, full-time and resident. The result was that Smith became, for the first time in his life, a prompt publisher: the first part of the illustrations of the collection was published in June 1816; part one of the catalogue appeared in August 1817, part two being ready then but never issued.

In preparing for these publications, Phillips began work immediately. Within three days of arriving in London he and Smith arranged shells according to the notions of Linnaeus, whose natural system of classification was based on organs of physiological importance. By February 1816 Phillips was helping Smith to make out even cornbrash fossils, on which Smith was the ranking expert. Phillips' contribution to the illustrations of Smith's fossil collection is difficult to estimate. *Strata identified*, published by James Sowerby, relied on his *Mineral conchology* and contained plates of fossils engraved by him. Later in life Phillips merely recalled that he drew many fossils for Smith and in general helped Smith with *Strata identified*.[22]

The lad's contribution to *Stratigraphical system*, the catalogue of Smith's fossil collection, is less ambiguous. If Smith was a capable drawer, he was totally inexperienced in 1815 in producing a catalogue *raisonné* of his fossil collection, especially a catalogue which tried to combine zoological classification with a stratified arrangement. He knew very little zoology so for him fossils were chiefly labellers of rock strata and nothing more: he was not interested in the living analogues of his fossils. In general he was sceptical about any help that collateral sciences, such as zoology, botany, chemistry, and mechanics, might be able to give to geology, which for him was an independent science with its own procedures. He was far more concerned with the empirically successful use of fossils in stratigraphy than with their biological significance. He could recognise fossils as old friends but not always knew their names. For the catalogue he needed help in conchology, which Phillips, who was then 'very deep in conchological studies', provided. Smith also needed aid in understanding works in Latin and French. The precocious Phillips provided both. It was Phillips who was responsible for all the descriptions, geological and geographical, of about 700 species of fossil shells, given in Smith's *System*.[23]

Phillips also drew up, under Smith's guidance, four tables, each of which showed the distribution of a particular group of fossils between various strata. Only one of these, of fossil echinodermata (sea urchins) was published in the *System*. This table showed which fossil species were peculiar to a particular stratum and which were repeated in other strata. It employed horizontal bands for strata, vertical columns for species, and stars for their occurrence. This table, showing the stratigraphical distribution of fossil species within a group of organisms, was the first of the kind ever published. It showed publicly that Smith was not a slave to the notion that each stratum contained exclusively certain so-called characteristic fossil species: he realised that some of them also appeared in other strata and had an interest in all species which occurred in a stratum. Two of the other tables, of fossil terebratulae and zoophytes, were never published; but in 1859, in his capacity as president of the Geological Society, Phillips reproduced the table of fossil ammonites. For young Phillips these tables were highly instructive in two ways. Firstly, he realised that the very idea of labelling strata by their allegedly characteristic fossil species was not straightforward. Secondly, he became aware of the proportionate prevalence of different groups of marine invertebrates in different rock strata, ie at different periods of time in the history of the earth. The table of fossil ammonites, for example, showed that the relative proportion of species varied from stratum to stratum, with far more species in the lias than in the other fifteen strata.[24] Such tables, drawn up by Phillips before autumn 1817, predate the similar attempts made in fossil botany in the 1820s by Adolphe Brongniart and the somewhat different statistical approach to fossils of tertiary strata taken by Charles Lyell in the 1830s. Phillips' juvenile tables eventually led to his mature efforts in statistical palaeontology.

In producing the catalogue of Smith's fossil collection, reference books were required. Some of them were so rare or expensive that they were borrowed from Banks and other supporters. Four types of works were consulted by Smith and Phillips. Firstly, there were six books published before 1800 which described and illustrated English fossils in general or from particular areas. They were: Edward Lhwyd's *Lithophylacii* (1699), the first work totally devoted to English fossils; *Fossilia Hantoniensia* (1766) by Gustus Brander and Daniel Solander who covered tertiary fossils; John Woodward's catalogue of English fossils (1728-9); and three regional studies, Robert Plot on Oxfordshire (1677), John Morton on Northamptonshire (1712), and John Walcott on Bath (1779), which mainly dealt with fossils from oolitic strata. Secondly, they examined four works on zoological classification. One was Linnaeus' famous general work . Another was the series of articles on the description and classification of fossil shells, mainly molluscs from the Paris area, published by Lamarck in the *Annales du Muséum d'Histoire Naturelle*. The other two works covered particular groups of animals, namely zoophytes and echinodermata by Ellis and Solander and by Klein respectively. The information derived from these two specialist zoological treatises was used in the tables showing the distribution of their fossilised ancestors in different rock strata. Thirdly, four recent standard works on palaeontology were consulted. Two of them, William Martin's book on Derbyshire fossils (1809) and James Sowerby's *Mineral conchology* (1812), were early attempts at applying to fossils the systematic Linnaean binomial nomenclature already used for living plants and animals. The other two works, James Parkinson's *Organic remains* (1804, 1808, 1811) and Townsend's *Character of Moses* (1813, 1815), are perhaps chiefly known for publicising Smith's discoveries; but at the time they were important reference works on fossils and stratigraphical palaeontology.[25]

The fourth type of work consulted concerned the tertiary fossils of the Paris region by Cuvier and Alexandre Brongniart. Their famous paper, first published in 1808 in a preliminary version, appeared in 1811 under the same title but in amplified form with map and detailed sections. The full memoir was reprinted unchanged in Cuvier's *Ossemens fossiles* (1812). It seems that Smith and Phillips consulted the 1811 version and not the English summary of 1810 and Farey's commentary on it which they probably knew. They no doubt welcomed as familiar the way in which Cuvier and Brongniart had worked out the stratigraphical sequence of strata on the basis of their fossils: the Frenchmen had applied a Smithian technique to strata above the chalk. But they had also gone beyond Smith by interpreting the origins of strata in terms of a contingent sequence of marine and freshwater periods and of periods of erosion. As Rudwick has stressed, Cuvier and Brongniart used fossils not only to reveal the succession of rock strata but also to reconstruct the history of the earth. They showed that there were strata formed in alternating marine and freshwater conditions, both being characterised by their fossil shells but the latter on occasion by fossils of terrestrial vertebrates. A rock stratum represented for them a particular period in the history of the earth, which was best characterised by its fossils. As the junctions between the marine and freshwater strata were abrupt, Cuvier and Brongniart concluded that long periods of geological repose were interrupted by sudden environmental changes which could cause extinction, that each marine incursion had led to the extinction of well-adapted terrestrial animals, that the Paris area was a basin which had been alternately a lake and a gulf of the sea, that the area had emerged as dry land between each submergence, and that the earth was very old.[26] Presumably Smith found it difficult to absorb Cuvier's historical geology and continued to see fossils as markers of strata; but Phillips, one assumes, welcomed the Cuverian approach to the history of the earth and of the succession of life on it. Certainly the main notions in Cuvier and Brongniart's essay were to become key elements in the view of the earth's history which Phillips was to develop later. Phillips presumably also derived from Cuvier an interest in the conditions under which fossils were deposited. For example Cuvier inferred that perfectly preserved and unmixed-up fossils occurring in regular and distinct beds of rock had been slowly deposited in a calm sea. Though Phillips was to present himself in the future as the pupil of Smith, it is indisputable that in helping his uncle with the catalogue of the fossil collection Phillips derived from Cuvier an interest in the conditions of existence of living things on the earth in the past and in the succession of life on it. By autumn 1817 Phillips, though only sixteen years old, was beginning to feel his way towards a synthesis of Smith's use of fossils in geology as ahistorical markers of strata and the historical palaeo-ecological approach of Cuvier. In older histories of the science, rival claims about priority were made for Smith and Cuvier. The case of the precocious nephew and assistant of Smith shows that the undoubtedly novel contributions of Smith and Cuvier, made independently, could be seen as complementary and not as antagonistic.

Phillips also helped Smith by drawing plans of coal and canal enterprises and by producing in summer 1817 two coloured lithographed geological sections of the strata of south London and north Wiltshire, which Smith sold for a shilling each. Based on information acquired by Smith in 1816 when he was investigating the water supply to a canal company, the Wiltshire section was stratigraphically important: for the first time the coral rag was recognised and distinguished as a distinct stratum, a discovery which was incorporated in amended versions of Smith's 1815 map.[27] Through these lithographed traverse sections Phillips learned to draw geological sections *à la* Smith, with most of the strata sloping in the same direction, each stratum having a different thickness. He acquired the latest stratigraphical knowledge, added to his competence in visual matters by developing a new skill as a lithographer, and had the satisfaction of helping his uncle financially.

1.3 Surveying with Smith

From autumn 1817 to the end of 1819 Phillips moved in part from the cabinet to the field when he accompanied his uncle on surveying expeditions. Until that happened Phillips had talked about geology with Richardson and Smith, drawn fossils from specimens, copied illustrations of them from books, helped his uncle with the catalogue of his fossil collection, used his facility in Latin to study older works on fossils and zoology, and exploited his fluency in French to read Lamarck on the classification of invertebrate marine fossils and Cuvier and Brongniart on the temporal history of the earth. He was also au fait with recent reference works in English on fossils and stratigraphy. And, of course, through daily contact in the same house Phillips' mind was in part moulded on Smith's. But the boy, though utterly familiar with Smith's approach, discoveries, and map, had not spent more than odd days in the field with his uncle. That situation changed in September 1817 when Phillips began to train under his uncle as an apprentice civil engineer and mineral surveyor. This period of Phillips' life, in which he helped his uncle on country surveys during expeditions from their London base, was painfully interrupted from June to August 1819 when Smith was imprisoned in London for eleven weeks for debt. Within three months of his release he was forced to give up Tucking Mill at Midford to his chief creditor, to relinquish the tenancy of his London home, and to submit to the sale of his furniture and books. He was able to keep for future use only his papers, maps, sections, and drawings. This disaster induced Smith in February 1820 to leave London, accompanied by his wife and nephew. This was a double renunciation of metropolitan ambitions. He had failed financially as a surveyor and engineer based in London; and he had been marginalised socially and intellectually by George Greenough, the leading figure in the Geological Society, who saw to it that Smith remained unrecognised by the Society and then in his own map of 1820 plagiarised from Smith's. At this time Smith began to carry a domestic cross: by 1820 his wife, not yet thirty years old and twenty-two years younger than him, began to be severely disturbed mentally. By the 1830s she was described in private as mad and insane. She died in 1844 having spent her last years in the County Lunatic Asylum at York.[28]

The journeys made by Smith and Phillips between autumn 1817 and late 1819 were undertaken for both professional and geological purposes. Smith's income was irregular and contracting so he was prepared to travel considerable distances to work on commissions as a civil engineer and surveyor. At the same time he was preparing the county geological maps which he began to sell in 1819. One journey was not intended to make money. In late 1818 Smith took Phillips to Churchill to see their relatives, leaving him there for Christmas (his eighteenth birthday) and the New Year.[29] The professional journeys on which he accompanied Smith were concerned with sea-defences, urban public amenities, lead and coal mines, and above all waterways and drainage. In October 1817 they inspected for the Bentinck family their sea banks at King's Lynn, Norfolk, which were in danger of being breached by high tides.[30] In February and March 1818 Phillips was in Monmouth, south-east Wales, with Smith who reported on its water supply, cleansing, lighting, and paving.[31] In May and December 1819 Smith examined lead mines in Swaledale, Yorkshire, which introduced Phillips to the mountain (or carboniferous) limestone of the county and enabled him then to begin to compare the Swaledale limestone strata with those he had already seen at Bristol, the Forest of Dean, and Pontypool.[32] In autumn 1819 Smith and Phillips travelled to the Forest to survey a colliery near Newnham. These mining surveys aroused Phillips' interest in carriages and railways used in mines.[33] He accompanied Smith on two waterways surveys. That on extending the navigable part of the river Waveney from Bungay to Diss on the Norfolk-Suffolk border was made in 1817 and published in 1818 but the scheme was not implemented.[34] The time-consuming second survey concerned the construction of a canal fourteen miles long which would join the river Aire at Knottingley to the river Don at Doncaster,

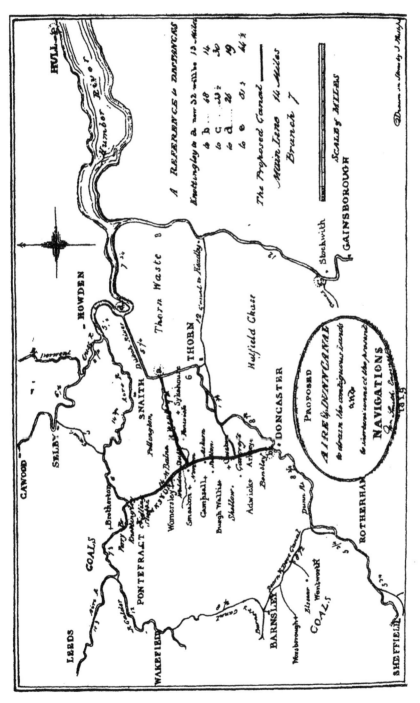

Fig 1.2 Phillips' lithograph dated 1819 of Smith's plan for a new canal between the Aire and Don rivers, running between Knottingley and Doncaster, with a branch running down the river Went. Phillips marked the [magnesian] limestone west of the northern end of the canal.

with a branch running eastwards down the river Went for seven miles from Norton to the Don north of Thorne (Fig. 1.2). This ambitious scheme involved draining contiguous land north and south of the river Went. Phillips, it seems, accompanied his uncle on two long trips made in autumn 1817 and summer 1818. The parliamentary bill for this canal was rejected by the House of Commons in April 1819 so that Smith lost the chance of being employed on its construction. Undeterred by this failure, in autumn 1819, after his release from prison, he worked solo on a third canal project, also unimplemented, for a Went canal to run eastwards for about twenty miles from the Barnsley canal at Cold Hiendley, four miles south of Wakefield, to the Don north of Thorne where it became the Dutch River.[35]

In these years Phillips learned much about techniques of surveying through experience in the field under Smith's guidance. He developed an eye for the lay of the land and became an accurate observer who knew how to use a theodolite, measuring chain, and spirit level. Later in life he told Edwin Chadwick, the sanitary reformer, that he was capable of following his uncle by writing a treatise on draining.[36] His journeys by coach as far north as Yorkshire and as far west as south Wales gave him a useful knowledge of English and Welsh topography and large-scale geological features. His uncle was adept at making important geological inferences and correlations on the basis of distant views of contours of hills. In 1794 on his first visit to Yorkshire Smith saw the Wolds from the top of York Minster; the view reminded him of the chalk downs of southern England so he concluded the Wolds were chalk. From a carriage in the Vale of York he saw the Hambleton hills, on the north-western edge of the north Yorkshire moors, and was reminded of the view of the oolitic Cotswold hills seen from the Vale of Gloucester. He concluded, again correctly, that the north Yorkshire moors belonged to the oolitic series of rocks.[37] Smith was a great talker and it is inconceivable that on coach journeys he did not comment on large-scale geological features to his nephew. The surveying work introduced Phillips in detail to rock strata which he had previously met only in books, maps, and cabinets. For example on the survey for the Aire-Don canal he saw for the first time the band of magnesian limestone which occurs on a north and south line between Knottingley and Doncaster. The proposed line of the canal ran consistently just east of the boundary of the porous magnesian limestone which Smith indicated on his plan of the venture.[38] While surveying for the canal Smith and Phillips had to confirm in detail where this limestone ended, perhaps using as a starting point Smith's 1815 map. This exercise, undertaken for commercial and engineering purposes, was probably Phillips' first experience of mapping the boundary of a rock stratum. These surveying trips also enabled him not just to see four different areas where there was mountain limestone, but to give him comparative experience of the Yorkshire version with others elsewhere. Thus began what became in the mid-1830s his important monograph on mountain limestone in which he took the Yorkshire variety as the reference type. Before 1817 Phillips had never seen Yorkshire; by early 1820 he had spent all told several months in a county which for him was intriguingly diverse in its scenery and geology. The Yorkshire surveying also provided occasional luxurious relief from the financial anxiety that Phillips shared with Smith. In early 1819 Phillips accompanied Lord Hawke, a promoter of the Aire-Don canal, in his carriage each way between London and Yorkshire.[39]

When not on professional surveys, Smith and Phillips undertook journeys to augment their geological knowledge and in particular to gather material for Smith's geological county maps. On occasion they covered long distances by foot over areas which were geologically revealing to Phillips. Late in 1819 they walked most of the way from Lincolnshire to Oxfordshire, which gave Phillips for the first time direct experience of the oolitic rocks in and between these counties. He also had the useful lesson of seeing Smith being unable to decide in the field whether a particular stone belonged to the forest marble or cornbrash stratum.[40] In order to gain details for the geological map of Yorkshire, Smith and Phillips visited Whitby and Scarborough in north-east Yorkshire in autumn 1817 to enable Smith to examine a second time the coastal geology. Phillips enjoyed his first sight of its spectacular exposures and examined the fossil collections of John Bird at Whitby and of Thomas Hinderwell and John Hornsey at Scarborough.[41] Presumably some of the specimens he saw reminded him of those he knew from south-west England and may have sown in his mind the idea of using fossils to correlate the strata of the Yorkshire coast with those elsewhere which were better known.

When living in Smith's house in London, Phillips continued to undertake some of his previous responsibilities but he also assumed new ones. As his uncle's assistant he remained a general amanuensis, copying out lists, drawing fossils, helping to produce part four of *Strata identified*, and writing out part of the unpublished portion of *Stratigraphical system*.[42] He also took on new work when helping Smith to colour geologically twenty-four maps of twenty-one English counties which Smith began to issue in 1819.[43] This novel experience of the problems of colouring geologically a topographical map complemented his experience in the field in mapping geological boundaries. In helping Smith with his surveying reports and plans, Phillips developed his skill as a lithographer, that is, in drawing and writing on stone and then taking impressions in ink from it. Invented in 1797, lithography was initially a crude process which produced blurred copies and at worst greasy daubs. It began to be used in the 1820s for illustrations in scientific books and journals: in 1824 the Geological Society of London adopted lithography in preference to engraving for the illustrations in its *Transactions*. It did so because lithography was much cheaper and it was better for depicting specimens and landscapes: it expressed texture and shading more effectively.[44]

Phillips' interest in lithography was economically driven. Early in 1817 he had thoughts about producing plates for a work on fossil conchology and began to explore lithography, using a German press, as a way of avoiding the expense involved in using an engraver such as Sowerby. Presumably he had learned from his uncle that the white lias stratum around Bath was the best British source for lithographic stone. For two years he made experiments on ink, paper, and modes of transfer, usually by trying various materials in various proportions. In summer 1817 he lithographed for sale a couple of geological sections made by Smith. The completion of the Aire-Don canal survey in September 1818 was a further spur. Phillips drew one plan which was printed lithographically by D.J. Redman but Phillips himself lithographed in 1818 and 1819 two more versions of it. It seems that at least 700 copies, apparently drawn on a stone from the lias stratum at Dunchurch, near Rugby, and on one from the magnesian limestone at Womersley, near Knottingley, were produced to publicise the proposed Aire-Don canal company and to accompany its prospectus. Phillips' ability to draw on stone and print from it had the attractions of cheapness and independence from lithographic printers. In February 1819, after discussion with Smith, Phillips devised a new lithographic press which he thought superior to the presses of Henry Bankes and Redman (Fig. 1.3). With Richardson's financial help and perhaps using a stone from the Bath lias, it was in action by July/August 1819. It was intended

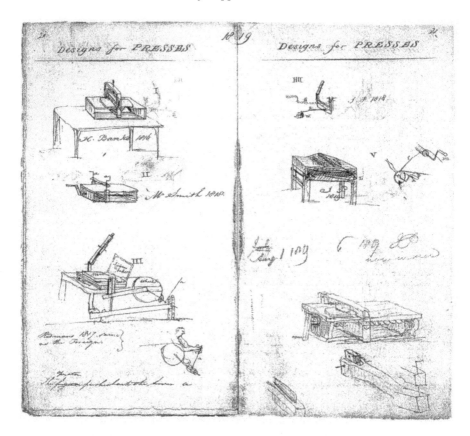

Fig. 1.3   **Phillips' drawings of 1819 of lithographic presses by: Bankes (1816),
I; Smith (1818), II; Redman (1817), III; Phillips (1818), IIII; Phillips
(1819, Feb-April) V; Phillips (1819, July/Aug, 'now in use') 6. Phillips
regarded his design as superior to the other three because it was
strong, simple, compact, and easily used.**

to help Phillips to survive while his uncle was in jail. Phillips issued an undated
lithographed circular from what he styled the 'Lithographic Office, 15 Buckingham
Street', ie his uncle's London home, soliciting lithographical work and offering what
would now be called duplicating services. He was ready to copy and print from stone
maps, drawings, designs, cards, circulars, and legal documents. He was also prepared
to produce prints for those who preferred to do their own drawing or writing on his
stone or prepared paper.[45] It is inconceivable that Phillips, assistant to Smith, would
have issued the circular under his own name, from his uncle's address and without
mentioning his uncle, if Smith had been living in his London house. Only on this
occasion in his London period did Phillips act independently of his uncle. This is
strong circumstantial evidence that the circular about Phillips' lithographic office
was issued when Smith was in jail for eleven weeks in summer 1819.

   Though it seems that Phillips published some plates of fossils in 1819, nothing
more is known about the way in which his lithographic press enabled him as an

eighteen-year-old orphan to survive in London for a desperate period during which he was effectively alone. But his venture into lithography was important in several ways. It was yet another mode by which he tried to help and repay his uncle. It gave him direct experience of the technical difficulties and possibilities of lithography. In later life, when he wrote scientific works, he was frequently his own sketcher and draughtsman and he knew what was technically feasible when he drew illustrations for others to lithograph. At the least Phillips' enthusiastic teenage venture into lithography, which endured for over a decade, heightened his awareness of the visual language of science and encouraged him in the future to give full expression to his remarkable visual imagination.[46] The lithographic venture confirmed his mania for gadgets. Indeed in 1819 the teenaged Phillips became the first in Britain, subsequently distinguished in science, to devise, construct, and use a lithographic press.

Though Smith's life from 1817 to 1820 was dominated by financial worries and calamities, he continued to attend the soirées held by Sir Joseph Banks in his London home. Sometimes Phillips accompanied Smith as his satellite. In summer 1818 Phillips enjoyed a rare and unforgettable experience: at Banks' home he saw and listened to Cuvier, who was visiting England. This incident strengthened Phillips' admiration of Cuvier and led to the conviction, held to the end of his life, that the Frenchman was the greatest of modern naturalists, the first and greatest of palaeontologists, and the master of comparative anatomy who had reconstructed from fossil bones the skeletons of extinct mammals. Banks also provided help as well as inspiration for Phillips. The ageing president of the Royal Society was so interested in plates of fossils and the table of fossil echinodermata which the teenager had produced that he lent Phillips rare and expensive books from his vast library devoted to natural history.[47]

Less pleasantly Phillips was a spectator at close quarters of the continued ostracism of his uncle by the Geological Society, epitomised by the hostility shown to Smith by Greenough, its first chairman and first president, who was again president 1818 20. Even after the publication of Smith's map and memoir (1815), his *Strata identified* (1816-9), and his *Stratigraphical system* (1817), he remained totally unrecognised by the Geological Society. Early in 1815, before Phillips joined Smith in London, Farey had published in the widely read *Philosophical magazine* a stinging attack on Greenough's conduct. As self-appointed champion of Smith, Farey defended Smith's priority as an investigator of England's strata and castigated Greenough for obtaining Smith's materials for unacknowledged use in his proposed geological map of England and Wales. Farey also exposed John Kidd's *Geological essay* (1815), dedicated to Greenough, for ignoring Smith's labours and the published tributes to them by Townsend and Parkinson, and for attributing Smith's discoveries to Greenough. Finally, Farey made the telling point that Greenough, like Smith, had published neither a geological map nor a book, so that criticism of Smith for tardiness applied a fortiori to the leisured and rich Greenough.[48] Whatever Smith thought then about being portrayed as victim, the peppery Farey represented him as such, with Greenough as his oppressor.

By November 1817 Smith found himself also pitted against Abraham Werner, the famous mineralogist who had taught for forty years at the Freiberg Mining Academy and had inspired many pupils. On the assumption that all rocks had been deposited as sediments in regular succession from a universal ocean, Werner had concluded from his studies of central European strata that there was a general order in their succession. As an investigator of the oldest strata, known as the primary, Werner had used mainly differences of mineral character, not fossil contents, to distinguish them. He had died on 30 June 1817. As he had been somewhat averse to writing, his disciples propagated his views in print, sometimes contentiously so. They were not uninterested in supporting their master's claims, especially just after his death.[49]

In 1817 William Henry Fitton, a Northampton physician who had been elected FGS only the previous year, launched his long career as an Edinburgh reviewer and commentator on English geology. Though Fitton had studied at the University of Edinburgh under Robert Jameson, professor of natural history and a vehement Wernerian, he argued in March 1817 that 'the importance of the study of organic remains [fossils] cannot be too highly appreciated by the geologist'. In November he gave equal credit to Smith and Werner as independent discoverers of the law of the general order of the succession of the strata, and portrayed geologists as candid and co-operative brethren of the hammer.[50] In this spirit of fair-minded adjudication, Fitton took the initiative in approaching Smith in order to give him 'undisputed possession' of what really belonged to him. On 30 November 1817, Fitton, Smith, and Farey met at a Banks' conversazione. Next day Farey drafted at his home a document about Smith's discoveries and reproduced it in the March 1818 issue of the *Philosophical magazine*. Farey prided himself on his 'fearlessly just severity' in supporting Smith's claims, but Fitton thought such denunciation was not just unhelpful but counterproductive. He preferred persuasion to confrontation.[51]

In February 1818, in a review of Smith's main publications, Fitton pleaded for justice to be done to Smith. Drawing on his conversation with Smith on 30 November 1817, Fitton portrayed Smith as an untaught and unassisted leader in geology who was deficient in publicising his discoveries and had used names which were new, local, and barbarous. Though Fitton argued diplomatically that fossils were useful assistants but not infallible guides in identifying strata, he called on the Geological Society to promote 'the welfare of an ingenious and deserving man, who is dependent solely upon his own exertions for support', and to give effect to his valuable observations by reworking his terminology. This plea was ignored, so Smith himself entered the fray by producing in June 1818 a full statement of his claims to 'the discovery and establishment of principles which have perfected [*sic*] the system of English geology'. Knowing the claims made by Fitton and Farey on his behalf, Smith decided to explain them himself. He gave chapter and verse showing the extent of the diffusion of his discoveries, beginning in 1799. He thought his achievements deserved 'national remuneration'. He represented himself as a victim of his own generosity: 'while all who could pick up the information thus profusely scattered thought themselves at liberty to publish it, I have been left to pursue, unrewarded and alone, the drudgery of more substantial utility, and my numerous papers, from want of means, are suffered to remain, from year to year unpublished'. Apropos Greenough and the Geological Society, he stressed that early in 1808 Greenough had twice seen in London Smith's fossil collection arranged by strata. He alleged that the early members of the Society had benefited by learning about his views purposely or imperceptibly. And he demonstrated that in the very first volume of its *Transactions* Parkinson had unambiguously acknowledged Smith's priority. As if to illustrate his poverty, this pamphlet was lithographed from a document written by Smith from his London home.[52]

In 1819 Greenough published his only book, *First principles of geology*. He believed in what Rudwick has called atomised empiricism, ie the diligent accumulation of facts untainted with theory. Accordingly he gave short shrift to

Smith's view that each stratum of rock could be identified by characteristic fossils, arguing that their utility had been greatly overrated because in different strata there were the same fossils and the absence of fossils in some strata was problematical. Having mentioned Smith just once and then only en passant, Greenough concluded that it was difficult to use fossils to determine the relative age of rocks. His scepticism attracted public and private criticism. Having struggled through it Robert Bakewell felt like 'John Bunyan's pilgrim after he had escaped from Doubting Castle'. Like Bakewell, Farey proclaimed that far more was known about the composition, structure, and past history of the crust of the earth, than Greenough admitted. Not unexpectedly he castigated Greenough for giving scant attention to Smith. Like Farey, Bakewell who was a mineral surveyor and textbook writer, never became an FGS. Greenough's friend, William Daniel Conybeare, an affluent vicar, told him privately that he had exaggerated the difficulties and uncertainties of the noble science of geology.[53]

In May 1820 Longman published Greenough's geological map of England and Wales, which appeared under the auspices of the Geological Society. From 1808 Greenough had assembled materials supplied gratuitously by his Geological Society colleagues and himself. Though he was a rich man, the Society underwrote his map to the tune of £1,000. In 1814 the appropriate committee of the Society decided that a new topographical map, to be coloured geologically, was desirable and appointed Thomas Webster, keeper of its museum and draughtsman, to produce it under Greenough's supervision. The final production cost of the geological map was £1,300. In contrast Smith worked solo, he was often near bankruptcy, and his map was not underwritten financially prior to publication. Inevitably Greenough's map improved on Smith's: it showed more geological detail, for instance of boundaries of strata, and more outcrops of igneous rocks. Though John Cary had constructed a special topographical map for Smith in 1813, the Webster/Greenough one showed much more topographical detail. Greenough's geological map was sold to the public at £6-6-0 but to fellows of the Society at £5-5-0, the price of Smith's. The result was that sales of Smith's map were killed off by Greenough's, at a time when Smith badly needed money. To add insult to injury Greenough's short accompanying memoir made scant acknowledgement of his debt to Smith. There is little doubt that Greenough did see and use manuscript maps, based on Smith's work, lent to Greenough by Farey on the assurance that Greenough's map would be merely an internal Geological Society affair, an assurance that was forgotten. For Smith Greenough's map was a ghost of his own and mocked his geological disappointments. For Greenough the stakes were high. His map was constructed on the principle of atomised empiricism, whereas Smith's was based on assumptions about characteristic fossils which Greenough questioned. Moreover Greenough had successfully led the early Geological Society, maintaining its independence against the hostility of Banks and the Royal Society. As a bachelor he cherished the Geological Society as his offspring and consequently regarded himself as the father of English geology. When that position was threatened, Greenough was prepared to behave unhandsomely against any competitor, especially one such as Smith who was a rival in mapping which Greenough regarded as his forte.[54]

For most of the time between autumn 1817 and early 1820 Phillips was in daily contact with his uncle. As Smith was a chatterbox, it is inconceivable that he did not discuss with his trusted assistant and substitute son the lack of recognition afforded him by the Geological Society. Presumably Phillips realised that in London there were helpful savants such as Banks and vindictive ones like Greenough, and that a body such as the Geological Society which was ostensibly devoted to the disinterested and cooperative pursuit of truth could be taken over by a particular party for its own ends. As he had contributed to Smith's works from late 1815, he shared Smith's resentment that they had been ignored or dismissed. Phillips also realised that there was a connection between financial circumstances and intellectual recognition: his uncle's poverty had led to tardy publication or no publication at all, which in turn fuelled priority disputes. As Phillips was to put in later, 'science, indeed, is a mistress whose golden smiles are not often lavished on poor and enthusiastic suitors'.[55] Finally Phillips' first experience of London had bright spots; but it was dominated by financial distress, culminating in his uncle's imprisonment, and the intellectual disappointment at the reception of his uncle's geological work. Like Smith, he quitted London in February 1820 in disgust. Henceforth he was very cautious about a career in the metropolis and fearful of its factionalism.

1.4   Wandering with Smith

From early 1820 to spring 1824 Smith and Phillips, now more his junior partner than an assistant or apprentice, tried to make ends meet by taking on mineral surveying jobs in the north of England. This work gave Phillips opportunities to develop practical techniques, such as making very detailed sections, and to further his interest in a variety of gadgets. Though Smith had suffered greatly in the past by spending his professional income on geology and was in desperate financial straits in summer 1821, between 1821 and 1824 he published nine more county geological maps to supplement the dozen published between January 1819 and February 1820. Phillips did much of the groundwork for those of Yorkshire (published 1821), Cumberland, Durham, Northumberland, and Westmorland (all 1824), sometimes in the company of his uncle and sometimes solo. In this period Smith encouraged Phillips to follow him in the perilous dual career of engineer/surveyor and geologist. Consequently Phillips did far more sustained fieldwork than before and enjoyed the freedom of making solo field trips which could last for four months. In this way he learned mainly on foot, and not from the top of a carriage, a great deal about the physical structure and stratification of the north of England and of the north midlands (Derbyshire, Nottinghamshire, and Leicestershire). By the end of the period he was such an accomplished geologist that he was capable of converting Smith to his own view about the occurrence of an important rock on the Yorkshire coast.

In these wandering years Smith and Phillips had no settled abode but stayed in no fewer than a dozen centres of operation in Yorkshire, Lincolnshire, county Durham, Westmorland, and Cumberland. They were generally cut off from polite scientific circles, though Smith was not forgotten by some leading gentlemen geologists in 1822. The reverend Adam Sedgwick, professor of geology at Cambridge, went out of his way to visit Smith in summer 1822 at Kirkby Lonsdale for guidance in collecting local fossils. Just before that meeting, Phillips had met Sedgwick by accident at High Force in Teesdale. In London, Fitton, a newly elected secretary of the Geological Society, continued his campaign to vindicate Smith: he declared his intention of proposing Smith as an honorary member of the Geological Society, with Greenough seconding Smith; but, fearful of the anti-Smith establishment, he withdrew his proposal. More importantly, in their classic work on the *Outlines of the geology of*

England and Wales, Conybeare and William Phillips gave to Smith the prominence that Greenough had denied him in 1819 and 1820. Though Conybeare admitted that on Smith's 1815 map some of the boundary lines for the lower oolitic strata were purely imaginary, in general he regarded Smith as simply 'great': the unassisted Smith was 'a great original discoverer' whose map was 'a great performance' which laid down 'great general views'. Conybeare stressed that from 1800 Smith's views were so freely and effectively communicated that most English geologists absorbed them as common property. He pointed out that Smith's terminology for the secondary strata was mainly unquestioned, unquestionable, and widely adopted. In fact in the *Outlines* Conybeare paid Smith the tribute of employing it. He hinted that Smith's great and real merit, in being the first to establish the order, position, and extent of the oolitic strata, had been unjustly suppressed.[56]

Such accolades were at odds with the lives then led by the homeless Smith and Phillips whose wandering precluded much contact with even provincial savants in the north of England. But they were not totally isolated. In 1821 in Keswick they met and learned from Jonathan Otley, who was about to publish his popular guide to the Lake District which contained a pioneering account of the divisions of the rocks and their geographical distribution. His detailed local knowledge was of service in the geological maps of Cumberland and Westmorland which Smith began to prepare in autumn 1821. Smith and Phillips were not alone in learning from Otley: in 1823, when Sedgwick wanted reliable guidance in his field work on the structure of the Lake District mountains, he turned to Otley as the best local expert and subsequently paid fulsome tribute to him.[57] For Phillips the contact with Otley aroused his interest in the slate rocks, then called greywacke, of an area which he had not previously visited. Otley's knowledge of the physical geology of the Lake District, including the ring shaped outcrop of mountain limestone on its periphery, and his views about how to distinguish stratification from cleavage in slate, were not lost on Phillips.

Late in 1821 Smith gave a lecture on the geology of Yorkshire to the Leeds Philosophical and Literary Society which had been established in 1818. It was not pre-planned but impromptu. En route on a surveying expedition, Smith and Phillips stopped in Leeds where on 16 November they attended a meeting of the Society at which Smith, regarded by Leodensians as a geologist of distinguished merit and reputation, offered to address it next day on the geology of Yorkshire. He did so using his own geological map of Yorkshire and one of the Leeds area drawn by Phillips who was in attendance. This was Phillips' debut in a scientific lecture theatre and his first contact with a provincial scientific society and its leading spirits. In particular he met John Atkinson, a Leeds surgeon who was the first curator and librarian of the LPLS, Edward Baines the Whig proprietor of the *Leeds mercury*, and Edward Sanderson George, a Leeds chemical manufacturer and a secretary of the Society who researched on the Yorkshire coal measures.[58]

As a mineral surveyor Phillips maintained his interest in lithography and chemical printing, and expanded that in surveying instruments and workshop tools. He was familiar with the graduation of what he called mathematical instruments and with elaborate portable surveying instruments such as a theodolite with a compass, a spirit level, a horizontal circle, and a vertical semi-circle, which could measure both horizontal and vertical angles. He was knowledgeable about lathes, both ordinary and oval ones, cutting screws, elliptographs (which drew ellipses), and devices for grinding the insides of tubes as well as lenses and mirrors. Thus he developed in this period his interest in workshop tools, some of which he seems to have made himself, and his competence in calibrating and using portable surveying instruments.[59]

Phillips learned important lessons from the surveying work. From borings made in the Yorkshire coal measures he became aware of their frequent faulting and daunting complexity. In a survey made for mineable copper ore at Conishead, near Ulverston, in north-west Lancashire, Phillips' meticulous quantifying spirit led him to record in one columnar section of strata, which depicted as a column their sequence and thicknesses, no fewer than 117 strata, of which the thinnest was a mere four inches. He also learned that miners in a particular area entertained notions at odds with his uncle's. In 1822 Smith and Phillips surveyed in county Durham a large estate at Haswell, which was situated on magnesian limestone. Knowing the dip of the strata, Smith concluded that there was coal at depth under the limestone, that therefore the land was valuable, and that it should not be sold. Local mining lore, in contrast, held that it was futile to search for coal under limestone. Smith's advice was rejected and no attempt made at the time to bore through the limestone for coal, though in 1833 after a successful sinking a colliery at nearby South Hetton was opened. Thus Phillips realised that local miners, who lacked wide stratigraphical knowledge, could not understand that coal was likely to be found in a new place only if typically accompanying strata were present. As a result of surveying with Smith the lead and copper mines of the Caldeck Fells in the eastern Lake District, Philips became interested in the extraordinary variety of minerals found there. On his return early in 1823 to Kirkby Lonsdale, he investigated these and other minerals, noting their colour, lustre, and translucency, measuring their specific gravity and hardness, and investigating their chemical composition by decomposing them by heat in a crucible. By early 1824 he was a competent mineralogist.[60]

Phillips' other main task in this period was to help Smith with his county geological maps. They used two methods of field work. The familiar one involved Smith and Phillips walking an area together to determine the lines of outcrop of strata to be marked on the map and the sequence of strata. Their unusual mode of mapping or of 'strata-hunting' frequently pressed Phillips for the first time into a role he was to assume more than once in later life: that of active subaltern under a commander-in-chief. Two prospectively rewarding routes of survey, through the country to be mapped geologically, were agreed. Smith then walked one of them, Phillips the other. From time to time Phillips left his lines of survey to rendezvous with the less mobile Smith and to report and discuss progress. In this mapping work, which established the boundaries of different rocks, Phillips often identified them by their fossils, of which he produced catalogues, but he did not disdain the help of mineralogy when appropriate: in delineating in Lincolnshire the oolite, lias, and cornbrash, with whose fossils he was familiar from south-west England, he also relied on mineral content. On these solo peregrinations, which occupied all told about six months, Phillips walked 2,000 miles on routes chosen for their geological interest. He explored in more detail some familiar territory, for instance around Doncaster, but he also extended his geological range. He went as far south as the Charnwood Forest, Leicestershire, whose ancient rocks were new to him, north-west to the Lake District, and north-east to Northumberland where between Haltwhistle and Hexham he walked for the first time the extensive basalt ridge. In 1822 he enlarged his knowledge of the magnesian limestone by tracing its boundaries in county Durham and by cataloguing its fossils he had collected for his cabinet. In 1821 he walked on the line of the Craven fault between Settle and Kirkby Lonsdale. By 1824 Phillips had attained a wide knowledge of the geology of northern England. It was derived, not from Greenough's map or Conybeare and W. Phillips' book, but from surveying on foot guided by his uncle's views and map of 1815. The width of Phillips' knowledge was shown in the large-scale traverse sections he drew and in the published geological county maps which incorporated his findings.[61]

The geological map of Yorkshire published in summer 1821 was different from Smith's other county maps. Measuring thirty-seven by forty-five inches it gave far more detail than his 1815 map. Its most remarkable feature was its pioneering attempt to divide and colour the coal district into groups of grits and sandstones with their accompanying coal seams.[62] In helping Smith to produce this map Phillips learned more about the geology of Yorkshire than of any other county. He also discovered that his uncle was fallible. On the map of Yorkshire Smith represented the alum shale of Whitby as being Oxford clay. Phillips, who had paid his second visit to the Yorkshire coast in 1820, presumably convinced Smith that this was a mistake: the Whitby shale belonged to the lias formation. Smith did not bother to publish a corrected edition of the map and exposed himself to criticism from Conybeare and W. Phillips. It was left to Phillips to produce a corrected version privately and on only a few copies.[63] Perhaps even more importantly by 1824 Phillips had convinced Smith that three particular fossils collected at Scarborough by Smith in 1820 and 1821 belonged to the Kelloways rock of Wiltshire. When Phillips first saw these fossils in 1821 he immediately claimed that they were characteristic of the Kelloways rock. He concluded somewhat impetuously that Smith had discovered Kelloways rock in Yorkshire and that it was possible to match these two occurrences over the large distance between Yorkshire and Wiltshire. Even though Smith had relied on these three fossils to identify Kelloways rock in Wiltshire, he was cautious about making a north-south correlation via them. It was only in spring 1824, when he re-examined the Scarborough cliffs and the stratigraphical position of the disputed rock, that he accepted the correlation Phillips had made three years earlier on the exclusive basis of characteristic fossils.[64] The two episodes show that qua geologist Phillips was no longer a pupil or assistant of Smith but a junior partner capable of making Smith change his mind about an important identification of a stratum.

Smith and Phillips spent much of 1823 at Kirkby Lonsdale where they finished the latest county geological maps, improved that of Yorkshire by introducing the slate rocks (later called Silurian), and enjoyed some measure of repose from working as mineral surveyors. Kirkby Lonsdale was not only an idyllic spot: it was geologically interesting. No fewer than five types of strata, slate, mountain limestone, millstone grit, coal measures, and red sandstone, occurred in the area and were all explored on foot by Phillips. He also traced the direction of diluvial detritus, contemplated the Craven fault which involved a vertical displacement of about 2,000 feet, and produced a section of Ingleborough with estimated thicknesses of its different strata. He noted those strata, later called the Yoredale series, which occur between the millstone grit cap of Ingleborough and the base of mountain limestone on which the hill rests.[65]

It seemed at this point in his life that Phillips would follow Smith in mixing the precarious professional career of a mineral surveyor and engineer with the endlessly fascinating study of geology. This expected destiny was disturbed in spring 1824 when the Yorkshire Philosophical Society, established in York in 1822, invited Smith out of the blue to give there not just a single performance but a course of lectures on geology. Phillips went to York to assist his uncle and faced what he later called 'the crisis' of his life. What skills did he take with him when on 26 January he left Kirkby Lonsdale to walk to York, not directly but going south-east to Doncaster before turning north to York (*c* 120 miles)?

By early 1824 Phillips, then twenty-three years old, had been trained as an engineer and surveyor by his uncle who had also employed him as a geological assistant. Smith had had other geological pupils and disciples in addition to Phillips but none of them worked so closely and for so long with him as his nephew. Through his connection with Smith, Phillips had learned many geological techniques. In the cabinet he had developed such basic curatorial skills as drawing, arranging, naming, and cataloguing fossils. In the field he could make the appropriate observations, measurements, and sketches, from which he could construct traverse and columnar sections. By surveying an area on particular lines, or routes, he could produce the data from which he could generate a geological map by colouring a topographical one. Thus Phillips had acquired under Smith a valuable apprenticeship in the field: he had learned through practice how to make the three chief visual representations used in geology, i.e., maps, traverse sections, and columnar ones. Guided by his uncle he could identify and correlate strata mainly by their fossils but without disdaining their lithology and stratigraphical position. Though an expert fossilist, he was a competent mineralogist as befitted one trained to be a mineral surveyor. He had acquired a wide knowledge of the topography and the broad geological structure of much of northern England, especially of Yorkshire. In two ways he had gone beyond Smith. Capitalising on his facility in French and Latin, Phillips had used the libraries of Richardson and Banks to develop an interest in the zoological aspects of geology. Encouraged by Smith, he nurtured an interest in gadgets, tools, and surveying instruments, some of which he designed and made for himself.

During this period the orphan Phillips was utterly dependent on his uncle financially. In return he aided Smith in his various enterprises, none of which gave his uncle security. It was financially impossible for young Phillips to sit at the feet of such professors of geology and natural history as Buckland at Oxford, Sedgwick at Cambridge, or Robert Jameson at Edinburgh. He was not able to learn from lectures at fashionable scientific sites in London, such as the Royal Institution, or at provincial scientific societies. Travel to the Continent, or even to Scotland, to see renowned geological phenomena was utterly out of the question. So, too, was hobnobbing with coteries of affluent gentlemen geologists. Particularly from 1820 he was, like his uncle, largely cut off from polite science. He had never taught anyone and on only one occasion had assisted a teacher, his uncle, in a lecture room; a situation that was to change decisively in spring 1824.

Notes

1. Phillips to J. Armetriding, 20 Oct 1827, PP.
2. Edmonds, 1982, pp. 141-3; Hunt, 1874; Phillips, notebook containing some incidents in the life of John Phillips and poems, University of London library, MS 517 is useful but not always reliable; William Smith (1769-1839), *DNB*; Anne Phillips (18 March 1803-29 May 1862). For information about excise officers I am indebted to William Ashworth.
3. Edmonds, 1982, pp. 143-5; Hunt, 1874; Mole, 1788; Simson, 1756.
4. Edmonds, 1982, pp. 143-6, 153-4; Hunt, 1874; Phillips, 1844a; Phillips to De la Beche, 18 March 1842, DLB P; Sir Joseph Banks (1743-1820), *DNB*, PRS 1778-1820.

5. On Benjamin Richardson (1759-1832), Jelly, 1832, Mitchell, 1872, Davis, 1943; Phillips, 1844a, pp. 27-31, 36-7; Sedgwick, 1831, pp. 271-9, drew on Phillips to Sedgwick, 5 Feb 1831, Se P, which reproduced and commented on Smith to Phillips, 24 Jan 1831; Sheppard, 1917, pp. 215-16; Joseph Townsend (1739-1816) *DNB*.

6. Cox, 1942, p. 13; Richardson to Smith, 21 Oct 1829, 11 Nov 1830 (q), SP; Edmonds, 1982, p. 146; Davis, 1943, pp. 132-3, 141-2; Richardson to Sedgwick, 10 Feb 1831, in Sedgwick, 1831, pp. 275-6; Smith, 1816-19.

7. Townsend, 1813; Buckland, 1840, p. 251; Conybeare and Phillips, 1822, p. 202; Lonsdale, 1832, p. 241; William Buckland (1784-1856), *DNB*, reader in mineralogy 1813-56 and geology 1818-56; William Daniel Conybeare (1787-1857), *DNB*, discoverer of ichthyosaurus, vicar of Brislington, Bristol, 1819-27, of Sully, near Cardiff, 1827-36, of Axminster 1836-44, dean of Llandaff Cathedral, Cardiff, 1845-57; William Lonsdale (1794-1871), *DNB* curator and librarian, Geological Society, 1829-42.

8. Jelly, 1832, p. 304 (q); on Harry Jelly (1800-1843) and his family, Torrens, 1975, pp. 97-9, Torrens and Winston, 2002; Phillips 1836a, p. xiii; Phillips, 1841a, p. vi; Phillips, 1871a, pp. 6, 411, 414-15.

9. Phillips to Lyell 26 Jan 1864, LP; Phillips to De la Beche, 18 March 1842, DLB P.

10. Edmonds, 1982, pp. 146, 153-4. Phillips presumably used Werner, 1809.

11. Phillips, draft preface to intended work on fossil conchology, 18 March 1819 (q), PP, box 51, folder 8; Phillips to Armetriding, 20 Oct 1827, PP.

12. Hunt, 1874, p. 597.

13. The standard works on Smith's career are Phillips, 1839a, 1844a, Sheppard, 1917, Cox, 1942, Eyles, 1969a and b, 1975, and Torrens, 2001, 2003; Knell, 2000, pp. 12-27 illuminates Smith's methods.

14. Porter, 1977, pp. 136-8.

15. On John Farey (1766-1826), *DSB*, see Torrens, 1994, p. 68 (q), 1998; Rudwick, 1976b, p. 169; Eyles, 1985, pp. 39-43; Torrens and Ford, 1989, pp. 1-28; James de Carle Sowerby (1787-1871), *DNB*, son of James Sowerby (1757-1822), *DNB*.

16. Fitton, 1832, p. 50; Smith, 1818a; Conybeare and Phillips, 1822, p. xlv (q); Warner, 1801; Parkinson, 1811; Farey, 1806, 1818a; Townsend, 1813; Richard Warner (1763-1857), *DNB*, a Bath curate 1795-1817; James Parkinson (1755-1824), *DNB*. The claim by Laudan (1976), that Smith's stratigraphy was not based on fossil-labelling, ignores much evidence to the contrary.

17. Porter, 1977, pp. 177-80; Smith, 1815a and b, p. 11; Rudwick, 1996.

18. On the sale, Eyles, 1967; Phillips, 1844a, pp. 78-9; Phillips to Harcourt, 17 Nov 1828, HP: Charles Dietrich Eberhard Konig (1774-1851), *DNB*, was keeper of natural history, British Museum, 1813-51.

19. Rudwick, 1996, p. 28.

20. Smith, 1816-19, 1817a and b. The section is reproduced by Sheppard, 1917, p. 144.

21. Smith, 1817a, p. vi.

22. Edmonds, 1982, pp. 147-8; Sowerby, 1812; Phillips to De la Beche, 13 Jan 1844, DLB P; Phillips to Sabine, 20 July 1854, Sa P; Cleevely, 1974, pp. 491, 549.

23. Eyles, 1967, pp. 201-3; Phillips to Lyell, 21 Dec 1840, LP (q); Farey, 1819a, p. 114; Phillips, 1844a, pp. 28, 130; Rudwick, 1976a, pp. 139-40.

24. Smith, 1817a, pp. iv, 119 (table); Phillips, 1844a, p. 79; Phillips to Ramsay, 22 Oct 1848, RP; Phillips, 1860a, pp. xxxciii, xli (table).

25. Smith, 1817a, pp. iv-v; Lhwyd, 1699; Brander and Solander, 1766; Woodward, 1728, 1729; Plot, 1677; Morton, 1712; Walcott, 1779; Linnaeus, 1758-9; Lamarck, 1802-9; Ellis and Solander, 1786; Klein, 1778; Martin, 1809; Sowerby, 1812; Parkinson, 1804, 1808, 1811; Townsend, 1813, 1815.

26. Cuvier and Brongniart, 1808, 1810, 1811; Cuvier, 1812; Farey, 1810; Rudwick, 1996, 1997; Georges Cuvier (1769-1832); Alexandre Brongniart (1770-1847).

27. Cox, 1942, pp. 9, 58-61; Eyles 1969b, pp. 98-9; Edmonds, 1982, p. 148; Phillips, 1844a, pp. 80-6.

28. On Mary Ann Smith, d 1844 aged 52, see Torrens, 1992a, 2003. Smith's first biographer mentions her just once: Phillips, 1844a, p. 94, alludes to her very manifest mental aberration in 1820.

29. Phillips, 1844a, pp. 90-1; Edmonds, 1982, p. 150.

30. Phillips, 1844a, pp. 86-8.

31. Smith, 1818b; Eyles, 1969b, p. 100.

32. Phillips, 1844a, p. 88; Phillips, 1836a, p. x; Phillips, Notes of tours and excursions 1819-1822, notebook 3a, PP.

33. Phillips, loose notes, Journey from Wentbridge to Dean Forest ... Oct-Nov 1819, notebook 2, PP; Phillips, notebook, Wheelcarriages, railways, roads, nd [c1819], Museum of History of Science, Oxford, Ms Museum 36.

34. Phillips, 1844a, p. 86; Cox, 1942, pp. 61-2; Eyles, 1969b, p. 100.

35. Smith, 1818-19; Edmonds, 1982, pp. 149-52; Eyles, 1969b, p. 101; Sheppard, 1917, pp. 153-4.

36. Edmonds, 1982, p. 152; Smith, 1806; Phillips to Chadwick, 9 May 1843, CP. Edwin Chadwick (1800-90), *DNB*, wanted to see any additions in manuscript Smith had made to Smith, 1806.

37. Phillips, 1844a, pp. 9-10, 13.

38. Phillips, 1844a, p. 88; Sheppard, 1917, p. 154.

39. Edmonds, 1982, p. 150; Edward Hawke (1774-1824) of Womersley Hall, ten miles north of Doncaster.

40. Phillips, 1844a, pp. 92-4.

41. Phillips, notes, 'Sept, Oct, Nov 1817', notebook 1, PP; John Bird (1768-1829) Whitby artist, first curator of the Whitby museum 1823-9; Thomas Hinderwell (1744-1825) Scarborough antiquarian; John Hornsey, Scarborough schoolmaster.

42. Cox, 1942, pp. 56-8.

43. Phillips to Sabine, 20 July 1853, Sa P; Eyles, 1969b, pp. 101-5.

44. Twyman, 1967, 1970; Rudwick, 1976b, pp. 154-7.

45. Phillips, notebook, 'System of lithography, 1819, PP, box 51, folder 16; Phillips, lithographed circular about his method of drawing and printing from stone, nd, watermark 1816, SP; Lee to Phillips, 5 Jan 1869, PP; Eyles, 1969b, pp. 99, 101; Edmonds, 1982, pp. 150-1; Twyman, 1972, pp. 11-12; Twyman, 1976, pp. xxxvii-xxxix.

46. In 1833 John Ford, (1801-75), headmaster of a Quaker school for boys (later Bootham School, York), noted with admiration that Phillips was a lithographer: Ford, 1877, p. 31.

47. Phillips, 1871a, pp. 230, 232; Phillips, 1855c, pp. 411, 433; Phillips to De la Beche, 13 Jan 1844, DLB P.

48. Farey, 1815; Kidd, 1815; John Kidd (1775-1851), *DNB*, professor of chemistry, University of Oxford, 1803-22, lectured on geology; George Bellas Greenough (1778-1855), *DSB*.

49. Laudan, 1987, is excellent on Abraham Gottlob Werner (1749-1817) and his followers but scornful of Smith.

50. Fitton, 1817a, p. 188; 1817b, pp. 71-4; William Henry Fitton (1780-1861), *DNB*, Robert Jameson (1774-1854), *DNB*, professor of natural history, Edinburgh, 1804-54.

51. Farey, 1818a, p. 173 (q); Farey, 1818b, p. 186 (q); Fitton, 1818, p. 312.

52. Fitton, 1818, p. 337 (q); Smith, 1818a, pp. 215, 218 (qs); Parkinson, 1811, p. 325.

53. Greenough, 1819, pp. 287-94; Rudwick, 1985, pp. 65-7; Torrens, 1990, p. 660 (q); Farey, 1819b; Conybeare to Greenough, 8 June 1819, GP; Robert Bakewell (1768-1843), *DNB*.

54. Greenough, 1820; Woodward, 1907, pp. 56-9; 'Greenough', *DSB*; Farey, 1820; Cox, 1942, pp. 41-3; Rudwick, 1963; Greenough, 'Smith's doctrine for which he received the Wollaston medal in 1831', nd, GP, box 17, resented the way in which Smith had become 'the darling and idol of the English school'; Thomas Webster (1773-1844), *DNB*, keeper of the museum and draughtsman, Geological Society, 1812-26, and secretary 1819-27.

55. Phillips, 1844a, p. 77.

56. Phillips, 1844a, p. 103; Clark and Hughes, 1890, vol. 2, p. 509; Torrens, 1990, p. 659; Conybeare and Phillips, 1822, pp. 218-19, xlvi-xlvii (qs), 165; Adam Sedgwick (1785-

1873), *DNB*, professor of geology, Cambridge, 1818-73; Fitton, a secretary, Geological Society 1822-4.

57. Phillips, 1844a, p. 98; Clark and Hughes, 1890, vol. 1, p. 251; Otley, 1823; on Jonathan Otley (1766-1856) see Ward, 1877.
58. Sheppard, 1917, p. 80; Phillips, 1844a, p. 101; LPLS, minute book of transactions, 16 Nov 1821; John Atkinson (1787-1828); Edward Baines (1774-1848), *DNB*; Edward Sanderson George (1801-30).
59. Phillips, notes and drawings 1819-22, Museum of History of Science, Oxford, Museum MS 34.
60. Phillips, notes of tours and excursions 1819-22, notebook 3a, PP; Phillips, 1844a, pp. 102-3, 106; Eyles, 1971; Phillips, characters of minerals and classification, with analyses made at Kirkby Lonsdale 1823, notebook 9, PP.
61. Phillips, 1844a, pp. 94-100; Phillips, notes of tours 1819-22, PP, notebook 3a; Cox 1942, p. 61; Phillips to Greenough, nd, fr 24 Nov 1836, GP; Phillips to Sedgwick, 20 May 1829, Se P.
62. Sheppard, 1917, pp. 155-60.
63. Phillips, 1844a, p. 96; Phillips, 1829b, p. xii; Conybeare and Phillips, 1822, p. 197.
64. Phillips, 1829b, pp. xii, 142-3.
65. Phillips, 1844a, 104; Phillips, notebook, 1822-3, notebook 8, PP.

Chapter 2

The Young Lecturer and Keeper

Phillips joined Smith in York in February 1824, primarily to assist with the eight lectures his uncle had undertaken to give for £50 to the Yorkshire Philosophical Society. In June Phillips made his first appearance at a YPS meeting as a guest of Smith. By summer 1829 he had secured total independence from Smith, who from 1826 lived in Scarborough and then near it. In 1826 Phillips assumed the post of first keeper of the Society's museum at a salary of £60 per annum, an emolument which he supplemented by fees from lecturing in York and in large towns in northern England. Elected a fellow of the Geological Society in 1828, he published next year his classic work on the geology of the Yorkshire coast (Fig. 2.1). For decades it was regarded as a leading prototype of regional stratigraphical treatises. How had the mineral surveyor transformed himself in just over five years into a geologist who combined curating, lecturing, research, and publication? How was it that Phillips, and not Smith, was able to establish himself as the leading expert on the coastal geology of Yorkshire, an area and a county on which Smith had worked for several years?

2.1 The Yorkshire Philosophical Society

When the president of the YPS, William Vernon Harcourt, invited Smith to give a course of lectures on geology in York in early 1824 he did so on the basis of Smith's reputation and of his geological map of Yorkshire which the Society had bought for its library in 1823. While the YPS was negotiating with Smith it bought his *Strata identified* and his *Stratigraphical system* for reference only. At this time Smith was not forgotten elsewhere. In March 1824 he was elected an honorary member of the Bristol Institution for the Advancement of Science, Literature, and the Arts, through Conybeare's influence, and in December 1824 honoured similarly by the Leeds Philosophical and Literary Society. In London Fitton more than kept Smith's name alive. In reviewing Buckland's *Reliquiae diluvianae* in 1824 he stressed that Smith's works formed the basis of Buckland's popular columnar section of British strata, published in 1818, and of all similar tables.[1] Next year Thomas Webster claimed that geologists should be grateful to Smith as the father of modern English geology.[2]

For Smith and the YPS a course of lectures was novel. Smith had talked geology endlessly but had never lectured. The YPS wanted to mount lectures as part of its aim of diffusing a taste for science, which as well as being valuable in itself might increase the membership, income, and resources of the Society. It did so first via geology, followed by chemistry and zoology.[3] This unusual choice was made because the YPS was not like other provincial societies which were devoted to the advancement and diffusion of science and literature in general: based in York, it was a scientific society with the ambitious aim of pursuing the geology of the large county of Yorkshire. This testing research programme had been set out by Harcourt, who acknowledged that

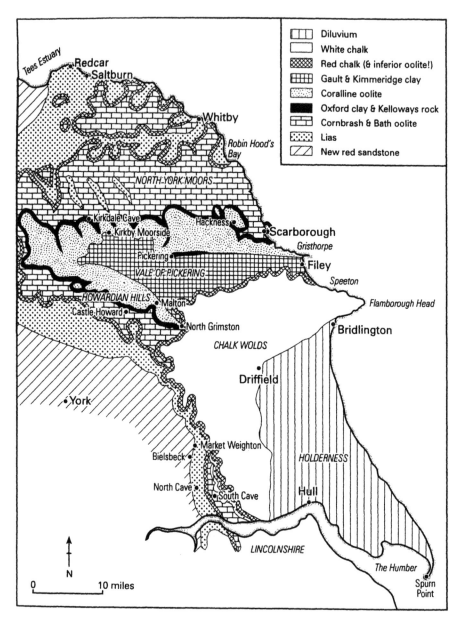

Fig. 2.1 Simplified geological map of north-east Yorkshire, based on that in Phillips, 1829b, and using his stratigraphical terminology. It shows the complicated geology of the area, the way in which inadequate colouring unintentionally conflated two different strata, and important geological sites in Phillips' lifetime.

he was the nurse, watcher, and promoter of the YPS but not its founder (Fig. 2.2). It was established by William Salmond, James Atkinson, and Anthony Thorpe on 7 December 1822 in York primarily to found a public museum devoted to the antiquities

Fig. 2.2   The reverend William Vernon Harcourt, from a bust by M. Noble, usually dated *c* 1871 but probably *c* 1851. It suggests his dignified firmness and tenacious dedication, which were amply shown in his management of scientific organisations.

of York and the geology of Yorkshire. All three had acquired specimens of bones from the recently explored cave at Kirkdale, near Kirkby Moorside, about twenty-five miles from York, and gave them to the museum as the nucleus of its geological collection. Though others possessed Kirkdale material, the three founders donated a sizeable and instructive collection of 245 specimens to the YPS. Their concerns were not uniform. Salmond, who possessed private means, had paid for excavation of the cave and mapped it in summer 1821, producing the plan used by Buckland when he explored it in December that year. Salmond thus facilitated Buckland's famous theory of 1822 that the cave was a hyena den. Atkinson, a local surgeon, was interested in comparative anatomy and its use in fossil osteology. Thorpe, a solicitor in York, was an antiquarian and in 1822 president of the York Subscription Library.[4]

Harcourt soon joined this trio and changed the focus of the Society from collecting, storing, and displaying geological and archaeological specimens to a programme of active research embracing the stratigraphical geology of Yorkshire and York antiquities. He had the social position, education, and leisure to suggest and to sustain this particular way of tapping local opportunity and local pride. He was the fourth son of Edward Vernon Harcourt, archbishop of York 1807-47. Intent on a clerical career, in 1807 he went to Christ Church, Oxford, where he graduated in classics. At Oxford he developed an interest in geology, attending some of Buckland's earliest lectures and meeting Conybeare. In 1814 his ordination launched him on a comfortable career as the incumbent of three livings near York, at Bishopthorpe 1814-24, Wheldrake 1824-34, Bishopthorpe 1835-8, and Bolton Percy 1838-61, all presented to him by his father. Favoured by extensive and powerful connections and a persuasive manner, Harcourt interested himself in philanthropic and scientific affairs in York and the surrounding area.[5]

His vision for the future of the YPS gave pride of place to the elucidation of the geology of Yorkshire. As the published aims of the Society made clear in early 1823, Yorkshire was interesting because of the great variety of strata which traverse it yet its 'geological relations' were imperfectly understood. Harcourt hoped to remedy this deficiency by adopting an approach advocated by Francis Bacon: workers in different areas would supply specimens of fossils and minerals, drawings and casts of fossils, report careful observations about boundaries of strata, and furnish traverse sections of strata to a co-ordinating central museum in York. On the basis of these minutiae, Harcourt hoped, general truths about Yorkshire's geology could be 'raised'. At the YPS meeting held on 6 January 1823, Harcourt dilated on its chief aim, which was to be research on the secondary strata, which he averred are almost all present as one moves east to west across the county, ie from the chalk to the mountain limestone. He pointed out that the geology of Yorkshire was in confusion: the maps seemed incorrect and eminent geologists differed about many aspects of its stratification. This confusion was to be banished by fossil-labelling of strata: fossils, he claimed, provided 'the most certain means of determining the order of succession of the strata'.[6] Thus the published aim of elucidating the geology of Yorkshire meant working on the stratigraphy of the secondary strata using characteristic fossils, not order of superposition and not lithology, as the main criterion.

These ideas of Harcourt's were derived primarily from Conybeare and, to a less extent, from Buckland, with both of whom Harcourt had been in correspondence. In a letter of 27 December 1822 Conybeare summarised the desiderata of Yorkshire geology; his recommendations were in part incorporated in the published aims of the Society, were read to it on 5 February 1823, and guided Harcourt for several years. In response to Harcourt's request for advice, Conybeare was generous with a 'few hints' to the Society of which he had already been made an honorary member. He pointed out that Yorkshire, with secondary strata ranging from mountain limestone up to chalk, was imperfectly explored and advocated the study of their fossils. He advised the Society to investigate the Whitby alum shales, which he thought belonged to the lias stratum, paying particular attention to their fossil saurians and crocodiles. He urged the YPS to try to illustrate the connection between oolitic rocks in Yorkshire and those elsewhere. Amplifying what he had just published in his *Outlines* about the problematic nature of Yorkshire oolite, Conybeare asserted that the top priority was field work on the terrain below the escarpment of the chalk Wolds, especially at North and South Cave, half-way between Hull and Goole, and at North Grimston, five miles south-east of Malton. He also recommended field work on the north, west, and south-west edges of the Vale of Pickering. Conybeare called for surveys of the new red sandstone, with its important local variations, of the belt of magnesian limestone, especially its relation to the coal measures, and of the millstone grit of the Pennine hills, in each case producing a full list of fossils, a map, and detailed traverse sections. He said little about the mountain limestone but arranged for J.S. Miller to donate to the YPS a copy of his book on crinoids to stimulate research on it. As in his book Conybeare stressed that authorities disagreed about basic points of stratification.

In his letter to Harcourt, Conybeare gently pinpointed some inadequacies of Smith. He regretted Smith's too frequent failure to give descriptive accounts of the evidence which ought to have been the basis of his maps. Apropos the Vale of Pickering he lamented that Smith did not distinguish the districts in which substrata were concealed by diluvium: for Conybeare any geological map had to show 'fundamental rocks'. He liked the first version of Smith's 1821 map of Yorkshire for its good scale and distinctions of strata; but he thought it wrong in mapping the Whitby alum shales as Oxford clay.[7] Conybeare was more severe privately and publicly on the recent account of the geology of north-east Yorkshire by Young and Bird. He confided to Harcourt that their coast sections were acceptable but their map was too small and general. In his *Outlines* Conybeare berated Young and Bird for forming general ideas by inspecting a limited district. Even worse they were decidedly and flippantly hostile to the idea of fossil-labelling of strata because they ascribed to the scriptural deluge the formation of most strata, the fossils of which were therefore so inter-mixed and dispersed that no single fossil could possibly be peculiar to just one stratum.[8]

Buckland gave useful advice about collecting fossils, collectors, and books, not only as an honorary member of the YPS but also in return for the help received in spring 1822 from Harcourt, who had sent drawings of specimens from Kirkdale cave. Buckland did not offer suggestions about interior locations for stratigraphic field work, but urged the YPS to organise the collecting of fossils from Whitby and Malton, of fossil plants from collieries, and of fossil shells and minerals from the lead mines in the mountain limestone area. A fortnight later Buckland's advice was still focussed on collecting. He advocated securing Whitby fossils through Young and Bird and inducing Hinderwell to bequeath his collection of Scarborough fossils to the YPS. Buckland also sent Harcourt a copy of his syllabus of lectures and gave a list of about two dozen geological books and maps, the most important being in his view Conybeare and Phillips' *Outlines*, of which two or three copies he thought should be available as the YPS's textbook.[9]

Harcourt took seriously the advice he had received from Conybeare and Buckland. He gave much attention to the YPS's museum which was opened to visitors in July 1823. That spring he visited London, not just to buy books, periodicals, and specimens but to learn from the mode of arranging and preserving fossils adopted in the museum of the Royal Institution, where he consulted Michael Faraday, and that of the Geological Society, where he questioned Webster. By the end of 1823 the YPS museum housed 2,000 specimens of minerals and fossils, mainly 'illustrative of the geology of Yorkshire', including fossils from the Oxfordshire oolite donated by Buckland, coral rag and lias fossils from Wiltshire and Somerset given by Conybeare and Miller, and Whitby fossils from Young. Harcourt himself donated almost 500 English fossil specimens.[10] In promoting the museum he was Bucklandian but in his geological field work, revealed to the YPS before Smith and Phillips appeared at it, Harcourt showed himself as a pupil of Conybeare. In March 1823 after reporting the donation by Conybeare of his *Outlines* and the fossil specimens of oolite and lias from south-west England, Harcourt stressed the analogy between some of them and a specimen donated that day from the Malton oolite. He went on to urge that the YPS could not only assemble the '*materials* of knowledge', such as specimens, but also advance it. To illustrate that advancing Yorkshire geology was feasible, he cited his own first experience of field work, on the junction between the new red sandstone and magnesian limestone in the area between Ripon and Boroughbridge. From the supposed order of stratification, he expected the former to lie on the latter but he found the line of junction concealed by vast diluvial deposits. He thus learned by experience the difference between superficial deposits and fundamental rocks as well as the difficulty of mapping boundaries of strata covered by diluvium. By October 1823 he had investigated South Cave, as Conybeare had suggested, specifically to discover which parts of the oolite series occurred in Yorkshire. He had also done field work on the clay at the northern and western edges of the Wolds, under the red chalk, and had concluded it was gault, of Kent and Sussex, on the basis of its fossil belemnites. Later that year Harcourt learned in the field about oolites in the Bristol area from Conybeare himself. By the time he invited Smith to lecture in York, Harcourt's commitment to geology had led to his being elected FGS in 1823.[11]

2.2   The geological lectures of Smith and Phillips

In accepting Harcourt's invitation to lecture at York, Smith regretted that his geological map of Yorkshire (1821) was imperfect, confessed that a complete account of the county's stratification was not easy, and that he and Phillips were preparing two traverse sections of it. He also made it clear that with Phillips' assistance and the use of the YPS library he could soon arrange the YPS collection of fossils both stratigraphically and zoologically, as he had done with his collection sold to the British Museum. He advised Harcourt that his *Stratigraphical system* would help in this task. For Smith's eight lectures, given in February and March 1824, he needed new geological maps, new traverse sections, and new drawings of fossils, all of which Phillips helped to provide. Smith devoted two lectures to Yorkshire, one on its economic geology and the other on its stratigraphy. In the latter Smith used fossil-labelling of strata to make long-distance correlations between Yorkshire and southern and south-western England. He even averred that the coast from Bridlington to Redcar gave the finest section of strata in Britain. Smith's lectures were so well attended that his fee was increased from £50 to £60, leaving the YPS with a profit of £23. They made a deep impression on Harcourt, who had Smith elected an honorary member of the Society on 9 March 1824. For Harcourt, Smith was *the* expert on Yorkshire geology so Harcourt advised him to publish by subscription his materials on it and he welcomed Smith's intention of revising his geological map of the county. Smith responded by staying in York, attending the next three meetings of the Society, and introducing Phillips as a visitor in June. Later that month Smith left York for Scarborough and rarely returned to it.[12]

Meanwhile Phillips had become a temporary employee of the YPS. In February 1824 he was paid £20 to arrange its fossils in stratigraphical order, a task which occupied him until early August. Before his lectures Smith had suggested that, with Phillips' help, he could arrange the collection. In his lectures Smith exhibited fossil specimens, chosen from the Society's collection, to illustrate his themes, many of which were stratigraphical. It soon became clear to Harcourt and to George Goldie, the senior secretary of the YPS, that Phillips' competence in stratigraphical arrangement of fossils was far greater than that of Salmond and J.B. Graham, the honorary curators of geology.[13]

This freelance employment enabled Phillips to use the Society's library which possessed decent runs of the main scientific and geological periodicals and reference only copies of Conybeare's *Outlines*, Cuvier's *Ossemens fossiles*, and Parkinson's *Organic remains*. In Phillips' wandering years access to publications was difficult. At York he could keep himself up to date from periodicals and could consult the books by Conybeare and Cuvier to which he was to be greatly indebted. His conservatorial work for the YPS had important intellectual consequences. Through arranging the Society's fossils Phillips apprehended that many of the strata of north-east Yorkshire could be identified, confidently and not tentatively, with well-known strata in the south of England. To illustrate this belief he compiled comparative catalogues of fossils which he used in his 1829 book. More generally Phillips began to think of devoting much of his time to fulfilling the YPS's chief aim of elucidating and illustrating the geology of Yorkshire. His growing loyalty to the Society was shown by two donations of fossils he made in May and July, and by his contribution as a visitor to its June meeting when he spoke about the way in which a single fossil genus occurred in many strata. In return he was elected an honorary member in August 1824.[14]

Phillips took two breaks from his cabinet work for the YPS. With Smith and Salmond he visited Kirkdale Cave and the Castle Howard area in March 1824. Next month he renewed his friendship with Edward George by walking from York to Leeds and then doing a week's field work with him in Nidderdale and Wharfedale. On this trip Phillips had at his disposal for the first time a mountain barometer, a portable mercury barometer which enabled its users to calculate their altitude from a barometric reading. It was not new in geological surveying. One of the best known types had been devised about 1750 by Jean André De Luc, a Swiss naturalist, and had been taken by De Saussure to the summit of Mount Blanc in his famous ascent of 1787. The mountain barometer was a particularly useful surveying instrument for a geologist who wished to construct coastal sections and to compare strata on the sides of valleys in the Yorkshire dales. Phillips was so impressed by George's instrument that he soon constructed his own pocket barometer.[15]

Having spoken in discussion at two YPS meetings, Phillips made his debut as a formal lecturer in September 1824 when he contributed three lectures on palaeontology inserted into a course of six on geology given by his uncle as a freelance venture in the town hall, Scarborough. Smith knew that palaeontology was becoming more important in geology, that his nephew was more expert on it than himself, and presumably trusted his nephew to lecture well. The stakes were high: lacking the protection of a learned society, Smith and Phillips depended at Scarborough on the admission fees of one guinea for the course and three shillings per lecture. Their experiment as joint lecturers was such a success that they repeated it in December 1824 at Hull in the Assembly Rooms under the auspices of the Literary and Philosophical Society, founded in 1822, which engaged them for £50.[16]

In autumn 1824 Smith and Phillips were based in Scarborough where they were working as surveyors for Sir John Johnstone who owned estates at nearby Hackness and Harwood-dale. By this time Phillips was doing solo field work on the geology of Yorkshire. In August 1824 on a walk from York to Scarborough he found Kimmeridge clay on the north side of the Vale of Pickering and then began sustained work on a geological section of the Yorkshire coast. He had at his disposal an excellent Englefield mountain barometer lent to him by Francis Cholmeley, then a vice-president of the YPS, through the good offices of Goldie. This instrument enabled Phillips to measure the heights of cliffs and exposures with accuracy, to make detailed sections at different places and to compare them; he could then confirm the order of succession of strata by comparing their fossils. Phillips took good care in late autumn to return to the YPS the specimens he had been allowed to borrow for the Scarborough lectures, to donate fossils collected at or near Scarborough, and above all to report twice to the Society the recent stratigraphical discoveries made there by Smith and himself. Harcourt was impressed by the coloured geological map which they sent and especially by their discovery of Oxford clay and of Kelloways rock at Castle Hill, Scarborough, and he was quick to realise that it was 'confirmed' by Fitton's recent work on the strata below the chalk in Sussex.[17]

In early 1825 Phillips remained in Hull, after his lectures there with his uncle, arranging the Society's fossil collection in stratigraphical order. At this juncture he received from Harcourt an invitation to give for £20 five or six lectures on geology to the YPS to supplement those given the previous year by Smith. Presumably Harcourt feared that Phillips might stay in Hull and certainly he wanted to maintain the YPS's contact with him. Phillips gave his first solo course of lectures in February 1825, four of the five being on the natural history of fossils. He devoted one entirely to vertebrates and concluded by reviewing the importance of fossils in the identification of strata. En passant he commented on the modes of existence of organisms in the geological past. The day before these lectures began, Phillips had delighted Harcourt not only with his comments on his drawing of a fossil crocodile recently found at Whitby but particularly with a long account of the recent research by Smith and himself on the geology of north-eastern Yorkshire. They had compared the succession of strata there with the general series of English strata using the three criteria of position, mineral character, and fossil content. Their results indicated that in the comparison mineral characters agreed 'pretty well' but the fossils did so 'remarkably'. Phillips revealed the postulated sequence of strata down from the Kimmeridge clay, through the coralline oolite, the calcareous grit, Oxford clay, and Kelloways rock, to the cornbrash; and he concluded that deposits of equal antiquity enclosed analogous but not identical fossils.[18]

After his York lectures Phillips returned for two months to Scarborough which he left for Sheffield where he and Smith gave in May 1825 a joint course of nine lectures in the Music Hall under the auspices of the Sheffield Literary and Philosophical

Society, founded 1822, for a fee of £60. This time Phillips used his York material to give four lectures on palaeontology compared with his three at Scarborough. The occasion brought to an end Smith's short career as a lecturer on geology. After his Hull lectures Smith was so stricken with rheumatic paralysis of the legs that on his journey to Sheffield he had to be carried in a sedan chair and lifted into coaches; while lecturing there to several hundred people he was confined to a chair. During his time in Sheffield Phillips visited the nearby mountain limestone area of Derbyshire.[19] He also turned down in May 1825 a sounding from George about being for nine months in the year paid sub-curator of the Leeds Society's museum, a post to which Henry Denny was appointed at a salary of £80 pa in spring 1826. Phillips felt he had to respect the wishes of Smith for whom he was making traverse sections in the coal measures, that a wandering life suited his health better, and that he should nurture his career as a lecturer on geology, his favourite subject.[20]

When the Sheffield lectures were completed Phillips went to Wakefield where in early June he gave in the Music Saloon as a private venture a course of six lectures on geology, including four on palaeontology. The experience was salutary: the course was not sponsored by a local literary and philosophical (lit-and-phil) society and it could not rely on a local taste for geology as developed as that at Scarborough, so it was not well attended. Phillips made little profit from those who paid 10/6 for the whole course or 2/6 per lecture. Having left his Wakefield host, the reverend Samuel Sharp, Phillips went to Leeds where he joined George in early July for some frustrating research on the coal measures. Initially Phillips was optimistic about correlating Leeds and Sheffield strata using a particular fossil as a marker in the scale of stratification; but, owing to attenuations of sandstones, the interposition of shales, the changing nature of coal seams, and the sudden faults in the rocks, he soon found it difficult to correlate even neighbouring strata. He realised that there was nowhere in the coal area equivalent to Scarborough castle hill with its fine exposure of a succession of strata. Henceforth, though he remained interested in coal measure plants, he was wary about working on the Yorkshire coal field per se. After George's death in 1830, his family consulted Phillips about the publication of his research on the Yorkshire coal field; but Phillips did nothing to secure the reputation of his friend whose paper was published belatedly and locally in 1837 in Leeds.[21]

2.3 Curating in York

In July 1825 Phillips reverted to mineral surveying when he examined an estate at Scotton, near Richmond, owned by Beilby Thompson, a vice-president of the YPS, to see if coal or lead ore occurred there. Characteristically he then geologised in Swaledale and Wensleydale, making sections which he used in his mountain limestone monograph. Next month he was working on the YPS collections partly to complete drawings for himself and partly because Harcourt had accepted his offer of assistance.[22] Fortune then favoured him. Adolphe Brongniart, the leading palaeobotanist in France, visited the museum in August and was shown some fossil plants from Whitby by Phillips. Brongniart realised immediately that they belonged to the middle series of his three divisions of fossil plants and confirmed his views about Höer in Denmark where he had seen the same genera and species and had inferred that the strata containing them were oolitic. Brongniart not only drew YPS specimens and received duplicate specimens but he also treated Phillips as a scientific equal.[23] This accolade by a distinguished foreigner presumably convinced Harcourt that the YPS would benefit by appointing Phillips as the first paid keeper of its museum. As recently as May 1825, in response to a query by Buckland, Harcourt had recommended Phillips for a post in a geological survey of the East Indies. By autumn 1825 Harcourt had decided that the growing museum needed more constant attention than its unsalaried curators could give in their spare time, and that without supervision the honorary curator of geology, J.B. Graham, could not cope. Phillips' arrangement of the fossil collection had been praised by no less than Buckland. Moreover Phillips had shown that he shared Harcourt's geological aims for the YPS. Phillips would be more than useful in the projected great northern museum, which Harcourt envisaged as a school of science and as a location for collections, books, lectures, and meetings. Goldie had long thought highly of Phillips as a curator. In December 1824 he wanted Phillips to sort the YPS collection of fossils and remove its numerous '*faults and dislocations*'. Goldie realised that Phillips' knowledge of conchology was the 'very best key to a knowledge of the fossil species'.[24]

On 11 October 1825 the YPS Council appointed Phillips as keeper of the museum and draughtsman from 1 January 1826 at a salary of £60 pa, provided half of it could be obtained voluntarily. That was soon raised with Harcourt and Goldie each giving £5. Next month Phillips' post was formally approved: he was to attend three days a week, from 10 am to 4 pm, for nine months of the year. Why did Phillips accept the York post when five months earlier he had rejected a similar one at Leeds? Through solo lecturing he had become a little more detached from Smith. No doubt he realised that some financial security was better than none at all. The demands on his time at York would not be unremitting: his conditions of employment gave him time to continue his lecturing and geological field work. As a country lad he liked York, an ancient ecclesiastical city and regional capital unbesmirched by industrialisation and, with a population of about 25,000, agreeably small. In contrast Leeds was dirty, smoky, and four times as big. By autumn 1825 he had contacts in York not matched in Leeds where he knew well only George and Atkinson. In old Ebor he was highly regarded by leading members of the YPS, such as Goldie, Salmond, Cholmeley, William Taylor, William Marshall, Eustachius Strickland, and especially by Harcourt. Perhaps Phillips sensed that Harcourt would be a more effective patron than George and that York was more likely than Leeds to become the scientific omphalos of north east England.[25] In celebration of his appointment Phillips became the leading speaker at the three remaining YPS meetings of 1825. He revealed in October the results of his latest research on the Yorkshire coast. He had just surveyed it from Saltburn to Bridlington, measuring the heights of cliffs with a telescope furnished with a tangent scale, and had produced an enlarged coastal section. He had identified strata by their fossils but made the caveat that the same fossils were not always found in the same strata. He noted that the Bridlington chalk contained twenty fossils, of which two-thirds were found in the chalk of southern England. Obsessed with 'analogous instances of the grouping of fossil shells', Phillips had taken censuses of fossils in order to define numerically the notion of characteristic fossils.[26]

Before Phillips took up his York post, he undertook his largest lecture course to date. For some time George and Atkinson had wanted him to lecture in Leeds, so they arranged for the Philosophical and Literary Society to pay him 40 guineas for ten lectures given in its Hall in December 1825 and January 1826. In the five devoted to palaeontology Phillips worked through successive types of fossils, ie plants, plant-like animals, shells, crustaceans, fish, reptiles, and the quadrupeds found by Cuvier in the gypsum of the Paris basin. Phillips was at pains to stress that palaeontology was not only useful in identifying strata but it also revealed the former conditions of the earth, including the antediluvian one. Like the York appointment, the Leeds lectures confirmed that, even though Phillips still described himself as a mineral surveyor, he had become totally independent from Smith occupationally, financially, and geographically. Phillips had assumed the role of lecturer quickly and successfully: the Leeds Society gave him a bonus of 10 guineas for his 'very admirable course'.[27] From September 1824 he had assisted Smith thrice by giving three or four lectures on palaeontology inserted into his uncle's courses. From February 1825 as a solo lecturer he had given three courses, the third being double the length of the first. From 1826 he could look forward to augmenting his salary as YPS keeper by lecturing locally on geology and palaeontology, without injuring Smith who was settled at Scarborough and indifferent to keepering or lecturing.

Now separated from Smith, Phillips was greatly indebted to his uncle. Smith's brief lecturing career had enabled Phillips to transform himself from an assistant mineral surveyor to an assistant and then solo lecturer. He had learned so much about field geology from his uncle that he was proud to style himself Smith's pupil. He admired his uncle's selfless devotion to geology, his calm and elastic mind, his fortitude, and his cheerfulness in the face of relentless adversity. Above all he was eternally grateful to his uncle's natural goodness of heart, which had literally kept him alive from 1807. At the same time Phillips had learned, by close contact with Smith, what to avoid. As a lecturer Smith was charming but unstructured and rambling. As a mineral surveyor Smith endured a fluctuating income, much of which he spent on geological mapping, and he lacked opportunities and indeed the taste for regular study and formal publication. He therefore suffered penury and saw his geological achievements ignored or disputed. Smith's sturdy independence and self-reliance precluded co-operating with others or learning from them. By 1826 his narrow range as a geologist was apparent and some of his views embarrassing. He was uninterested in sciences collateral with geology, such as zoology and botany, and by force of circumstances and conviction confined his stratigraphical correlations to Britain.[28]

The post of paid keeper or curator of a museum in Phillips' time could involve drudgery and overwork without the compensation of a decent salary. Knell has shown that some geological curators were slaves to their employers who gave neither security nor the prospect of advancement. At the Geological Society of London Lonsdale received £200 pa as its full-time curator, librarian, and assistant secretary from 1829. By 1836 his workload, which on occasion involved seventeen hours a day, had made him ill. When he resigned in 1842 he was replaced by Edward Forbes who held office for only two years, tired of being seen as a mere servant of unreasonable fellows. At the Bristol Institution Miller was paid £150 a year as curator from 1824: a victim of an eleven-hour day he died exhausted in 1830. In Yorkshire two curators received pittances. At Whitby the Literary and Philosophical Society appointed as its keeper in 1837 a local lad, Martin Simpson, an expert on ammonites. Four years later he was sacked because the Society could not afford his salary of £20 a year. In Scarborough John Williamson curated the museum of the Literary and Philosophical Society for £30 pa from 1829 to 1838 when £10 was granted to his wife as his assistant. In 1848 the Society forced him to become part-time on £20 a year until he retired in 1853.[29]

As keeper of the YPS's museum Phillips was not particularly well paid but he was not exploited and certainly not a slave to its collections. Some of his responsibilities were humdrum. He was not uplifted by stamping books in the YPS's library for four days. Copying out the odd scientific paper given to the YPS was a mere secretarial chore. Attending some YPS members in the museum was not always inspiring. Labelling specimens and making entries in catalogues could be tedious. On the other hand, some of his tasks which look boring were useful and interesting to him. He mended damaged specimens, such as a fossil crocodile head, which stimulated his interest in comparative anatomy. In identifying and arranging specimens in geology and natural history, he made himself even more familiar with the works of Cuvier. He assisted the honorary curators and learned from three of them. In helping William Marshall to arrange the mineralogical collection, he used the terminology for crystals created by Haüy, whose theory linked their external form and their supposed internal structure. Phillips' interests in entomology and antiquities were stimulated by his respective contacts with Thomas Backhouse, a Quaker nurseryman and banker, and with Charles Wellbeloved, a Unitarian minister who was principal of Manchester College which, despite its name, was located in York from 1803 until 1840. Phillips was responsible for recording verbal communications made at YPS meetings and thus became familiar with the interests of its active members and broadened his own. He arranged swaps of duplicate specimens with other institutions and sent donations of YPS duplicates to such individuals as William Bean in Scarborough, Adolphe Brongniart in Paris, Count Sternberg in Prague, and De la Rive in Geneva. Through duplicates Phillips maintained and created valuable contacts, especially in the field of fossil botany. Though negotiating with local collectors about specimens could be irritating, it permitted him to shape the museum's holdings in line with his interests and increased his familiarity with private collectors and the contents of their collections. Receiving donations was mundane but in some cases he had the satisfaction of knowing that his own suggestion had been fruitful. Though the YPS did not mount a large lecture programme, Phillips assisted visiting lecturers such as William Scoresby. From time to time Phillips acted as a secretary to Harcourt, the president, and Goldie and William Gray, junior, the leading secretaries. He was thus involved in the policy-making and running of the Society, becoming a trusted colleague of Harcourt and Goldie, and developing a close friendship with Gray, a local lawyer and meteorologist.[30]

As keeper Phillips received distinguished visitors from home and abroad who found it profitable not only to examine the collections but also to hear about his research. August De la Rive, the Swiss naturalist, and Wilhelm Haidinger, the Austrian mineralogist, came to inspect specimens. On their long tour of England and Scotland the German mining officials, Heinrich von Dechen and Carl von Oeynhausen, came to learn from Phillips' section of the Yorkshire coast. The three most important British visitors were: William Hutton of Newcastle-upon-Tyne, and Henry Witham of Barnard Castle, both leading fossil botanists; and in June 1826 Roderick Murchison, an affluent former soldier who began to devote himself from 1824 to geology which he conveniently combined with field sports. Murchison was making himself felt in the Geological Society, of which he had just been elected a secretary (Fig. 2.3). Introduced by Harcourt, Murchison not only looked at the YPS's specimens from the unusual vantage point of the floor on which he lay. He also learned from Phillips about the geology of the Yorkshire coast before going to Sutherland, Scotland, to settle the geological age of the Brora coalfield: Phillips showed Murchison his coast section and gave him lists of strata and coal plants and sketches of the cliffs. These resources helped Murchison to show that the Brora coalfield belonged to the lower oolitic rocks of Yorkshire and not to the older carboniferous strata; and to launch his illustrious and imperially minded career, which included a knighthood in 1846 and being director general of the Geological Survey 1855-71.[31]

Fig. 2.3   The bemedalled Sir Roderick Impey Murchison, an engraving of 1851 of W.H. Pickersgill's oil painting of 1849. Proclaimed King of Siluria in 1849 by bishop Wilberforce, Murchison attained the beau-ideal of the independent gentleman dedicated to science.

As keeper to the YPS, Phillips knew that a scheme for a new building to house the collections, then lodged in rented rooms in Low Ousegate, had been formally launched in March 1825. The plan was to erect a spacious and imposing home for the YPS on a semi-derelict site of three acres known as Manor shore, which adjoined the river Ouse and contained the ruins of St Mary's Abbey. The architect initially chosen was Richard Sharp of York who produced in April 1825 a plan for a modest single-storeyed building costing £3,000. The immediate success of the building fund induced Harcourt to think by early 1826 of a bigger and more dignified building which would be a 'great Northern Museum'. He turned to the fashionable London architect, William Wilkins, who was certain that Grecian and not Gothic was the only style suitable for the locality or the purpose of the building. By summer 1827 Wilkins had designed a completely new exterior, with an impressive two-story façade, but he retained much of Sharp's plan for the interior (Fig. 2.4). The estimated cost was £5,650 with a further £2,000 needed for fencing, a road, and furnishings. In October 1827 the foundation stone of the Yorkshire Museum was laid by the Archbishop of York. The digging of the site for it revealed important medieval remains which Charles Wellbeloved preserved as a 'memorial of departed splendour' and described on behalf of the YPS. From this time archaeology became more important to it. Another cause of delay concerned the acquisition of the Manor shore which belonged to the Crown. After negotiations lasting three years, Harcourt eventually prevailed on the government to pass through parliament the appropriate legislation which enabled the YPS to gain the site from the Crown at an annual rent of £1-0-8. This remarkable

Fig. 2.4   The membership ticket for the first meeting of the British Association for the Advancement of Science held in York in 1831 showed the Gothic ruins of St Mary's Abbey and the recently opened neo-classical Yorkshire Museum of which Phillips was keeper.

example of Harcourt's stamina and skill as a lobbyer of government, in which he had relatives, was surely not lost on Phillips. The Yorkshire Museum was not fully occupiable until February 1830 when it was officially opened.[32]

Phillips' contribution to the new Yorkshire Museum was humble but useful. As keeper he was never a member of the YPS's building committee, established in March 1825 and chaired by Harcourt, but he attended some of its meetings by invitation in July and August 1827. Before that time he had offered advice about the cases for the ever-increasing specimens and drawn a plan of the site for Harcourt to send to Wilkins. In summer 1827 he discussed the use of the basement storey with Sharp, who acted as assistant architect and site superintendent for Wilkins. Phillips was particularly keen that the design of the zoology room of the new Museum should permit a gallery to be put in subsequently in order to accommodate another tier of cases. In summer 1829 he was deeply involved in the transfer of some 24,000 specimens from Low Ousegate to the new Museum and in the design of movable brackets for shelves.[33]

2.4 Performing: home and away

Phillips did not confine himself to curating the YPS's collections. Through giving papers to it, contributing to discussions at its meetings, and lecturing to it he soon became its leading scientific performer. Most of his papers were on geology. Many of them rehearsed material which ultimately appeared in his two monographs on the geology of Yorkshire. Three of them, on the distribution of erratic rocks in Yorkshire, the geology of Ferrybridge, and that of the north side of the Vale of Pickering, were published in the *Philosophical magazine*, which spread the reputations of himself and the YPS. Two of them involved co-operation with YPS members. In the work on erratic rocks Phillips organised four members, Marshall of Tadcaster, J.B. Graham of York, Jonah Wasse of Boroughbridge, and the reverend Charles Vernon Harcourt of Rothbury, to supply illustrative data about the directions of diluvial currents which Phillips believed had carried the rocks far from the nearest place where they occur in situ. On the two research trips made in May 1826 and April 1828 to Ferrybridge, Phillips was accompanied by William Bulmer, a local clergyman.[34] Phillips' interest in gadgetry was revealed to the Society in accounts of two inventions of his, a new lithographic press (1826) with roller and scraper combined in a strong, small, cheap frame, and a new mountain barometer, designed to remedy defects of the common types, which he had used in June 1828 when surveying the Yorkshire coast. This was by no means the sum of his interest in mechanical contrivances. In 1827 he improved his turning lathe, made a blowpipe for the YPS, a balance, and a hair-spring hygrometer; and also ground an achromatic lens for a telescope.[35] In zoology he developed an interest in entomology through his association with Backhouse, Backhouse's employee Henry Baines, a Unitarian, and William Hewitson, whom he accompanied on excursions to propitious local sites such as Stockton moor. Having collected insects for ten years in the York area, Phillips gave his collection to the YPS in 1837. His interest in entomology extended beyond bug-bagging: he was capable of expatiating to the YPS on the compound respiration system of scorpions and on the noises made by insects.[36]

Phillips' belief in co-operative endeavour was also shown in meteorological research he began in 1828. It was stimulated by the request made by the Royal

Society of Edinburgh in 1827 to record hourly meteorological observations on 15 January and 17 July each year, and guided by hints on procedure given in his own journal by David Brewster, its secretary. In response Phillips organised Jonathan Gray, William Gray jnr, William Taylor, and William Bulmer to make the appropriate observations in January 1828 under his direction. Phillips' particular contribution was to use four different types of thermometer, including one he had devised for measuring dewpoint, in compiling a meteorological register every hour of the day and night. The aim of the York quintet was to assist the Edinburgh Society by generating data from which it was hoped to compare the meteorological features of different places at the same time. Brewster, who had donated to the YPS library in July 1827 a copy of his hourly temperature observations made at Leith in 1824 and 1825, was delighted by the YPS's meteorological observations which were repeated in July. This episode showed Brewster that the YPS was effective and it confirmed that Phillips believed that science could usefully be collaborative when data from many sources were required to form empirical generalisations from which general laws, he hoped, could be produced.[37]

Phillips was also involved in the evening conversation meetings which were established apparently in 1829 to supplement the YPS's monthly meetings held on a Tuesday at 1.00 pm. This was an inconvenient time for those who were neither leisured nor leisurely. With his varied attainments Phillips was central to the evening meetings and gathered round him an inner circle of enthusiastic and extremely able local members such as Taylor, William Gray jnr, Wellbeloved, Robert Davies, John Kenrick, William Hincks, and Baldwin Wake, but not Harcourt, who met at various homes to dispute conversationally about scientific topics. Phillips' papers and lectures to the YPS as well as his keepership gave him many valuable contacts; the evening meetings gave him cronies.[38]

As well as giving papers to the YPS and contributing extensively to discussions, Phillips gave three courses of lectures to it in 1826, 1827, and 1829 (see Appendix 1). They were useful to him financially, generating £36, £30, and 30 guineas for eight, six, and six lectures respectively. Those of 1826 covered the familiar topics of geology and the natural history of fossils but included two novelties: for the first time in a lecture Phillips displayed his section of the Yorkshire coast from Bridlington to Redcar; and in his discussion of fossil plants he considered them in order of stratigraphical position going upwards through the strata. Those of 1827 and 1829 showed Phillips' serious interest in zoology. That of 1827 covered thirteen classes of invertebrates, ranging from microscopic animalcula to cephala (squid, octopus), illustrated by Phillips' own drawings or where possible by specimens. His classification drew on such standard authors as Linnaeus, Cuvier, and Lamarck; but, unusually for a geologist, Phillips learnt from Latreille's mode of classifying insects. The 1829 course, given in the lecture theatre of the new Yorkshire Museum, considered aquatic animals of the York area, from animalcula to otters. In these zoological lectures Philips was at pains to stress that the adaptations of creatures to their mode of existence via ingenious and admirable contrivance indicated the wise and well-ordered plans of a benevolent God who filled all space and blessed all his works. Though the theological concern was muted in the 1826 lectures, there is no doubt that for Phillips lecturing was a sacred act. In March 1826 he prayed: 'Oh! God! While I study thy works assist – while I explain them inspire me'.[39]

As well as lecturing to the YPS, Phillips began to give lectures to the York Mechanics' Institute which was run and supported mainly by Wellbeloved, Kenrick, and Hincks, his Unitarian clerical friends. Founded in 1827, its early years were discouraging and difficult not least because it was frowned on by Tories and Anglicans as subversive of social order. Wellbeloved and his Unitarian colleagues disagreed. In their view it was a Christian duty to remove ignorance and barbarism and to make the means of mental improvement available to all. For them ignorance was certainly not the mother of devotion. It was knowledge which led to virtue and humility when it was cultivated from pure motives and in a benevolent and cautious way. Such self-improvement by artisans was to be attained by reading books from the Institute's library, to which Wellbeloved gave no fewer than eighty-nine volumes when it was launched, and by listening to lectures on the sciences of nature and of man. Wellbeloved was keen for astronomy, meteorology, geology, geography, botany, and zoology to be taught for two reasons. Firstly, when pursued in a Baconian inductive way these sciences undercut dogmatic monopolies of knowledge and required caution in order to gain truth gradually. Secondly these sciences could be used, via the arguments to and from design, to prove the existence of the power, wisdom, and goodness of a unified creator, an all-wise disposer, and gracious providence, but no more.

Phillips was not formally a member of the Institute until 1831 but he heard and regarded as excellent Wellbeloved's address of March 1828 on knowledge leading to diffidence and humility. Next month he heard Wellbeloved dilate at the Institute on the advantages of studying natural history and single him out for praise. In March 1829 Phillips gave two gratis lectures to it, one on the structure of the earth and the other on the geology of the Yorkshire coast. Presumably he was invited to do so by Wellbeloved who regarded geology as an important and not subversive science which asked demanding and great questions about the formation of the earth and its subsequent changes. The way in which Phillips soon allied himself with the Institute and continued to support it, indicates his broad sympathy with the views of the Unitarian clergymen who ran it. For several years Phillips was the only notable Anglican savant in York, and exponent of Baconian induction and co-operation, who supported the Institute. He did so because he believed in non-dogmatic, unzealous, and irenical practical Christianity which encouraged co-operation between sects outside church and chapel. As an orphan who had improved himself, he advocated self-improvement for all. And he admired Wellbeloved's chief characteristic, which was benevolence practised both privately and publicly. Wellbeloved was a Unitarian version of the Anglican Richardson in that respect.[40]

From his secure base in York Phillips accepted invitations from the local lit-and-phils to lecture in the familiar haunts of Hull (twice), Leeds, and Sheffield. He also made his debut at the Royal Manchester Institution founded in 1823 to promote science, literature, and the fine arts (see Appendix 2). These performances outside York, which he slotted into his YPS commitments, were welcome financially: they generated £40 in 1826, £60 in 1827, and £60 in 1828. In these years Phillips earned each year about £150-£180, of which at most two-thirds came from lecturing engagements in York and northern towns. Of the five courses a minority covered geology and fossils. The majority, on the zoology of invertebrates, drew on and enlarged that given at York in spring 1927. Indeed, Phillips shuttled between York and Leeds while lecturing in both places at that time. The Sheffield course was peculiar in that it was given in three parts, presumably to fit in with Phillips' commitments. It is clear that he was locally in demand as a lecturer, not just as a geologist and fossilist but also as a zoologist, a role which befitted his post as keeper at York. It was also a matter of intellectual conviction: by 1826 Phillips was convinced that zoology and chemistry were essential in geology because these sciences were connected.[41]

On occasion Phillips thought that lecturing outside York gave him insufficient financial reward and was therefore a waste of time. Generally, however, he welcomed not just the cash but also the contacts and opportunities that such lecturing brought him. At Hull he became particularly friendly with William Hey Dikes, curator to the Hull Society; John Edward Lee, later a close friend and a well-known antiquarian; the reverend Christopher Sykes of Rooss, near Hedon; William Stickney of Ridgmont, Hedon; and John Crosse, an antiquarian. They provided a base for his research on Holderness and assisted him with it. In Manchester Phillips stayed with G.W. Wood, the Unitarian businessman who had heard about Phillips from the York Unitarians and booked him to lecture at the Royal Institution. During this engagement Wood quickly came to revere Phillips as a geologist and urged his friend William Henry, a wealthy chemist, Unitarian, and textbook writer, to introduce himself to Phillips. Through Henry Phillips then met Edward Holme, the well-known Unitarian physician, and John Dalton, the famous Quaker creator of the chemical atomic theory. Phillips was particularly impressed by the dissenters. He wished to emulate Holme's wide knowledge, Henry's excellent taste, well-regulated mind, and warm desire for the advancement of science, and Dalton's 'infantine simplicity of heart'. From his Manchester base Phillips made geological excursions to a salt mine at Northwich and to Castleton and Clitheroe, at both of which he studied the relations between the mountain limestone and the millstone grit. Phillips' lectures at Leeds and Sheffield confirmed and extended his acquaintances. In Leeds, where he also gave papers in 1827 (twice) and 1829, his main friends remained George, Atkinson, and Hamilton, to whom he added Denny and William West, a Quaker chemist. In Sheffield there were no equivalents to Phillips' chums in York, Leeds, and Hull; but he had good relations with Luke Palfreyman, a Unitarian solicitor, Edward Barker of the Sheffield Lead Works, and especially Robert Younge, a Unitarian silver-roller, all of whom were officials in the local lit-and-phil.[42]

2.5　Harcourt as patron

In his new life as keeper at York Phillips was from time to time melancholic. He felt he was still a wanderer without purpose, he lamented the separation from his sister whom he had last seen in 1815, and he was acutely aware of the vicissitudes of his life after he became an orphan. Yet generally he was buoyant and convinced that the romance of his life had just begun. He appreciated that he had received much disinterested kindness which proved to him that humans were not degraded. Moreover he believed in an omnipresent and omniscient God who continued to bless all His wonderful works, including Phillips himself: the bounteous author of his life had enabled him to exercise his God-given reason in studying the natural world and, through the knowledge gained about adaptation, design, contrivance, and the constancy of natural laws, to enjoy the delight of adoring the beneficent supreme and his unceasing watchfulness. Phillips was certain that, compared with 'fettering dogma', nature showed in 'clear *unsullied* splendour' God's benevolence. In contemplating in June 1828 his 'upward way' on the arduous paths of science, Phillips decided that his ambition should be modelled on God's benevolence and that his own self-interest should lead him to try to make others happy so that his own happiness might be augmented ten-fold.[43]

As an employee of the YPS Phillips had the satisfaction of not only agreeing totally with its aims but also secured through it essential means of instruction previously not available to him. That became particularly clear as early as summer 1826 when the YPS gave Phillips leave of absence to enable him to accept a proposal of George's that they and others should visit Scotland for about seven weeks. In the event George brought along two LPLS friends who were clergymen, Richard Hamilton, a dissenter, and William Bathurst, an Anglican. Phillips invited Goldie, a Catholic who had been keen for a year to accompany him on a geological excursion. As a non-sectarian Anglican, Phillips was happy to join a geological party distinguished by Christian harmony and 'fun, frolic, and philosophy'. For years he had wanted to examine Scotland's rocks and to see the geological phenomena made famous by James Hutton and John Playfair, who had claimed that unstratified rocks such as granite and basalt were igneous in origin. Accordingly, Phillips examined Arthur's Seat, Salisbury Crags, and Calton Hill in Edinburgh, he marvelled at the big sheets of basalt and the conglomerate rocks around Oban, he was not disappointed by the basalt columns of Fingal's cave on the Island of Staffa, he was excited by Glen Tilt, near Blair Atholl, where he saw veins of red granite shooting into the surrounding schists, and he was fascinated by the many dykes, some crossing and branching, which run out to sea from the Isle of Arran. He also visited the famous but puzzling parallel roads of Glen Roy near Spean Bridge.

This Scottish tour enlarged his geological horizons: previously he had seen mainly undisturbed secondary sedimentary strata, from the chalk down to the old red sandstone, his experience of igneous rocks being confined to the whin sill of northern England and outcrops of granite. Though he wished as a Baconian to keep his mind free from theoretical views, he was intrigued by Hutton's section on Salisbury Crags, where both the 'greenstone' [as he called it] and the sedimentary sandstone were much altered at the place of contact. Such alterations produced in surrounding rocks by basalt, pitchstone, porphyry, and granite, persuaded Phillips that they were at one time fused and of igneous origin. The decisive experience occurred on the Isle of Arran which he visited solo for a week expressly 'to try the theories of geology in the open court of nature'. Near Lochranza he saw that granite veins had passed through slate, had enclosed pieces of it, and had divided themselves. He concluded that apropos the origin of granite the Wernerians were heretical so he 'embraced the true geological faith' of the Huttonians. Except on just one day Phillips did not examine a fossil during seven weeks in Scotland: he devoted himself to non-palaeontological structural geology, with particular reference to junctions between igneous rocks and sedimentary strata, the relations between igneous rocks, and veins and dykes.[44]

Through the YPS he soon acquired a large circle of friends, mainly in York, Leeds, and Hull, who liked and respected him. Through it he developed a new career as keeper and lecturer, and gained a measure of independence which contrasted with his previous privations: in September 1826 he was able to send £10 to Smith. Through the YPS he gained enough security to make York his home, in lodgings in Gillygate, which was conveniently close to Low Ousegate and even closer to the Manor shore. He also had the satisfaction of agreeing with Harcourt's theological views about the museum and science. The former was not just devoted to accumulating and displaying specimens; by diffusing knowledge about the works of creation and the wisdom of nature, it contributed to the noble and exalted understanding of God.

TERTIARY STRATA
(divided by Lyell into pliocene, miocene, eocene)

Formations	Subdivisions
Crag	
Freshwater marls	
London clay	

SECONDARY STRATA

Cretaceous system

Chalk	Upper chalk
	Lower chalk
Green sand	Upper green sand
	Gault
	Lower green sand

Oolitic system

Wealden	Weald clay
	Hastings sands
	Purbeck beds
Upper oolite	Portland oolite
	Kimmeridge clay
Middle oolite	Upper calcareous grit
	Coralline oolite/Coral rag
	Lower calcareous grit
	Oxford clay
	Kelloways rock
Lower oolite	Cornbrash
	Forest marble
	Great/Bath oolite
	Fullers' earth
	Inferior oolite
Lias	

New red sandstone strata

New red sandstone
Magnesian limestone

Carboniferous system

Coal measures
Millstone grit
Carboniferous limestone
Old red sandstone

TRANSITION STRATA

Silurian system

Ludlow rocks
Wenlock rocks
Caradoc rocks
Llandeilo rocks

Cambrian system

Plynlymmon rocks
Bala limestone
Snowdon rocks

Fig. 2.5   Table of the succession of British formations and their classification, as widely adopted by the late 1830s.

Harcourt's ideology of science was equally high-minded. He admitted that science had useful practical applications and developed good intellectual habits such as minute and accurate observation. But for him the chief justification for science was the elevating moral effect of investigating the laws of nature and tracing the hand of creative wisdom in processes and contrivances. If pursued in this way, 'science, with a secret moral charm, allays the animosity of parties, and pours a friendly feeling over the most discordant minds'. Thus Harcourt and Phillips shared a theological view of the YPS's museum and an irenic ideology of the science it promoted.[45]

Phillips knew that Harcourt esteemed him publicly. The four successive annual reports of the YPS for 1826 to 1829 alluded to Phillips as valuable, indispensable, meritorious, and knowledgeable. He was praised for his ardour for science and his happy talent for explaining it. His salary of £60 pa was denounced as inadequate. Privately Harcourt made Phillips welcome at his home, promoted Phillips' social life, and was happy to invite him to meet John James Audubon, the famous ornithologist. Harcourt's constant and considerate kindness and encouragement ensured that at York Phillips was never put in the ignominious and demeaning position that was the lot of some keepers elsewhere.[46] Moreover, when the occasion arose, Harcourt exercised his influence for Phillips' benefit. In September 1827 Phillips decided that he wished to become a fellow of the Geological Society, basing his application on his latest research on the slate rocks of the Lake District and the Ingleton area which he had visited on foot with Thomas George the previous month. In November 1827 Harcourt wrote on Phillips' behalf to Murchison, a crony and a secretary of the Geological Society, who was happy to propose Phillips as a 'most eligible' FGS and to secure backers for his candidature. His paper on slate rocks from Kirkby Lonsdale to Malham was read on 21 December 1827 and 4 January 1828. Its novel account of faults in the area and of erratic blocks at Norber and its awareness of the difficulty of distinguishing between stratification and cleavage planes impressed Buckland, another crony of Harcourt and immediate past president. On 4 January 1828 Phillips was elected FGS which cost him, as a non-resident, 10 guineas admission fee. Through the support of Harcourt, Murchison, and Buckland, Phillips had acquired a useful qualification still denied to his uncle. Unlike the sturdily independent Smith, Phillips was prepared to exploit patronage networks: he sent an interesting paper to the Geological Society, thanked Murchison for his assistance, and represented himself as a provincial enthusiast who would welcome geological hints and commands from headquarters in London.[47]

Phillips appreciated that Harcourt did his best to encourage and support his uncle. It was through Harcourt and Phillips that Murchison came to know and value Smith in 1826. When Murchison, as a self-styled foreigner, wished to be informed about the geology of the Yorkshire coast, he learned not just from Phillips in the museum at York but also from Smith who accompanied Murchison by boat from Scarborough to Whitby and on foot under the cliffs between Robin Hood's Bay and Whitby. Murchison learned from Smith directly that Sedgwick's paper of May 1826 on the Yorkshire coast was deficient in that it had not mentioned Oxford clay, Kelloways rock, cornbrash, and great oolite, which Smith had established as having 'representatives' on that coast (Fig 2.5). Without Smith's recent but unpublished results Murchison could not have claimed in his Brora coalfield paper of 1827 that the Yorkshire coast was the connecting link between Brora and the oolitic rocks of south-west England.[48]

Smith had taken advantage of living in Scarborough by colouring geologically Knox's map of the area. Knowing Harcourt's interest, in October 1826 Smith spoke to the YPS about this map, on which he had represented his latest discoveries, and presented it to the Society. Smith told Harcourt and Johnstone that he hoped to complete and publish in the near future a new large geological map of Yorkshire, costing 3 guineas and based on Cary's topographical map, and an accompanying memoir costing 2 guineas. Harcourt was so convinced that Smith would produce that year this geological map and book, to be published by subscription, that he secured from Murchison a promise that those London geologists, who esteemed Smith highly, would support it. Sadly Smith did not complete either the map or memoir. Harcourt also pressed Murchison and Buckland to launch a campaign for an annuity for Smith to stop him dying in a poor-house. Though Murchison regretted Smith's distressed state, he did not think that his fellow geologists would subscribe to an annuity fund for Smith as well as to a book by him; and he warned Harcourt that in some quarters Smith had the reputation of being an improvident man who would squander speedily any money given to him. By late 1827 it was obvious that there was insufficient support for the annuity scheme.[49]

In fact Smith was employed from May 1826 for about nine months in improving the water supply to Scarborough. In a hill above the town he excavated in rock a subterranean reservoir to dam the water from a small spring. It fed a lower, new, large, covered, and mainly underground reservoir built entirely with bricks and located in the town. This system, Smith thought, would solve the problem of the shortage of water in summer when Scarborough was full of visitors. Superintending these waterworks precluded Smith from attending to his intended publications on

Yorkshire geology, but he did find time to write a short paper on the new Scarborough water supply. He despatched it to the YPS where it was read in March 1827, and then revised by Phillips before Harcourt secured its publication in June in the *Philosophical magazine*.[50]

Harcourt's concern for the welfare of Smith, whom he regarded as ill-rewarded for his successful labour in the service of science, was alleviated by Johnstone. In February 1828 Johnstone, who had married one of Harcourt's sisters in 1825, appointed Smith as his land-steward on his Hackness estate. He did so presumably out of charity, perhaps prompted by Harcourt. As President of the Scarborough Philosophical Society (founded 1827) Johnstone also appreciated Smith's valuable contribution to the design of its museum formally opened in August 1829. The unusual circular plan was suggested by him so that the local strata could be depicted in colour at the base of the dome and below the geological specimens could be arranged on sloping shelves, passing beneath each other in the order of strata. Above all Johnstone wanted to give Smith anything which would enable him at last to publish all his geological work. Sadly Smith procrastinated: apart from his remarkable geological map of Hackness, on a scale of six inches to the mile, he ignored Johnstone's entreaties.[51]

Harcourt exerted himself as a patron but he was also a practising geologist who appreciated from 1824 the importance of work done by Smith and Phillips on the Yorkshire coast. Prompted by Conybeare, Harcourt realised that it needed to be supplemented by research on the interior, especially around Cave and Malton. When Phillips became keeper to the YPS he was pulled away from the coast and especially Scarborough, which he had praised in 1825 as 'the finest spot for a geologist that the whole earth contains'. From 1826 he learned from Harcourt the importance of field work at places which Conybeare had highlighted. At the same time Harcourt was delighted to have a skilled mapper, stratigrapher, and fossilist as his companion on geological excursions. In 1826 they made three trips together. On Harcourt's initiative in March they examined the strata below the chalk at Cave which Harcourt had investigated in 1823. With Phillips' section and coloured map of the area at his disposal, Harcourt quickly published their finding that the strata there were identical with some of those found on the Yorkshire coast, namely coral rag, Kelloways rock, inferior oolite, and lias. In June they spent two days in the west Howardian hills and inferred correlations between the strata there and some of those at Cave and on the coast. In September they went to Scarborough to negotiate on behalf of the YPS with fossil collectors and to collect fossils themselves. On the way Phillips took care to make a section at North Grimston, south-east of Malton, and confirmed the succession of chalk, clay, and oolite which Conybeare had postulated. All these results were incorporated into Phillips' 1829 book. Having benefited from Phillips' aid in the field, in late 1826 Harcourt attacked the Castle Howard area solo and reported in December to the YPS the relation of its geology to that of Cave and the coast. Harcourt had obviously grasped one essential of the approach taken by Smith and Phillips and even lectured Murchison on it, namely that in correlating strata, which had been provisionally identified from their appearance, both 'stratical position' and characteristic fossils were important. Murchison was impressed: in late 1826 and early 1827 he urged Harcourt, aided by Phillips' section of the Yorkshire coast, to produce the authoritative account of it. Knowing his limitations as a surveyor and fossilist, Harcourt did not even try to do so.[52]

2.6   The coastal geologist

Until July 1828 Phillips' aim, it seems, was to publish his section of the Yorkshire coast and to develop his geological knowledge of its interior. Then he heard that Webster intended to survey the Yorkshire coast. Given Webster's long-established reputation as a coastal geologist and as an expert on the upper oolitic strata, this was a threat. Determined to secure prior publication for his section, to oust Young and Bird as authorities, and to give substance to his belief that it was noble to compile the natural history of Yorkshire, Phillips began composing his 1829 book in July 1828. It was written so quickly that the prospectus of 6 November 1828 anticipated that it would be published on 1 February 1829.[53] Pressure of time explains why Phillips did not describe in detail the fossils he himself drew on stone with pen and ink, an omission which lead John Dunn to accuse him of failing to acknowledge that Phillips' Scarborough friends had discovered many of them.[54] The odd mistake on the map, such as Hedon instead of Hessle, and its ungenerous scale at about ten miles to the inch were probably the results of hasty composition. The map was confusingly coloured: the red line, representing red chalk adjacent to the white chalk, was indistinguishable from the narrow stripe of 'full orange', representing the inferior oolite, which ran between the lias and the cornbrash/Bath oolite.

Phillips was well equipped to write the book. Its first chapter, covering the principles of geology, drew on his public lectures. For fossil specimens Phillips enjoyed access to the YPS collection in his care, to public collections in Leeds, Sheffield, Hull, and Whitby, and to above twenty private collections. He was particularly indebted to the fossil hunters of Scarborough, such as William Bean, John Dunn, and John Williamson, who generously sent specimens to the YPS and reported their latest discoveries to it.[55] Phillips also had available his section of the coast which was constructed from many and repeated measurements. His first attempt, made in 1824 and 1825, had been twice revised: in October 1827 he had surveyed on foot the coast from Redcar to Scarborough and the Vale of Pickering using an Englefield barometer lent to him by Cholmeley; and in June 1828, using his own improved mountain barometer, he had covered the whole of the coast. He knew that his section had been warmly approved by Murchison, Oeynhausen, Dechen, and Brongniart, and that this sort of detailed surveying was beyond the capacity of Harcourt and the ageing Smith. For some time friends had mentioned to Phillips the desirability of producing illustrations of Yorkshire geology. In 1825 Bean offered to combine his labours with Phillips in illustrating Yorkshire fossil species, and in early 1828 Witham deplored to Phillips the lack of 'illustrative fossils of the oolite'. In any event it was entirely appropriate that the keeper of the YPS, dedicated to illustrating the geology of Yorkshire, should publish a monograph on it, especially as in April 1828 his paper on the Vale of Pickering had shown that in this interior location there were eight strata which occurred in the same order on the coast.[56]

The part of the book which gave him most trouble was the introduction. How was he to do justice to Smith who had obtained by 1824 'a correct view of the stratification of the whole coast', without appearing to be yet another plagiarist of his mentor and uncle? And how was he to secure his own reputation without denigrating Smith? Initially Phillips thought of trying to deal with the vexed subject of Smith's general claims, which would have involved correcting what Phillips regarded as mistakes made in the 1818 account of Smith by Fitton who in autumn 1828 as president of the Geological Society was recruiting subscribers for Phillips' book. He rejected this plan as impracticable. Instead, in an introduction vetted by Harcourt, he gave a historical view of previous work on the geology of the Yorkshire coast, which enabled him to praise his uncle, to bring out his own originality, and to give less attention to the larger question of Smith's achievement. Privately Phillips showed his sympathies by ensuring that the first copy of the book to be sent out went to Fitton, who was very pleased with it.[57]

Phillips' book was published by subscription in April 1829 at £1. Many of the 400 subscribers lived in York, Leeds, Sheffield, Hull, Whitby, and Scarborough, where he was well known. In the south-west of England Miller recruited subscribers in Bristol and took their money as did the faithful Richardson in Bath with six takers. In London Fitton organised the subscription: he was interested in the Yorkshire coast and he had tried for over a decade to vindicate Smith. He distributed copies of the prospectus at the Geological Society where he secured the support of Murchison and the administrative help of John Taylor, its treasurer, and of Lonsdale. Fitton realised that Phillips had not named a London bookseller on the cover of the book and, though suspicious of John Murray, eventually approached him. By late 1830 Murray had sold about fifty copies and sent £44-2-0 to Phillips.[58]

His book accelerated the rehabilitation of Smith to whom it was dedicated by his affectionate nephew and grateful pupil. Its first page asserted that Smith was the first in England who studied and taught others to study the earth inductively because methodologically he had discovered how to identify strata from their embedded fossils and doctrinally he had produced the notion of general geological formations, ie that there was a widespread conformity or analogy in the ordering of the strata. The former was not for Phillips a routine claim: privately he believed that geology had become *the* Baconian inductive science mainly through Smith's labours. Phillips changed the title of his book shortly before publication from the *Geology of the Yorkshire coast* and *A description of the strata of the Yorkshire coast* to *Illustrations of the geology of Yorkshire*. He may have had in mind John Playfair's *Illustrations of the Huttonian theory*, because he was to Smith what Playfair was to Hutton. In lucid prose Playfair had defended and modified the ideas of Hutton whose publications were prolix. Phillips presumably knew and liked Gideon Mantell's *Illustrations of the geology of Sussex*, a recent work in which local geology of general resonance had been produced by a local who offered figures and descriptions of fossils. An illustration also meant an examination and distinct representation. In that sense Phillips offered three sorts of illustrations all drawn by himself: a coloured map; a coloured section of 100 miles of coast on the generous horizontal scale of one and a half inches to the mile and vertical one of 400 feet to the inch; and fourteen plates depicting 400 fossil species in order of the strata to which they belonged. Finally the new title emphasised the wealth of Baconian data from which he had induced his generalisations.[59]

In his book Phillips represented Smith and Werner as the leaders of the 'modern school of geology' so, in a delicate balancing act, he found in Yorkshire only two geological formations (assemblages of rocks having some common feature) associated with named individuals, that is, Smith's clay vale formation and Werner's new red sandstone formation. Phillips saw Smith and Werner as independent and simultaneous stratigraphical pioneers: to acknowledge one discoverer did not mean dismissing the other. Of course, Phillips' book *was* an exercise in Smithian geology in its approach and broad results. Phillips' prospectus stressed that he had 'strictly examined' Smith's law about the regular distribution of fossils and then applied that law to determine the relations of the Yorkshire strata to those known elsewhere. In the inscription in the copy of the book which Phillips gave to Smith he described it as Smith's own work and a proof that he had not slighted his uncle's instructions. Phillips showed what Smith had discovered but not published, namely the presence and succession in Yorkshire of strata already known in the south-west of England particularly from Smith's own work. In Phillips' coast section he used Smith's colours for the oolitic rocks. Generally he followed Smith in the importance he attached to characteristic fossils. This was crucial for identifying in Yorkshire the cornbrash and lower calcareous grit, both of which were variable mineralogically but well characterised by fossils. In his attitude to fossil-labelling of strata, Phillips drew on his uncle's insights, published in 1817, that sometimes a particular stratum did not contain a small number of fossils exclusive to it and that in such a case the stratum could be identified only on the basis of its assemblage of fossils compared with those of the strata immediately above and below. Apropos the generally thin but important marker stratum of the cornbrash, which was no thicker than five feet in Yorkshire, Phillips reported that it was not identifiable by the presence of particular species of fossils found in no other stratum but by the '*occurrence together*' in it of some fossils present in the rocks above it and of other fossils found below it. Lithology and stratigraphical position offered little aid in identifying what Phillips suspected was a Yorkshire version of the cornbrash, so he relied on comparing its 'whole suite' of fossils with that of the cornbrash elsewhere.[60]

In his book Phillips presented himself as a calm and non-speculative inductive geologist, but he was bold in some of his postulated correlations between Yorkshire and south-west England, especially with respect to the Cave and Gristhorpe oolite which he referred to the great or Bath oolite. This correlation was briskly challenged in the 1860s by Thomas Wright who referred these deposits to the inferior oolite which he divided into three zones, each labelled by a particular ammonite and its own fauna. Though Phillips was aware that strata could be subject to local variations, he thought it worthwhile to try to identify and correlate widely separated rocks even if they differed totally in their lithology. Thus Phillips' 1829 book, in which his homage to Smith was more than lip service, showed the power of Smith's palaeontological stratigraphy and that it had not been effectively bypassed in the 1820s. On the contrary, as a long 1831 review of the book stressed, Phillips had confirmed and developed Smith's opinions in a remarkable degree.[61]

Phillips went beyond his mentor in two respects. Firstly, he traced the extent to which some Yorkshire strata were identical with or analogous to those in several parts of the European continent. Such lateral correlations, over big distances and across a sea, were not new but never made by Smith. In his famous paper of 1814 on the Isle of Wight, Webster had compared the sedimentary basins of the Isle of Wight, London, and Paris. In the 1820s Conybeare and W. Phillips had routinely given details of foreign localities where English strata occurred, De la Beche had demonstrated the general accord in the nature and order of succession of strata of opposite coasts of England and France, and above all Adolphe Brongniart had established correlations between Yorkshire and Scandinavia. Secondly, Phillips showed that a creative and not slavish exponent of Smith's palaeontological stratigraphy was also a Cuverian. In a book dedicated to Smith, the epigraph was taken from the last sentence of Cuvier's *Ossemens fossiles*, which had tried to give 'un premier coup d'oeil jeté sur ces immenses créations dans anciens temps'. Certainly Phillips' approach was Cuverian in its concern with conditions of deposition of fossils, violent geological disruptions, mass extinctions, and an overall progressive direction to the history of life as revealed by the fossil record.[62]

The general Smithian character of the 1829 *Illustrations* becomes even clearer if it is compared with three previous works on the Yorkshire coast. Nathaniel Winch's survey, published in 1821, had relied heavily on Bird's local knowledge and made no attempt to establish analogies with southern England. In 1822 George Young and John Bird gave a general and not detailed coast section; they illustrated the most remarkable fossils, which were ordered zoologically and not by the strata containing them; they considered the local but not country-wide relations of the strata; and they so vehemently rejected fossil-labelling of strata that they made no reference to Smith's work and had therefore not realised the importance of the cornbrash and Kelloways rock strata. Moreover Young was and remained a vigorous scriptural geologist. For him the geology of the Yorkshire coast confirmed the sacred scriptures. He believed that there had been a Deluge, previously assumed by him to have lasted a year, which was responsible for the formation of most strata. It followed that the lowest strata were not of the highest antiquity and that fossils were so churned up that no fossil could be restricted to one stratum. The era of the Deluge, a historical event recorded by Moses, was the key to Young's general theory of the earth; it played the same role in William Eastmead's 1824 account of Kirkby Moorside and Kirkdale cave. In their second edition Young and Bird went out of their way to praise Granville Penn's exposition of Mosaic geology without disturbing the first edition structurally or methodologically.[63]

In contrast to Young and Bird, Phillips thought it ill-advised to try to compare a 'chronology of nature' with the records of history; but he had to explain east Yorkshire's huge erratic blocks and the vast deposits of diluvium (clay and gravel) which made up Holderness. Drawing on Buckland's diluvial theory, Phillips' answer was that there had been an irresistible and huge flood flowing from the north-west. It had carried erratic blocks of granite 100 miles from Shap to Flamborough and it had created Holderness by depositing diluvium there. It had happened after the stratification of the earth had been completed but a long time ago because in Holderness there were lake deposits, containing skeletons of deer and elk, which rested on the diluvium and were themselves not covered. Compared with Young and Bird's single age of the deluge, Phillips had clearly in his mind a triple distinction between antediluvian, diluvian, and post-diluvian periods. He also suspected that the layering of diluvial deposits indicated that, besides a mighty flood, there had been quiet waters and local currents.[64]

The most formidable competitor for Phillips was Sedgwick, who in 1826 had published on the coast of his native county. Though Sedgwick had made some correlations with the south-west of England, his field work was so hastily done that he did not offer plates of fossils, a section, a map, and data on the thicknesses of strata. He took Young and Bird's order of strata as correct and never ventured inland. Sedgwick's field work was so limited that he could not feel cheated if others developed his hints. He had started in print a particular hare but, as so often in his career, had not caught it. As president of the Geological Society, he responded generously to Phillips' book both privately and publicly: he averred it was one of the best geological works in English and would secure for Phillips a European reputation.[65]

Phillips' monograph was well received. In York the YPS proudly reported that it was the first fruit of the Yorkshire Museum and its geological collection. In Leeds the Philosophical and Literary Society rejoiced that its extraordinary merits had been noted on the Continent as well as in Britain. On the Yorkshire coast it inspired in the 1830s the work of Louis Hunton and William Crawford Williamson on zoning within strata, using the limited range of certain fossils. In London Fitton, the immediate past president of the Geological Society, praised it publicly and privately; and he lent a copy to Leonard Horner, the warden of University College, London, who liked it. Both Murchison and Buckland eulogised it. Conybeare judged its discussion of oolitic rocks to be very valuable, though he was mistaken in asserting that Murchison's Brora research suggested Phillips' oolite work: he was misled by Murchison's failure to acknowledge in print his debt to Phillips in 1826. The greatest accolade came from De la Beche who deemed Phillips' book 'very excellent' and adopted it quickly as the standard work on the oolitic strata of north-east England.[66] By summer 1829 the keeper of the Yorkshire museum was recognised as an expert geologist by Buckland, Sedgwick, Fitton, Murchison, Conybeare, and De la Beche, who had all come to the fore in the Geological Society in the 1820s. Phillips' reputation had been local and regional; from 1829 it became national.

Notes

1. YPS Council minutes, 9 March 1824; *YPS 1823 report*, p. 40; *YPS 1824 report*, p. 30; Conybeare to Smith, 27 March 1824, SP; George to Phillips, 25 Dec 1824, PP; Fitton, 1824a, p. 217; Buckland, 1818, 1823; Rupke, 1983, pp. 122-7.
2. Webster, 1825, p. 39; Torrens, 1975, pp. 88-92.
3. Phillips, 1844a, p. 107; *YPS 1824 report*, p. 4.
4. Orange, 1973, pp. 7-13; Pyrah, 1988, pp. 14-24; Boylan, 1981. For rival views of Salmond's centrality in 1822 see Salmond to Milton, 16 Nov 1831, Wentworth Woodhouse muniments, G 32/5, Sheffield Public Library; and Harcourt to Milton, 11 Nov 1831, Fitzwilliam papers, Northamptonshire Record Office; Atkinson speech, YPS meeting 2 Feb 1830 reported in *Yorkshire gazette*, 6 Feb 1830. On his cave theory see Buckland, 1822, 1823, and Rupke, 1983, pp. 31-50, and on cave geology Shortland, 1994. William Salmond (1769-1838), James Atkinson (1759-1839), Anthony Thorpe (1759-1829).
5. William Venables Vernon Harcourt (1789-1871), *DNB*, known until 1830 as William Venables Vernon; Edward Vernon Harcourt (1757-1847), *DNB*, archbishop of York 1807-47.
6. *YPS, 1824*, pp. 6-8; YPS general meetings minutes, 6 Jan 1823; for Harcourt's debts to Francis Bacon (1561-1626), *DNB*, see Morrell and Thackray, 1981, pp. 267-72; on Baconianism, Yeo, 1985.
7. Conybeare to Harcourt, 27 Dec 1822, HP; YPS general meetings minutes, 5 Feb 1823; Conybeare and Phillips, 1822, pp. xlv, xlvii, 164, 189, 197, 216, 230, 270, 311-12, 390-1; Miller, 1821; John Samuel Miller (1779-1830) curator Bristol Institution 1823-30.
8. Conybeare to Harcourt, 27 Dec 1822, HP; Conybeare and Phillips, 1822, pp. xiii, xiv; Young and Bird, 1822, pp. 300-4; George Young (1777-1848), *DNB*, a Presbyterian minister at Whitby.
9. Buckland to Harcourt, 15 Dec 1822, HP, 29 Dec 1822 in Melmore, 1942, pp. 322-3.
10. Pyrah, 1988, pp. 31-3; *YPS 1823 report*, p. 16.
11. YPS general meetings minutes, 5 March and 14 Oct 1823; Buckland to Harcourt, 18 Oct 1823, HP; Harcourt to YPS, 3 Nov 1823 in Melmore, 1942, pp. 320-2; Harcourt to William Danby, 10 Dec 1823, Danby Papers, Bradford district archives.
12. Smith to Harcourt, 3 Jan 1824 in Harcourt, 1891, pp. 194-5, and 12 Jan 1824, HP; *Yorkshire gazette*, 28 Feb 1824; *York courant*, 24 Feb 1824; Edmonds, 1975a, pp. 375-8; YPS general meetings minutes, 9 March, 13 April, 11 May, 8 June 1824; YPS scientific communications, 9 March 1824.

13. YPS Council minutes, 8 March 1824; John Baines Graham, an Anglican minister; George Goldie (1784-1853), a Catholic Scotsman, physician to York dispensary and York county hospital, secretary YPS 1823-31.

14. *YPS 1823 report*, pp. 37-40; Phillips, 1829b, p. xiii; Phillips, 1836a, p. xi; YPS general meetings minutes, 11 May, 8 June, 13 July, 10 Aug 1824; YPS scientific communications, 8 June 1824.

15. Edmonds, 1975a, pp. 377-9; Phillips, 1828a, p. 243; George to Phillips, 26 May 1824, PP; Phillips to George, 21 July 1824, Devon Record Office, 138 M/F 1293; Phillips, notes on York to Leeds and Greenhow, 21-9 April 1824, notebook 11, PP; Phillips, 1836a, p. xi; McConnell, 1990.

16. Edmonds, 1975a, pp. 380-4.

17. Phillips to Goldie, 7 Jan 1825 in Melmore, 1943, pp. 23-4; Phillips, 1829b, p. xiii; Phillips, notebook on Yorkshire coast 1824-7, 13 August 1824, notebook 12, PP; Phillips 1828a, p. 243; Phillips to Goldie, 1 Nov 1824 and nd [early Dec 1824] in Melmore, 1943, pp. 21-3; YPS general meetings minutes, 9 Nov, 14 Dec 1824; YPS scientific communications, 9 Nov 1824; Fitton, 1824b; Sir John Vanden Bempde Johnstone (1799-1869) of Hackness Hall, MP Yorkshire 1830-2, Scarborough 1832-7, 1841-69; Francis Cholmeley (1783-1854) of Brandsby, fifteen miles north of York, wealthy antiquarian and meteorologist; Henry Charles Englefield (1752-1822), *DNB*, published details of his portable barometer in 1806 and encouraged Webster's research on the geology of the coasts of the Isle of Wight and Dorset.

18. Edmonds, 1975a, pp. 384-5; Harcourt to Phillips, nd [5 Jan 1825], PP; *Yorkshire gazette*, 19 Feb 1825; YPS scientific communications, 8 Feb 1825.

19. Edmonds, 1975a, pp. 386-9.

20. Phillips to George, 24 May 1825, J.B. Murray collection, Yale University Library.

21. Edmonds, 1975a, pp. 389-90; Phillips to Goldie, 14 May, 2 July 1825, and Phillips to Harcourt, 10 June 1825, in Melmore, 1943, pp. 26-9; Samuel Sharp (1773-1855) vicar of Wakefield 1810-55. On 17 Dec 1824 George read to the LPLS Phillips' paper on coal plants and origin of coal, minute book of transactions, LPLS; George, 1837; T.W. George to Phillips, 15 March 1830, PP.

22. Phillips, notebook, Yorkshire coast 1824-7, July 1825 notebook 12, PP; Phillips, 1836a, pp. xii, 40; Phillips to Harcourt, 10 June 1825, in Melmore, 1943, pp. 28-9; Paul Beilby Thompson (d. 1852), of Escrick Park, near York, MP Wenlock 1826-32, East Yorkshire 1832-7.

23. *YPS 1826 report*, pp. 6-7; Brongniart to Goldie, 17 Oct 1825, in Melmore, 1943, pp. 150-1; YPS scientific communications, 9 Nov 1825; Brongniart, 1825; Launay, 1940, pp. 157-8; on the debts of Adolphe Théodore Brongniart (1801-76), *DSB*, to Phillips see Brongniart, 1828a, pp.x, 115-18, 167-8, 225, 263, 304-5, 377.

24. Orange, 1973, p. 27; *YPS 1825 report*, pp. 6-7; YPS general meetings, 9 Nov 1824; *YPS 1826 report*, p. 9; Goldie to Phillips, 29 Dec 1824, 27 June 1825, PP.

25. YPS council minutes, 11 Oct, 7 Nov 1825; for the contrast between York and Leeds see Armstrong, 1974, Feinstein, 1981, and Fraser, 1980; William Taylor (1790-1870) Anglican minister in York and inventor; William Marshall, from 1833 styled William Hatfeild, (1799-1844) of Newton Kyme, near Tadcaster, honorary curator of mineralogy YPS 1823-44; Eustachius Strickland (1787-1840) affluent York antiquarian and multiple office-holder in YPS.

26. YPS scientific communications, 11 Oct, 9 Nov, 13 Dec 1825; Phillips, 1829b, p. xiii.

27. George to Phillips, 25 Dec 1824, PP; Edmonds, 1975a, pp. 395-7; LPLS Council minutes, 27 Jan 1826.

28. Phillips, 1844a, pp. 109-10, 113-14, 127-30, 133, 143-4.

29. Knell, 2000, pp. 92-111; Edward Forbes (1815-54), *DNB*; on Martin Simpson (1800-92) see Sheppard, 1922, and Hemingway, 1946; on John Williamson (1784-1877) see Williamson, 1884.

30. Phillips, day book, Feb 1826-Nov 1828, YPS; *YPS 1826 report*, p. 11; YPS Council minutes, 16 Feb 1826; George to Harcourt, 9 Jan 1826, YPS Letter book 1822-8; René Just Haüy (1743-1822), *DSB*; Thomas Backhouse (1792-1845) curator of entomology 1823-32; Charles Wellbeloved (1769-1858), *DNB*, curator of antiquities 1823-58; on William Bean (1787-1866), Scarborough fossil collector, see McMillan and Greenwood, 1972; Kaspar Maria von Sternberg (1761-1838), *DSB*, palaeobotanist of European reputation; August Arthur De la Rive (1801-73), professor of physics, Geneva Academy; William Scoresby (1789-1857), *DNB*, an Anglican clergyman then at Bridlington.

31. Phillips, day book, 1-2 June 1826 (Murchison), 16 Nov 1826 (Haidinger), 20 March 1828 (Witham); Hutton to Copsie, 5 Feb 1828, YPS miscellaneous correspondence 1821-50, on his 1827 visit; Phillips, diary 21 Aug 1826-July 1827, 11 May 1827 notebook 16, PP (Dechen and Oeynhausen); Geikie, 1875, vol. 1, pp. 130-1; Murchison, 1827; Wilhelm Karl Haidinger (1795-1871), *DSB*; Heinrich von Dechen (1800-1889), *DSB*; Carl von Oeynhausen (1795-1865); William Hutton (1798-1860), *DNB*, a Newcastle insurance agent and leading palaeobotanist; Henry Thomas Maire Witham (1779-1844), *DSB*, an innovative Catholic palaeobotanist of Barnard Castle; on Roderick Impey Murchison (1792-1871), *DNB*, as imperial geologist and geographer see Secord, 1982, and Stafford, 1989.

32. Orange, 1973, pp. 18-23, 31; *YPS 1825 report*, pp. 8-9, with inserts about Sharp's design; *YPS 1826 report*, pp. 20-2; *YPS 1827 report*, pp. 7-9, 29-38; *YPS 1828 report*, pp. 9-10; *YPS 1829 report*, pp. 1-2, 13-14; Forgan, 1986, pp. 98-9; Addyman, 1981; Wellbeloved, 1829, p. 9; Richard Hey Sharp (1793-1853); on the York work of William Wilkins (1778-1839), *DNB*, see Liscombe, 1980, pp. 170-4.

33. Phillips, day book, 4 April, 5 May, 5 July, 20 July, 2-10 Aug 1827; YPS minutes of building committee, 1825-9, 5 July 1827; Phillips to Harcourt, 4 Aug 1827, 29 July 1829, HP; Phillips to Murchison, 11 June 1829, MP.

34. Phillips, 1827, 1828a and b, given as papers to YPS on 7 Nov 1826, 1 Jan 1828, and 4 July 1826/4 Nov 1828 respectively, YPS scientific communications; Jonah Wasse, a physician; Charles George Venables Vernon Harcourt (1798-1870) Anglican vicar of Rothbury, Northumberland, brother of William Vernon Harcourt; William Bulmer (1779-1869) Anglican vicar in York.

35. YPS scientific communications, 7 March 1826, 1 July 1828; Phillips, daybook, 8, 16-19 May 1827; Phillips, diary July 1827-March 1829, Sept 1827, notebook 17, PP.

36. Phillips, diary Aug 1826-July 1827, 28 March, 6 April 1827 notebook 16, PP; *YPS 1837 report*, p. 2; Phillips, text of noises of insects, 1 April 1828, YPS miscellaneous correspondence 1821-50; YPS scientific communications, 6 Feb 1827; Henry Baines (1793-1878) sub-curator to the YPS 1829-70; William Chapman Hewitson (1806-78), *DNB*, expert on diurnal lepidoptera, then resident in York as an articled land surveyor.

37. Brewster, 1826, 1827; *YPS 1827 report* p. 63; *YPS 1828 report*, pp. 23-4; YPS scientific communications, 21 Jan, 5 Feb, 7 Oct 1828; Brewster to Phillips, 2 May 1828, PP; Brewster, 1830, pp. 324-5; David Brewster (1781-1868), *DNB*, secretary of the Royal Society of Edinburgh 1819-28; Jonathan Gray (1779-1837), York lawyer and meteorologist, treasurer to the YPS 1823-37, was father of William Gray junior, a secretary to the YPS 1827-38.

38. Phillips, diary 1827-9, 28 Feb 1829, notebook 17, PP; Robert Davies (1793-1875), *DNB*, a solicitor, town clerk of York 1827-48, antiquary, and treasurer to the YPS 1837-49; John Kenrick (1788-1877), *DNB*, Unitarian clergyman, tutor in classics and history from 1810 and professor of history 1840-50 at Manchester College, active in YPS from 1823 to death; William Hincks (1794-1871), *DNB*, Unitarian clergyman, tutor in natural philosophy at Manchester College 1827-39, curator of botany for YPS 1828-39; Baldwin Wake, a York physician.

39. Phillips, bound syllabuses, 1824-52, folder 12 PP; Phillips, diary 1825-6, 3 and 8 March, 16 April 1826, notebook 13, PP; YPS Council minutes, 7 March 1826, 1 Jan 1827, 5 June 1829; Pierre André Latreille (1762-1833), *DSB*, used numerous external features as the basis of classification of insects.

40. Book listing lectures, classes, number of members of Institute 1827-51, York City archives, TC 49/1:15; Peacock, 1981; Kenrick, 1860; Wellbeloved, 1828; Alborn, 1986, esp pp. 54-76, is excellent; Phillips, diary 1827-9, 27 March, 29 April 1828, 3 and 10 March 1829 notebook 17, PP.
41. Phillips, financial accounts, diary 1827-9, notebook 17, PP; Phillips, bound syllabuses, 1824-52, PP; Phillips, diary 1825-6, speech at YPS dinner, 7 March 1826, notebook 13, PP; on the Royal Manchester Institution see Bud, 1974, and Kargon, 1977, pp. 16-20.
42. Phillips, 1836a, p. xii; Phillips, diary on Wolds and coast 1828, 16-28 May, notebook 19, PP; Phillips, diary 1826-7, 9 Oct 1826, 11-29 June 1827, notebook 16, PP; Dikes to Phillips, 14 Jan 1828, PP; Wood to Phillips, 18 May 1827, 21 July, 14 Sept 1828, PP; Sykes to Phillips, 25 July, 8 Nov 1828, PP; Watson to Phillips, 14 Nov 1828, PP; Younge to Phillips, 20 April 1829, PP; William Hey Dikes (1792-1864); John Edward Lee (1808-87), *DNB*, archaeologist then living at Hull; Christopher Sykes (1774-1857); George William Wood (1781-1843), treasurer of Manchester College 1808-43, close associate of Wellbeloved and Kenrick, active in Manchester Literary and Philosophical Society, Royal Manchester Institution, and Manchester Mechanics' Institution, Foxite Whig MP South Lancashire 1823-5, Kendal 1837-43; William Henry (1774-1836), *DNB*; Edward Holme (1770-1847), *DNB*; John Dalton (1766-1844), *DNB*; on William West (1792-1851) see Morrell and Thackray, 1981, pp. 444-7; Luke Palfreyman, secretary to the Sheffield LPS 1823-31; Edward Barker (1798-1832) was curator of mineralogy; Robert Younge secretary 1826-31. George persuaded Phillips to give papers to the LPLS on the mechanical effect of rain, rivers, and the sea, and on the materials which compose the surface of our planet, 20 April 1827; on sensation in animals, 7 Dec 1827; on digestion in animals, 18 Dec 1829; *Transactions of Leeds PLS* (1837), vol. 1, pp. xiii, xv.
43. Phillips, diary 1825-6, 13 Feb, 3 March, 6 and 16 April 1826 notebook 13, PP; Phillips, diary 1826-7, 26 June 1827, notebook 16, PP; Phillips, diary 1827-9, 30 Nov 1827, 8, 22, and 23 June 1828 (qs), notebook 17, PP; Phillips, diary Wolds 1828, 8, 22 (q), 23 (q) June 1828, notebook 19, PP.
44. Phillips, diary 1825-6, 7 March 1826; YPS Council minutes 3 July 1826; Phillips, tour of Scotland 1826, notebook 14, PP; Phillips, tour in Scotland, 1826, Mitchell Library, Glasgow, 5 and 9 August (qs); Phillips to Harcourt, 31 July 1826, HP; Phillips, 1829b, p. 27-8; Richard Winter Hamilton (1794-1848), *DNB*, Congregational minister in Leeds; William Hiley Bathurst (1796-1877) minister, Barwick in Elmet, near Leeds; James Hutton (1726-97), *DNB*; John Playfair (1748-1819), *DNB*.
45. Phillips to Armetriding, 20 Oct 1827, PP; Phillips, diary 1825-6, 9 Sept 1826, notebook 13, PP; *YPS 1827 report*, p. 14; *YPS 1828 report*, p. 13 (q).
46. *YPS 1826 report*, p. 11; *1827 report*, p. 12; *1828 report*, p. 18; *1829 report*, pp.9, 13; Phillips, day book, 26-7 April 1827; Phillips, diary April-Dec 1829, 23 April 1829, notebook 21, PP ; John James Audubon (1785-1851), *DAB*, was in Britain recruiting subscribers to his *Birds of America*.
47. Phillips, diary 1827-9, 2 Sept, 13 and 15 Nov 1827, notebook 17, PP; Murchison to Harcourt, nd [late 1827], nd [early 1828], HP; Phillips to Murchison, 21 Jan 1828, MP; *PGSL* (1834), vol. 1, pp. 38-9; Phillips, 1829c; Thomas George, a relative of E.S. George.
48. Phillips, 1844a, p. 112; Geikie, 1875, vol. 1, pp. 131-2; Murchison to Harcourt, 20 Nov 1826, HP; Sedgwick, 1826.
49. YPS scientific communications, 3 Oct 1826; *YPS 1826 report*, pp. 15-16; Sheppard, 1917, pp. 188-9; Harcourt to Murchison, nd [1826], MP; Murchison to Harcourt, 20 Nov 1826, 6 March, 30 June 1827, HP.
50. Phillips, 1844a, pp. 112-13; Smith, 1827; Sheppard, 1917, pp. 187-91; YPS scientific communications, 6 March 1827; Phillips, day book, 16-17 March 1827; Smith to Phillips, 8 March 1827, in Melmore, 1943, pp. 28-9.
51. Phillips, 1844a, pp. 113-14; Johnstone, 1840; Sheppard, 1917, pp. 168-74, 206-207; Theakston, 1841, pp. 56-9.

52. Phillips to Goldie, 11 April 1825, in Melmore, 1943, pp. 25-6 (q); YPS scientific communications, 4 April, 5 Dec 1826; Phillips, day book 13-17 March, 5-6 June, 6-11 Sept 1826; Harcourt, 1826; Phillips 1828a, p. 243; *YPS 1826 report*, pp. 15, iii, v; Harcourt to Murchison, nd [1826], MP; Murchison to Harcourt, 20 Nov 1826, 6 March 1827, HP.

53. Phillips to Sabine, 20 July 1853, Sa P; Phillips, diary 1827-9, 8 Dec 1827, notebook 17, PP; Phillips 1829a; Phillips, diary Wolds 1828, 26-7 July 1828, notebook 19, PP; Fitton to Phillips, 30 Oct 1828, PP.

54. Dunn to Phillips, 25 June 1831, PP; Theakston, 1841, p. 120; John Dunn (1790-1851) Scarborough surgeon, fossil collector, then secretary of the Scarborough LPS. The plates were lithographed by Inchbold of Leeds because there was no adequate lithographic printer in York.

55. Phillips, 1829b, pp. xiii; YPS scientific communications, 6 Nov 1827, Phillips reported Bean's discovery of fossil plants at Gristhorpe near Scarborough; 1 Jan 1828 Phillips reported Williamson's letter about Bean's discovery of fossil starfish also found by Williamson; Dunn to Phillips, 15 Dec 1826, PP, donated Oxford clay fossils from Scarborough.

56. Phillips, diary 1827-9, 21 Sept, 4-20 Oct 1827, notebook 17, PP; Phillips, 1829b, p. xiii; Phillips to Brongniart, 2 Nov 1827, PP; YPS scientific communications, 11 Oct 1825; Witham to Phillips, 26 Jan 1828, YPS miscellaneous correspondence; Phillips, 1828a.

57. Phillips, 1829b, p.xii (q); Phillips, diary 1827-9, 16 Jan, 19 March 1828, notebook 17, PP; Phillips to Harcourt, 17 Nov 1828 HP; Fitton to Phillips, 6 Dec 1828, 23 March 1829, PP.

58. Miller to Phillips, 27 Feb, 6 March 1829, PP; Phillips to Miller, 31 March 1829, American Philosophical Society, Philadelphia; Fitton to Phillips, 30 Oct, 6 Dec 1828, 23 March, 23 May 1829, PP; Lonsdale to Phillips, 25 Feb 1830, PP; Phillips to Murchison, 11 June 1829, MP; Goldie to Phillips, 13 April 1829, PP; Phillips to Murray, 20 Dec 1830, Mu P; John Taylor (1779-1863), *DNB*, Unitarian mine owner, treasurer Geological Society 1823-43; John Murray (1778-1843), *DNB*.

59. Phillips, diary 1828 Wolds, 26-7 July 1828, notebook 19, PP; Phillips, 1829a; Phillips, 1829b, pp. xi, xvi; Hutton, 1795; Playfair, 1802; Mantell, 1827; Gideon Algernon Mantell (1790-1852), *DNB*, a Lewes surgeon.

60. Phillips, 1829b, pp. xvi, 3-4, 32-3, 44-5, 59, 75, 87, 114-15, 137, 145-6 (qs); Phillips, 1829a; Phillips to Smith, 30 March 1829, Hawkins collection, University of Reading Library, printed by Sheppard, 1934, pp.171-2; contrast Laudan, 1987, pp. 164-8.

61. Phillips, 1829b, pp. 20, 152-3, 158-60; Phillips, 1858a, p. 85; Phillips, 1860a, p. xli; Wright, 1860; Thomas Wright (1809-84), *DNB*, Cheltenham surgeon; contrast Rupke, 1983, pp. 192-3, and Laudan, 1976, pp. 222-4, with Anon, 1831.

62. Phillips, 1829b, pp. 28-30, 42, 117, 166; Webster, 1814; Conybeare and Phillips, W. 1822; De la Beche, 1822; Phillips, 1830; Cuvier, 1825, vol. 5, part 2, p. 487.

63. Phillips, 1829b, pp. xv, xvi; Winch, 1821; Young and Bird, 1822, pp. 4-5, 179, 289, 300-4, 311, 324; Young and Bird, 1828, pp. 355-6; Young, 1817, vol. 2, pp. 790-2; Eastmead, 1824, pp. 32-40; Penn, 1822; Nathaniel John Winch (1768-1838), *DNB*, secretary to Newcastle Infirmary, primarily a botanist; William Eastmead (d 1847), *DNB*, dissenting minister, Kirkby Moorside; Granville Penn (1761-1844), *DNB*.

64. Phillips, 1829b, pp. 16-19 (18q), 22-3, 48, 51, 55, 61, 68-9, 104-5; Buckland, 1823, pp. 173, 192-4.

65. Sedgwick, 1826, pp. 339, 341, 351; Phillips, 1829b, p. xiv; Sedgwick to Phillips, 11 April 1829, PP; Sedgwick, 1830, pp. 199-200.

66. *YPS 1828 report*, p. 8; *YPS 1829 report*, p. 6; LPLS, report 1828-9, minute book of transactions 1821-41; Hunton, 1837; Williamson, 1837, 1896, p. 12; Torrens and Getty, 1984; Fitton, 1829, p. 117; Goldie to Phillips, 13 April 1829, PP; Phillips to Murchison, 11 June 1829, MP; Buckland to Phillips, 12 April 1829, PP; Conybeare, 1833, p. 378; De la Beche, 1830, p. 82; Louis Hunton (1814-38) belonged to an alum working family

at Loftus on the Yorkshire coast; William Crawford Williamson (1816-95), *DNB*, of Scarborough, son of John Williamson; Leonard Horner (1785-1864), *DNB*, *DSB*, warden of University College London 1827-31, elected FGS 1808, president Geological Society 1845-7 and 1860-2.

Chapter 3

The Spreading Reputation 1829-1833

During these years Phillips gained greater financial security by increasing his regular income three-fold. In 1831 the YPS raised his salary as its keeper from £60 to £100 pa. Next year he was appointed the first assistant secretary of the British Association at £100 pa. Until 1834 he was on balance content to be a provincially based savant with a national role and international reputation. From his York base he made at last a couple of continental tours in 1829 and 1830. The year 1831 was eventful. Phillips contributed to the final vindication of his uncle who was canonised from the presidential chair of the Geological Society as the father of English geology. In spring he considered but eventually rejected the chair of geology at University College London. In autumn he was a key figure at the first meeting of the British Association for the Advancement of Science held in York. Late in 1833 he applied for the chair of mineralogy and geology at the Royal Dublin Society but then withdrew. In this chapter I shall first consider his activities in York and then discuss the episodes alluded to above. I shall argue overall that Phillips' situation at York, which appeared solitary and limited when viewed from London, gave him sufficient opportunities to develop his career and pursue his research so that for the time being he repressed his metropolitan ambitions.

3.1 The York savant

After the publication of his 1829 book in April and before the Yorkshire Museum was formally opened in February 1830, Phillips showed his commitment to the YPS by becoming an elected subscribing member, which cost him £5 admission fee and £2 annual subscription. He became eligible for honorary office-holding and in February 1830 was elected a secretary, joining his chums Goldie and William Gray, junior. In his double capacity as keeper and secretary to the YPS, Phillips became even more the trusted right-hand man of Harcourt, who as president had seen to it that from October 1829 Phillips was assisted by a sub-curator, Henry Baines, who was paid 52 guineas a year. In Harcourt's speech, made at the anniversary meeting of 2 February 1830, he invoked Francis Bacon's authority to argue in Phillips' presence that the next major aim of the YPS was to pay its scientific servants decent salaries and expenses to enable them to work full time, to bring to a focus the scattered efforts of part-time honorary curators, and to push the Society towards realising its objects.[1]

At this celebration Phillips was praised by Goldie as a friend, keeper, and philosopher whose profession was natural science. He heard himself described by Dunn as a highly gifted 'cyclopaedia of science'; and noted that Johnstone and Dunn from Scarborough regarded the YPS as the great heart of Yorkshire science whose blood nourished other societies in the county. Best of all, he heard Harcourt propose a toast to Smith, as the father of English geology, to which Smith responded. That sobriquet had previously been conferred on Smith by Johnstone and Dunn in August 1829 at a public dinner to celebrate the opening of the Scarborough Museum; but Phillips was absent abroad then. In 1830 in his presence the paternity of English geology was conferred on Smith, at a ceremony which was sufficiently important to be reported nationally. Moreover the stone from which the Yorkshire Museum was built was Kelloways rock, which had been quarried at Johnstone's Hackness estate and donated by him. For Phillips and Smith it was a highly appropriate choice because it was a key stratum in their elucidation of the stratification of the Yorkshire coast.[2]

As keeper of the Yorkshire Museum, Phillips was well aware that it had cost the YPS £9,800, leaving a debt of £1,500. In compensation for this financial constraint, the new premises were far superior to the Society's previous rented accommodation. They offered a library, a lecture theatre with tiered seats, three rooms lighted by glass skylights for the geological, mineralogical, and zoological collections, and in the upper storey a room for himself. The new Museum proved to be popular, with 4,000 recorded visits made in the first nine months. The YPS also aimed to be instructive so Phillips was heavily engaged in arranging and cataloguing specimens. In order to lighten his extensive and still extending labours he was aided by eight honorary curators, an arrangement which encouraged participation and specialisation. As a reward for devoted service, his salary was raised in February 1831 to £100 pa even though the Society's annual surplus of income over expenditure was hardly enough to pay the interest on its debt. Harcourt still saw Phillips as central to his visionary aims for the YPS, 'the erection of a well-supported SCHOOL OF SCIENCE in this northern metropolis, and the execution of such a HISTORY OF YORKSHIRE as the antiquary and the natural historian may be contented to possess'. Harcourt also had another purpose. He wished to resign the presidency of the YPS in 1831 to Lord Milton, Whig MP for Yorkshire, who had wide scientific interests. He was already a patron of the YPS and a trustee of its Museum to which he and his father had donated £350. In his successful approach to Milton, Harcourt stressed that Phillips was well qualified for conducting all the material parts of the YPS: he had 'very superior scientific attainments' and was 'modest, sensible and popular, well contented with science and £100 a year'.[3]

Phillips' main problem as keeper was coping with ever-increasing collections. By early 1833 the Museum contained 20,000 geological, mineralogical, and zoological specimens, of which only 13,800 were properly exhibited. The zoology collection was a particular problem. In order to house it adequately, a subscription with a target of £450 was launched but at the end of 1833 only £179 had been raised. The zoological collection was so unsatisfactory that the YPS confessed that it could not guarantee the safety of new donations. Some zoological specimens were, of course, big: in 1832 Phillips prepared for display a skeleton, twenty-eight feet long, of a beached whale donated by his friend Christopher Sykes. Phillips was so keen to have such specimens properly exhibited that in March 1834 he gave £20, half the proceeds of a YPS lecture course on the natural history of animals, to help the subscription meet its £450 target.[4] Perhaps in compensation for the difficulties posed by dead specimens, Phillips proposed in July 1830 that a limited collection of live animals, paid for by private subscription, be set up in the Society's grounds. The menagerie, which contained monkeys and a bear, was short-lived. In late 1831 the bear proved troublesome and was offered for sale to the Zoological Society of London which accepted him and advised that he travel by stage coach from York to London as an outside passenger. By November 1833 the YPS welcomed the complete extinction of the menagerie.[5]

More happily Phillips was involved in two new ventures of the YPS, a laboratory and an observatory. The former was endowed in 1833 by the Earl of Tyrconnel, a loyal member of the Society, who offered to it his large collection of chemical and electrical apparatus. As keeper and as an expert on instruments, Phillips went to Kiplin Hall, near Catterick, to inspect it and pronounce favourably on it. Previously the YPS had owned only a miscellaneous collection of donated philosophical instruments. Tyrconnel's liberality enabled the Society to launch, on a small scale, research and lecturing in chemistry. Though Harcourt had learned the science from no less than William Wollaston and Humphry Davy, he had not encouraged it because it did not fit the Society's dedication to Yorkshire's natural history.[6] Phillips was a competent mineralogist and not uninterested in electrical apparatus, but made little or no use of the laboratory.

The Society's observatory was also externally stimulated but only partly endowed. At the meeting of the British Association in York, several eminent astronomers urged the YPS to establish an astronomical observatory. One of them, William Pearson, donated a clock, a transit instrument, an achromatic telescope, and the conical roof of his own observatory in Leicestershire, leaving a committee of the Society to launch a subscription of £300 to build the observatory. Though contributions fell well below the target, the building was begun in 1832 and completed in July 1833. Phillips was not on the observatory launch committee, whose members were William Taylor, Jonathan Gray, and Thomas Donkin, but he was a member of a committee of subscribers which in 1833 supervised its construction. Phillips was by then more competent in astronomy than most geologists. From 1826 he had made his own telescopes. In 1832 he had used James Atkinson's telescope to observe and draw sun spots. In his garden he had a stone tablet on which he had drawn a meridian line. He was capable of doing the appropriate calculations which gave the altitude of the sun from measurements made with a sextant. Even so he felt he needed advice about the YPS observatory and in June 1833 consulted at the Cambridge BAAS meeting George Airy, then professor of astronomy there and future astronomer royal. He urged that the transit instrument be used to correct the clock, that the telescope be mounted equatorially, and be used by intermittent observers to observe those intermittent and irregular phenomena which were troublesome for established observatories because they involved breaks from routine. The telescope, which had a focus of six feet and an aperture of just over four inches, was mounted equatorially and in July used by Phillips. In September 1833 the YPS appointed a committee of seven observers, of whom Phillips was one, to run the observatory which was initially open only to subscribers. Their leading spirit was William Lewin Newman, an actuary to a York insurance company. Phillips, too, had high hopes for the small observatory: in conjunction with the improved zoological collection, it would confirm the YPS's position as the leading scientific institution in provincial England. The episode showed Phillips' growing interest in astronomy, the queen of the sciences, and confirmed his readiness to contribute to co-operative endeavour.[7]

Phillips continued to be easily the most prolific performer at the YPS's monthly meetings. As keeper he reported on the museums he had seen in summer 1829 in the Low Countries, France, and the Rhineland and those visited in spring 1831 in the south of England. He particularly liked Mantell's purpose-built museum at Lewes because it perfectly illustrated the geology of Sussex. He kept the Society up to date with his latest research on the carboniferous geology of north-west Yorkshire. In May 1833 he revealed the important discovery that the sandstones and shales, which locally divided the mass of limestone in three north-west dales, varied in thickness. The previous year Phillips expounded and quickly published his Cuvierian conclusion about the coal measures in the Halifax area: on the basis of the fossil evidence there had been periodical alternations of marine and freshwater deposits, a view at odds with that of Conybeare. Of course Phillips reported interesting geological results obtained during his continental tours. He rejoiced in the stratigraphical correlations between Le Havre and south England. He expatiated on his firm belief that violent, turbulent, and transient floods were responsible for certain aspects of Swiss geomorphology. Phillips also commented on a specimen of jaw-bone, from the slate of Stonesfield, near Oxford, which he had found in the collection of Christopher Sykes. Having declined Sykes' offer of it to himself, Phillips persuaded him to place it in the Yorkshire Museum in 1831. It was important because only three other specimens were known and because Cuvier had concluded that they belonged to a marsupial mammal called didelphis. If they did and if the Stonesfield slate belonged to the oolitic formation, then it followed that contrary to general opinion mammals existed well before the tertiary period and that there had been no exclusive or necessary sequence from simple to complicated life in the past. 'In the full spirit of Cuvier', Phillips did not view the Stonesfield fossil quadruped as a puzzling anomaly: he welcomed it as a valuable guide to the succession and conditions of life on the earth in the past.[8]

In 1829 Phillips was involved in the excavation of mammal bones found in a marl pit at Bielsbeck, two miles south-west of Market Weighton and slightly nearer Hull than York. For him it was 'the most important thing known about osseous remains' and for the YPS the most remarkable geological phenomenon ever investigated by its members. The episode not only contributed to the natural history of the antediluvian world: it also showed the YPS, as the county philosophical society, exploiting and appropriating the discovery. On 25 July 1829 William Dikes, curator of the Hull LPS, wrote to his friend Phillips about his discovery in the black marl of the bones, which were accompanied by fresh water and land shells. Dikes neither consulted Phillips nor invited him to the site; but Phillips decided that he could identify the shells better than Dikes and that therefore a York trio of Phillips, Harcourt, and Salmond (an expert on cave excavation and fossil bones) should see for themselves the highly important remains of bear, elephant, rhinoceros, deer, ox, and horse. On 31 July the three Yorkists spent a day at Bielsbeck and decided that Harcourt would write a paper about it for rapid publication in the September issue of the *Philosophical magazine*, with Salmond covering the bones and Phillips the shells. Knowing that Dikes was a non-publisher, Phillips disingenuously urged him yet again to publish independently or be 'hooked into ours'. Though the paper acknowledged Dikes' priority of discovery, Harcourt's interpretation depended on Phillips' section and his conclusion that all the shells of molluscs in the black marl were identical with those still living in Yorkshire. Above the black marl was a layer of gravel, containing chalk and white flint, which for Harcourt and Phillips had been laid down by a mighty current after the marl had been deposited. They concluded that mammals, adapted to a temperate climate, lived in Yorkshire where their remains and the shells were deposited in the black marl. Harcourt added a theological gloss to these conclusions in that he argued that the flood recorded by Moses, and not high tides and ancient inundations of the river Humber, explained the scale and position of the gravel deposits.[9]

In autumn 1829 Harcourt organised a new excavation at Bielsbeck on behalf of the YPS which paid £22 for it. The bore, twenty-six feet in depth, revealed in downwards order: sand; gravel; grey marl containing elephant, horse, and rhinoceros bones, and erratic rocks from north-west England but no shells; and black marl containing wolf, elephant, horse and bison bones, and many unbroken shells of land and freshwater creatures. Aided by Phillips, who visited the site twice, Harcourt concluded that, whereas the black marl had been deposited by tranquil water, the grey marl and gravel had been produced by violent floods. Having given these results first to the YPS, Harcourt published them quickly in the January 1830 issue of the *Philosophical magazine*. Drawing on Phillips' Cuvierian approach to the reconstruction of the creatures and climate of Yorkshire in the past, and in particular on his conclusion from the condition of the bones and shells that they had not been transported far, Harcourt claimed again that some mammals, not then alive in Yorkshire, had adapted to a temperate climate there and with the molluscs had been contemporary inhabitants of Yorkshire in antediluvian times. As before, Harcourt added a theological gloss. He hoped that further excavations would fix the chronology of the Deluge, which he believed had followed creation 'at no very considerable interval of time', was relatively recent, and had preceded the appearance of human beings. For Harcourt the confirmation and extension of the links between natural and civil history were central to geology.[10]

Harcourt's papers, based on Phillips' expertise in the field and in interpretation, were favourably received in London. Lyell welcomed Harcourt's results, about past inhabitants of Yorkshire living together in a temperate climate, as being of the greatest importance. As president of the Geological Society, Sedgwick praised Harcourt for revealing 'the extreme link of a great chain binding the present order of things to that of older periods' in which considerable extinction occurred. Harcourt not only gained intellectual credit for the Bielsbeck excavation: he also appropriated for the YPS the best ensemble of specimens. He persuaded William Worsley, a long-standing member of the YPS and owner of Bielsbeck, to donate the excavated materials to the Yorkshire Museum, leaving duplicates for the Hull museum. Though Dikes was content to let Harcourt and Phillips explain the significance of his finds at Bielsbeck, he was incensed at losing the best specimens to York. Phillips, who kept Dikes well informed about the excavations, was caught in the cross-fire. He claimed that he was ignorant of Harcourt's intention for the specimens: his own contribution was 'anatomical not diplomatic' because it was concerned with the classification and not the disposal of the bones.[11]

As a researcher Phillips revealed to the YPS his new interests in electricity, magnetism, and meteorology, in all of which his approach was dominated by instruments and measurements made with them. From 1830 he designed, made, and published the details of his modification of Volta's electrophorus, which generated statical electricity. Phillips also designed and made in 1831 a new type of dipping needle, with a double balancing system which eliminated vibration because it made the needle's centre of gravity coincide with its axis of movement. It determined the dip of the earth's magnetism by measuring the angle which the needle made with the horizon. This was familiar work for Phillips in that as a surveyor he knew about the dip of strata, i.e. their slope as measured by their angle of inclination to the horizon. In 1832 Phillips did not report to the YPS his failure to make a type of Hansteen needle with which he hoped to measure changes in magnetic intensity produced by auroras.[12]

Phillips' interest in meteorology was driven by his belief in Baconian induction, i.e. that regular, systematic, discriminating, and accurate observations made in different places using exact instruments under standard conditions would generate appropriate data; these, it was hoped, would lead to empirical laws, which summarised the phenomena in what he called 'scientific generalisation', and thence to precise theories from which these laws could be deduced. All his meteorological work was focussed on measurement with instruments, some of which he designed and made for himself. It was stimulated and reinforced but not created by the British Association. Its first meeting gave a sensible impulse to the YPS which set up a meteorological committee of Jonathan Gray, William Gray junior, and Phillips.[13] The research on rainfall, begun in late 1831 mainly by the two younger men on the comparative quantities of rain falling at the top of York minster and near its base, was suggested by Phillips to the Association which then recommended it be pursued under its name. The aims were to determine the general law about the different amounts of rain collected on buildings and on the ground near them and from that law to test the belief that raindrops grew bigger by condensation as they fell through moist air. Accordingly they measured rainfall with three gauges, one on the ground near the Yorkshire Museum, one on its roof, and one on a nine-foot pole above the battlements of the central tower of York minster. They tried to eliminate instrumental error and the distorting effect of winds, especially at the top of the tower. Their provisional result, using gauges which were no more than tin boxes and funnels, was announced in 1833 first to the YPS and then to the BAAS: rainfall diminished with height above ground in the same place according to a particular empirical formula; and their probable inference, as opposed to plausible speculation, from this empirical law was that raindrops did increase in size as they fell. Phillips was well aware that data from just one place were not enough, that eddying winds and local physical geography were distorting, and that his method of using average results was open to objection. Even so they revealed a meteorological paradox which impressed no less than John Herschel: rainfall in the hills was greater than in the plains, as Phillips knew from experience, but rainfall at a low situation was greater than at an elevated one in the vicinity.[14]

Also in late 1831 Phillips and Gray began research on the hourly oscillations of barometric pressure in order to establish one element of the local climate of Yorkshire. Their ambitious aim was to draw 'the curves of horary oscillations', ie. to produce graphs showing them. Using a barometer designed and made by themselves, they failed to do so but they did produce for the YPS tables showing the hourly variation of barometric pressure, reduced to standard temperature, for 1832 and 1833. They themselves recorded the pressure on the hour eight times during the day and from 1833 employed an early riser to record it at 4 am. Such 'multiplied and intensive fagging' was modelled on similar work undertaken at Edinburgh from 1827 by James Forbes, who had spoken about it at the 1831 meeting of the British Association and rapidly become a close friend of Phillips.[15]

Phillips also encouraged YPS members to form 'a grand association of weatherglass people over Yorkshire' to record features of its climate. This project was based on the meteorological observations made at Ackworth in 1830 and reported to the YPS in March 1831 by Luke Howard, the well-known Quaker meteorologist and classifier of clouds. By 1833 meteorological registers at six different places were being kept and summarised for the YPS; but, significantly, not one table was ever published except that for monthly rainfall of 1834. The activities of these Yorkshire weatherglass people, involving gentry, clergymen, teachers, and Phillips himself, produced merely raw data unguided by any hypothesis, a result which did not bother Harcourt who believed in provincial participation by all capacities. But such activity was disappointing to Phillips and castigated in 1833 by William Whewell as antiquarian meteorology in which endless and mere observation produced not knowledge but despair.[16]

Phillips tried to reduce the labour of meteorological recording by inventing, making, and revealing to the British Association in 1832, the first successful self-registering maximum thermometer. It relied on an air bubble to separate a small portion of the column of mercury. To make it Phillips melted the end of the sealed tube of an ordinary thermometer, of fine bore and full of mercury, and plunged it instantly into the flame of a spirit lamp. This sort of skill in designing and constructing a new measuring device, which was delicate, exact, and durable, was not common among his geological contemporaries. Phillips' maximum thermometer enjoyed a long life: in its commercial form it was still widely used well after his death.[17]

At the meeting of the British Association held at Cambridge in 1833 the phenomena of the aurora borealis were extensively discussed and it was agreed that Phillips should compile instructions about observing them. These guidelines were addressed to the general membership who were to report their results to Phillips. The aim was to establish a system of contemporaneous observations which would give data pertinent to the magnetism of an aurora and the height of its arch. Though Phillips was highly regarded as a meteorologist by such leading mathematicians as George Peacock and John Lubbock, his instructions were strongly criticised by experts. Richard Potter of Manchester objected to Phillips' assumption that auroras were only a few miles high and thought miscellaneous displays a false basis for calculations. Airy objected to the use of divided instruments and not using the eye on the whole phenomenon. Above all John Dalton, who had contributed to the auroral discussion at Cambridge, told Phillips that his instructions repeated or ignored what was commonplace. Phillips, it seems, did not know that Dalton himself had published long ago his determination of the height of the arch as being 150 miles and his theory that the aurora was a magnetic phenomenon, with the beams governed by the earth's magnetism. In his enthusiasm for the Association's aim of promoting collaborative inductive research and for Forbes' notion that the infant science of meteorology needed total revision, Phillips had ignored the major meteorological work of the revered Quaker whose support had been so vital to the early Association. Significantly Dalton did not contribute to the tabular conspectus of the aurora of 17 September 1833, collated by Phillips and containing his own observations. In the auroral work Phillips showed an uninformed and tactless impetuosity which on occasion overcame his usual prudence. Certainly in late autumn 1833 his enthusiasm for auroral observation and comparison led him to talk about them to three successive meetings of the YPS.[18]

As a lecturer Phillips was in demand. Though urged by friends to lecture in Edinburgh, Birmingham, and Preston, he declined. He remained loyal to the YPS, giving two courses on geology, two on zoology, and one on meteorology (see Appendix 1). At the YPS's request he performed in a new genre, that of a gratuitous public lecture given in the evening during the spring assizes in the theatre of the Yorkshire Museum. Covering geology, zoology, and meteorology in 1830, 1831, and 1832 respectively, they attracted a fashionable audience of 300 including 'the sages of the magistrates box and the belles of the ballroom'. Phillips' concern to popularise science was also shown in his continuing association with the York Mechanics' Institute to which he gave two gratuitous lectures on geology in January 1831.[19]

Outside York Phillips confined himself as a lecturer mainly to the familiar haunts of Leeds, Manchester (twice), and Hull, where he expatiated on zoology as well as on geology (see Appendix 2). These visits consolidated friendships and gave him new contacts. In 1831, for example, he stayed a week in Manchester with William Henry at whose home he met the young chemist William Gregory and the redoubtably omniscient William Whewell; and he dined with Wood, his chief patron there. On his trip to Hull in 1832 he did field work around Louth in north Lincolnshire with Lee and Dikes and stayed for a time with Sykes, all old chums, but he also resided with James Alderson, a Hull doctor who rose to be physician extraordinary to Queen Victoria.[20] Phillips ventured to just one new locale, Halifax, where in autumn 1831 his lectures on geology constituted the first course sponsored by the recently founded Halifax Literary and Philosophical Society. His performances received rapturous local acclaim and led to friendship with Christopher Rawson, its president, who facilitated Phillips' research on the alternations of marine and fresh-water deposits in the coal seams of the area. Phillips clearly became selective about lecturing. In one year (1833) he undertook no engagement outside York. When he did his standard rate by 1831 for twelve geology lectures was £80 and for six £50.[21] His income from external lecturing remained important to him: in 1831 and 1832 it generated £100, a sum equal to his York salary and from 1832 to that as assistant secretary to the British Association.

In his zoological lectures Phillips continued to stress the correspondence of structure and function in animals and their exact and beautiful adaptation to their environments. In those on geology he distanced but did not entirely divorce himself from scriptural geologists. At Halifax he gave his views on what he called the narratives of creation and the Deluge. The former showed the eternal providence and forethought of God who created the world and human beings. Genesis was written to induce religious contemplation and to be intelligible to Jews: it was entirely concerned with the general circumstances attending the origins of humans, God's final creation. The point of the narrative of the Flood was to show that the Author of the world continued to watch over it, with power to destroy and save. The Flood, which was controlled by God and used natural law-like means, was a warning to posterity and the last great violent occurrence on the earth. Phillips concluded that geology was not incompatible with biblical accounts of creation and of the Deluge by inspired writers. At York in 1831 he argued that there was a general accord between scripture and geology because geology, if pursued freely and without reference to religious doctrine, could be the handmaid of religion in that it provided evidence of physical changes of the same kind as those recorded in the Bible. Thus the three scriptural periods, antehistorical, antediluvian, and post-diluvian, and two events, creation and the Deluge, agreed broadly with geological epochs. Phillips believed that the scriptural reference to the appearance of dry land corresponded roughly with the great mountain elevations following the tertiary period. Though he was not a scriptural geologist in that he did not try to make geological results conform to scripture, he was a passive reconciler: he was not unhappy when geologists, pursuing their science free from religious constraint, produced conclusions which broadly and loosely were not incompatible with scripture.[22]

3.2   Geological raids on the Continent

The eighteenth-century Grand Tour of Europe was assiduously cultivated by rich
Britons partly to indulge in hedonistic pleasures free from prying eyes at home
and partly to celebrate Britain's superiority to absolutism and Catholicism. After
1815 affluent geologists avoided the first temptation but still succumbed to the
second. They also wished to see famous geological phenomena, especially those
which did not occur in Britain, to make cross-channel comparisons, and to meet the
continental brethren of the hammer. In 1828, for example, Murchison had spent six
months in Europe, partly with Lyell who subsequently preached to his companion
that travelling was the first, second, and third requisite for a modern geologist, given
the adolescent state of the science. Murchison obeyed: in 1829 he and Sedgwick
devoted five months to research in the Austrian Alps. Lyell himself followed his own
sermon. From 1818 he had undertaken continental tours, culminating in his major
expedition to France and Italy from May 1828 to February 1829, which confirmed
the uniformitarian views he expounded in his *Principles of geology*.[23]

Such lengthy and expensive tours abroad were impossible for Phillips who
generally had neither the leisure nor the money for them. In 1829 and 1830, however,
he was able to make two continental trips, after which he never left Britain until he
visited the Auvergne in 1855. The first tour was stimulated by an invitation from
Charles Daubeny to contribute the greater part of the article on geology to the
Encyclopaedia metropolitana. Armed with a letter of introduction from Buckland,
who thought Phillips' book on the Yorkshire coast was excellent, Daubeny spent a
whole day in April 1829 in York with Phillips who accepted the proposal in early
June. Phillips felt that his article should not deal just with Britain. He hoped to
compare distant regions with familiar British areas and to meet the eléves of other
schools of geology, in order to avoid dogmatism based on merely local results. The
YPS backed his plan by giving him leave of absence for August and September.
He received useful advice on his itinerary from Daubeny and Buckland, who also
gave him introductions to Jacob Nöggerath, professor of mineralogy and mining at
Bonn, and to Cuvier. Indeed Buckland was about to send to Cuvier for identification
Phillips' drawings of Yorkshire coal measures specimens of supposed serpents which
Buckland thought were fish [24]

Accompanied by his close friend the reverend William Taylor, Phillips travelled
via London where he met Lonsdale at the Geological Society and Faraday at the Royal
Institution. On the Continent he moved eastwards from Calais to the river Rhine via
Dunkirk, Lille, Ghent, Antwerp, Malines, Brussels, Namur, Liège, Maastricht, and
Aix-la-Chapelle. He then ascended the Rhine from Cologne to Bingen, returned to
Koblenz, and cut across the Eifel to Liége. He returned to Namur and Brussels before
reaching Paris via Mons, Valenciennes, and Cambrai. He left France from Le Havre,
the geology of which fascinated him. Geologically he focused on the mountain
limestone of Namur, the cretaceous strata of Maastricht, the slate rocks of the Rhine
between Bonn and Bingen, and the volcanic rocks and so-called 'transition limestone'
of the Eifel. He inspected museums at Lille, Ghent, Brussels, Namur, and Bonn, and
dutifully evaluated them for Harcourt. Ever mindful of the YPS he bought German
works for its library. Besides renewing at Bonn his acquaintance with von Dechen, he
met for the first time some well-known savants. In Namur, Jean-Baptiste d'Omalius
d'Halloy, an expert on the French stratigraphic column, made him welcome. In Bonn
he met Nöggerath, the authority on the older rocks of the Rhineland. The climax of

his tour was his stay in Paris where he made contact for the first time with Brongniart senior, Valenciennes, Laurillard, and above all Cuvier, and renewed his friendship with Brongniart junior.[25]

Phillips was not entirely impressed by the Brongniarts. Alexandre denounced Sowerby's delineations in his *Mineral conchology* as careless yet Phillips thought that many of the fossils in Alexandre's cabinet were wrongly labelled so that his recent *Tableau des terrains* was vitiated. In Phillips' view Brongniart was too busy with his job at the royal porcelain works to attend to his cabinet properly. Though Phillips enjoyed a field trip with Adolphe, he found to his surprise that Adolphe's collection of fossil plants was neither large nor splendid so that in his publications he was dependent on others. In contrast Cuvier had no feet of clay. He was gracious and friendly to Phillips who gave him a copy of his Yorkshire coast book. Through Cuvier he met Achille Valenciennes and Charles Laurillard, who were his aide-naturalistes. Moreover Cuvier gave Phillips access to the anatomy galleries of the Muséum national d'histoire naturelle. Socially Cuvier looked after Phillips by inviting him to a soirée and to a meeting of the Institut de France, at which he saw François Magendie, the experimental physiologist, and heard Étienne Geoffroy St Hilaire, then warming up for his acrimonious controversy with Cuvier in 1830. All this was heady and inspiring stuff for Phillips who had never set foot in the Royal Society of London.[26]

The 1829 excursion was heady in another way. In August Phillips' sister, Anne, was then, it seems, living in Brussels where he met her by arrangement after fourteen years of separation. He invited her to come back with him to York to be his housekeeper. This dyad of elder brother and younger sister was often idealised at the time into the powerful and protective brother and the dependent and caring sister. Unlike marriage such a partnership was generally based on the innocence of childhood and was unsullied by sexuality. That was true of the Phillips siblings whose mutual affection was also based on the shared sorrows of being orphans. For Phillips the new arrangement with his sister was attractive. He gained domestic convenience, sympathetic companionship, and an accomplished hostess who impressed his scientific friends. She was also prepared to become a helpmeet geologically so she aided her brother just as Charlotte Murchison and Mary Buckland helped their husbands. She was put into geological service quickly: her first day in the field under her brother's tutelage was spent at Le Havre on their way home.[27] For her the new arrangement gave her a home and income. Presumably she realised that, if Phillips were to marry, she might lose both whereas he would not. He remained a bachelor so they looked after each other from 1829 until 1862 when she died.

The instigator of the 1830 tour to Switzerland was G.W. Wood who had on several occasions wanted Phillips to instruct him on a geological tour. In the event Wood, his wife, and their son William accompanied Phillips and may well have covered his expenses of around £40. Phillips wished to learn still more about continental geology for his article for the *Encyclopaedia metropolitana*. In pursuit of his developing interest in English mountain limestone, he wished to increase his knowledge of continental limestones. To complement his visit to the Eifel in 1829, he decided in 1830 to inspect the adjacent Ardennes, especially around Namur. Around Strasbourg he wanted to make reconnaissance surveys of the Vosges and the Black Forest. In Switzerland he proposed to see for himself two sorts of famous phenomena: the huge contorted structures of the Alps; and alpine glaciers and their effects. He also hoped, on his way from Lyons to Paris, to visit the Auvergne, a classic site for volcanic

phenomena. As in 1829 the YPS was happy to grant him two months' leave of absence.[28] The July revolution in France, in which Charles X was replaced by Louis Phillipe, disrupted Phillips' itinerary in only one respect: he felt it unwise to visit the Auvergne. Phillips regarded the absolutist and clericalist proclivities of Charles as odious so he was not alarmed by the heavy tread of the National Guard's sentinels in Strasbourg in early August, especially as he thought that all respectable citizens had joined it. The Guard guaranteed the transition to constitutional monarchy, the freedom of the press, and a very tranquil journey for Phillips from Thionville to Strasbourg just after 'the three glorious days' of 27-29 July.[29]

Phillips was well armed with introductions given in 1829 by Sedgwick and von Dechen to Phillipe Voltz in Strasbourg. The latter told Voltz that Phillips was 'un geologue distingué, un excellent connoisseur des petrifactions et un definateur parfait'. Buckland again provided introductions, this time to Genevan savants. Though Phillips again met d'Omalius d'Halloy in Namur and on his way home again attended a soirée given by Cuvier in Paris, the social highlights of the tour were provided by Strasbourg and Geneva. In the former Phillips was impressed by the museum run by Voltz, a mining engineer knowledgeable about stratigraphical palaeontology. Phillips was amazed to discover that the fossils from the Kelloways rock were properly labelled. Voltz was an expert on belemnites and aroused in Phillips an enduring interest in them. Through Voltz he met in Strasbourg Édouard de Billy, a mining engineer and geological mapper. Phillips found de Billy so congenial that when the Frenchman visited Britain in 1833 they spent several days together on field work in Wensleydale, Swaledale, and Teesdale, the results of which appeared in Phillips' 1836 monograph on mountain limestone. In Geneva Phillips found the public museum and the private collection of Jean de Luc to be rewarding; but above all he admired the way in which the professors there, though on small salaries, enjoyed high status and influence not least through giving open lectures which produced an educated citizenry.[30] This tour, like that of 1829, was not only illuminating geologically: it also boosted Phillips' self-esteem. He had been welcomed by leading French, Swiss, and Belgian geologists who valued his book on the Yorkshire coast.

3.3 Vindicating William Smith

Between 1831 and 1833 Phillips welcomed and contributed to the final rehabilitation and vindication of Smith. In February 1831 his uncle became the first recipient of the Wollaston medal of the Geological Society and was dubbed by Sedgwick, its president, the father of English geology (Fig. 3.1). Next year the government awarded Smith a pension of £100 pa. In 1833 Fitton issued his last defence of Smith, and ended a campaign lasting fifteen years, when he published a revised version of his 1818 account.

The Smith cult was *not* launched in 1831 at the Geological Society of London by a group of young Turks of whom Sedgwick was the spokesperson.[31] In 1829 and early 1830 Smith had been publicly recognised in Yorkshire as the father of English geology and his nephew's monograph had all but conferred that title on him. Nor was he ignored in London. In summer 1830 Lyell's *Principles of geology* praised Smith for classifying the secondary strata on the basis of their fossils, for his generosity in responding to queries so that his views became widely known, and for his 1815 map which threw into 'natural divisions the whole complicated series of British rocks'.

**Fig. 3.1   The reverend professor Adam Sedgwick, from an engraving of 1833
of T. Phillips' oil painting of 1832. A brilliant orator with an explosive
yet generous character, he was widely idolised as the first of men.**

Lyell denounced Smith's terminology as provincial and barbarous but admitted that
its wide use showed Smith's priority. Lyell also made two mistakes, both favourable
to Smith: according to Lyell, Smith's table of strata, compiled in manuscript in 1799,
was published in 1790. No wonder that Phillips liked Lyell's honourable mention of
Smith! Sedgwick, too, admired Smith's achievements. In 1826 he had published his

verdict that Smith's labours on the secondary strata were incomparable and had laid the foundation of all that was known about them. Three years later, in his pioneering paper on the magnesian limestone, Sedgwick praised Smith's county maps for their excellent detail and averred that they gave incomparably the best delineation of this formation.[32]

In autumn 1830 Phillips was still concerned about Smith's reputation which he thought needed a helping hand to maintain its just station. He was thinking about making something of notes on Smith's life when Sedgwick informed him that in mid-January 1831 the Geological Society had decided to award its Wollaston medal to Smith for being the first in England to use and to teach fossil-labelling of strata. The award was contentious: it seems that Sedgwick and Fitton, president and immediate past president, presumably supported by Murchison, the next president, secured it for Smith against the opposition of those such as Greenough who were suspicious of Smith's use of fossils. Sedgwick thought that as an act of public justice the Geological Society should correct its false position about Smith so he turned to Phillips for evidence about Smith's priority in identifying strata by fossils. Phillips, who was overjoyed that Smith's disappointment, neglect, and forced exclusion from the world of science were to end, quickly provided for Sedgwick appropriate documentation secured from his uncle, offered his own comments on it, and hinted that Sedgwick should consult Richardson.[33]

In his effusive presidential speech of 18 February 1831 Sedgwick used unpublished materials provided by Smith, Phillips, and Richardson to justify to the assembled fellows of the Geological Society the resolution of the Council that Smith should receive the Wollaston medal because he was indeed 'the first, in this country, to discover and to teach the identification of strata, and to determine their succession by means of their imbedded fossils'. Sedgwick traced the origin of Smith's achievement as far back as 1787, stressed that by 1791 Smith had proved 'the continuity of certain groups of strata, by their organic remains alone, where the mineral type was wanting', and tabled supporting documentary evidence. Though Sedgwick thought that in fossil-labelling only large groups were reliable, and particular genera and species were useless, he alluded to Smith in an act of public justice and gratitude not just as the father of English geology but as 'our master'.[34] This elevation of Smith, previously ignored by the Society, was important for Phillips. Henceforth he assumed a higher and unique place in English geology as the nephew and chief pupil of Smith. In any future disputes in the Geological Society and elsewhere about the relation of palaeontology to geology, it was inevitable that he would be seen as the inheritor of the Smithian mantle.

At the first meeting of the British Association held in September 1831, which Smith attended, Murchison went out of his way to call Smith the father of English geology. Afterwards Murchison spent two days with Smith at and near Hackness and devised a scheme of benefiting Smith which involved government paying him to colour geologically the published ordnance maps. Murchison broached the plan first to Phillips who agreed that such work was Smith's forte and suggested that his uncle should colour the maps under the direction of the Geological Society in order to ensure that he adopt a regular system and to assure government that the scheme was feasible. After it had been backed by the Council of the Geological Society Murchison discovered that Lord Morpeth had already approached Lord Lansdowne, a cabinet minister and committed FGS, about a government pension for Smith. Murchison seconded this application and added to it that his own scheme of employing Smith on state service had the advantage of satisfying the Humists who advocated financial retrenchment and opposed sinecure pensions. In late December Murchison learned that Harcourt and Johnstone had doubts about his scheme which Phillips still supported warmly: they thought that Smith was too independent to work under anyone's direction and that in the manual labour of colouring maps he depended on Phillips' skilful assistance. As Murchison did not withdraw his plan to help 'our good old father in geology', government was faced with two proposals. In January 1832 it awarded a pension of £100 per annum to the ageing Smith, then in his early sixties. A pension avoided the danger for government of establishing a new type of post which might in the future provide a precedent for the creation of further geological positions paid for by government.[35]

Phillips welcomed Smith's pension because it supplemented his uncle's income as land steward at Hackness. Illness prevented Phillips attending the 1832 meeting of the British Association at which the Wollaston medal was physically presented to Smith. But his concern for his uncle's reputation was still so strong that in late 1832 he lent unpublished maps and documents to Fitton who wished to revise his 1818 account of Smith and to add to a suppressed 1821 version of it. Fitton's chief aim was to strengthen the defence of Smith's originality as proclaimed by Sedgwick. Like Phillips, whom he regarded as the best authority, Fitton viewed the period 1790-1803 as crucial in the development of Smith's notions, the diffusion of which was so widespread as nearly to amount to publication.[36]

3.4   London overtures

In 1831 Phillips made his debut as a lecturer in London when he gave the first ever solo course of geology lectures at University College. This was a preliminary move in a scheme devised by Murchison to procure for Phillips the professorship of geology, combined with a secretaryship of the Geological Society. Unlike the metropolitan gentlemen who backed his venture, Phillips was not too affluent to consider appointment to the chair. His course of lectures was well received but by June 1831 he had decided that the scanty but certain salary of his York post, the resources it provided, and his many local and northern attachments, were worth preserving so he ended his pursuit of the UCL chair. At the end of 1831 Murchison and Lyell concocted a similar scheme whereby Phillips would replace Lyell as professor of geology at King's College, London, and also become a secretary of the Geological Society. Phillips was so enamoured of this plan that he gave formal notice to the YPS of his possible move to London. Nothing came of it because Lyell did not resign until 1833.

University College, known until 1836 as the University of London, was founded in 1826 by religious dissenters and their sympathisers, mainly Whig alumni of the University of Edinburgh. In its organisational characteristics and intellectual emphases it was indebted to the Athens of the north. It offered a wide range of subjects under the heads of liberal education, professional training, and ornamental accomplishment. Most of the professors of science were connected with medicine but from the start geology was seen as ornamental accomplishment and not as part of the mind-training given through liberal education. University College, London was a secular institution, with no chapel and no religious tests for admission, which mounted lectures to non-residential students who could enrol for whatever courses they chose.

The professors received only small or non-existent salaries so they depended for income on class fees paid by students. The warden of UCL, Leonard Horner, was not in holy orders and was the first scientist to head a British university.[37]

The chair of geology and mineralogy was not a top priority at UCL. In 1827 Samuel Hibbert-Ware, an expert on Shetland geology who had lectured on geology in Manchester, offered his services. Having received no reply to his letters, in 1829 he castigated UCL and Horner for violating 'the most common rules of polished society' and withdrew his candidature. That year Lyell ruled himself out of consideration for the chair: UCL had no dignity or reputation to confer, its classes were precarious, and no guaranteed salary was available. In 1829 UCL formally revealed that the chair was suspended so as a temporary measure geology and mineralogy were covered by three professors in their own lecture courses. Edward Turner, chemistry, included mineralogy; John Lindley, botany, discussed fossil botany; and Robert Grant, comparative anatomy and zoology, expatiated on animal remains. By spring 1831 UCL had not seen any opportunity of having the chair ably filled.[38]

That situation changed because George Goldie, who was in London enquiring about a medical chair for himself, told Horner in January 1831 about Phillips' 'rare talent and felicity as a lecturer'. Horner already admired Phillips' book on the Yorkshire coast as 'most excellent', he was impressed by Phillips' conversation when they met in 1830 in London, and he knew that Murchison and Sedgwick valued Phillips highly. Horner took the bait and concocted a scheme for Phillips to give a trial course of lectures at UCL as a prelude to his election to the chair of geology in November 1831. Horner also wanted Phillips to become senior secretary of the Geological Society in succession to Murchison and claimed that De la Beche was prepared to occupy the post temporarily and then to vacate it in favour of Phillips.[39]

Phillips took considerable care about the arrangements for his lectures because he knew that he would be going on trial for the chair, that lecturing to a learned London audience containing his masters in geology would be different from talking to his York friends, and that as the chair would be without salary he needed to judge the probable size of his audience and amount of class fees. Accordingly he was careful to induce Horner and Murchison, as president of the Geological Society, to distribute copies of his printed syllabus of twelve lectures. At the same time he was not sanguine about London or UCL. The visit, lasting about six weeks, would be his longest stay in the capital since he left it in 1820. When he passed through London in 1830 he noted its hollow smiles and solid vexations. Phillips knew that at UCL Lindley, Grant, and even the flamboyant Dionysius Lardner (natural philosophy) could not recruit a class of decent size so he had low expectations about his own success.[40]

On his way to London Phillips stopped at Stamford where he was joined by Murchison. Together they explored the oolitic rocks of Collyweston and Ketton, but failed to find another jaw of didelphis to match the YPS specimen that Phillips brought. In London Murchison quickly arranged for Phillips to make his debut in person at the Geological Society with a paper on atmospheric erosion. At this meeting in late April he met new faces. His paper was read by De la Beche, as secretary, and in the ensuing discussion Phillips listened to Greenough, Fitton, James Yates (a Unitarian minister in London), and two UCL professors, Grant and Anthony Thomson (materia medica). No doubt he enjoyed the racy eloquence of Sedgwick whom he had met only once. He also learned that Lyell had just been elected to the chair of geology at King's College, London. Phillips' lectures at UCL attracted on average an audience of twenty-five, including at various times Lord King, Murchison, Lyell, Turner, Horner and Yates, and from Oxford Buckland and Daubeny. The lectures were unusual in that, after his discussion of primary rocks, Phillips considered the sedimentary strata chronologically from the oldest to the most recent. They were novel in that he grouped them into five systems, the carboniferous, saliferous (magnesian limestone and new red sandstone), oolitic, cretaceous, and tertiary or marino-lacustrine. By system he meant a period in the earth's history during which certain strata, with their distinctive fossils, were deposited Europe-wide. Accordingly Phillips discussed the sepulture and conservation of fossils. He also gave attention to diluvial deposits, which he explained by the violent action of extensive floods caused by subterranean movements which occurred after the tertiary era. Phillips' lectures charmed Horner, Turner, Grant, and Thomson who united in wanting Phillips to take up the chair and to lecture in 1832 for no more than two months of the year. The only critic was Lyell who was 'vastly taken' by the opening lectures but then cryptically found that 'things do not exactly fit'.[41]

Throughout May 1831 Phillips was feted as a geological star. He attended two meetings of the Geological Society where he met, as well as old friends, for the first time the young James Forbes, Charles Babbage (professor of mathematics at Cambridge and scourge of the Royal Society of London), and Webster. Murchison, Yates, and Horner invited him to breakfasts and dinners, at one of which he met for the first time William Conybeare. At one of Murchison's soirées he made the acquaintance of Lonsdale, Curtis, and Peter Roget, secretary of the Royal Society of London and later author of the famous *Thesaurus*. In this giddy metropolitan whirl, Murchison and Horner went out of their way to make Phillips welcome. So, too, did Lyell who gave Phillips an introduction to Mantell whom he visited at Lewes when his lectures were ended. Having inspected Mantell's rich museum, Phillips visited Oxford at Daubeny's invitation, went to Stonesfield, and attended a lecture by Buckland. With these Oxonians, James Forbes, then staying in Oxford, and Conybeare, whom he found agreeable, he geologised on Shotover and Brill hills north-east of Oxford. Phillips capped this enjoyable field day with dinner at the country home of Sir Alexander Croke, the well-known judge.[42]

Before he left London Phillips thought he had gained reputation from his lectures but he decided not to push for the chair and not to offer more lectures. London, of course, had its attractions for him. His visit had proved that he would have enlarged contacts with London scientists, be welcome at the Geological Society, and be better appreciated than in York. He knew that he could make money in London by journalism and by lecturing at such fashionable venues as the Royal Institution where he had heard a feeble performance on geology by Webster. On the other hand there were aspects of London he disliked. It was so expensive he would have needed an income of £300 pa to maintain his York style of life. It was also dirty: 'the very grass is the colour of the chimney sweep, the flowers are all paper'. Its men of science were too often divided into parties and sects, particularly after the ructions in 1830 at the Royal Society of London where there had been a contested election for the presidency. As Phillips noted, 'the jealousy among the men of science here is wonderful and you feel to walk on a cavity, and to be grasped by a hand of friendship no firmer than a ghost's shadow'. At UCL the prospects were not alluring. He had attracted only a small audience to his lectures, the geology chair carried no salary, and there would be hot competition from Lyell at King's College. Horner, Phillips' chief backer at UCL, was so hated by some professors for his autocracy that he resigned as warden effective summer 1831. Generally UCL was tottering financially, it was vilified as a utilitarian conspiracy, it was denounced as the godless house in Gower Street because it gave no religious teaching, it was ridiculed for being a joint stock company masquerading as a university which could not award degrees, and its professors were powerless as the sacking in July 1831 of Granville Pattison from the chairs of anatomy and surgery confirmed.[43]

At York Phillips no doubt sometimes felt that he was thrown away because cut off from London's acclaim and resources. On the other hand York was small, relatively quiet, affordable, and clean with its 'green grass to walk on with gay flowers to delight the eye and glad songs of birds to charm the ear'.[44] The ancient city was a convenient centre for his ongoing research on Yorkshire's geology. At the YPS he enjoyed modest but secure reward, honest friends, good fellowship, the patronage of Harcourt, and the satisfaction of having helped to make it into one of Britain's leading provincial institutions. Moreover, as keeper, lecturer, and publisher, he was in York an unrivalled hyperborean star.

In December 1831 the delicate balance in favour of York was seriously disturbed. Lyell was thinking of resigning from his chair at King's College, London, where he had not yet lectured, and with Murchison concocted another scheme whereby Phillips would replace Lyell at KCL in 1832 and also become secretary of the Geological Society. Lyell and Murchison thought Phillips the best man and wished to promote his geological career. They were sure that, if he found KCL too narrow, he would still prosper in London because in their view he was so widely accomplished. Lyell was even prepared to subsidise Phillips financially for illustrations and specimens. Phillips responded eagerly to the 'benevolent vigilance' of Murchison and Lyell, acknowledging his solitary and limited position in York and stressing that it was only in London, 'the field of action', that ' knowledge may be gained and when gained made useful'. He confessed that interest, duty, and desire combined to induce him to accept their proposal. He was so serious about his possible move to live in London that in January 1832 he gave formal notice to the YPS that he might leave in summer 1832. Though York was still a scene of scientific enjoyment and the YPS still fosteringly protective, Phillips was prepared to sacrifice them in favour of the opportunities offered by the KCL chair and the prospect of being at the centre of geological action as secretary of the Geological Society. The scheme came to naught because Lyell did not resign his chair until October 1833.[45]

3.5   The early British Association

Phillips was a key figure in the British Association from its inception in 1831 until his death in 1874. As assistant secretary from 1832 he was for thirty years its most enduring officer so he was publicly identified with it more than anyone else. The Association was founded to give a stronger impulse and greater direction to research, to promote contact between all those interested in science, to obtain greater public attention to science, and to remove public obstacles to its progress. Avoiding competition with existing societies, it was a peripatetic body which met annually in summer in different places for just one week of the year. Its meetings, which combined festival, feasting, and spectacle, received extensive press cover. It had a relatively open membership policy and offered various modes of participation so it claimed to be the most representative institution of the British body scientific. As it existed to advance science, its leaders took care to promulgate their definition of it, to demarcate proper science from unacceptable endeavours, and to organise its meetings around sections devoted to particular sciences. The Association contributed four remarkable innovations to British science. It followed the Geological Society of London in making discussion of papers a normal practice. From 1833 it gave research grants from its own funds, a unique role until 1850 when the Royal Society of London followed suit though using government money. Again from 1833 the Association became a successful lobbier of government. Compared with the Royal Society, it was activist; compared with specialist societies, it mustered greater scientific weight. Lastly, its annual *Reports* not only summarised the proceedings of the sections. They contained commissioned reports on the state and progress of particular sciences or topics drawn up by experts.[46]

The originator of the BAAS was David Brewster who, in a review of October 1830 of Babbage's *Decline of science*, suggested that to remedy the depressed state of British science an association of nobility, clergy, gentry, and philosophers should be formed. In late February 1831 he wrote to Phillips, as secretary of the YPS, giving him the surprising news that the arrangements for the first meeting of a British Association of men of science were being made and that it was hoped that it would be held at York 'as the most centrical city for the three kingdoms'. He asked Phillips to enquire about accommodation, to gauge the support likely to be given by the YPS and local worthies, and told him that the meeting would probably be held in mid or late July. Not knowing that Brewster was given to fantasising about his private panaceas, Phillips took his letter at face value, consulted the YPS and the Lord Mayor, and in March replied favourably on their behalf. Next month Brewster advertised in his journal a great scientific meeting to be held at York in July with John Robison, secretary of the Royal Society of Edinburgh, as interim secretary. Aware of the excited political atmosphere generated by the Reform Bill, Robison soon developed cold feet: he thought September 1831 a better date than July and consulted Phillips about postponement until 1832. Meanwhile in London Murchison was so dissatisfied with such procrastination that on 25 May 1831 he issued from London an anonymous circular about the proposed meeting at York. He settled the starting date as 26 September 1831 and urged all interested to write to Phillips. He alluded to these decisions having been taken by a committee of the principal scientific societies of London and Edinburgh; but, the very wording of the circular shows that it was drafted by Murchison in conjunction with Phillips who was then in London, seeing Murchison regularly, and happy to undertake its distribution with him.[47]

Phillips arrived back in York in mid-June 1831 and presumably reported to the YPS the reactions of savants in Edinburgh, London, and Oxford to the proposed meeting. It responded in early July by setting up a committee under Harcourt, with Phillips as secretary, to make arrangements for visitors to the meeting. This committee resolved to print a further circular identifying the forthcoming meeting with neither London nor Edinburgh but with York and the YPS. This circular, dated 12 July 1831, was not anonymous but signed by Harcourt and Phillips. Wishing to avoid the moral disgrace of failure, they distributed over 400 copies of the circular, inserted an advert in the August *Philosophical magazine*, and issued a second circular on 7 September.[48]

In the four weeks before the meeting Harcourt worked out the title, aims, and constitution of the proposed Association, which were mostly adopted when it met, while Phillips was concerned with promoting and organising the meeting of which he was acting secretary and factotum. His engaging and indefatigable efficiency, noticed by friends and strangers, helped to make the meeting, attended by 353 gentlemen, an unignorable success. On the day before the meeting he entertained to dinner the Edinburgh contingent of Brewster, Robison, Forbes, and James Johnston. On the first evening a conversazione was held in the Yorkshire Museum where Phillips became the Association's first performer. To welcome, entertain, and instruct the visitors, he delivered in his 'usual style of modest perspicuity' an extempore lecture on the geology of Yorkshire. With Murchison, Phillips launched the Association's practice of discussing papers when they commented on William Hutton's view about the whin sill of Cumberland and Northumberland.[49]

After the meeting Phillips, who had been appointed York secretary of the Association with William Gray, junior, told Buckland that he was elected president of the next meeting to be held at Oxford and helped Harcourt to compile the *1831 Report* which was published in February 1832. Though Phillips intended to attend the Oxford meeting, 'spasmodic respiration' prevented him from doing so. In his absence he was elected assistant general secretary, at a salary of £100 pa, to help Harcourt who was appointed honorary general secretary. The joint appointment ratified formally the valuable contributions they had made to the Association as its architect and clerk of works. But it was also important in another way because at Oxford the fledgling Association was taken over by a coterie made up of academics from Cambridge and Oxford, London gentlemen, and the odd aristocrat. At the York meeting London savants, except some geologists led by Murchison, were mainly absent; the University of Oxford was represented by the solitary Daubeny; while the universities of Cambridge, Edinburgh and Glasgow, and the two London colleges, held themselves aloof. At York it was resolved that the Association should always be provincial in its places of meeting. At Oxford it became non-provincial in its governance because a Council, dominated by London gentlemen and Camford academics, was quietly created to run the Association fifty-one weeks of the year. Yet the Association was British and depended on support from the provinces. The election of Harcourt and Phillips as secretaries, who sat on Council ex officio, signalled that provincial interests were still recognised.[50]

Phillips was happy to accept the paid post of assistant secretary. Though it involved much administration, including the heavy editorial responsibility of helping to compile the annual *Report*, it had several advantages. Phillips knew from experience that he could work effectively and harmoniously with Harcourt who took on the ill-defined role of secretary on the condition that Phillips be paid to help him. Phillips' appointment, he was told, was widely popular and free from jealousy. He hoped that his greater financial security would obviate his spending three months a year anxiously preparing a new course of lectures to make money and that he would therefore be able to devote more time to the Yorkshire Museum. Though he was aware that maintaining the right tone of the Association would be difficult, his new post consolidated his allegiance and fidelity to it. As a new ship fairly afloat, it needed good steerage from Harcourt and himself.[51]

His Association post gave him a national role through which he built up a wide array of contacts usually denied to the isolated provincial. That became obvious at the next meeting held at Cambridge which was attended by about 850 people. Having given Whewell, the senior local secretary, some very useful suggestions, Phillips was invited by Whewell to stay in Trinity, Whewell's college, so that he would be conveniently near Sedgwick, the president, Whewell, and Harcourt. Others staying in Trinity were William Hamilton, astronomer royal for Ireland, John Lubbock, and Thomas Chalmers, the famous presbyterian minister. Among many new friendships, four stood out. Though Phillips had met Sedgwick briefly in 1822, they did not meet again until May 1831 at the Geological Society. In 1833 as BAAS officers and as students of the geology of north-west England they became close friends. Phillips had met Whewell only once. At Cambridge Whewell was impressed by Phillips' good-humoured administration and his wide scientific interests. Through Whewell Phillips met Lubbock who immediately regarded him as a most valuable person. Finally, Phillips met an important group of Irish savants, the reverend Humphrey Lloyd, Hamilton, James Macartney, and Thomas Robinson, the Armagh astronomer. Phillips was also visible scientifically. He was a secretary of the geology section, gave a meteorological paper to that for mathematics and physics, and at an evening meeting along with Sedgwick, Buckland, and Whewell, he hunted down Henry Boase, a Penzance physician who believed that mineral veins were formed at the same time as the rocks in which they were found. Thus the early British Association provided for Phillips a national platform through which he furthered his reputation as an organiser, editor, speaker, and researcher.[52]

3.6 The Royal Dublin Society

Late in 1833 several savants whom Phillips had met through the British Association wanted him to be a candidate for the chair of mineralogy and geology at the Royal Dublin Society. Founded in 1731 to improve Ireland's economy by promoting applied science and agriculture, the Society was Ireland's oldest enduring scientific institution. By the early 1830s it was mainly a teaching body whose professors, salaried by government, had traditionally given free lectures. Its sumptuous premises at Leinster House accommodated a large museum and library, and it ran a botanical garden at Glasnevin. From the 1790s it received government grants which were so generous that in 1836 a select committee questioned what they had achieved. Its record in geology and mineralogy was not negligible. In 1812 it had appointed as its mining engineer and lecturer on geology and mining Richard Griffith, a pioneer geological mapper of Ireland. When he resigned in 1829 no successor was appointed. In 1813 Charles Giesecke was elected professor of mineralogy, with responsibility for the RDS's museum, at a salary of £300 pa. His salary was halved by government in 1832 and, as an ad hoc measure, he was granted £150 pa as keeper of the museum. When he died suddenly in March 1833 it was decided to convert his chair to one of mineralogy and geology.[53]

The post was drawn to Phillips' attention by Airy and Sedgwick. Phillips had never visited Dublin and had no geological confrères there analogous to those in London but presumably he knew that the city was Britain's third largest and that the RDS was the Irish equivalent of London's Royal Institution where Michael Faraday flourished as a lecturer to fashionable audiences. The few Dublin savants he knew were not established acquaintances: he had met most of them at the 1833 BAAS meeting, the exception being the reverend Bartholomew Lloyd, provost of Trinity College, whom he had met in York in 1831. On the assumption that the salary of the post was £300 pa and its duties light, he applied for it by 21 November 1833. Well aware that he was two months late in the field for the chair, Phillips was keen to gain it. He thought that Ireland, with its carboniferous limestone, was geologically interesting and he wished to apply himself 'heart and soul to his *metier* without being obliged for bread to study and dilate de omnibus rebus'. Having approached Sedgwick, Murchison, Greenough, and Whewell for testimonials, by late November he withdrew his candidature. He learned that the salary was £150 pa, only £50 more than that of his YPS post, that audiences at all RDS lectures had declined because the government had introduced class fees, and that he would have been responsible for the RDS's museum. Phillips acted so impetuously in applying for the chair, without knowing its salary and duties, that he was the only one of ten candidates who withdrew before John Scouler was elected to it in February 1834. The episode shows that on occasion Phillips felt frustrated at York. It also showed him that his reputation had spread to Dublin where the Lloyds, William Hamilton, James Macartney, and Robert Hutton, all supported his application. Even so he judged that there was no point in exchanging York for the Irish capital: an increase in salary of just £50 pa was insufficient compensation for the upheaval, the loss of his York chums, and the onerous responsibilities of the RDS chair.[54] Thus by late 1833, in spite of three metropolitan encounters, Phillips' base was still York.

Notes

1. *YPS 1829 report*, pp. vii, viii, 3, 17, 39; *Yorkshire gazette*, 6 Feb 1830.
2. *Yorkshire gazette*, 6 Feb 1830 (q), 5 Sept 1829; *Philosophical magazine*, 1830, vol. 7, pp. 213-23; *YPS 1829 report*, p. 2.
3. Orange, 1973, pp. 21-2; *YPS 1830 report*, pp. 1, 4, 12, 16 (q); YPS Council minutes, 12 Jan 1831; *YPS 1829 report*, pp. v, vi, 44; Harcourt to Milton, 18 Jan 1831, Fitzwilliam correspondence, Northamptonshire record office (q); Charles William Wentworth Fitzwilliam, Lord Milton, third Earl Fitzwilliam from 1833 (1786-1857), *DNB*, president of BAAS 1831, president YPS 1831-57. The eight curators covered geology, mineralogy, comparative anatomy, ornithology, entomology, botany, the library, and antiquities and coins.
4. *YPS 1832 report*, pp. 4, 18; *YPS 1833 report*, pp. 2, 3; YPS Council minutes, 17 March 1834.
5. YPS General meetings minutes, 6 July 1830; *YPS 1830 report*, p. 9; Orange, 1973, pp. 41-2; Vigors to Phillips, 19 and 26 Dec 1831, PP; Nicholas Aylward Vigors (1785-1840), *DNB*, secretary of the Zoological Society of London 1826-33.
6. Tyrconnel to Phillips, 27 Oct 1833, PP; YPS Council minutes, 27 Nov 1833; *YPS 1833 report*, p. 3; John Delaval Tyrconnel (1790-1853), fourth earl; William Hyde Wollaston (1766-1828), *DNB*; Humphry Davy (1778-1829) *DNB*.
7. YPS, *Proposal for the erection of an observatory in the grounds of the Yorkshire Museum*, 20 March 1832; *YPS 1831 report*, p. 15; *YPS 1832 report*, p. 6; *YPS 1835 report*, p. 3; J. Gray to Phillips, 19 Sept 1832, Taylor to Phillips, 23 June 1833, Airy to Phillips, 5 July 1833, all PP; Phillips to Danby, 20 March and 17 July 1833, Danby papers, Bradford District Archives; YPS Council minutes, 18 Sept 1833; Phillips, notes on astronomy, PP, box 93, file 5, for sun spots drawing 25 Jan 1832; YPS Scientific communications, 4 Dec 1832; Ford, 1877, p. 31; William Pearson (1767-1847), *DNB*, Anglican minister in Leicestershire and key member of the Astronomical Society; George Biddell Airy (1801-92), *DNB*, professor of astronomy Cambridge 1828-35, astronomer royal 1835-81. A transit instrument was used to observe the passage of an astronomical object across the meridian of the observer. An equatorially mounted telescope enabled the observer to keep an object in view for a long time. An achromatic telescope minimised distortions of image and colour.
8. YPS Scientific communications, 6 Oct 1829, 5 July 1831 (museums), 7 May 1833, 2 Oct, 6 Nov 1832, 1 Dec 1829 (Le Havre), 5 Oct 1830 (Switzerland), 5 April 1831 (didelphis), *YPS 1831 report*, pp. 11-13, Rupke, 1983, pp. 162-4; Phillips, 1837, 1839, vol. 1, pp. 96-

8 (q). Two of these YPS communications were published: Phillips 1830 (Le Havre), and 1832a (Halifax ganister coal) which disagreed with Conybeare and Phillips, 1822, pp. 344-6.

9. Sheppard, 1934, p. 174 (q); *YPS 1829 report*, p. 7; Dikes to Phillips, 25 July 1829, Phillips to Harcourt, 29 July 1829, HP; Harcourt, 1829; for the importance of Bielsbeck, see Knell, 2000, pp. 180-92.

10. Harcourt, 1830, p. 9; *YPS 1829 report*, p. 22

11. Lyell, 1830, p. 96; Sedgwick, 1830, p. 197; Sheppard, 1934, p. 178 (q); William Worsley (1792-1879) of Hovingham, fifteen miles north of York.

12. YPS Scientific communications, 5 Dec 1830, 2 April 1833 (Phillips, 1833a), 1 March 1831, reported to BAAS 1834 but unpublished, *BAAS 1834 report*, p. xlvi; Phillips to Forbes, 19 Jan 1832, FP; Christopher Hansteen (1784-1873), *DSB*, devoted his life to terrestrial magnetism.

13. *YPS 1832 report*, pp. 7-8; YPS Scientific communications, 7 Feb 1832(q); *BAAS 1831 report*, p. v.

14. *BAAS 1831 report*, pp. 49-50; *BAAS 1832 report*, p. 116; Phillips, 1833b; Phillips, 1834b; YPS Scientific communications, 5 March 1832, 5 March 1833; *YPS 1834 report*, pp. 5-7; Phillips to Forbes, 5 Dec 1831, 16 Feb 1832, FP; Phillips to Lubbock, 18 Sept 1833, Lub P; Middleton, 1965, pp.169-70; Herschel, 1861, pp. 103-4.

15. YPS Scientific communications, 7 Feb 1832(q); *YPS 1831 report*, p. 15; *YPS 1832 report*, p. 7-8; *YPS 1833 report*, p. 3-4; *YPS 1834 report*, p. 5; Phillips to Forbes, 10 March 1832, FP; Whewell to Murchison, 10 Oct 1831, MP (q); Forbes, 1832a; James David Forbes (1809-68), *DNB*, professor of natural philosophy, University of Edinburgh, 1833-60, principal of United College, University of St Andrews 1859-68.

16. Phillips to Forbes, 5 Dec 1831, FP (q); *YPS 1831 report*, p. 23; *YPS 1832 report*, p. 15; *YPS 1833 report*, p. 12; *YPS 1834 report*, pp. 5, 13; *BAAS 1831 report*, p. 21; *BAAS 1833 report*, pp. xx-xxii; Luke Howard (1772-1864), *DNB*; reverend William Whewell (1794-1866), *DNB*, fellow of Trinity College, Cambridge, 1817-41, master 1841-66.

17. Phillips, 1832b; Sydenham, 1979, p. 288; Middleton, 1966, p. 155; Scott, 1887, pp. 26-8.

18. *BAAS 1833 report*, pp. ix, 486-9; Lubbock to Whewell, 22 Aug 1833, WP; Peacock to Phillips, 18 Oct 1833, PP; Potter to Phillips, 23 Oct, 31 Dec 1833, Airy to Phillips, 21 Oct 1833, Dalton to Phillips, 8 and 29 Oct 1833, in Phillips, papers on aurora borealis, 1833-41, PP, box 94; Forbes, 1832b; Dalton, 1793, pp. 54-60, 65-75, 153-94; YPS Scientific communications, 1 Oct 1833; YPS General meetings minutes 4 Nov, 2 Dec 1833; reverend George Peacock (1791-1858), *DNB*, fellow of Trinity, Cambridge, from 1836 professor of astronomy and mathematics; John William Lubbock (1803-65), *DNB*, banker and treasurer of the Royal Society of London 1830-5, 1838-47; Richard Potter (1799-1886), *DNB*, Manchester merchant and natural philosopher.

19. Witham to Phillips, nd, fr 26 Dec 1831, Robison to Phillips, 28 Oct 1832, PP (Edinburgh); Gilbertson to Phillips, 13 Oct 1832, PP (Preston); Goldie to Phillips, 1 May 1831, PP (Birmingham); YPS Council minutes, 8 March 1830, 28 Feb 1831, 8 and 19 March 1832; Ingham to Murchison, 24 March 1830, MP (q); Book listing lectures, classes, number of members of York Mechanics' Institute and classes, 1827-51, York City archives, TC 49/1: 15, 12 and 19 Jan 1831.

20. Phillips to Anne Phillips, 25 Aug 1831, 12 Nov, nd fr 29 Nov 1832, PP; William Gregory (1803-58), *DNB*, professor of chemistry, University of Aberdeen, 1839-44, University of Edinburgh, 1844-58; James Alderson (1794-1882), *DNB*.

21. *Halifax LPS 1831 report*, p. 30, *1832 report*, pp. 55-6; *Halifax and Huddersfield express*, 15, 22, 29 Oct 1831; Phillips, 1832a; Rawson to Phillips, 12 July 1831, 2 March 1832, Alexander to Phillips, 29 Dec 1831, PP; Phillips to Anne Phillips, 18 Oct 1831, PP; Christopher Rawson (1777-1849), Halifax banker and principal founder in 1830 of Halifax Literary and Philosophical Society of which president 1830-42; Phillips to Winstanley 21 April 1832, Royal Manchester Institution archives, M 6/1/51, Manchester Public Library.

22. Phillips, notes for Halifax lecture, 26 Oct 1831, PP, box 111, f. 4; draft for York lecture, 23 March 1831, PP, box 103, f. 1.

23. Geikie, 1875, vol. 1, pp. 148-62; Lyell to Murchison, 12 Jan 1829, in Lyell, 1881, vol. 1, pp. 232-4.

24. Buckland to Phillips, 12 April, 17 and 29 July 1829, PP; Phillips, diary April-Dec 1829, notebook 21, 22-3 May, 3 June 1829, PP; Phillips to Daubeny, 3 June 1829, PP; Phillips, tour in Scotland 1826 and of continent 1829, notebook 14, 9 Aug 1829, PP; YPS Council minutes, 7 July 1829; Daubeny to Phillips, 24 June, nd fr 4 Aug 1829; Charles Giles Bridle Daubeny (1795-1867), *DNB*, professor of chemistry, University of Oxford, 1822-55, and expert on volcanoes: Daubeny, 1826; Johann Jacob Nöggerath (1788-1877).

25. Phillips, notebook 14, PP; Phillips, Foreign tour 1829, Library Magdalen College Oxford, Ms 336; Phillips to Harcourt, 1 Sept 1829, HP; Phillips, 1830; Michael Faraday (1791-1867), *DNB*; Jean Baptiste Julien d'Omalius d'Halloy (1783-1875), *DSB*; Achille Valenciennes (1794-1865), *DSB*; Charles Léopold Laurillard (1783-1853).

26. Phillips, diary 8 Sept-*c* 19 Oct 29 notebook 19a, 12, 17, 21, 23, 25 Sept 1829, PP; Adolphe Brongniart to Phillips, 20 and 24 Sept 1829, PP; Brongniart, 1829; François Magendie (1783-1855), *DSB*; Étienne Geoffroy St Hilaire (1772-1844), *DSB*.

27. Phillips, Foreign tour 1829, Magdalen College Library, dedicated to Anne as a memorial of past and a pledge of future affection, 18-19 Aug 1829; Phillips to Harcourt, 16 Oct 1829, HP; Charlotte Murchison (1788-1869); Mary Buckland (1797-1857); Davidoff and Hall, 1987, pp. 348-53.

28. Wood to Phillips, 21 May 1830, PP; Phillips, Foreign tour 1830, Library Magdalen College Oxford, Ms 337, 3 Aug 1830; YPS Council minutes, 5 July 1830.

29. Phillips to Harcourt, 6 Aug 1830, HP; Phillips to Anne Phillips, 3 Aug 1830, PP.

30. Sedgwick to Phillips, 12 July 1829; von Dechen to Voltz, 27 Aug 1829 (q); Buckland to Phillips, 10 July 1830; Phillips to Anne Phillips, 3, 21 Aug, 21 Sept 1830, all PP; Phillips, Foreign tour 1830, 3 Aug, 7 Sept 1830; Phillips, 1835a, pp. v-vi; Phillips, 1836a, pp. xiii-xiv; Phillipe Louis Voltz (1785-1840), *DSB*; Édouard Louis Daniel de Billy (1802-74); Jean André de Luc (1763-1847).

31. Contrast Rupke, 1983, pp. 192-3 and Laudan, 1976, pp. 222-4.

32. Lyell, 1830, pp. 70-1; Phillips to Anne Phillips, 26 Nov 1830, PP; Sedgwick, 1826, p. 360; Sedgwick, 1829, p. 43.

33. Phillips to ?, 22 Nov 1830, Ms 967, f. 176, National Library of Scotland; Phillips to Anne Phillips, 29 Sept 1830, PP; Sedgwick to Phillips, 24/5 Jan 1831, 29 Nov 1847, PP; Forbes to Brewster, 24 July 1831, FP; Sedgwick to Smith, 28 Jan 1831, SP; Phillips to Sedgwick, 23 Jan, 5 Feb 1831, with copy of Smith to Phillips, 24 Jan 1831, Se P.

34. Sedgwick to Phillips, 9 Feb, 3 March 1831, PP; Sedgwick, 1831, pp. 271-9 (274, 278 qs), 204; Phillips, 1844a, pp. 114-17.

35. *York courant*, 4 Oct 1831; Murchison to Buckland, 12 Oct 1831, MP; Murchison to Lansdowne, 20 Dec 1831, Murchison to Harcourt, 29 Dec 1831, HP; Murchison to Phillips, nd [Oct 1831] and nd [Dec 1831] in Geikie, 1875, vol. 1, pp. 190, 193-4 (193 q); Phillips to Murchison, 31 Dec 1831, MP; Johnstone to Phillips, 24 Jan 1832, PP; George William Frederick Howard, Lord Morpeth (1802-64), *DNB*, of Castle Howard, near Malton, a patron of the YPS, attended the 1831 BAAS; Henry Petty Fitzmaurice, third marquis of Lansdowne (1780-1865), *DNB*, a cabinet minister as Lord president of the council; Humists were disciples of Joseph Hume (1777-1855), *DNB*, an advocate of financial retrenchment.

36. Phillips to Smith, 29 Sept 1832, SP; Fitton to Phillips, 10 Sept, 9 Oct 1832, 25 Jan 1833, PP; Fitton, 1833.

37. Bellot, 1929; Harte and North, 1978.

38. Hibbert to Brougham, 24 March 1827, BP; Hibbert to Horner, 29 Oct, 11 Dec 1827, UCL applications for chairs, geology, 1827, UCL archives; Hibbert to Horner, 10 July 1829, UCL College correspondence (q), UCL archives; Lyell, 1881, vol. 1, pp. 178, 257-8; Bellot, 1929, pp. 140-1, Samuel Hibbert-Ware (1782-1848), *DNB*; Edward Turner (1798-

1837), *DNB*, professor of chemistry 1828-37, a secretary of the Geological Society 1830-5; John Lindley (1799-1865), *DNB*, professor of botany 1829-60 and a leading British fossil botanist; on Robert Edmond Grant (1793-1874), *DNB*, see Desmond, 1984.

39. Goldie to Phillips, 31 Jan (qs), 7 Feb 1831, PP; and generally Edmonds, 1975b; De la Beche was indeed a secretary 1831-2.
40. Phillips to Horner, 5 March, 2 and 16 April 1831, UCL College correspondence; Phillips to Murchison, 16 April 1831, MP; Phillips, Foreign tour 1830, Magdalen College Library, 20 July 1830; Phillips to Anne Phillips, 28 April 1831, PP; Dionysius Lardner (1793-1859), *DNB*, professor of natural philosophy UCL 1827-31.
41. Geikie, 1875, vol. 1, pp. 181-2; Phillips, Collyweston 1831, Ketton and Collyweston, notebooks 25 and 22, PP; Phillips, 1831; Edmonds, 1975b, pp. 168-72; Phillips to Anne Phillips, 11 May 1831 (q), PP; James Yates (1789-1871), *DNB*, a supporter of UCL; Anthony Todd Thomson (1778-1849), *DNB*, professor of materia medica UCL 1828-49; Peter King, seventh Lord King (1776-1833), *DNB*, another UCL supporter.
42. Phillips to Anne Phillips, 28 April, 6 and 18 May, 10 June 1831, PP; Phillips, notebook 22; Phillips to Mantell, 31 May 1831, Ma P; Phillips, Stonesfield 1831, notebook 24, PP; Shairp, 1873, p. 69; Charles Babbage (1792-1871), *DNB*, inventor and professor of mathematics, University of Cambridge 1828-39; John Curtis (1791-1862), *DNB*, entomologist; Peter Mark Roget (1779-1869), *DNB*, physician, secretary of Royal Society 1827-49; Alexander Croke (1758-1842), *DNB*.
43. Phillips to Harcourt, 2 June 1831, HP; Phillips to Anne Phillips, 28 April, 11 and 18 (q) May, 10 June 1831, PP; Bellot, 1929, pp. 190-212; Granville Sharp Pattison (1791-1851), *DNB*, professor of anatomy UCL 1827-31 and surgery 1830-1.
44. Phillips to Anne Phillips, 18 May 1831, PP.
45. Phillips to Murchison, 8 and 31 Dec 1831 (q), MP; Lyell, 1881, vol. 1, p. 359; Lyell to Murchison, 22 Dec 1831, MP; YPS Council minutes, 2 Jan 1832.
46. Morrell and Thackray, 1981.
47. Brewster, 1830; Brewster to Phillips, 23 Feb 1831, Davies to Phillips, 9 March 1831, Robison to Phillips, 25 March and 29 April 1831, all BAAS foundation volume, Bodleian Library; YPS Council minutes, 28 Feb 1831; *Edinburgh journal of science*, April 1831, vol. 4, p. 374; Geikie, 1875, vol. 1, p. 185; Phillips to Harcourt, 2 June 1831, HP; and generally Morrell and Thackray, 1981, pp. 35-94; John Robison (1778-1843), *DNB*, gentleman inventor and secretary Royal Society of Edinburgh 1828-39.
48. YPS Council minutes, 4 July 1831; York Reception committee proceedings, 5 and 6 July, 2 and 5 Sept 1831, Bodleian Library; *BAAS 1831 report*, pp. 5-6, 8, reproduced the two circulars; Morrell and Thackray, 1981, pp. 67, 544-5; *Philosophical magazine*, 1831, vol. 10, p. 150.
49. Geikie, 1875, vol. 1, pp. 185-6; Johnston, 1832, pp. 4, 7, 17-18; *York courant*, 4 Oct 1831 (q); Phillips to Thomson, 17 July 1871, KP; *BAAS 1831 report*, pp. 56-8, 77, 83; James Finlay Weir Johnston (1796-1855), *DNB*, chemist and protégé of Brewster.
50. Buckland to Phillips, 1 Oct 1831, PP; Phillips to Forbes, 29 Jan 1832, FP; Phillips to Murchison, 24 May 1832, MP; Harcourt to Phillips, 2 July 1832, PP; Phillips to Harcourt, 2 (q) and 6 July 1832, HP; Buckland to Phillips, 26 July 1832, PP; BAAS Council minutes, 26 June 1832; on the BAAS coterie, Morrell and Thackray, 1981, pp. 297-370.
51. Phillips to Harcourt, 8 and 21 July 1832, HP; Harcourt to Phillips, 2 July 1832, PP.
52. Whewell to Harcourt, 12 June 1833, Whewell to Phillips, 10 July 1833, Lubbock to Whewell, 22 Aug 1833, all WP; *BAAS 1833 report*, pp. xxxix, 401-12; Whewell and Henslow, 1833, p. 73; on the socialising at Cambridge, Morrell and Thackray, 1981, pp. 165-75; William Rowan Hamilton (1805-65), *DNB*, an undistinguished astronomer whose forte was mathematical physics; Thomas Chalmers (1780-1847), *DNB*; the reverend Humphrey Lloyd (1800-81), *DNB*, professor of natural philosophy, Trinity College, Dublin, 1831-43, provost 1867-81; James Macartney (1770-1843), *DNB*, professor of anatomy there; the reverend Thomas Romney Robinson (1792-1882), *DNB*, director of Armagh observatory 1823-82; Henry Samuel Boase (1799-1883), *DNB*.

53. Berry, 1915, pp. 154-69; Meenan and Clarke, 1981, esp pp. 154-66; Carl Ludwig Giesecke (1761-1833), professor of mineralogy, Royal Dublin Society, 1813-33; on Richard John Griffith (1784-1878), *DNB*, mining engineer and professor of geology, RDS, 1812-29, see Davies and Mollan, 1980.

54. Phillips to Sedgwick, 15 and 20 Nov 1833, H. Lloyd to Sedgwick, 22 Nov 1833, all Se P; Phillips to Murchison, 20 Nov 1833, MP (q); Phillips to Mantell, 21 Nov 1833, Ma P; Phillips to Greenough, 28 Nov 1833, GP; Phillips to Whewell, 7 Dec 1833, Whewell to Phillips, 19 Dec 1833, both WP; Hamilton to Phillips, 3 Dec 1833, Macartney to Phillips, 8 Dec 1833, Sedgwick to Phillips, 18 Dec 1833, all PP; Phillips to Hamilton, 7 Jan 1834, Ham P; *Proceedings of the RDS*, 1834, vol. 70, pp. 32, 95-6; reverend Bartholomew Lloyd (1772-1837), *DNB*, provost of Trinity College Dublin, 1831-7; Robert Hutton (1785-1870) Dublin merchant and fossil collector, Dublin MP 1837-41. The better known candidates for the chair were: William Francis Ainsworth (1807-96), *DNB*, a surgeon; Whitley Stokes (1763-1845), *DNB*, lecturer in natural history, Trinity College, Dublin from 1816, Regius professor of medicine 1830-43; John Scouler (1804-71), *DNB*, professor of geology, natural history, and mineralogy, Anderson's University, Glasgow, 1829-34, professor of mineralogy and geology, RDS, 1834-54; William MacDonald (1797-1875), Campbeltown doctor, professor of natural history, University of St Andrews, 1850-75; John Murray, junior, (1798-1873), Edinburgh doctor and lecturer; and John Vaughan Thompson (1779-1847), *DNB*, zoologist, district medical inspector, Cork, 1816-35.

PART II
MAKING A CAREER 1834-1853

Chapter 4

The Provincial Base

From 1834 Phillips continued as keeper of the Yorkshire Museum on a salary of £100 pa and as honorary secretary to the YPS until in late December 1840 he resigned from the former post. Though he retained his base in York, he distanced himself steadily from the YPS as his engagements proliferated and new opportunities arose. He added to his reputation as a geological researcher with his monograph of 1836 on the mountain limestone of Yorkshire. It was not only a work on regional or topographical geology which fulfilled one of the leading desiderata of the YPS. It was also a pioneering account of much of the carboniferous system of rocks which occurred well down the stratigraphical column, below the new red sandstone and above the old. He continued as assistant secretary to the British Association, which in 1834 raised his salary from £100 to £200 a year in recognition of his centrality to it and his increased responsibilities. He sustained his lecturing commitments, reaching in 1838 a peak of seven courses, of which three were given in London and four in the provinces. Though they paid well, they were so exhausting that he decided to reduce their number in future.

In addition to these established roles Phillips embarked on three new activities. Firstly, in 1834, an *annus mirabilis* for him, he was elected FRS and assumed the chair of geology at King's College, London, from which he resigned in August 1839. Secondly, he launched himself as an expert author in a flurry of general works on geology, of which four were commissioned contributions to encyclopaedias. Clearly Phillips hoped that the enduring printed word would add to his pedagogic reputation previously based on the evanescent spoken word. Thirdly, from 1836 he became more and more involved with the Geological Survey, which had been established under De la Beche in 1835. Initially Phillips worked voluntarily for it for two years. Then in spring 1838 the Treasury agreed to pay him a fee of £250 plus expenses to complete his work for the Survey on the fossils of Devon and Cornwall. In December 1840 De la Beche secured unwritten agreement from government that Phillips would be employed on field work for the Survey for eight months in the financial year 1841-2 at a salary of £300 pa plus expenses. It was this attractive prospect which induced him to sever his most enduring connection with York science. Late in 1840 he looked forward to a regular income of £500 pa (£200 BAAS, £300 Survey), which was a considerable improvement on the £200 (YPS £100, BAAS £100) of early 1834 and the £300 (YPS 100, BAAS £200) from mid-1834. In this chapter we shall examine the continuities and novelties in Phillips' roles in the provinces, leaving for the next chapter his London-based activities and general publishing and for chapters six and seven his contributions to the Geological Survey.

4.1   YPS keeper

Phillips withdrew from the York keepership in two stages. Firstly, in 1837 he negotiated a new contract with the YPS which, desperate to retain him, accepted conditions favourable to him but unfavourable to it. There is little evidence that Phillips was deeply dissatisfied then. In 1834 he dedicated his *Guide* to the YPS and rejoiced that Louis Agassiz found its geological collection essential for his important work on fossil fish. Next year the young princess Victoria and her mother, the duchess of Kent, examined the collections in the Museum of which they became patronesses and gave £50 to the library. In 1836 Phillips welcomed the donation by George Fox of bones of the Irish elk, an extinct species of deer with huge antlers, from which he constructed an almost complete skeleton of which he produced a drawing which was commercially published. This elk skeleton was the first in any English museum and showed that Phillips had learned from Cuvier in reconstructing it from bones. The Irish elk was highly problematic because though extinct it was found in peat bog, a recent deposit. Phillips was sufficiently loyal to the YPS to donate his whole collection of British insects to it but he found it impossible to fulfil his duties for the stipulated period or times so in March 1837 he offered to resign as keeper at the end of the year. He felt overworked, he knew he would be away from York for five months on BAAS work, on general lecturing, and on professing at King's College; and he wanted to be free from his YPS bridle for the sake of his health, his research, and his prospects in science. In response the YPS was so keen for him 'to continue the general direction of the scientific affairs of the Society' that it asked him to do what he could for the Museum in effect for three to four months a year whilst still being paid £100 pa. In June 1837 Phillips accepted what he knew was a handsome proposal which maintained his connection with the YPS on his conditions and with no financial loss. He still took pleasure and pride in the flourishing condition of the YPS of which he was also senior secretary.[1]

In these roles Phillips was not inactive between 1837 and 1840. He organised the YPS subscription to buy from Thomas Allis his private osteological collection and negotiated the cost of £350 with him. Phillips was a leading supporter early in 1838 of two proposals which broadened the YPS's attraction for the public. One was to give free admission on certain days to the Garden and Museum to benefit and improve the taste of the poorer citizens of York without injury to the YPS. The other was to hold a horticultural exhibition in the Society's grounds. Both ventures were so successful that they were regularly repeated. Phillips was also deeply involved in two schemes pertaining to buildings acquired by the YPS in January 1837 when it bought from government for £2,500 the rest of the Manor shore site between the Museum and the river Ouse. The Society acquired not only five and a half acres but also the semi-derelict buildings on it. One of these was the hospitium of St Mary's Abbey. In November 1838 William Etty, the well-known artist and a native of York, urged that it be restored and used in part as a school of drawing and for art exhibitions. As an accomplished draughtsman Phillips was elected provisional secretary of a fine arts committee of the YPS which backed Etty's scheme and launched a public subscription which raised £490 to which Phillips contributed £5. In February 1839 he was put on an organising committee of subscribers. A year later the plan for restoring the hospitium and using it as a drawing school had made no progress because very little of the sum raised for the restoration had reached the YPS. At this stage the YPS changed the project. It decided to preserve, but not restore, the hospitium and to use it to further 'the objects of the Society' and not to promote the fine arts in York. By early 1841 it had raised from its members a subscription sufficient to repair the shell of the building.[2]

Phillips was far more successful in his scheme for the medieval gatehouse of St Mary's Abbey, another building acquired by the YPS in 1837. In April 1839 he began to restore this house, which he had begun to rent from the YPS. Later that year he and his sister were installed in their new home which he christened St Mary's Lodge (Fig. 4.1). Though the property must have caught his eye before 1839, he made no moves before then. In 1837 he had recovered from the bleeding in the lungs which afflicted him in May 1836 but he was still too much occupied with 'scientific *business*' at the expense of '*productive* industry'. He felt that his 'unhappy brains' were being pushed to 'near destruction'. He was so unwell that before the 1837 meeting of the BAAS he escaped to the Isle of Man and after it to the Lake District to restore his health. In 1838 he knew he was lecturing to excess, mainly to deliver himself from pecuniary wants. He was so tired in mind and exhausted in spirit that in June his sister threatened to leave their home because of his bad moods, rough temper, and hasty words. He acknowledged to her that he had suffered from a clouded mind and that he was to blame. He implored her not to desert him, especially as he had no notion of marrying, and to be his companion for the rest of their lives, never to be 'disunited till God awards a final separation, final at least for earth'. He was terrified at the prospect of being left on his own so he assured his sister

I love you as much (*it can not be more*) as ever: I am able to maintain you in comfort, I have solemnly vowed to be your protector in life, I have laboured hard for no other object than to gain the means of a quiet and peaceful *evening of life with you*: I have gained these

Fig. 4.1 St Mary's Lodge, Marygate, York, drawn by Phillips in 1839 when he renovated it, was his charming medieval residence to which he was greatly attached, not least as a retreat when his career was stressful.

means: our health is yet uninjured: we have the reasonable hope of God's blessing on pure acts and wishes: past errors we may amend and future happiness we may almost ensure ... with you as my companion, I shall be more even than I have been, a happy man. Firmly do I believe that with your brother you shall be a happy woman. Separated we shall be wretched. *We will not be divided.*[3]

By October 1838 their relations were sufficiently improved for Phillips to propose to the YPS that he rent the Marygate gatehouse and adjoining ground from it and that, using the extra income generated by his lectures that year, he would reconstruct the roof and interior at his expense. He wanted to exchange his modest accommodation in Penley Grove Street, where he had lived for about five years, for a detached and grander building with its own garden and with facilities and an ambience appropriate to his ambition and reputation. It was agreed that he reconstruct the gatehouse, built about 1470, as his scientific home, restore its architectural features, and at the end of a thirty years' lease at £15 pa resign the renovated building to the Society. The YPS agreed to his gaining possession in April 1839 but Robert Davies, the YPS treasurer and friend of Phillips, reduced the rent to £7-10-0. The YPS also treated him well with respect to his privacy. In summer 1839 it spent £123 on a new road, which included the gatehouse in its grounds and enlarged his garden. It diverted a path so that the gatehouse archway became unusable as an entrance to the YPS grounds. For £720 the YPS bought adjacent properties in Marygate to safeguard its north-western boundary, a purchase which benefited Phillips greatly because his new home became part of a conservation area owned and managed by the Society. For such vital advantages Phillips was asked in October 1839 to pay an increased rent of only £10 pa and, given the money he had spent on a YPS property, he and his sister as joint tenants were given free rent until October 1840.[4]

In remodelling the gatehouse Phillips indulged his taste for Gothic and neo-Gothic architecture. Though he had assured the YPS that its original architectural character would be scrupulously respected, he added a low stone parapet which concealed a flat lead roof designed for meteorological observations. He altered the ground and first floors, putting in new internal walls and ceiling flags to carry the weight of fossils above it. His dining room, twenty feet long, was built for entertaining. He remodelled all the basement windows, and all windows on the south-west side of the building, in Tudor style. He also designed domestic equipment. His boiler/stove was like a modern Aga cooker. His loo-bolt and window were copied from those of Babbage's home in London and the latch locks were approved by that irascible inventor. As Gideon Mantell testified, St Mary's Lodge was a charming residence in a charming spot which made it a charming retreat.[5]

Phillips had spent just over a year in his new home when he resigned in late December 1840 from the post of keeper which he had occupied for fifteen years. He did so primarily because of the alluring prospects in the Geological Survey which De la Beche had dangled before him. Of course on occasion he deplored the inability of the YPS to buy either important specimens illustrating Yorkshire's geology, or new cases and drawers for its geology room. He was well aware that the systematic display of fossils was incomplete: some were disposed in vertical cases in order of the strata, but there was no separate exhibition of fossils arranged zoologically. He was unhappy with the library's lack of modern reference works in zoology and botany. From time to time he felt that the York system of a salaried keeper aided by honorary curators had not given him unfettered control. But he showed his regard for the YPS by withdrawing from office-holding in stages: he continued as secretary until 1842, and as honorary curator of the geological collection, the Society's largest, until 1845.[6]

As a performer at the ordinary meetings of the YPS, Phillips was quite active until 1837. Solo or accompanied he reported annually on the meetings of the British Association. Of three geological papers, those on an ancient forest in Holderness and on measurements of subterranean temperature in Monkwearmouth colliery were soon published. Phillips also reported his continuing research on auroras, particularly their magnetic and electrical features; and he determined the height of one of them as 184 miles. From 1837 the YPS ordinary meetings were poorly supplied with papers and poorly attended, in part because Phillips contributed so little other than routine reports every October on the British Association meetings. Over the whole period 1834-40 Phillips made little contribution to the YPS's declining collaborative work on meteorology and astronomy, which he had previously supported so enthusiastically. His continuing research on rainfall, in collaboration with Gray until 1835 and then solo, was not reported to the YPS but to the BAAS. His growing interest in anemometers, which recorded various features of the wind, and his continued observations on changes in barometric pressure never surfaced at YPS meetings. Phillips was unable to set up an effective working partnership in the YPS with John Ford, its leading meteorologist from 1834. In astronomy Phillips, it seems, did not use the YPS's observatory; and, having revealed to the YPS in 1834 details of an achromatic telescope he had made, he kept quiet about the achromatic refracting telescope of two and a half inches aperture and thirty-seven and a half inches focus which he bought in the late 1830s from his friend, Thomas Cooke, a York schoolmaster who began to make optical instruments commercially in 1837. Advised by Phillips, his first patron, and financially indebted to William Gray, junior, Cooke prospered as the leading maker in Britain of high-quality refracting telescopes for amateurs. It seems that from 1834 Phillips made no attempt to promote the YPS observatory in conjunction with William Newman, the Society's keenest astronomer.[7]

As a lecturer to the Society Phillips was far less prominent than before. In seven years he gave only two courses of lectures, one on the natural history of inferior aquatic animals and one on the contents of the Museum (See Appendix 1). In contrast he supported strongly the evening discussion meetings of an inner circle of YPS members to whom he explained recent important scientific discoveries by foreigners as well as Englishmen. For instance in 1839 and 1840 he discussed Darwin's recent account of the results of his voyage on the Beagle; he summarised Quetelet's work on asteroids, shooting stars, and meteors; he dilated on terrestrial magnetism; he epitomised the views, given to the British Association in 1840, of Liebig an organic chemistry and of Espy on storms; he commented on the importance of footsteps of quadrupeds and even birds in the new red sandstone; and from the existence of upright fossil trees, discovered in 1840 at Bolton in the coal measures, he inferred that plants were made into coal in situ and not transported. Phillips even sent to York his sketch of a raised beach at Barnstaple to initiate discussion at a 'much beloved evening conversazione'.[8]

As keeper of the Yorkshire Museum Phillips had some power over its specimens. For instance in 1839 he lent to Richard Owen its specimen of the jaw of didelphis from the slate at Stonesfield which Owen had inspected at York with a hand lens. It helped him to quell doubts raised in 1838 that the animal was not a marsupial mammal but a reptile. As the scientific director of the YPS Phillips encouraged its members to publish works on the natural history of Yorkshire and the antiquities of York. In 1840 Henry Baines, Phillips' assistant keeper, produced his *Flora of Yorkshire*, which was indebted to herbaria of YPS members and to William Hincks who revised much of the text. It was a Baconian local survey which aimed to bring together exact data from many sources. It carried an introductory essay by Phillips on the physical geography of Yorkshire in which he related the nature of the flora to eight geological regions into which he had divided the county. Two years later his friend Wellbeloved, the curator of antiquities, published his book on York under the Romans. Its origin was a course of lectures, given to the YPS in 1840 on the specimens in the Museum and on the spate of Roman finds made in the 1830s. Wellbeloved had been prodded into action by Phillips who had suggested the lectures and then approved their plan. Wellbeloved's approach was not unlike that of Phillips in inferring from their remains the modes of life of previous inhabitants of the earth: from the 'relics', ie physical remains of Roman York, Wellbeloved drew conclusions about 'the history, customs and arts of those who founded Eburacum'. Moreover Phillips had some experience of practical archaeology. In 1836 he and Jonathan Gray inspected and opened tumuli near Kirkby Moorside and found fissure graves.[9]

Though Phillips was a local scientific star with a national reputation and a professorship in London, he continued to support the York Mechanics' Institute. He never served on its organising committee but he proclaimed his public commitment to it by joining Kenrick in 1834 to form the local committee in York of the Society for the Diffusion of Useful Knowledge. More importantly between 1834 and early 1838 he gave six gratuitous lectures to the YMI, covering lower invertebrates (2), physical geography (2), coal mines, and rain. These were difficult years for the YMI because patrons and audiences were not easily recruited. Many of the middling classes thought that the diffusion of sound knowledge to inferior classes was undesirable and improper, and were indifferent or adverse to checking frivolity, dissipation, and vice. Artisans, for whom the YMI ostensibly existed, were unenthusiastic about its programme for their mental and moral improvement. There was, it seems, no great interest in either technical instruction, in a city where there was no staple manufacture until the railway boom of the 1840s, or in lectures on science unless they contained experiments. In September 1838 the YMI changed its name to the York Institute of Popular Science and Literature in order to create more patrons and to reflect the social fact that much of its audience was made up of tradesmen, clerks, and shopmen. Like so many of its counterparts elsewhere, it broadened its aims to include rational amusement of its members, the cultivation of their tastes, and raising the popular idea of enjoyment. Presumably Phillips accepted these revised purposes because in late 1838 and 1839 he was asked to lecture to the York Institute and he remained loyal to it in the 1840s. Of course his absences from York prevented him from being an activist or regular performer in it; but, like his friends Wellbeloved, Kenrick, and Hincks, he lectured gratis to it when it was in the doldrums, thus saving fees of outside lecturers and releasing funds for the library to which he donated the odd book. His involvement in his local mechanics' institute was unusual among leading research geologists and showed his strong commitment to the ideology of the diffusion of useful knowledge.[10]

4.2 The mountain limestone monograph

In 1835 and 1836 Phillips issued two books which elucidated the geology of Yorkshire and thus accomplished the principal purpose proposed by the YPS when it was formed. The 1835 volume was a second edition of his 1829 monograph on the Yorkshire coast. By citing more continental works and collections, by discussing the Bielsbeck excavations, and by giving a much expanded table of fossils which included continental localities, the second edition strengthened but did not change the main conclusions of the first.[11]

The 1836 monograph had been a long time in the making. In his early work with Smith he had acquired a general knowledge on foot of the mountain limestone, then also known as metalliferous and carboniferous limestone, of Yorkshire and adjacent areas (Fig. 4.2). His stay at Kirkby Lonsdale with Smith was particularly interesting and profitable. In 1824 he explored Greenhow hill with his friend Edward George, discovered how useful the mountain barometer was, and became interested in fossil-labelling of the limestone shales below the millstone grit. By late 1827 he had done field work on the limestone of Derbyshire (twice), the Isle of Arran, and around Clitheroe in Lancashire; and he had examined again the Craven faults. In that year he showed his growing interest in Yorkshire's limestone by giving two papers to the YPS, one on the 1,800-foot fault between the millstone grit and the mountain limestone around Settle and the other on the geological relations between limestone and the slate on which it rests, also near Settle. In 1829 he published his important paper on the two Craven faults between Kirkby Lonsdale and Malham, with appropriate sections and a coloured map showing slate rocks, old red sandstone, scar limestone, various strata later called the Yoredale series, millstone grit, and new red sandstone. This map, on a scale of three miles to the inch, covered an area which was to be crucial in the 1836 book. On the basis of this paper Phillips began to compare systematically different limestone districts. In 1829 and 1830 he saw for himself the limestone of Namur and Luxembourg and renewed his acquaintance with the Mendips and the Avon Gorge. He also inspected important cabinets of fossils in all those areas. Simultaneously the YPS called for 'closer research and a more copious induction of facts' pertinent to the mountain limestone and shale of north-west Yorkshire, which it believed to be the lowest of fossil-bearing strata in England.[12]

By January 1832 Phillips had done enough research to issue the prospectus for the second part of illustrations of the geology of Yorkshire. It would focus on north-west Yorkshire, be adorned by sections and a coloured map, and offer numerous original figures of fossils all drawn by Phillips. He called for subscribers to pay £1-10-0 for the volume which he hoped to publish in 1833 after he had completed his field work in 1832. Subsequently Phillips complained that 'want of leisure, which so fatally retards the progress of men devoted to science' was responsible for publication being postponed until March 1836. But the delay enable him to write a book which was more than a companion to his volume about the Yorkshire coast. In 1829 he correlated the strata of the Yorkshire coast with those known in south-west England. In 1836 he claimed that the Yorkshire mountain limestone was the 'general type or standard of comparison for the limestone deposits of Europe' including Britain.[13]

Between January 1832 and March 1836 Phillips solicited and received various types of help in assembling the materials of his mountain limestone book. Like Murchison, who was then working on his *Silurian system*, Phillips was greatly indebted to residents in key localities; unlike Murchison, he gave them due

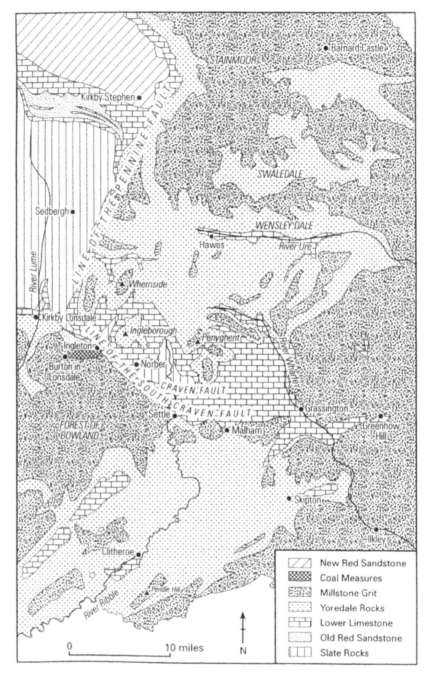

Fig. 4.2 Simplified geological map of north-west Yorkshire, based on that in Phillips, 1836a, and using his stratigraphical terminology. It shows his interest in faults, as well as boundaries between strata, and important geological sites in Phillips' lifetime.

acknowledgement. For data on the important coalfield at Ingleton and Burton in Lonsdale, Phillips relied entirely on Thomas Hodgson, owner of the Burton colliery. For facts about lead mines and mineral veins he was indebted to their owners and to such well-known mining surveyors and industrialists as Thomas Sopwith, Westgarth Forster, and Hugh Pattinson. On Phillips' first trip to Ireland in 1835 he learned about its limestone from Griffith, Scouler and Joseph Portlock, all of whom were leading experts on Irish geology. Some locals opened their collections to Phillips, others lent to him important specimens of fossils, and some even gave him specimens. His greatest obligation was to William Gilbertson, a Preston apothecary who had amassed a fine collection of local fossils. He was happy to send to Phillips at convenient intervals the whole of his collection, pertinent books, his own drawings of crinoids, and advice about classification. Gilbertson provided for Phillips no fewer than 258 of the total of about 420 species mentioned in Phillips' book and his specimens formed the chief source of the 642 figures which Phillips drew for it. In a remarkable act of self sacrifice, Gilbertson enabled Phillips to expound in 1836 the meanings of his collection just one year before he himself published a catalogue of it.[14] Phillips was also indebted to some national experts in two ways. De la Beche, Murchison, Viscount Cole, and Sir Philip Egerton allowed Phillips to examine their valuable collections; and shortly before his book went to press it was subjected to informal peer-group review. After the meeting of the British Association held in Dublin in 1835, several geologists met at Florence Court, Viscount Cole's mansion at Enniskillen, where Sedgwick, Murchison, Griffith, Agassiz, and Sir Philip Egerton talked with Phillips about the content of his forthcoming volume.[15]

In his field work from 1832 Phillips was not always alone. In 1833 he traversed Wensleydale with de Billy, his guest from Strasbourg, and explored Teesdale with Salmond and Witham. Next year he examined the basalt and limestone of Northumberland with Hutton. His main work, however, was done solo and required eight different excursions on foot to different parts of north-west Yorkshire. He measured the thicknesses of strata by making over a thousand observations with a mountain barometer and by early 1833 was certain that some strata thinned out in one location and thickened in another. Later that year he surveyed Whernside from three different directions, a procedure which enabled him to confirm the direction of attenuation and thickening of strata. In a remarkable diagram in his book, Phillips summarised the variations of thickness in the main strata in a north- west/south-east direction (Fig. 4.3). This discovery was associated with his conclusion about the lateral variation of mountain limestone, which he called his 'method of variation'.[16]

Phillips regarded his 1836 monograph as his best work so he not only supplied copies, printed in York, directly to subscribers but also sent fifty for John Murray to sell in London. Though it ostensibly covered the mountain limestone district of Yorkshire, it said relatively little about the slate rocks, in deference to Murchison's forthcoming volume, and about the coal measures. It focussed mainly on the alternating beds of limestone and red sandstone above the old red sandstone, the lower scar limestone, the upper limestone called by Phillips the Yoredale series (Yoredale was Uredale, usually called Wensleydale) and the millstone grit. Having examined no fewer than eight different areas where mountain limestone occurred, Phillips chose the Yorkshire version as the standard type because it was the most comprehensive, lithologically and palaeontologically, and the most accessible to exact research. He gave fifty pages of description of mountain limestone fossils, arranged zoologically, which were illustrated on twenty-two plates engraved by J.W. Lowry. Phillips was

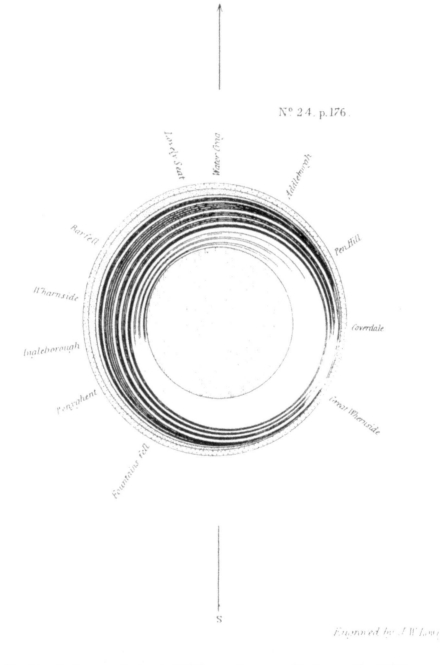

N.° 24. p. 176.

Fig. 4.3 **A diagram, drawn by Phillips and engraved by Lowry, for Phillips, 1836a. It showed the different thicknesses of the lower limestone and of sandstones/shales in eleven localities. The former, represented in white, was thickest in the south-east of the area considered; the latter, in fine dots, thickest in the north-west.**

proud of the large number (392) of species he himself had figured and jubilant that 80% of them (322) were new and named by him. Though his book was not a treatise on the carboniferous system, it was a *magnum opus* on the palaeontology of the mountain limestone, a key stratum in it. In contrast his traverse sections and map, which he drew, were disappointing because he had not the money to produce better versions. Though he had made many detailed sections at key locations, he published only three, not coloured but hatched and stippled, on an unspecified horizontal scale of about three miles to the inch. The solitary map showed eleven different strata, distinguished not by colour but by hatching and stippling, on a scale of five miles to the inch. It also depicted the lines of dykes and veins, the lines of faults including the long and bent Pennine fault, the shorter Craven fault and its two main branches, several anticlines (elliptical dome structures), and some remarkable dipping of strata with no angle given.[17]

Phillips' 'method of variation' had some important consequences. He argued that the carboniferous system was characterised not by the ubiquity but by the prevalence of coal, limestone, and red sandstone. He claimed that the mountain limestone was not sharply separated from the old red sandstone beneath it but merged into it via alternating beds of limestone and red sandstone. Similarly he showed that between the lower scar limestone and the millstone grit above it there were up to five intermediate beds which he called Yoredale limestone. Thus the lines of division between major strata were arbitrary and could be merely local in their application. It followed that any attempt to make stratigraphical correlations over large distances, specially if the areas in between had not been preserved, was hazardous. Moreover in making such correlations geologists relied on data which Phillips knew were generated in part by personal judgement. Generally Phillips presented himself as a rigorous Baconian who recorded exact facts from which he inferred the laws of phenomena, which he then explained using theory. But he acknowledged and valued the tacit skill and personal interpretation involved in creating data. To the Baconian geologist

> gradations and variations are often known too minute for description yet necessary to the train of argument, and influencing rightly his own conviction; the relative value of the observations has due weight with him in clearing up discrepancies and correcting results; and thus data are made available which would be too incomplete or apparently disagreeing for other men to employ with safety.[18]

Phillips was so conscious of the significance of the distribution of fossils in various strata that he compiled a table showing, for each of five strata, the total number of fossil species and their distribution between nine zoological classes (Fig. 4.4). This Smithian exercise showed that the millstone grit contained sixteen species and the lower scar limestone 390, giving an obvious contrast between these two strata. Yet Phillips noted that all sixteen species of the grit recurred in the limestone so that none of them was specific to the grit. There was also a good deal of overlap between the Yoredale rocks and scar limestone in that of 100 species in the former seventy-two occurred in the latter. But there was also a striking difference between these strata in that the scar limestone contained forty species of crinoids and the adjacent Yoredale rocks just one. Thus in 1836 Phillips was moving towards a statistical palaeontology, inspired by his uncle and based on the number and distribution of species arranged by zoological class.[19]

Strata of the carboniferous system	Polyparia	Crinoidea	Echinida	Plagimyona	Mesomyona	Brachiopoda	Gasteropoda	Cephalopoda	Crustacea	Total
Coal measures	-	1	-	1	2	3	-	7	-	14
Millstone grit	6	-	-	1	5	4	-	-	-	16
Yoredale rocks	8	1	2	11	5	29	9	34	1	100
Lower scar limestone	40	40	3	26	25	96	91	61	8	390
Alternating limestone and red sandstone	3	-	1	-	-	-	-	2	-	6

Fig. 4.4. **Table of the distribution of fossil species, grouped zoologically, in five carboniferous strata; adapted from Phillips, 1836a, p. 244.**

His book also contained contributions to igneous geology and to what William Hopkins in 1835 had called physical geology. In a discussion of basalt whin sill and its relation to dykes, Phillips made the important methodological point that it was dangerous to suppose that there had been only one flow of basalt which had only one opening through which it flowed. He concluded that the sill belonged to the same geological period as its attendant strata, whereas the dykes were independent and posterior to it, though derived from the same plutonic focus. More remarkably he made eighty-nine observations of the directions of long fissures in rocks and showed, in another telling diagram, that they were not random but occurred mainly in two particular directions at right angles to each other (Fig. 4.5). Phillips' muted interpretation of this empirical law was that some general cause had controlled the consolidation of the rocks in a sort of crystallisation which resulted in their splitting into large masses in two particular directions. Phillips was happy to discover that, independently of William Hopkins, he had implemented two of the three procedures which together constituted for Hopkins physical geology: he had produced geometrical laws about a geological phenomenon and assigned them to a distinct mechanical cause; but he had not tried to deduce them by considering mathematically a stratum under tension, as Hopkins had. It is clear that Phillips, unlike many geologists, combined a good geometrical eye for phenomena with an interest in mechanical and physical principles. Indeed, by early 1837 he thought that physical geology ought to be an English speciality.[20]

Phillips gave even more attention to subterranean forces which had broken and greatly displaced consolidated strata to produce the Pennine and Craven faults. He was particularly impressed by the maximum relative displacement of strata on the two sides of a fault (3,000 feet), the great length of disruption (fifty-five miles) of the

SYMMETRICAL STRUCTURES OF ROCKS.

The following diagram represents the result of the investigation by breadths of shade corresponding to the frequency of long joints or fissures, parallel to each radius.

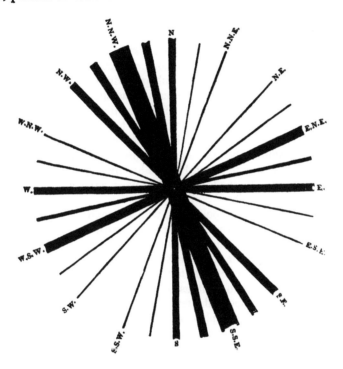

Fig. 4.5   **A diagram drawn by Phillips for Phillips, 1836a. As part of a concern with physical geology and the symmetrical structures of rocks, it showed the relative frequencies of the directions of long joints in the secondary rocks of Yorkshire, the most frequent being north-north-west/south-south-east.**

Pennine fault, the pervasiveness of steeply dipping strata near the lines of the faults, the existence of numerous parallel flutings along the dip of one fault as if produced by an enormous planing machine, and the general absence of igneous rocks on the lines of the faults. He concluded that the faults were produced by sudden and violent displacements caused by the contraction of the earth as it cooled. For Phillips the evidence of sudden mechanical fracture, produced by inconceivable violent force, was overwhelming and totally incompatible with Lyell's view that large effects were explicable by invoking ordinary causes of present-day intensity acting over long periods of time.[21]

Phillips' monograph was generally well received. As president of the Geological Society, Lyell welcomed it as excellent, particularly for its description and figures of fossils of the mountain limestone. In May 1836 Thomas Torrie congratulated Phillips on producing the first proper monograph on the limestone which would serve as a reference work for all in Britain interested in carboniferous rocks. In summer 1836 Joseph Prestwich regarded and used it as an excellent reference work to identify many fossils from the Coalbrookdale coal measures and thereby determined their age. Phillips' account of the carboniferous uplands in Yorkshire was unchallenged until the Geological Survey mapped it at six inches to the mile thirty years later. In France Élie de Beaumont was very pleased with it. In Belgium Laurent de Koninck, an expert on carboniferous fossils, welcomed and used Phillips' 'remarquables recherches' on Yorkshire's mountain limestone. There were, however, two British geologists who in private were unhappy with Phillips' book: Murchison and Sedgwick agreed that Phillips was clever but had shown bad taste. Murchison envied Phillips' facility in writing and his skill in describing and drawing fossils. But he deplored to Sedgwick the paucity of sections which were not even coloured, some objectionable new nomenclature, and Phillips' coxcombry in giving his own view about which strata constituted the Silurian system instead of espousing Murchison's. Murchison dismissed the way in which Phillips had measured the average specific gravities of common rocks, failing to realise that, when a mineral surveyor, Phillips had met the problem of averaging.[22]

Sedgwick, a dedicatee of the book along with Harcourt, had more cause to be aggrieved. Though he and Phillips had separately explored the same ground, Sedgwick thought that Phillips gave insufficient acknowledgement of his debts to others. In Phillips' 1827 paper on the slate rocks of Yorkshire it was Sedgwick, the referee, who insisted that Phillips give due credit to Otley. In 1836 Sedgwick found his own views, of which abstracts were in print, used by Phillips without acknowledgement. Sedgwick was particularly upset by the fate of his paper on the carboniferous chain between Penyghent and Kirkby Stephen, read to the Geological Society in January and March 1831, published in abstract in 1834 and in full in 1835, and of his remarks on that topic made in his presidential address to the Society in February 1831. In these he acknowledged Phillips' valuable work on part of the Craven fault and then took up the subject where Phillips had left it. In the abstract version and his address, Sedgwick established the existence of the Pennine fault, dated it as occurring just before the period of the new red sandstone, used the notion of anticlines, and realised that some strata thickened and others attenuated. In the full paper of 1835 Sedgwick produced a classification of the limestone which was broadly like that of Phillips. Sedgwick did not venture east of Settle, offered no map, and said nothing systematic about fossils, but he gave ten sections of three kinds which revealed his mastery of structural geology. Phillips told Sedgwick that his book was partly printed before he read the full version of Sedgwick's paper. It remains curious that in his preface dated 1 March 1836 Phillips praised Sedgwick as the best researcher on the west border of the limestone district of Yorkshire but did not cite the key paper published in 1835 by one of his two dedicatees or the earlier versions. Perhaps he was influenced by Sedgwick's insistence in February 1836 that he should delete the description in the dedication of Sedgwick's research on Yorkshire as 'splendid'. For his part Sedgwick told Phillips he had written a great book and the fossil plates were delightful, but he indicated he was peeved by Phillips' cavalier attitude to his own previous research and remained very critical of Phillips' account of the relation between the old red sandstone and the mountain limestone.[23] In his book Phillips showed that he was so keen to increase his own reputation as a research geologist on secondary strata that he took the risk of alienating, albeit temporarily, two of his most loyal supporters. It paid off: by 1836 he had two monographs to his credit but they had none.

In 1839 Murchison published by subscription his *Silurian system*, copiously illustrated with 112 woodcuts, five maps, thirty-one plates of fossils, and nine coloured sections. In this great work, which was epoch-making metaphorically and literally, Murchison was ambivalent about Phillips' mountain limestone monograph. Murchison claimed that Sedgwick, as the expert on the structural geology of northern England, provided a framework to which Phillips was indebted. On the other hand Murchison acknowledged three important debts to Phillips' book. It was a model for Murchison's in that it combined geological description and zoological proof, ie it offered detailed stratigraphy and figures of all known fossils in each group of deposits. Secondly, Murchison recognised that Phillips was an authority on carboniferous fossils. For instance, Phillips demonstrated to Murchison the vital point that Silurian crinoids differed from those in the carboniferous system. Lastly, Murchison was very impressed by Phillips' chapter on long symmetrical joints and fissures in rocks: for his own field work on this topic he not only took advice from Phillips but also in summer 1836 at Brecon and Ludlow learned from him how to make the appropriate measurements and calculations. Indeed, it was precisely to help the innumerate Murchisons of the geological world that in 1836 Phillips devised his cardboard geological intersector which allowed one to read off directly the direction and dip of a joint plane with respect to the plane of stratification, without having to calculate them.[24]

An important consequence of Phillips' 1836 monograph was the founding in 1837 of the Geological and Polytechnic Society of the West Riding of Yorkshire, a part of the county devoted to coal-mining and iron- making. In his book Phillips stressed that he had concluded his research on the geology of Yorkshire, thus leaving the coal measures to others. He also pointed out that in them the continuity of particular seams for forty miles 'admits of satisfactory demonstration' which implied that the succession of strata was well known or knowable. This wild remark presumably attracted the attention of Thomas Wilson, Henry Hartop, and William Thorp, three subscribers to Phillips' book and future activists in the YGS. They did not know, however, the extent of Phillips' reluctance to attack the coal measures. On the death in 1830 of Edward George, his family had put his unpublished materials on the Yorkshire coalfield into the hands of Phillips for publication but he did nothing with them and was not involved in the appearance in 1837 of George's long account of the Yorkshire coalfield. Phillips' unpublished and sporadic research on the coal measures had confirmed Conybeare's judgement that generalisation about them was hazardous because they were so disordered and disarranged. Privately Phillips thought that coalfield research should be left to practical men who were acquainted with minute local details. In his solitary published paper on the Yorkshire coal measures Phillips was not interested in stratigraphical correlations but in the alternation of freshwater and marine deposits in the lower measures.[25]

It was practical men involved in the coal and iron industries who on 1 December 1837 in Wakefield decided to form a society to collect and record geological and mechanical information concerning the local coal measures. For advice Wilson, as chairman, turned to Phillips who attended the next meeting two weeks later and pointed out the 'proper objects' of the Society. Those present followed his views so that, in a circular widely distributed that month, its title was the Geological Society of the West Riding of Yorkshire. The circular revealed that Phillips, who had been elected with William Smith as the first honorary members, might provide 'the hand of science' to methodise in a Baconian way the information which the practical miners would accumulate.[26] Wilson tried hard to lure Phillips to address the next meeting (March 1838) on the general objects of the Society but he was unable to attend. In the meantime Wilson had been taking advice from James Johnston who recommended that the Society should devote itself to the practical working of coal mines, their products, and ancillary industries, as well as to the exclusively geological programme advocated by Phillips. Faced with these two different agenda, Wilson and Hartop prepared a prospectus of the Society to be known as the Geological and Polytechnic Society of the West Riding of Yorkshire. Their prospectus, approved at the meeting of March 1838, dropped the exclusive emphasis on coal geology and the plaudits given to Phillips, both of which were prominent in the circular of December 1837.[27]

Having appeared at the first formal meeting of the Society, Phillips could not be cajoled into attending any of its meetings in its crucial opening three years. He confined his involvement to advising it about its museum and its proposed section across the Pennines to connect the Lancashire and Yorkshire coalfields. In short, he refused to be the great Baconian synthesiser for the YGS. He knew that the coal trade, which was notoriously secretive, would be reluctant to give accurate local information. He was aware that the YGS's programme of correlating coal seams was difficult owing to the bending of coal seams, their varying thicknesses, and the faults which dislocated them. The coal measures offered few revealing natural exposures and no site analogous to Scarborough Castle hill or Wensleydale, which had given Phillips the key to the geology of the Yorkshire coast and dales. There was less possibility of spiritual refreshment in the coal measures where the landscape was neither spectacular nor elevated. Though Phillips reduced his commitment to the YPS in 1837, he remained loyal to it: the YGS was economically driven by its desire to create geological knowledge applicable to coal-mining and iron-making, but it overlapped with the YPS which specialised in the polite geology of the whole county and collected pertinent materials in its museum of which he was still keeper. Given Phillips' relative detachment from the YGS, Thorp assumed the mantle of geological director and in 1844 proposed to bring together his publications on the coalfield in a quarto volume entitled *Illustrations of the geology of the Yorkshire coal district* which would be a continuation of Phillips' 1836 monograph and be printed uniformly with it. He failed to recruit enough subscribers so his book remained a dream; but in about 1847 he published a series of sections of the Yorkshire coalfield which remained the best available until replaced in the late 1860s by those of the Geological Survey.[28]

4.3 Carboniferous expert

As the ranking expert on Yorkshire's geology, including most of the carboniferous strata, Phillips was invited to undertake paid consultancy work, to which in principle he was not opposed. Apropos coal-mining he thought that colliers were often deluded and superstitious, and that their lore encouraged trials for coal which geologists deemed foolish. On the other hand geologists could make feasible predictions, contrary to miners' expectations, about where coal might be found. Between 1834 and 1840, however, Phillips was so consumed by multifarious activities that he did little consultancy unless it paid well for either little or highly interesting work. In 1840 he received two invitations, neither of which he accepted. One was to report on the prospect of finding coal near Sandwell, Birmingham. The other was to report to the Yorkshire Agricultural Society (founded 1837) about the relation between the county's agriculture and geology. Though the Society had £75 to spend on such a survey and though Phillips in his publications had routinely justified geology as useful in agriculture, he agreed to do the survey and then withdrew. The task was undertaken by William Thorp, who was a keen practitioner of applied geology.[29]

Phillips undertook three consultancies, two of which were not time- consuming. In 1835 he advised John Yeoman and Henry Belcher, two Whitby solicitors and entrepreneurs, that the results of borings for coal at Newtondale, north of Pickering, were discouraging.[30] Two years later he was involved in a celebrated case concerning the famous and public old 'sulphur' spring at Harrogate. In 1835 Joseph Thackwray, a hotel owner, sunk a well on his own premises, only eighty- two feet from the old spring. It was alleged by owners of other hotels that Thackwray's well injured the spring and their trade. He was indicted under the Knaresborough Enclosure Act and tried at York Assizes in March 1837. Each side lined up scientific men, some of whom had produced written statements, to give evidence in court. The Crown summoned two geologists, Phillips and Smith, and two chemists, Dalton and Daniell; but the defence called on a wider range of experts, including Newcastle mining engineers (Thomas Sopwith, John Buddle), doctors who were experts on mineral waters (William Clanny, Adam Hunter, Peter Murray), two chemists (James Johnston, William West) and a geologist (W.C. Williamson). The experts for the prosecution argued that the dip of the strata had created a basin, containing the sulphur water, on which all the wells and springs drew, and that the small differences in the chemical composition of the waters from wells and springs were insignificant. Those for the defence claimed that local geological disturbances had made each outlet independent and that the variations in chemical composition, though small, were significant. Given the opposing views of such eminent authorities, the judge urged a compromise, accepted by Thackwray, that his well be made public, and the jury found him not guilty. The packed court was thus deprived of the roaring of contending scientific lions.[31]

Phillips' biggest consultancy was undertaken in the winter of 1836-7 for the Lancaster Mining Company which was established in July 1836 to find coal in the area. One member of its provisional committee was William Whewell, who had bought £50 of shares to help to promote the prosperity of his native town. He persuaded the committee that Phillips should survey the Lancaster area for coal and other minerals. By late September the company had accepted Phillips' terms: a fee of 150 guineas, which included travel and personal expenses, and £150 expenses for plans and sections, his report to be produced by the end of March 1837. Phillips was keen to accept the commission. The fresh air and exercise involved in field work were, as always, welcome. He was happy to survey the Bowland district as an appendix to his 1836 book. He thought that new red sandstone covered coal measures at Ingleton and Burton in Lonsdale, he knew that there were three collieries working coal in the millstone grit south-west of Ingleton, and he was aware that around Garstang, 10 miles south of Lancaster, the grit dipped westwards. On his trips made between November 1836 and March 1837, which occupied about four weeks, Phillips did not use boring rods or drive levels to look for coal: he worked as a geologist, mapping the area, making a general section, and noting local variations. Using mainly natural exposures, he explored the structural geology of Bowland, concluding from his study of anticlines that it was a large geological basin. His findings were economically discouraging: there was no workable coal, other than that already exploited in the millstone grit. It was possible that, if the strata were continuous, coal measures might occur deep below the new red sandstone only around Garstang where there was an unconformity; but otherwise Phillips argued that geology forbad boring for coal in the Lancaster area. Phillips' report, received by the Company in May 1837, was published that year, accompanied by a coloured columnar section and coloured traverse sections, but not by the map which it withheld. Faced with such depressing conclusions the Company was wound up in 1838.[32]

In 1840 Phillips was involved as an unpaid adviser to another mining project in north-west England. The small Ingleton coalfield was worked at two collieries, one at Ingleton owned by G.J. Serjeantson and the other only three miles away at Burton in Lonsdale owned by Thomas Hodgson, a Quaker who had reopened it in 1837 after consulting Phillips. In 1839 Hodgson again consulted Phillips, whom he thought infallible in geology, about the prospects for the whole coalfield. Phillips advised him that each colliery alone would have difficulties but that if they could be worked together under a single company its future would be so bright that he would like to join it. Hodgson, who had subscribed to the Lancaster Mining Company and seen it fail, proposed to form a Lonsdale and Gretadale Mining and Railway Company which would lease both collieries. He envisaged four shareholders, with himself having half the shares and Phillips, who had promised £300, having a twelfth of them. Sedgwick refused Hodgson's invitation to be a shareholder so, as the sole philosophical member of the company, Phillips was to be the supreme arbiter and director of the operation of the united collieries, if the shareholders differed, and was to be paid for such professional services. In spring 1840 Serjeantson agreed that the scheme for united collieries was feasible but in summer he withdrew so they continued to be run separately and Phillips' negotiations with Hodgson ceased.[33]

4.4   The philosophical carnival

As assistant secretary to the British Association, Phillips developed an unrivalled knowledge of its machinery which enabled him to become an important adviser and policy- maker while still remaining its general factotum. His centrality was recognised in 1834 when his salary was raised from £100 to £200 a year, thus binding him to the Association by a stronger and more responsible bond. Henceforth he was required to visit the places of meetings to supervise the local arrangements prior to them and to pay his expenses from his increased salary.[34] Between 1834 and 1840 he attended all the meetings from beginning to end, except that of 1839 at Birmingham which he left early in order to be with his uncle who died in Northampton two days after it had begun. When Murchison replaced Harcourt as senior general secretary in 1837, Phillips became even more involved with correspondence about research grants, research committees, and contributions to the annual *Reports* of the Association which he edited solo from that time. He continued to perform in the Association's popular section C and in the powerful section A, devoted to geology and mathematical and physical science respectively. For a time he joined the Association's magnetic crusaders, thus showing the strength of his commitment to *physique du globe*.

For about two weeks before each meeting began Phillips stayed in the locality to supervise the local arrangements in line with any decision Council might have made and his own experience and aims. Sometimes he paid a preliminary visit to meet the local secretaries and to examine the facilities available. Accommodation was vital. Each town visited had to offer big rooms for general meetings of the Association, for its General Committee, for evening lectures, and for promenades; and also up to seven rooms for the scientific sections which met simultaneously. Phillips advised the locals about which sections should go where, about the admission of local members, about registration of members, about the local subscription raised to receive the Association, about excursions and exhibitions, about eating arrangements, and about lodgings for visitors, which were always a source of difficulty and anxiety. Phillips was particularly exercised by the arrangements for the issue of ladies' tickets. Though women were usually excluded from scientific societies in the 1830s, they increasingly penetrated the Association's sections until in 1840 they were formally admitted to all of them though still not allowed to be members of the Association. Local members paid dearly for the entertainment of their ladies: in the late 1830s they subscribed to the local fund £2 for one lady's ticket, £5 for two, and £10 for three. Phillips connived at this financial exaction because it went some way to controlling the scramble for ladies' tickets and dangerous overcrowding in the section rooms. Phillips also dreaded the evening meetings because of the difficulties of striking a balance between promenades and formal meetings, including lectures, and of finding suitable lecturers. On three occasions Phillips gave useful advice before a meeting to key speakers. In 1835 he set out the aims and means of the Association for William Hamilton, a local secretary. In 1839 he laid out the ways in which Harcourt, as president, should deal with the geology and scripture question. Next year he helped Murchison to draw up his secretarial address.[35]

During the week of the meeting Phillips stayed with a local savant, sometimes a local secretary of the Association, or in a hotel in order not to be tarred with any local political party brush. In this way he extended the range of his contacts. In 1835 for instance his host was Humphrey Lloyd who became quickly a close friend. As each meeting had two to four local secretaries, Phillips met many leading local savants scattered over Britain and often dined with their friends. In counterbalance to such pleasures the increasing size of the meetings which peaked in 1838 with 2,400 male attenders, caused Phillips great anxiety.[36] Furthermore in the late 1830s he was working for the first time with the pre-emptive and autocratic Murchison, the senior general secretary. With Harcourt, the architect of the Association, Phillips as the clerk of works had an easy relation though his post was indefinite; with Murchison there was for two years no tacit trust. Phillips was pained by Murchison's improvisation of arrangements and the bitter dispute at the 1838 meeting between Murchison and Babbage about the running of the Association. Phillips even contemplated retirement from his post but was saved for it by outdoor recreation soon after the meetings. In 1837 he left Liverpool in great haste for the Lake District where 'it was only by boating on Coniston Water, climbing the Old Man, and beating stones like a mason for a month that I got over my horreur d'assemblie'. After 'the hot work' of the 1838 meeting he left Newcastle for four days of hammering, again in the Lake District, to cool down. In 1839 Phillips faced another source of anxiety: it seemed that the Chartist disturbances at Birmingham might wreck the meeting or cause it to be postponed. Racked by uncertainty Phillips yearned for quiet, peace, and good order which for him favoured the expansion of a philosophical spirit.[37]

There were particular features of the meetings which he liked. One was harmony among savants. He hoped that 'those who had once met in friendship would never after exhibit in their controversies that acrimony and pertinacity which has sometimes disgraced the cause of science and weakened the efforts of its disunited battalions'. Another was the diffusion of science. Phillips was particularly pleased by the 'upwards movement of the standard of mind among the ordinary members of the Association'. Indeed he thought that since 1836 the Association's chief function had been to diffuse new results and ameliorate old knowledge. At the personal level Phillips welcomed the Dublin LL D which his uncle received at the 1835 meeting while realising that Smith, like John Dalton, was a token provincial of symbolic importance wheeled on to the Association's stage to show that assiduous humbly born men could make their scientific mark.[38] There were also aspects of the Association he disliked. The philosophical carnival did not visit smaller towns and stimulate their scientific life because they had neither the money nor the accommodation to receive the Association. Another was that it did not harness and guide the zeal of provincials as Harcourt and he had hoped. He was disappointed by the failure of the Association to give specific and feasible jobs to provincial institutions who would combine their powers. A third defect of the Association was that by the mid-1830s many research committees had become dead letters. It was easy enough to suggest desiderata but very difficult to persuade scattered individuals to satisfy them collaboratively between meetings.[39]

If Phillips regretted the Association's indifference to Baconian participation and collaboration, he wanted its sections to be rich in scientific interest. Though he was busy with administration during meetings, he found time to perform not just in the geology section but also in that devoted to mathematics and natural philosophy. In his geological contributions four themes were prominent. Firstly, he stressed the difficulty of correlating strata using characteristic fossils. In 1835 in commenting on the joint paper by Sedgwick and Murchison on their newly enunciated Cambrian and Devonian systems, Phillips proclaimed that a stratum should not be identified by any particular genus but by a combination of coexisting genera. In 1838 he sounded further notes of caution. Having stressed that it was difficult to identify both living and fossil species, he urged that in classifying strata chemical and mechanical features should be used as well as fossils. Secondly, he tilted at Lyell's views. In reporting in 1835 a newly discovered tertiary deposit found at Bridlington, he commented on Lyell's way of classifying tertiary deposits according to the proportionate number of recent species of molluscs in them: for Phillips this method was valuable but had such wide limits of error that it should be used cautiously. Next year Phillips analysed the distribution of erratic rocks from the Lake District. He concluded that Lyell's iceberg theory and ordinary water were inadequate explanations because the distribution of the erratics had been affected by the general present-day configuration of the land. He stressed that whatever carried the erratics was not generally capable of overcoming natural obstacles because they never passed over these obstacles except at comparatively low points. More speculatively Phillips argued contra Lyell that the greater regularity of physical structure in older rocks was the consequence of the greater intensity of the earth's central heat. Thirdly, in 1836 and 1837 he went out of his way to praise Hopkins' papers on physical geology, because in their use of mathematics they showed how geology might eventually become an exact physical science. Lastly, Phillips was interested in two aspects of the coal measures, namely the certain existence of concealed coalfields underneath the magnesian limestone and the new red sandstone, and the uncertainty surrounding the origin of coal. With characteristic caution Phillips was agnostic about whether coal had been formed from plants *in situ* or drifted from a distance.[40]

Phillips' research on subterranean temperatures straddled sections A and C of the Association. The subject was part of *physique du globe* and also of interest to geologists because it was related to the questions of whether the earth was cooling and whether its nucleus was still molten. Phillips' work was stimulated by the Association which in 1834 established a committee, including him, to supervise the recording of the temperature of the earth below its surface. While lecturing in Newcastle in autumn 1834 Phillips was told by William Hutton that a pit at nearby Monkwearmouth, at 1,500 feet below sea level, was the deepest coal mine in Britain. Hearing that W.L. Wharton, another friend, was preparing to record underground temperatures in it, Phillips took charge of a group of five, including two Durham professors and one reader, which in November 1834 measured the temperature at 1,500 feet depth in three ways (of rock by bores, of gushing water, and of bubbling gas) to minimise modifying effects. The rough result, quickly published, was that there was an increase in temperature of $1°$ F for every sixty feet of descent. This figure was in accord with the rate of increase found in the 1820s by Louis Cordier, the French mining engineer, and convinced Phillips that the cooling earth was a fact and not a theory. Accordingly he organised from 1835 the Association's work on subterranean temperature, using standard thermometers supplied by him at its expense to individuals such as Buddle, James Forbes, Robert Were Fox, Eaton Hodgkinson, and himself. They used standard procedures designed to eliminate known sources of experimental error and to neutralise those which were unknown. Their nation-wide results, ideally recorded on a standard form, confirmed that there was a general increase of temperature with depth below the earth's surface. For Phillips they fitted neatly into Joseph Fourier's theory of the progressive cooling of the earth, in which he was a firm believer for the rest of his life.[41]

In 1836 and 1837, perhaps as a change from topographical geology, Phillips extended his interest in earth science by working on terrestrial magnetism under the aegis of Humphrey Lloyd and Edward Sabine, who were key figures in the magnetic crusade pursued in and through the Association in the late 1830s and subsequently. Before that time Phillips' interest was confined to measuring magnetic dip with his own needles, on which he dilated at the 1834 meeting. There he met François Arago, France's leading magnetician, who suggested improvements to them. Next year at the Dublin meeting of the Association the magnetic crusade began to take shape. Lloyd gave an account of the research on direction and intensity of magnetism in Ireland done by himself, Sabine, and James Ross, the discoverer of the north magnetic pole in 1831. In February 1836 Lloyd invited Phillips to join himself and Sabine in his projected contribution to the Göttingen Magnetische Verein founded in 1834 mainly by Friedrich Gauss to promote simultaneous magnetic observations in Europe. Flattered by this recognition and stimulated by what he had learned in Dublin, in spring 1836 Phillips began research on the magnetic dip in Yorkshire and constructed isoclinal curves which linked places of equal dip. His aim was to discover how far any flexing or breaks in these curves depended on local topographical configuration, such as hills. His provisional results, revealed to the 1836 BAAS, were that in flat areas the curves were bent to the south and on hills to the north. These conclusions were attacked by William Scoresby and William Ritchie who argued that the differences of dip were within the range of experimental error; but defended by Lloyd who claimed they were not and gave some of his own results.[42]

Phillips realised he was an inexperienced magnetic observer and not au fait with the literature, but he had set his heart on expanding his research to include intensity as well as dip measurements. Accordingly in May 1837 he began these in north Yorkshire. At about this time Lloyd and Phillips agreed to combine their English results under Lloyd's superintendence and name. As a magnetic novice Phillips expected Lloyd to 'inwardly bemoan the folly which induced your friend to meddle with the cold iron of a magnetic needle'. In July 1837 Lloyd, having learned that Sabine proposed to magnetise in England, preferred to work with his old associates Sabine and Ross. On hearing this Phillips released Lloyd from their hasty engagement, agreed to throw his observations into a common stock, and made up a magnetic quintet of Lloyd, Sabine, Ross, Fox and himself. Phillips had already observed in July 1837 on the Isle of Wight, where he stayed with Harcourt, and in August on the Isle of Man. He was given the job, as a magnetic underling under Lloyd and Sabine's superintendence, of measuring dip and intensity in the north of England as part of the huge task of recording them for the whole of Britain in 1837. Though Phillips welcomed such Baconian co-operation and the way the British Association had encouraged it, he quickly met problems: using three different needles and two modes of measurement, he found it difficult to produce consistent results for magnetic intensity, his Manx work was severely criticised by Lloyd, and it was not easy to correlate his York results for intensity with those of London, the base station. Even so, Phillips observed at twenty-two stations covering Yorkshire, the Lake District, Newcastle, Birmingham, and the Isles of Wight and Man.[43]

His results for dip and intensity were recorded as tables in the report by Sabine on Britain's magnetism published in the Association's 1838 *Report*. Phillips surveyed magnetically at more stations than either Sabine or Lloyd but, knowing his limitations as a magnetician, took no part in their attempts to perfect instruments, their research on the earth's magnetic intensity, or the Association's successful lobby of government in 1839 for a system of fixed magnetic observatories and an Antarctic magnetic expedition. He was happy to have contributed data, recorded in tables, about the actual state of magnetism in Britain in 1837, which would not only form the basis of the laws of magnetism he hoped would be discovered but would also permit long-term variations of magnetism in Britain to be traced. Presumably he agreed with Sabine that this magnetic work provided a model national survey, worthy of imitation. But he also believed that, in the study of terrestrial magnetism, frequent and simultaneous observations over wide areas of the earth's surface for long periods of time were necessary and best conducted by government. The role of individual exertion was for him therefore limited to initiating research and he was content to have joined for two years a Baconian enterprise which via co-operative survey and measurement produced for the British Association a pioneering magnetic map of Britain.[44]

In the mid and late 1830s Phillips was recognised as zealous and able meteorological observer who took a great interest in instruments. Baden Powell, professor of geometry at Oxford, admired Phillips' rain-gauge work carried out at York and confessed that Oxford could produce no equivalent. In 1838 the meteorological committee of the Royal Society of London established a sub-committee of Sir John Herschel, Daniell, and Phillips to draw up instructions and registers for meteorological observations. Two years later Phillips was the meteorological expert to whom the Newcastle Literary and Philosophical Society deferred and he sat on a BAAS committee appointed to decide the fate of 120,000 meteorological observations made at Plymouth under the Association's auspices by William Snow Harris.[45]

Phillips believed that meteorology was like terrestrial magnetism in that measuring instruments were fundamental to its progress which required exact, numerous, and corresponding observations on a uniform plan and at great expense, so that the role of the individual was limited to initiating projects or maintaining them as pilot schemes. Phillips therefore continued with his rain-gauge work which he reported regularly to the Association. In 1835 he felt he had sufficiently reliable data to maintain the view that raindrops grew while falling and to produce empirical equations which summarised the dependence of rain fall on height, temperature, and humidity at different times of the year. Three years later his results were attacked in section A by Alexander Bache who had been provoked by Phillips' paper of 1833 to it. Bache claimed that his own experiments, with four gauges at the corners of a tower, showed that eddy winds were not secondary as Phillips had assumed. Bache was supported by Lloyd, John Stevelly, a Belfast professor, and Thomas Robinson, the director of the Armagh Observatory, Ireland, who suggested that a globular gauge might eliminate the capricious effects of eddies. The fiercest criticism of Phillips' work came from Herschel who asserted that all rain gauges were useless because of eddy currents and that 'mere individual exertion' was little better than inaction. Next year Whewhell argued to section A that rain gauges were so defective that at best they gave mediocre results.[46]

At the 1840 BAAS Phillips reported that in response to Bache's objections, which he thought minor, he had not only begun to rotate his gauges but had devised a new compound or azimuthal-and-inclination gauge, which had five equal funnels and tubes, one with an horizontal opening, and four with curved tubes and vertical openings directed to the four quarters of the horizon (Fig. 4.6). With this apparatus he still claimed that rain drops grew as they fell. James Forbes thought these precautions gave valid results but Stevelly remained unconvinced. At the 1841 meeting Phillips confessed that he could not obtain consistent results from a funnel gauge and Robinson's globular gauge which he had begun to use in 1839. He called for a three-way comparison of these two gauges and his own compound one, a call to which he himself did not respond. Phillips' rain-gauge work, pursued persistently from 1832 to 1841, led to his being given a drubbing qua meteorologist but it also enabled him to secure a moral triumph. In 1841 Robinson praised Phillips as a true philosopher who had given 'a remarkable example of the ennobling effect of a true love of science' because he had tested his own opinions and produced results which seemed to tell against them.[47]

As the solitary paid and permanent officer of the Association Phillips was deeply involved in its government. He was a member of the General Committee, nominally its sovereign body, which convened only during its meetings, and of the Council which until 1836 ran the Association between meetings. Then it also deliberated during them and expanded its powers. He was also a member of the Committee of Recommendations, established in 1834; it met only during meetings and dealt with research grants and desiderata. Though the Association's leaders proclaimed it was the British Parliament of Science, it was managed by an inner cabinet which promoted its own interests, often at the expense of provincials whom it regarded as sciolists. Yet the Association could not entirely disregard the provincials: it met in the provinces and depended on provincial support and money. After Harcourt's resignation as general secretary in 1837 Phillips was the leading provincial in the inner cabinet,

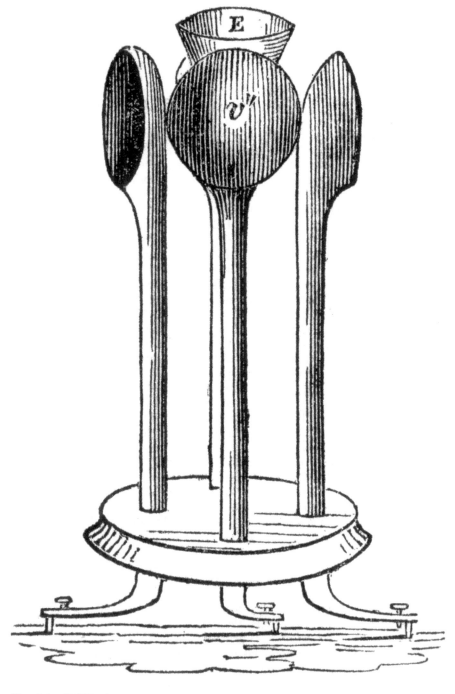

Fig. 4.6 **Phillips' compound rain gauge. Though it was criticised at the British Association in 1840, it received honourable mention at the Great Exhibition in 1851 when it was displayed in modified form.**

which was dominated by London gentlemen and Oxbridge academics, but he rarely defended provincial interests. On three constitutional matters he initiated changes which extended the power of the cabinet and reduced the importance of provincials.

His first campaign concerned the General Committee which, with up to 200 members, was in Phillips' view too large for a governing and partly executive body. His aim was to ensure that its membership was restricted to 'the elite of our corps scientifique'. In spring 1837 he was a member of a Council committee whose recommendations excluded from the General Committee representatives of all the Yorkshire lit-and-phils except the YPS. By September Phillips realised that this draconian proposal would outrage provincial sensibilities so he persuaded the Council to soften its exclusiveness. At the 1837 meeting the new regulations were not adopted but suspended until that of 1838. In March 1838 the Council appointed Murchison, Sabine, and Phillips to produce new rules for membership of the General Committee, which were adopted at the 1838 meeting. Henceforth the provincial input into the General Committee was restricted to three representatives from philosophical institutions in places of meetings, past and present, and three from any publishing lit-and-phil. Thus Phillips at the least connived at the exclusion from the General Committee of delegates from the societies at Sheffield, Hull, Halifax, Whitby, and Scarborough, all of which had nurtured his early career.[48]

The second constitutional matter concerned annual membership. The Association quickly became a booming financial business, its main source of income from 1833 being the sale of annual membership costing £1 to locals where the Association met. Large numbers provided, it seemed, the visibility essential for the Association's role as a lobbier of government and the money to fund research grants. Yet for Phillips these numbers provided administrative turmoil and a loss of scientific and social respectability. At the 1838 meeting income from annual members exceeded £2,000, but Phillips felt the scientific sections were too full of idlers in the form of annual members, of whom few compounded for life membership which cost £5. Such locals produced crowded meetings in which deliberation was impossible. Their presence also meant great expense for hiring large rooms, for fittings, and for refreshments. Phillips knew that, though the locals had raised a local hospitality fund of £3,100 in 1838, the BAAS had been forced to top it up by £500. In early 1839 Phillips took the initiative in bringing before Council his 'reform budget' which tried to change the terms of membership in order to limit the size of meetings without depriving the Association of money. Phillips wanted to rid it of those locals who had no intention of continuing to support it, or to tax them heavily, so he proposed that annual membership be abolished and that every new member pay £5 admission fee in exchange for receiving the *Reports*. Council accepted the main thrusts of Phillips' proposals but decided it was impolitic to abolish annual membership: instead, annual subscribers had henceforth to pay an admission fee of £1 in addition to the membership fee of £1. The 1839 meeting was the last under the old arrangements and confirmed for Phillips the evils caused by £1 members. He was therefore very happy to double the sum required from annual members, mainly provincials, as part of his aim of rendering the Association more exclusive.[49]

The third matter concerned the hospitality fund raised by the locals to receive the Association in style. In 1838 and 1840 the Novocastrians and the Glaswegians raised £3,100 and £2,350 respectively, but both of them had to be subsidised by the Association. Phillips deplored the way in which locals spent money on lavish display and spectacle and disregarded the Association's own purposes. He was embarrassed by the financial bribery of ladies' tickets, on the sale of which local funds were dependent. In December 1840 Phillips and Sabine, a general secretary from 1839, were chiefly responsible for a new scheme which enabled the Association to go anywhere without eating up the land and to accept invitations from towns as well as conurbations. Henceforth the permanent officers of the BAAS, using its funds, took over much of the local administration, leaving the local fund to pay for rooms. Big public dinners were discouraged. In an attempt to curb local display the BAAS subsidy to local funds was abolished. In future ladies' tickets would cost £1 per ticket per attending member and the proceeds would go to the Association.[50] Thus Phillips gained more control over the locals apropos the arrangements of meetings, an aim which was in harmony with his restricting access of some locals to the General Committee and his doubling of the annual subscription. Though he believed that from the mid-1830s the Association's chief role was the diffusion of knowledge, he put its scientific respectability before its popularity. Though he was financially the poor boy in the BAAS' governing coterie and a provincial who had risen the hard way, he rarely defended provincial interests. Instead, as a co-opted member of that coterie, he risked appearing to dictate to the provincial masses by going out of his way to control local involvement.

Between the meetings Phillips undertook heavy correspondence on behalf of the Association about research desiderata, committees, grants, and reports, using cyclostyled letters whenever possible.[51] From the mid-1830s he edited solo the Association's annual *Reports*, which were then unique in Britain because they contained reports on research undertaken at the Association's request, and often aided by its money, as well as reports on the state of various sciences. Phillips enjoyed the contacts with expert reporters and though their reports made himself familiar with all the sciences deemed acceptable by the Association. The *Reports* also contained notices and abstracts of miscellaneous communications to the scientific sections of the Association. In the late 1830s Phillips concluded that these notices were troublesome to him, useless because they had been previously published in journals such as the *Athenaeum*, and were so expensive to print that they made the price of the *Reports* unacceptably high. On one occasion he wanted to raise the level of the notices by including a long geological paper, replete with map and section, but was ordered by Murchison and Peacock, the two general secretaries, not to do so. Subsequently Phillips failed to convince the Council that the notices be scrapped.[52] Had he won his case he would have been saved the great anxiety of a vehement plagiarism dispute between Alexander Nasmyth, a London dentist who looked after the teeth of the Queen and the prince consort, and Richard Owen, the ambitious comparative anatomist employed by the Royal College of Surgeons, London.

In the scientific sections of the Association priority and originality could be claimed in a spoken short paper, which was followed by full publication. In August 1839 Nasmyth gave to the medical section a paper on the microscopic structure of teeth, their enamel covering, and their pulp. Owen, whose career had already been advanced by conspicuous patronage from the Association, did not hear Nasmyth's paper because he was president of the zoology and botany section. It seems that Nasmyth's abstract of his paper was approved by Phillips, as editor, who then sent it to the printer. In early June 1840 Owen asked Phillips to suspend the publication of Nasmyth's abstract on the grounds that Nasmyth had allegedly plagiarised from Owen's paper on teeth read by Owen in Paris in December 1839. Phillips replied that he could neither suspend nor suppress the abstract. It was revealed to Phillips in late

June 1840 that Nasmyth *had* made additions to his abstract, which was set up for printing but not printed off, thus depriving Phillips of the power of substantiating its authenticity. In response to Phillips, Nasmyth claimed he made only minor alterations. In July 1840 Owen complained to the Association's Council about Nasmyth's alleged plagiarism and asked for his abstract to be postponed. The technical issue was whether at the BAAS meeting Nasmyth described cellular ivory of teeth as calcified pulp. With Yates, the secretary to the Council, the general secretaries, and most of the Council unavailable, in August 1840 Phillips took unilateral action by suspending publication of Nasmyth's abstract pending an enquiry by Council. Nasmyth was so outraged by what he regarded as persecution by Phillips, who had arbitrarily passed and executed sentence before either charge or trial, that he complained to the Council about Phillips' behaviour. In September 1840 it ended Phillips' anxiety by supporting his action and, having made two enquiries, it eventually concluded in summer 1841 it could not judge the fidelity of Nasmyth's abstract not least because he would not produce the original version.[53]

The affair was publicised extensively in *The Lancet* which was happy to have Owen in its sights. It characterised Owen as a literary depredator and a practised interpolator who exploited the delay between spoken and published versions of a text. Nasmyth remained so bitter that in his 1841 book on teeth, based on his paper of 1839, he attacked the BAAS Council to which in private he was still complaining. This time *The Lancet* castigated the Council, characterising it as a feeble and slippery 'conclave of drones' which was in Owen's pocket. It christened the members of the two investigatory panels of medical experts who had reached no conclusion, 'professors Goggle and Smudge'. Faced with such attacks, the Association printed nearly all the relevant documents in its *1841 Report*.[54]

Privately Phillips had some sympathy with Nasmyth who had accused him of acting on Owen's advice. James Yates had more. He resigned as secretary to the Council after nine years' service because he thought the Council had treated Nasmyth badly and favoured Owen. In Yates' view Owen's charge had no foundation and more importantly was a slur on Nasmyth's character. For his part Owen deplored in 1840 the use of the notices part of the *Reports* as a trap to catch whatever discovery might fall in it between a meeting and the publication of the appropriate *Report* almost a year later. Yet next year the inconsistent Owen exploited that time gap to make crucial changes to an exposition given at the 1841 meeting so that it appeared from the *1841 Report*, published in April 1842, that he had invented the dinosaur in 1841.[55]

4.5 The burden of lecturing

Phillips' involvement with BAAS meetings, usually held in August and September, did not facilitate foreign trips. Indeed, it was the 'grievous tyranny' of the Association which prevented him in 1837 from touring in the Auvergne with R.H. Greg and in 1839 from accepting from the reverend William Bilton a free trip to Norway.[56] Phillips' lecturing engagements constituted a further obstacle to distant travel (see Appendix 2). He undertook these to supplement his regular income. For six lectures his fee was usually £50 and for eight 60 guineas. In Manchester, where he was much in demand, he embarked on a new experiment of giving concurrently the same course of lectures to three different audiences. In June and July 1836 he gave two lectures a week for six weeks to the Royal Manchester Institution, the Manchester

Athenaeum, and the Manchester Mechanics' Institute, for which he received £160. When not performing in Manchester, he gave one lecture a week during the same six weeks to the Liverpool Royal Institution. His receipts from this stint of lecturing roughly equalled his annual salary from the British Association. In 1838 he gave his second triple course at Manchester, receiving 120 guineas for a total of twenty-four lectures given in a month. Two years later he gave yet another triple course, lecturing in Manchester to the Royal Institution and the Athenaeum and in Liverpool to the Mechanics' Institute, which generated 100 guineas for eighteen lectures all told. He lectured most in Manchester because there he could give double or triple courses. He ventured to the new locations of Newcastle, Liverpool, Birmingham, Bristol, and Chester, usually lecturing on geology because after all from 1834 he was a professor of geology, could adopt and sell his *Guide* as a textbook, and always had materials ready. In contrast with the past he lectured little in Yorkshire because he could make more money elsewhere. York, where he lectured twice, was the only place where he did not pursue financial gain: in 1834 he devoted half the proceeds to help house its zoological collection properly and in 1839-40 his main purpose was to raise awareness of the value of its zoological and geological collections.[57]

Phillips' lecturing in the provinces exacted high personal costs. Whether staying in lodgings or with hosts and however much he was wined and dined, he often felt homesick. He was away from home for long periods. For instance he spent much of autumn 1834 in Newcastle giving twenty lectures at the rate of three a week. He also mounted there a parallel practical class in geology involving five field excursions and ten lectures on the application of geology to mining, draining, engineering, and agriculture. It was tiresome to have to lug around maps, sections, models, fossils, and drawings of fossils and scenery. By the late 1830s Phillips feared the long labour and anxiety of yet another lecture course. At Manchester in 1838 he enjoyed the company of new friends such as William Langton, William Fairbairn, and Eaton Hodgkinson; but he endured sleepless nights in a noisy hotel in that 'furnace of a place' and his lecturing left him hoarse and exhausted. In 1834 he had told Thomas Winstanley, his chief patron in Manchester, that lecturing in geology was essential because it made him keep up to date in a subject which was changing rapidly and augmenting wonderfully. By the late 1830s Phillips' view had changed. Lecturing was desirable financially to augment his regular income and to give him the financial stability his uncle had rarely enjoyed. But he was aware that he lectured to excess in 1838, that he undertook engagements not always to his taste, and that the trials of lecturing were destroying his 'repose of mind'.[58] It became clear to him that provincial lecturing, however lucrative, was a punishing and unreliable way of increasing his scientific reputation; and it was an exhausting diversion from advancing science.

Notes

1. *YPS 1834 report*, p. 2; Agassiz, 1833-43; *YPS 1835 report*, pp. vi, i, 5, 17; *YPS 1836 report*, pp. 4, 12-13; Rudwick, 1997, pp. 159-64; Rudwick, 1976a, pp. 167, 172-3; *YPS 1837 report*, pp. 2, 5 (q); Phillips to Harcourt, 16 March 1837, HP; Phillips to Forbes, 28 March 1837, FP; YPS Council minutes, 6 March, 3 April, 11 May, 19 June 1837. Jean Louis Rodolphe Agassiz (1807-73), *DSB*, Swiss naturalist; George Lane Fox (1793-1848), a member of the YPS, of Bramham Park, near Tadcaster.
2. Phillips to Harcourt, 11 Dec 1837, 11 Feb 1838, HP; *YPS 1838 report*, p. 3; YPS General meetings minutes, documents dated early Jan, and 1 Jan 1838; Orange, 1973, pp. 43-5;

YPS 1836 report, pp. 1-2; *YPS 1838 report*, pp. 7, 17-18; *YPS 1839 report*, pp. 11, 16 (q); Prospectus of a subscription for restoring the hospitium of St Mary's Abbey, York, 1839; Phillips, printed circular about Etty's lecture on 5 Nov 1838, PP; Thomas Allis (1788-1875), superintendent of the Friends' Retreat, York, and expert on comparative osteology; William Etty (1787-1849), *DNB*.

3. Phillips to Forbes, 20 May 1836, 28 March 1837 (q), FP; Phillips to Harcourt, 10 Aug 1837, 7 July 1838, HP; Phillips to sister, 1 July 1838 (q), nd (franked 2 July 1838) (q), PP.

4. YPS Council minutes, 1 Oct 1838, 8 Feb, 6 May, 10 June, 7 Oct 1839; YPS General meetings minutes, 5 Feb 1839; *YPS 1838 report*, pp. 9, 16; *YPS 1839 report*, pp. 11-12.

5. Phillips, Considerations on Gothic architecture and On Gothic arches, windows, and buttresses, given to YPS evening conversazione, 31 Oct 36, 14 Jan 1839, PP, lecture scripts (26); Phillips to sister, 23 Feb, 1, 6, 7, 12-14, 16, 22 March 1839, PP; Phillips, Drawings and details of St Mary's Lodge 1839, notebook 40, PP; Royal commission on historical monuments, 1975, p. 17, plate 18; Phillips to Whewell, 4 Nov 1839, WP; Curwen, 1940, pp. 257, 286.

6. *YPS 1839 report*, pp. 8-10; *YPS 1840 report*, pp. 11-12; Goldie, 1841; YPS Council minutes, 4 and 18 Jan 1841, 3 Jan 1842.

7. YPS General meetings minutes, October 1834-9; YPS Scientific communications, 4 March 1834 (Holderness), 2 Dec 1834 (Monkwearmouth), Phillips, 1834c and d; YPS Scientific communications, 15 Jan, Dec 1835, Jan 36 (auroras); *YPS 1837 report*, p. 4; for Phillips' interest in Whewell's anemometer, Whewell to Phillips, 28 Dec 1834, PP, and 2 Feb 1835, WP, Phillips to Whewell, 12 Jan 1835, WP; for his results on pressure changes, Phillips to Whewell, 18 Apr 35, WP; for short-lived collaboration with Ford (1801-75), a Quaker schoolmaster in York, *YPS 1840 report*, p. 22; YPS Scientific communications, 1 July 1834 (telescope); for Thomas Cooke (1807-68) and Phillips, McConnell, 1992, p. 50, *YPS 1868 report*, p. 12, Phillips to Whewell, 4 Nov 1839, WP, Chapman, 1998, pp. 234-7, and Pritchard, 1868-9.

8. YPS evening meetings minutes, 21 Oct, 18 and 25 Nov 1839, 5 Oct, 16 Nov 1840; Phillips to sister, 29 March 1840, PP; Darwin, 1839; Lambert Adolphe Jacques Quetelet (1796-1874), *DSB*, Belgian natural philosopher; Justus von Liebig (1803-73), German chemist, on whom see Brock, 1997; James Pollard Espy (1785-1860), *DSB*, USA meteorologist.

9. Owen to Phillips, 30 Jan, 19 April 1839, PP; Richard Owen (1804-92), *DNB*, comparative anatomist, Royal College of Surgeons, London; on didelphis, see Rupke, 1994, pp. 142-7 and Desmond, 1989, pp. 306-21; Baines, 1840, pp. i-xvi; Wellbeloved, 1842, pp. iv-v; Phillips, 1853a, p. 210.

10. Book listing lectures ... of ...YMI, York City archives, TC 49/1:15, 15 Dec 1835, 12 Jan and 29 March 1836, 23 Jan 1838; YMI minute book of committee, York archives, TC 49/1: 1, 31 Oct 1838, 28 March 1839; YMI minute book of monthly meetings, York archives, TC 49/1:2, 18 March 1834, 31 Mar 1835; *YMI annual reports*, for 1833-40.

11. Phillips, 1835a, pp. v-vi, 142-70.

12. Phillips, 1836a, pp. x-xiii; YPS Scientific communications, 3 July, 2 Oct 1827; Phillips, 1829c; Sedgwick to Phillips, 28 May 1829, PP; *YPS 1830 report*, p. 6.

13. Prospectus, FP; Phillips to Murchison, 24 May 1832, MP; Phillips, 1836a, pp. xvi (q), xx, x (q).

14. Phillips, 1836a, pp. xiv-xvii, 125-30, 203, 241; Hodgson to Phillips, 9 Dec 1835, PP; Gilbertson to Phillips, 4 Aug 1829, 18 Oct 1832, 27 Dec 1833, 12 Nov 1835, PP; Gilbertson, 1837; Thomas Sopwith (1803-79), *DNB*; Westgarth Forster (1772-1835); Hugh Lee Pattinson (1796-1858), *DNB*; Joseph Ellison Portlock (1794-1864), *DNB*; William Gilbertson (1789-1845).

15. Phillips, 1836a, p. xv; Viscount Cole (1807-86), later third earl of Enniskillen, collected mainly fossil fish; Philip de Malpas Grey Egerton (1806-81), *DNB*, studied them.

16. Phillips, 1836a, pp. ix (q), xiii-xvi, plate 24, diagram 24; YPS Scientific communications, 7 May 1833; Phillips, Yorkshire June-Nov 1832 and 1833, notebook 28, and Yorkshire 12 Aug-Sept 1833, notebook 27, PP.

17. Phillips to Murchison, 10 Oct 1835, 2 Feb 1836, MP; Phillips to Murray, 16 Feb 1835, 26 March 1836, Mu P; Phillips, 1836a, pp. ix-x, 241-3; for his figures of fossils, Phillips to Mantell, 28 Sept 1833, Ma P, Phillips to Murchison, 16 Nov 1833, MP, Phillips to Harcourt, 15 June 1835, HP, Phillips to Hutton, 8 Feb 1836, Hu P, Phillips to Buckland, 8 March 1836, Bu P.

18. Rudwick, 1985, pp. 146-7; Phillips, 1836a, pp. 11-12, 15, 18, 35, 58, 174-7 (174 q), 194-5.

19. Phillips, 1836a, p. 244.

20. Phillips, 1836a, pp. 84-5, 90-98 (diagram p. 98), xix on Hopkins; Smith 1985, pp. 73-6; Hopkins, 1835; Hopkins to Phillips, 19 Oct 1836, PP; Phillips to Forbes, 28 March 1837, FP; Phillips to Whewell, 15 March 1836, WP; William Hopkins (1793-1866), *DNB*, a private crammer at Cambridge.

21. Phillips, 1836a, pp. 99-107, 193.

22. Lyell, 1837, pp. 488-9; Torrie to Phillips, 18 May 1836, PP; Prestwich to Phillips, 9 July 1836 inserted in YPS, Scientific communications 1822-38; Prestwich, 1840; J.B. Pentland to Phillips, 26 June 1836, PP; Koninck, 1842-4; *QJGS*, (1875), vol. 31, p. xxxiv (Koninck when receiving Wollaston medal); Murchison to Phillips, spring 1836, in Geikie, 1875, vol. 1, pp. 232-3; Murchison to Sedgwick, 17 March 1836, Se P; Thomas Jameson Torrie (d 1858) assistant editor of *Edinburgh new philosophical journal*; Jean Baptiste Armand Louis Léonce Élie de Beaumont (1798-1874), *DSB*; Laurent Guillaume de Koninck (1809-87), *DSB*, Liège chemist and palaeontologist.

23. Sedgwick to Murchison, 7 April 1836, MP; Sedgwick, 1835b, was read spring 1831 and abstracts published in *PGS* (1834), vol. 1, pp. 247-9, 318-20; Sedgwick, 1831, pp. 285-6; Phillips to Sedgwick, 15 March, 2 April 1836, 28 Aug 1837, Se P; Sedgwick to Phillips, 8 Feb, 28 March, 7 April 1836, 24 Aug 1837, PP; Phillips, 1836a, pp. xviii-xix.

24. Murchison, 1839, p. 4; pp. 580-1; pp. 161, 558, 586, 648, 671-5; pp. 244, 247, 339, 730; Phillips, 1836b, pp. 170-2.

25. Phillips, 1836a, pp. xx, 134 (q); T.W. George to Phillips, 15 March 1830, PP; George, 1837; Phillips to Harcourt, 3 Feb 1836, HP; Conybeare and Phillips, 1822, pp. 325, 347-8, 364; Phillips, 1832a; Thomas Wilson (1800-76), Barnsley mine owner and Cambridge graduate; Henry Hartop (1786-1865); William Thorp (1804-60), Anglican minister at Womersley near Pontefract and pupil of Sedgwick at Cambridge, Thorp to Sedgwick, 28 June 1830, Se P.

.26. YGS minute book, 1, 14, 16 Dec 1837; circular of 16 Dec 1837 reproduced in Davis, 1889, p. 6; and generally on early YGS, Morrell, 1983, 1988b.

27. Wilson to Phillips, 10 Feb 1838, Wilson P; prospectus, 15 March 1838, in Davis, 1889, pp. 12-14.

28. Wilson to Phillips, 23 Oct, 1 Nov 1839, PP; Thorp's prospectus, Se P, 1E93a; Davis, 1889, p. 228.

29. Phillips, 1837, 1839, vol. 2, pp. 288-96; G. Lloyd to Phillips, 24 Oct 1840, PP; C. Howard, secretary of YAS, to Phillips, 30 March 1840, PP; Hall, 1987, pp. 56, 61; Thorp, 1841; Davis, 1889, pp. 80, 215-27.

30. Belcher to Phillips, 31 Jan 1835, PP; the borings were undertaken in connection with the Whitby to Pickering railway, opened 1836.

31. *Leeds mercury*, 18 March 1837; Grainge, 1871, pp. 121-2; Grainge, 1864, pp. 22-5, for Phillips' statement; Phillips to Harcourt, 16 March 1837, HP; John Frederic Daniell (1790-1845), *DNB*. professor of chemistry, King's College, London; John Buddle (1773-1843), *DNB*; William Reid Clanny (1776-1850) *DNB*, Sunderland physician and inventor; Adam Hunter (1794-1843), Leeds physician; Peter Murray (1782-64), Scarborough physician; William West (1792-1851), Leeds analytical chemist.

32. Phillips, 1837; Phillips, 1836a, pp. 72-3, 126-9; T. Mason, secretary of company, to Phillips, 5 and 30 Sept 1836, 24 May 1837, PP; Whewell to Phillips, 20 Sept 1836, PP; Phillips to Whewell, 26 Sept, 4 Oct, 22 Nov 1836, WP; Phillips to Harcourt, 11 Nov 1836, HP; Phillips to sister, nd [fr 23 Nov 1836], 22 March, 7 April 1837, PP; J. Heywood to Phillips, 21 July 1838, PP.

33. Hodgson to Phillips, 14 Dec 1839, 9 March, 2 July 1840, PP; Hodgson to Sedgwick, 27 Jan, 10 Feb 1840, Se P; Harris, 1968; Thomas Hodgson (1799-1869).

34. *BAAS 1834 report*, p. xxviii; Phillips to sister, 15 Sept 1834, PP.

35. For pre-meeting preparations, Phillips to Harcourt, 3, 5, 8 Aug 1836, 10, 14 Aug 1837, HP; on dealing with ladies, Morrell and Thackray, 1981, pp. 148-57; as adviser, Phillips to Hamilton, nd [1835], Ham P, 1835 (509), Phillips to Harcourt, 20 April, 12 June 1839, HP; Phillips to sister, 16 Sept 1840, PP.

36. Phillips to sister, 23 Sept 1834, 1 Aug 1835, PP; Phillips to Harcourt, 11 Aug 1839, HP.

37. Morrell and Thackray, 1981, pp. 189-90, 252; Phillips to Currie, 7 Nov 1837 (q), Fitzwilliam Museum, Cambridge; Phillips to Harcourt, 11 Feb, 7 July, 10 Aug, 17 Sept 1838, HP; Phillips to sister, 27 Aug 1838 (q), 11 Aug 1839, PP; William Wallace Currie (1784-1840), a local secretary in 1837 and first mayor of the reformed Liverpool Corporation.

38. Phillips to Hamilton, nd [1835] (q), Ham P; Phillips to Harcourt, 17 Sept 1838, 29 Sept 1840 (q), HP; Phillips, 1844a, pp. 120-1; Davies, 1969, pp. 27-30.

39. Phillips to Harcourt, 17 Sept 1838, HP; Phillips to Hamilton, nd [1835], Ham P; Morrell and Thackray, 1981, pp. 503-4.

40. Phillips to sister, 23 Sept 1834, PP; *Literary gazette* (1835), p. 535; *Athenaeum* (1838), p. 599; *BAAS 1835 report*, p. 62; *BAAS 1836 report*, pp. 87-8; *Literary gazette* (1836), p. 565; *Athenaeum* (1836), pp. 624-5; *Athenaeum* (1837), pp. 750-1; *Athenaeum* (1836), p. 632; *Athenaeum* (1838), pp. 599, 610; *Athenaeum* (1840), p. 775.

41. Phillips to sister, 16 Nov 1834, PP; *BAAS 1834 report*, p. xix; Wharton to Phillips, 8 Nov 1834, PP; Phillips, 1834d; instructions and forms, PP, box 98, f. 4; *BAAS 1835 report*, p. xix; *BAAS 1836 report*, pp. 291-3; *BAAS 1837 report*, p. 37; Buddle to Phillips, 6 Feb 1836, PP; Phillips to Forbes, 20 May 1836, FP; Phillips to Fox, 2 April 1836, Fox Papers; Pierre Louis Antoine Cordier (1777-1861), *DSB*, found an increase of 1° C per 30-40 metres of depth and assumed that the earth was fluid below 5,000 metres; Robert Were Fox (1789-1877), *DNB*, measured in Cornwall; Eaton Hodgkinson (1789-1861), *DNB*, retired Mancunian pawnbroker measured in Lancashire; Jean Baptiste Joseph Fourier (1758-1830), *DSB*, French mathematical physicist. The Monkwearmouth five were: Johnston, professor of chemistry, Durham; Charles Thomas Whitley (1808-95), reader in natural philosophy; Temple Chevallier (1794-1873), *DNB*, professor of mathematics and astronomy; William Lloyd Wharton (1789-1867), high sheriff of County Durham and pupil of Sedgwick; and George C. Atkinson, active in the Natural History Society of Northumberland, Durham and Newcastle upon Tyne.

42. *BAAS 1834 report*, p. xlvi; Phillips to Whewell, 26 Oct 1834, WP; Lloyd, 1835; Lloyd to Phillips, 17 Feb 1836, PP; Phillips to Forbes, 20 May 1836, FP; *BAAS 1836 report*, p. 31; *Athenaeum*, 1836, p. 649; Edward Sabine (1788-1883), *DNB*, a general secretary BAAS 1839-52, 1853-59, president BAAS 1852, president Royal Society of London 1861-71; Dominique François Arago (1786-1853), *DSB*, a leading physical scientist and bureaucrat of science; James Clark Ross (1800-62), *DNB*, arctic explorer; Carl Friedrich Gauss (1777-1855), *DSB*, director of Göttingen Observatory; William Scoresby (1789-1857), *DNB*, arctic scientist; William Ritchie (1790-1837), *DNB*, professor of natural philosophy, University College London, 1832-7; for BAAS magnetism and the magnetic crusade, Morrell and Thackray, 1981, pp. 353-70, 523-31.

43. Phillips to Lloyd, 4, 23 July, 5 and 21 Aug (q) 1837, Terr Mag P; Phillips to sister, 21 July, 17 Aug 1837, PP; Phillips to Sabine, 4 Aug, 4 Nov 1837, Sa P; Lloyd to Sabine, 27 July 1837, Sa P, PRO; Lloyd to Phillips, 3, 28 Aug 1837, PP; Sabine to Phillips, 2 Dec 1837, 1 Apr 1838, PP; Phillips, magnetic observations 30 May 1837-, notebook 37, PP.

44. Sabine, 1838, pp. 50, 70-4, 144-7; Sabine to Harcourt, 7 Aug 1839, HP; Phillips to Sabine, 8 March 1845, *BAAS 1845 report*, pp. 37-8.

45. W.H. Sykes to Phillips, 18 Oct 1835, Powell to Phillips, 15 Nov 1835, PP; Phillips to Herschel, 26 March 1839, Her P; Daniell to Phillips, 11 Apr 1839, Wailes to Phillips, 5 Feb 1840, PP; *1840 BAAS report*, p. xxiii; reverend Baden Powell (1796-1860), *DNB*; John Frederick William Herschel (1792-1871), *DNB*; William Snow Harris (1791-1867), *DNB*.

46. Phillips to Harcourt, 6 July 1837, HP; Phillips, 1835b; *Athenaeum* (1838), p. 620; (1839), p. 643; Alexander Dallas Bache (1806-67), *DSB*; John Stevelly (1794-1868) professor of natural philosophy, Royal Belfast Academical Institution and then Queen's University, Belfast; reverend Thomas Romney Robinson (1792-1882), *DNB*.

47. Phillips, 1840a, 1841b; *Athenaeum* (1840), p. 793, (1841) pp. 623-4.

48. Morrell and Thackray, 1981, pp. 304-6.

49. Morrell and Thackray, 1981, pp. 310-12.

50. Morrell and Thackray, 1981, pp. 221-1.

51. Phillips to Whewell, 26 and 28 Sept, 3 Oct, 8 Nov 1836, WP.

52. Phillips to Harcourt, 25 Aug 1837, HP; Murchison to Harcourt, 18 July 1838, 16 Nov 1839, HP; Phillips to Forbes, 26 Oct 1839, FP; BAAS Council minutes, 1 Dec 1839; reverend George Peacock (1791-1858), *DNB*, professor of astronomy and geometry, Cambridge, and a general secretary of BAAS 1837-9.

53. *Athenaeum* (1839), p. 707; *BAAS 1841 report*, pp. 1-23; BAAS Council minutes, 15 Sept, 19 Nov 1840, 8 Jan, 27 Feb, 29 and 30 March, 28 May, 5 and 28 July 1841; Phillips to sister, 16 Sept 1840, PP; Desmond, 1989, p. 320; Rupke, 1994, p. 114; Alexander Nasmyth (1789-1849).

54. *The Lancet* (1839-40), vol. 2. pp. 376-8, 486-93, 567-75, 623; (1840-1), vol. 2, pp. 841-3; Nasmyth, 1841, pp. iii-xvi.

55. Phillips to sister, 10 Apr 1841, PP; Yates to Nasmyth, 12 Aug 1841, Richard Taylor collection, St Bride's Printing Library, London; Torrens, 1992b, 1997.

56. Greg to Phillips, 8 March 1837, Bilton to Phillips, 1 and 30 April (1839) (q), PP; Robert Hyde Greg (1795-1875), *DNB*, Manchester Unitarian businessman and antiquarian; William Bilton (1798-1883) lived in London and Bideford.

57. Phillips, syllabuses of lectures, folder 12, PP; Winstanley to Phillips, 7 Jan 1835, 18 Feb 1836, 28 Feb, 3 May 1838, Royal Manchester Institution archives, M6/1/49/2; Phillips to Winstanley, 17 Feb 1838, RMI archives M6/1/51; Lynelt to Phillips, 1 Sept 1840, W.B. Hodgson to Phillips, 6 Oct 1840, PP.

58. Phillips to sister, 27 Oct 1834, 6 July 1837, 1, 2, 6, 10 (q), 20 July 1838, PP; Phillips to Winstanley, 28 Jan 1834, RMI archives, M6/1/52; Phillips to Harcourt, 7 July 1838, HP; Phillips to De la Beche, 20 Oct 1840, DLB P; William Langton (1803-81), *DNB*, financier and statistician; William Fairbairn (1789-1874), *DNB*, structural engineer.

Chapter 5

The Professor and Popular Writer

5.1 King's College, London

Phillips became a fellow of the Royal Society of London in 1834, a move which was suggested to him the previous September by John Lubbock, then its treasurer, who had been impressed by Phillips at the meeting of the BAAS that year. Initially Phillips was deterred by the high cost of being FRS, which at maximum was £50 (£10 entrance fee, £40 life subscription), but when Lubbock persisted in wishing to propose him he agreed to the financial ordeal of spending £14 (entrance £10, £4 annual subscription). In November he approached Sedgwick and Murchison to support Lubbock's proposal. Phillips' appointment as professor of geology at King's College, London, in January 1834 highlighted the importance of his being FRS, to which he was elected on 10 April 1834 just eleven days before he gave his first lecture at King's. Of his thirteen supporters, eight were leading geologists, most of whom also knew him well through the British Association (Sedgwick, Murchison, Buckland, Greenough, De la Beche, Daubeny, John Taylor, William Clift). Three non-geologists knew him only through the British Association (John Gray, Peacock, Powell). The other two supporters were professors at University College, London, who had been impressed by Phillips' lectures there in 1831 (Lindley, Turner).[1] For many years Phillips took little part in Royal Society affairs but was fast off the mark in putting the three magic letters, FRS, behind his name on the title pages of his books and on advertisements for his lectures. At the age of thirty-three the orphan who had never attended a university was doing well: he was professor Phillips, FRS, FGS.

King's College, London, was established in 1828 as a rival to University College, a secular institution founded mainly by whigs two years before. The history of the foundation of King's gives plausibility to the view that it was a defensive bastion of church and state. After all, its initiator was George D'Oyly, rector of Lambeth and a high Anglican, who wrote in February 1828 to Robert Peel, who had joined Wellington's ministry the previous month as home secretary. D'Oyly deplored the absence of religious teaching at UCL as a moral and spiritual omission and proposed a second college in London to rectify it. D'Oyly's proposal was supported by his great friend, Charles Manners-Sutton, archbishop of Canterbury, who recruited bishops, royalty, and high tories to the cause. Peel was keenly in favour of the religious basis of higher education and obtained Wellington's support. Thus King's was backed by leaders of state and church; hence its name.[2]

Yet it also makes sense to view King's as an early manifestation of alert conservatism and liberal Anglicanism. Peel distrusted rigorously unrelenting toryism: having resigned as home secretary in 1827 on account of his opposition to Catholic emancipation, in March 1829 he introduced the bill granting the measure as a desirable reform endorsed by public opinion. A similar approach to the issue was taken by Charles Blomfield, bishop of London from August 1828. In 1829 he opposed Catholic emancipation but when it became law accepted it. With D'Oyly and Edward Copleston, dean of St Paul's Cathedral, Blomfield formed a clerical triumvirate which dominated the governing Council of King's. It was Blomfield who preached a sermon in the chapel in October 1831 when King's began to admit students. His theme was the duty of combining Christianity and secular knowledge. Blomfield argued that, provided the respective provinces of science and revelation were properly allocated,

> we may join in the praises which are lavished upon philosophy and science, and fearlessly go forth with their votaries into all the various paths of research, by which the mind of man pierces into the hidden treasures of nature ... and removes the veil which ... obscures the traces of God's glory in the works of his hands.

Thus Blomfield wished to harness the march of intellect for Christian purposes so that 'the wholesome nutriment of science' would not be converted into 'the poison of unbelief or immorality.' Indeed the motto of King's, Sancte et sapienter, reflected his liberal Anglican views.[3]

At King's the principal was a clergyman and most professors were Anglicans. Regular students were required to attend chapel and to know about Anglican doctrines and the evidences of natural and revealed religion. Otherwise King's was rather like UCL in the wide range of subjects it offered, the way in which science was taught as part of liberal education and medical training, the opportunities given to occasional students (for whom there was a conscience clause which enabled dissenters to attend), the precarious financial position of many professors who were mainly or entirely dependent on class fees, and general adversity in its opening years. In one respect King's took the initiative. Whereas UCL had no regular professor of geology until 1841, King's showed alert conservatism and liberal Anglicanism in appointing no less than Charles Lyell in 1831 as its first professor of the subject. His uniformitarianism and his view that geology required a vast time scale in comparison with human history were known to Copleston who wanted an assurance from Lyell that his geological views did not contradict belief in the Noachian deluge. Though privately Lyell wanted to free geology from Moses, he gave such an assurance to Copleston who thanked him for his candid statement. Whether Lyell's behaviour was uncandid or hypocritical, King's attitude was clear: by not electing a Mosaic geologist, dedicated to making geology conform to scripture, it acted on its liberal Anglican principles. Lyell sought the chair because he thought it would be agreeable, influential, and financially worthwhile. But even before he gave his first course he wanted to relinquish it because publication of a readable book was a superior way of enlightening the public and securing fame. However in 1832 he lectured to an audience of eighty, including women, fellow geologists, but very few King's students. Next year, when women were banned, only fifteen attended his lectures so he resigned in October.[4]

In 1831 Lyell had thought of Phillips as his successor but in autumn 1833 had lost interest in the scheme of having Phillips resident in London as professor at King's and secretary to the Geological Society. The initiative for Phillips' appointment came from Blomfield, who failed to persuade Sedgwick to replace Lyell temporarily and then consulted Buckland who warmly recommended Phillips. Though Buckland revealed to Phillips that the KCL chair carried no salary and that the class fees would pay for only six weeks' residence in London, he urged Phillips to take it in order to promote his future welfare. In January 1834 Phillips was elected to the professorship and next month received the good news from Blomfield. It was soon agreed that he would give his first course of eight lectures at the rate of one a week in spring 1834. From the start Phillips never contemplated permanent residence in London but he was prepared to visit it for a few weeks of the year in order to be and to style himself a professor of geology and thus become like Sedgwick and Buckland. Non-residence also meant that he would not be a leading and full-time figure at King's, such as the chemist Daniell: he would be a visiting teacher, gaining cachet but avoiding responsibility. Moreover the chair seemed compatible with his proliferating engagements. Unlike Lyell, who was secretly anti-clerical, Phillips did not have to dissimulate. He was an Anglican, many of his associates were priests, both Anglican and dissenting, and he agreed with King's liberal Anglicanism. As for London there were esteemable people there but the place in 1833 had struck him as having no sympathy and was pervaded by 'din and dust and smoke'.[5]

Phillips's course of eight lectures, given in spring 1834 at one a week for seven weeks, generated £29 from the class fee of 1 guinea (see Appendix 2 for his lecturing in London in the 1830s). He focussed on the classification of stratified rocks treated historically from most to least ancient, on subterranean heat which had made the earth fluid, and made no mention of Mosaic geology.[6] Next year he wanted to give nine lectures in three weeks but reluctantly agreed to deliver eight in less then four weeks. He then told KCL that he was so hard at work on his 1836 monograph and BAAS affairs that he wished to postpone his lectures to the following academic year. In response Henry Smith, secretary to the College Council, told him that interests of the public and the College were paramount so that he had to make a personal sacrifice. Phillips dutifully delivered his eight lectures, which examined the important question of the interior temperature of the earth in ancient and modern times and reached the anti-Lyellian conclusion that the cooling of the earth had been responsible for huge and violent dislocations of land and sea. Though the class numbered only a dozen, it included Baron Parke and the historian Henry Hallam [7]

In April 1836 Phillips told King's that he had no time to give long courses and that, as his chair was merely ornamental, he was forced to adapt his few lectures to extra-collegiate auditors. Later that month he withdrew from his proposed course on fossils, because of slight bleeding in his lungs, while confessing that he intended to visit London on other business. It is surprising that by June he was fit enough to give a lucrative quadruple course of lectures in Manchester and Liverpool.[8] In 1837 he reduced his lecturing at King's to six lectures given in sixteen days in May but took the precaution of advertising them more widely than before. His popular and philosophical review of the study of fossils attracted only thirteen payers but drew clerics such as D'Oyly and William Otter, first principal of King's, and such leading geologists as Lyell, Murchison, Buckland, Whewell, Lonsdale, Samuel Pratt, and Robert Hutton, who were keen to hear Phillips' views on the analogies and differences of present-day and ancient organisms, the succession of life on the earth, and the successive conditions of its surface.[9] Next year Phillips gave in May and June an elementary course of just eight lectures.[10]

In late 1837 and 1838 various schemes, which involved Phillips playing a less peripheral role at King's, were mooted. Daniell and John Royle, professor of materia medica, wanted Phillips to give a long and elementary course which would include much mineralogy which they regarded as the ABC of geology. By April 1838 Thomas Bell, professor of zoology, had secured Phillips' 'cordial co-operation' in a proposed scheme for systematic instruction in natural sciences. The plan, which was warmly backed by Daniell and Royle, involved the teaching of seven subjects by a strong team: chemistry (Daniell), geology (Phillips), botany (David Don), zoology (Bell), experimental philosophy (Charles Wheatstone), general mechanics and astronomy (Henry Moseley), and natural products as applied to the arts (Royle).[11]

Phillips soon forgot about this scheme because in autumn 1838 King's launched a successful class in civil engineering and mining, which involved an extensive course by Phillips on geology, mineralogy, and the theory and practice of mining. King's College, London, expected and announced in its prospectus of October 1838 that Phillips would teach this course to civil engineers in early 1839 as well as his usual course on geology in early summer. Though Phillips had complained about his chair being peripheral, he was unenthusiastic about it becoming less so. In late autumn he pleaded that he had other avocations, imperative duties, and overwhelming heavy and anxious pursuits. He claimed that it had been agreed in spring 1838 that he would give a limited course of twelve lectures to civil engineers who would hear nothing about the working of mines and the art of mining because he had neither the time nor the practical knowledge. Ignoring his training as an engineer and mineral surveyor under William Smith, he stressed to King's that he had studied mines but only as a geologist. He revealed that 'pure geology is my subject; courses of arranged *lectures* suit my taste; but I have neither health, nor time, nor habits of mind for *continued teaching*'. Having made this surprising professorial confession, Phillips advised King's that the new course should be split into three parts: geological mineralogy; geology in relation to the theory of mining to be taught by himself; and mining operations by a practical man. Two of these options were implemented in session 1838-9: James Tennant taught geological mineralogy; and in spring 1839 Phillips delivered twelve lectures, thrice weekly, mainly on geology with reference to its applications in mining, architecture, and engineering.[12]

Phillips was aware that his attitude to the new venture would be construed as unsatisfactory and that his own position as professor was unsustainable. Though Daniell wanted Phillips to stay at King's and drop other things, in February 1839 Phillips told him, Wheatstone, Thomas Hall (the instigator of the engineering and mining scheme), and John Lonsdale, the new principal, that he intended to resign. Phillips' lectures to his 'mining class' did not pay his expenses but they attracted thirty-two students so Lonsdale tried to persuade him to change his mind. In August 1839 Phillips formally resigned from his King's chair, adducing as his main reason overexertion caused by too many widely extended employments. He preferred temporary repose in the country, with the Geological Survey, to London; he wished to recover his strength and to devote more effort to the advancement of science. Though Phillips was aware that his chair was not a bed of roses in 1839 when theological contests raged, the main reasons for his resignation were anxieties about lecturing, money, and his health. Like Lyell, his predecessor at King's, he cut the cockneys in order to devote himself more to research and to address his geological peers with publications. Again like Lyell he usually gave eight to twelve lectures per year. Neither approached the heavy teaching load undertaken by Phillips' successor, David Ansted.[13]

In 1838 and 1839 Phillips spent more time on freelance lecturing in London than on professing at King's. He used his post there, where his lectures were always given in the minimum period, as a platform for lecturing at two important metropolitan venues, the Royal Institution founded in 1799 in Mayfair and the London Institution established in 1805 in the City. In 1838 he dovetailed his eight lectures at King's with twelve on geology which he gave in afternoons at the Royal Institution from late February to mid-June at the rate of one a week, with gaps when he was lecturing in Bristol. In April and May he also gave six evening lectures on geology to the London Institution at the rate of one a week. These invited performances were so successful that in spring 1839 he returned to the Royal and London Institutions. As before he dovetailed the performances there with his course at King's. Indeed on Fridays in March 1839 he lectured at 3 pm at King's and at 7 pm at the London Institution. To the Royal Institution he gave six lectures on his favourite topic of the heat of the earth as a cause of geological change and to the London a survey course on geology. These lectures brought him cachet, decent financial rewards of 80 and 50 guineas for his courses at the former, and the friendship and advice of Faraday with whom he had been on amiable but not close terms from 1833 via the British Association. It was Faraday, an exemplar of unremitting work, who stressed to Phillips in 1839 that for the sake of science and his health he should reduce his lecturing. Like others Phillips revered Faraday for his moral purity and straightforwardness. In contrast Phillips was presumably puzzled by Darwin who went out of his way in March 1839 to attend just one of Phillips' lectures at the Royal Institution.[14]

On his visits to London as a lecturer at King's College, the Royal Institution, and the London Institution, Phillips enjoyed its intellectual and social opportunities. Naturally he gave preference to the Geological Society, attending its meetings and its anniversary dinner when he could. Like a long absent relative Phillips came up occasionally from the provinces and was not overawed. Though he gave no paper, he liked the vigorous discussion of papers, then unique in London's scientific societies. Such urbane geological skirmishing, conducted with manly vigour and usually tempered by good will and manners, appealed to Phillips so he contributed to it. For instance after Darwin's paper of March 1838 on earthquakes, volcanoes, and the gradual elevation of mountain ranges, Phillips contended for greater pristine forces than Darwin or Lyell accepted and was bombarded by Darwin who opened up his whole battery of arguments. Phillips also heard Darwin's famous paper on coral reefs given to the Geological Society in May 1837. When in London Phillips was invited to breakfasts, dinners and soirées, mainly by geological cronies and acquaintances such as Murchison, Yates, Lyell, Gray, Babbage, and Robert Hutton. He liked such events because at the least they showed he was respected as an author by London's leading geologists; at best they were rich in contacts and friendships. For instance, at one of Lyell's soirées in 1835 he met Babbage, Mantell, De la Beche, William Saull, Lord Cole, Francis Baily, Sir Philip Egerton, Thomas Weaver, William Lonsdale, and Sir Woodbine Parish. In 1839 he dined at Murchison's with Buckland, Lyell, Hopkins, the sculptor Francis Chantrey, Charles Stokes, a financier and collector, and the Earl of Selkirk.[15]

Phillips took advantage of his fellowship of the Royal Society of London, without giving much to it. He attended the fashionable presidential soirées given by the Duke of Sussex and the Marquis of Northampton. In 1839 Phillips heard two famous papers given to it. He liked Darwin's theory of land elevation in his account of the parallel roads of Glen Roy and was intrigued by William Talbot's new invention of 'photogenic drawing'.[16] Phillips also found time to listen to a few lecturers in London. At the United Service Club he heard William Harris and at the Royal Institution his friend Mantell, whom he thought flowery and a malcontent. In March 1839 he attended more than one lecture on scripture and geology given by John Pye Smith, a Congregational minister. Generally Phillips was impressed by Smith's courage in defending geology against its detractors. Not unexpectedly little of Phillips' social life was centred on King's College, though he did dine with D'Oyly and attended the farewell dinner for Otter.[17]

5.2   The popularisation of geology

From 1834 Phillips published a series of geological works which were primarily addressed to a general audience though they were not uninteresting to his geological peers. They established his reputation as an expert capable of popularising syntheses of his subject. The *Guide*, an independent work published by Longman, appeared in 1834 and was so well received that second and third editions followed in 1835 and 1836. The remaining works were all published as encyclopaedia articles. In 1835 Phillips' long article on geology, co-authored with Daubeny and written for Edward Smedley's *Encyclopaedia metropolitana*, was separately published. In 1837 and 1839 Phillips produced his two-volume *Treatise* written for Lardner's *Cabinet cyclopaedia*. In 1838 there appeared another *Treatise* which was composed for the seventh edition of the *Encyclopaedia britannica*. From 1838 to 1843 Phillips published anonymously a series of short articles on geology in the *Penny cyclopaedia* of the Society for the Diffusion of Useful Knowledge. While professor at King's Phillips not only established himself as a major writer of geological books alongside Lyell and De la Beche but also, as a prolific expositor of the science in no fewer than four different encyclopaedias, he surpassed every other British geologist. In these publications he publicised himself: Darwin did not like the way in which Phillips referred to them, as if saying you must, ought, and shall buy everything I have written. Phillips was thus an important contributor to the new practice of popularisation developed in the 1830s just when the term 'scientist' was created. Such popularisation, which was not orientated to an existing syllabus or to passing examinations, as text books were, took advantage of the mechanisation of book manufacture which led in that decade to a drastic cheapening of books, to mass medium publication, and what Secord has called 'steam reading'.[18]

Phillips revealed his expert synthetic ability not only in words but in his Index geological map of the British Isles, engraved by Lowry and published commercially in 1838. Not a hack compilation, it involved decisions about geological boundaries and the colours to be given to different geological formations. Phillips knew that the forthcoming second edition of Greenough's large-scale map would be expensive and would cover only England and Wales so he compiled a cheap, convenient, and simplified map of the British Isles on a scale of about thirty miles to the inch. It drew not only on his own observations and on published maps but also on the private communications he had solicited from other fellow geologists. For Ireland he was greatly indebted to Griffith who allowed him to inspect the unpublished third edition of his geological map of Ireland. Phillips' Index map was praised for its general accuracy and was affordable: whereas Greenough's map of 1840 sold at £5 to the public, Phillips's cost 10/- on paper, 13/- if mounted in a case, 16/- if varnished on black-rollers, and 18/- on mahogany rollers. Phillips updated it via several editions, the last of which was published in 1862.[19]

Phillips had been encouraged by friends who were impressed by his lectures to write an elementary book on geology. He knew that Lyell's *Principles* was not an introductory text. It was concerned with the regulative notions of geology. Published in three volumes between 1830 and 1833, it cost 47 shillings. There was scope, therefore, for an-up-to date and cheap account of the subject which Lyell had raised to a new level of sophistication. The *Outlines* by Conybeare and William Phillips was beginning to show its age but it was becoming sadly clear that Conybeare was incapable of revising it. In contrast Bakewell's popular *Introduction to geology* had been revised: the fourth edition, running to over 600 pages and costing a guinea, had appeared in 1833; but it was indifferent to palaeontology and the reconstruction of the earth's history. De la Beche's *Geological manual*, which Phillips thought an unrivalled reference work, was first published in 1831 and cost 18 shillings. It was so popular that a third edition, occupying more than 600 pages, appeared in 1833. Addressed more to the student than the accomplished geologist, it was strong on palaeontology where Bakewell was weak and on the secondary strata which Lyell had virtually ignored. In these respects Phillips' *Guide* followed De la Beche's *Manual*. It differed in being much shorter, smaller, and cheaper: published as a portable volume, its 139 pages fitted into a pocket and cost 5 shillings. It capitalised on Phillips' experience as a lecturer in expounding the elements of his subject and was in part designed to be used as a textbook by members of his classes. It offered a dense summary of the ascertained inductive results of geology and was judged to be the first successful 'grammar of geology' in English. In revising the first edition Phillips received detailed useful advice on style from his friend John Kenrick and was indebted to Murchison for many useful suggestions. With a print run of 1,000 for the first edition and 1,500 for the second, the *Guide* was a popular work which sold 100 copies in Newcastle in autumn 1834 when Phillips was lecturing there, and in Edinburgh it was in such demand among natural history students that it sold out. De la Beche generously regarded it as a standard work. So, too, did the publisher, John Murray, to whom Phillips had offered it unsuccessfully before turning to Longman. The *Guide* aided the career of Lowry as a geological illustrator. The frontispiece of the second and third editions showed in three dimensions the geology of the Isle of Wight, drawn by Phillips and lithographed by Lowry who had done the plates of the first edition (Fig. 5.1). Its origin was Lowry's and Phillips' plaster model of Wight, geologically coloured and divided transversely to give a coloured geological section through it. Through this collaboration Lowry was confirmed as Phillips' favourite lithographer and Phillips became the geologist closest to Lowry.[20]

The *Encyclopaedia metropolitana*, edited mainly by Smedley and designed according to a plan originally suggested by the poet Coleridge, was published in twenty-nine volumes from 1817 to 1845. Charles Daubeny, who was an expert on active and extinct volcanoes, was invited to write on geology; but, feeling his knowledge was limited to igneous rocks, he asked Thomas Webster in May 1828 to collaborate with him. When Webster withdrew from the arrangement, Daubeny approached Phillips whom he met in York in April 1829, having been introduced by letter by Buckland. Phillips' monograph on the Yorkshire coast, just published, showed that he was an expert on fossils, sedimentary strata, and the history and principles of geology. In June 1829 Phillips agreed to be Daubeny's co-author not least because he believed that learning should not be shut up in cloisters and confined to one class of society. The two men soon became close friends: Phillips stayed with Daubeny in Oxford in June 1831 and worked with him in the early days of the British Association (Fig. 5.2). In the event many of the 280 quarto pages of the text of their joint article were by Phillips, with Daubeny writing only on igneous rocks. The

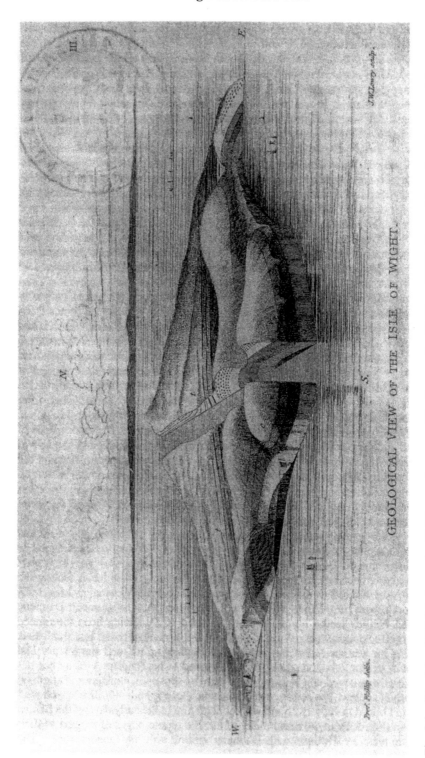

Fig. 5.1 The Isle of Wight, drawn by Phillips and engraved by Lowry, for Phillips, 1836b. It combines cleverly an aerial topographical view, a coastal section, and via a north-south slice a traverse section.

article was up to date, with references to mid-1834, and it covered not just Britain but Europe. It devoted many pages to secondary strata to which Lyell had given only three pages in the third volume of his *Principles*. It contained many big tables

Fig. 5.2 Charles Giles Bridle Daubeny, an engraving of 1836 by M. Haughton. Known affectionately as 'Little Dubs', Daubeny has too often been underestimated as a steadfast though temperate reformer at Oxford.

of fossils of each geological formation. It offered seven plates engraved by Lowry. Three were devoted to volcanic phenomena, two to fossils, one to the possible fates of strata, and one drawn by Phillips to veins of igneous origin cutting through strata. The article was completed by two geological maps constructed by Phillips, one of the British Isles and the other of Europe. Both were dated 1833, not coloured, engraved by Lowry, and lacked sections.[21]

It was presumably Phillips' success as a contributor to the *Encyclopaedia metropolitana* which led to his composing treatises for Lardner's *Cabinet cyclopaedia* and then for the seventh edition of the *Encyclopaedia britannica*. In writing for the former, Phillips was one among many: when completed in 1846 this huge publishing venture of Longman, begun in 1829, systematised scientific information in no fewer than 61 titles. Presumably Phillips met Lardner through the British Association in which Lardner was prominent as a vigorous disputant and flamboyant lecturer. By 1838 they were sufficiently close for Phillips to be listed as a principal contributor to the *Monthly chronicle* which Lardner and Edward Bulwer-Lytton ran until 1841. Generally Lardner wanted what he called profound men to write intelligible and philosophical works even more popular than contributions to the *Encyclopaedia metropolitana* but Phillips did not fully oblige him. His *Cabinet cyclopaedia treatise*, which ran to over 600 octavo pages and cost 12 shillings, was a comprehensive and in parts quite technical exposition of considerable interest to accomplished geologists as well as to beginners. It offered fuller discussion with more illustrations and diagrams than the *Encyclopaedia britannica treatise* which was half as long.[22] Both Phillips and Adam Black, the publisher of the *Britannica*, wanted separate publication. Phillips did not wish to see his piece on geology buried in the *Britannica* while Black hoped that a volume costing six shillings would prosper as an Edinburgh University textbook.[23]

The *Penny cyclopaedia* was one of the main publishing ventures of the Society for the Diffusion of Useful Knowledge which had been founded in 1826 by Henry Brougham and, though avowedly non-partisan, was very much a whig venture. Among the sorts of knowledge which it deemed useful was science, a form of learning which it tried to spread to middle and lower class readers. It aimed to raise the standards of popular publishing in general; and in its *Penny magazine* and *Penny cyclopaedia*, both began in 1832 and published weekly at one penny, tried to remove any financial barrier to the acquisition of useful knowledge. It seems that Fitton and then De la Beche were lined up to write on geology for the SDUK's Library of Useful Knowledge but, given their failure to deliver, the Society turned to Phillips to write for the *Penny cyclopaedia*. He accepted readily. Politically he was not hostile to the Society's whiggery and he supported its aims. In 1834 he had joined his close friend Kenrick to form the local committee of the SDUK at York. Their commitment to the Society was so strong that they remained the York committee until its demise in 1846. They saw the SDUK as analogous to the York Mechanics' Institute and the British Association in that all tried to spread knowledge and to raise the standard of mind of the public. Phillips was therefore happy to write for the *Penny cyclopaedia* a twenty-four page article on geology (1838) and short but telling pieces on organic remains (1840), palaeozoic series (1840), the parallel roads of Glen Roy (1840), rocks (1841), the saliferous system (1841), the Silurian system (1842), stratification (1842), and the geology of Wales (1843). He also contributed palaeontological pieces on goniatites (1838), polythalamacea (1840), and turrilites (1843). All these articles, especially those dealing with the terminology of geological systems, were written

without a trace of *de haut en bas* and therefore demanded some prior knowledge
on the part of the reader. They satisfied the desiderata of the editor, George Long,
who wanted self-contained reference articles which included general views and
gave sound information in 'clear and perspicuous language'. Phillips' articles were
unsigned but he did not conceal that he was their author. By 1841 he was citing
some of them as showing *his* priority of discovery. Charles Knight, the publisher of
the *Penny cyclopaedia*, regarded Phillips as a most valuable contributor. For his part
Phillips presumably welcomed its steady sale of no fewer than 20,000 copies to the
biggest audience he was ever to reach.[24]

5.3 Lyell's principles of geology

There were many common themes running through the general works which Phillips
published between 1834 and 1840. Most but not all of these themes were reactions,
not always favourable, to the first edition of Lyell's *Principles*, the last volume of
which appeared in 1833. It makes considerable sense to see Phillips' general works
as codifying and strengthening the opposition of many leading British geologists to
some of the main tenets advanced by Lyell (Fig. 5.3). Like Sedgwick, Conybeare,
Murchison, Greenough, De la Beche, and Fitton, Phillips acknowledged that Lyell
had put geology on a new footing but also tried to ensure that there would be few
out-and-out Lyellians.[25] In these general works Phillips also revealed his view of his
uncle's achievement, his attitude to fossil-labelling of strata and geological systems,
and not least his position as an Anglican on the controversial Genesis-geology
question.

The subtitle of Lyell's *Principles* proclaimed that his book was 'an attempt to
explain the former changes of the earth's surface by reference to causes now in
operation'. His chief concern was therefore methodological, i.e. with the regulative
principles of reasoning which would make the new subject of geology truly scientific.
In pursuing this aim Lyell assumed that the only proper causes to be employed in
geological argument were those which in his time were visibly acting. He thus
confined causation to just those agents which humans had observed. In addition
to this restriction of kind he proposed one of degree when he proclaimed that only
the present- day intensity of observed causes should be used to explain past events.
Having rejected as uncontrolled speculation such notions as paroxysms to explain
large effects, Lyell invoked agents of present-day intensity which had produced a
succession of small effects over a very long time. He made the huge assumption, as
Sedgwick pointed out, that 'the physical operations now going on are not only the
type, but the measure of intensity of the physical powers acting on the earth at all
anterior periods'.[26]

In order to illustrate his recommended principles of method, Lyell offered many
examples and discussed major geological questions. Thus the *Principles* was
construed as a work of doctrine as well as method. Some of his tenets were highly
controversial. For instance, Lyell argued that the fossil record was incomplete. If so
the absence of fossils of mammals and birds in the oldest rocks did not mean that
they had not lived in the most distant times: it meant that no remains of them had
survived. Lyell used his argument, about the necessary incompleteness of the fossil
record, to criticise the view that it showed there had been a general progressive

**Fig. 5.3   Sir Charles Lyell, a lithograph of 1851 by T.H. Maguire. Even his
critics acknowledged that Lyell had an unrivalled synoptic and
penetrating view of his subject.**

development or continual improvement from simple to complicated forms of life: he
rejected 'progression' in the history of life on the earth.[27]

His approach to the question of past climate on the earth also contained interlocking
elements. Lyell accepted that in the past there had been warmer climates, as in the
carboniferous period when tropical plants flourished. Rejecting on methodological
grounds the theory of the cooling earth, he invoked changing distributions of land
and sea to explain changes of past climates. He was therefore critical of the notion
that there was a direction to the history of the earth. It is not surprising that Lyell's
opposition to progression in the history of life and to directionalism in the history
of the earth was construed as implying that he was advocating a steady-state or
cyclical system of geology. After all Lyell seemed to be arguing in favour of random
fluctuations about a mean when he proclaimed 'however constant we believe the

relative proportion of sea and land to continue, we know that there is annually some small variation in their respective geographical positions'.[28]

Lyell's concern to connect the present with the past was also manifested in his pioneering work on tertiary strata (those above and newer than chalk). Assuming that the rate of change in the living world had been uniform and using tables and identifications of fossil molluscs made for him by Paul Deshayes, whom he paid, Lyell dated the various deposits according to the percentage of living species they contained. Using this novel mode of fossil-labelling, he divided the tertiary period arbitrarily into the eocene, miocene, older and newer pliocene. The eocene contained about 3% of living species, the miocene 18%, the older pliocene between 35 and 50%, and the newer pliocene 96%. These numbers show that Lyell did not see these four epochs as contiguous: he viewed them as short isolated moments between which there were long periods of time of which no record had been preserved. Thus Lyell's classification of tertiary strata emphasised the incompleteness of the fossil record.[29]

5.4   Phillips' principles of geology

Apropos Lyell's prescriptions for geological argument Phillips agreed with Sedgwick and Conybeare that Lyell was too much an advocate and exponent of a priori reasoning. They all doubted whether the agents of the past were like present-day causes in intensity as well as kind. It seemed to them that an explanation in geology should invoke 'forces adequate to the effects and coextensive with the phenomena'. Yet, as Sedgwick stressed, 'Lyell will admit no greater paroxysms than we ourselves have witnessed – no periods of feverish spasmodic energy, during which the very framework of nature has been convulsed and torn asunder. The utmost movements that he allows are a slight quivering of her muscular integuments'. Phillips agreed: on the question of the intensity of past agents Lyell assumed what had to be discovered. Arguing from the nature of conglomerate rocks, the wide distribution of large erratic blocks, huge diluvial accumulations, and strata which were inclined, bent, shattered, and even retroflexed (as in the Malvern hills), Phillips inferred the existence of violent convulsions. His clinching piece of evidence was provided by faults in Britain, one of which (the Pennine fault) involved large dislocations 3,000 feet deep and 100 miles long. They showed no trace of gradual accumulation of many small movements: there were no fissures filled with an aggregate ground from materials on their sides. It seemed to Phillips that sudden and short-lived violent convulsions had occurred on several occasions in the past. Indeed he regarded as important Élie de Beaumont's research on the timing and direction of great convulsions in the earth's past, which was a far cry from Lyell's quiet earth approach.[30]

With respect to the fossil record, Phillips agreed with Lyell that for terrestrial vertebrates it was fragmentary and therefore unreliable; but for marine invertebrates it was a tolerably good if incomplete guide because each present-day genus was represented in it. On the question of overall and broad progression in the forms of past life, Phillips was more sceptical than many of his contemporaries. Like Lyell he welcomed the discovery of didelphis: it showed that mammals existed much earlier than had been supposed and destroyed the notion that an age of mammals had followed an age of reptiles. On the basis of the relative completeness of the fossil record of marine invertebrates, Phillips argued that it was a gross error, and a speculative violation of inductive geology, to assume that all earlier forms of life were less complex. The example of belemnites showed that, in a group of creatures of different geological age, curiously and highly organised forms existed in the past but not now. Phillips therefore concluded that 'the whole notion of a gradual amelioration or enrichment of the animal organisation may be dismissed as a mere illusion of the fancy of a finite being, who vainly transfers to the work of the Almighty the pattern of his own limited labours'.[31]

His own view was that systems of life were always adjusted to the actual and changing conditions of land and sea which had been providentially planned by God. These systems of life were part of God's scheme of successive creations, the last great one being of humans who had probably appeared, Phillips thought, in post-tertiary times. On the positive basis of the marine fossil record, he offered a palaeo-ecological interpretation which viewed past creatures, including the meaner forms of every geological age, in relation to their conditions of life which a caring God had arranged. He was keen to deny progression probably because, like Lyell, he felt that the notion led to belief in transmutation or evolution of species. Unlike Lyell, Phillips did not attack French transformism at length but he took pot shots at it. He rejected transformation, as expounded by St Hilaire and Lamarck, as a plausible but poetical conjecture: the fossil record showed that the appearance and extinction of species were sudden and not gradual. Given the overwhelming evidence in favour of adaptation in both past and present-day forms of life, Phillips believed in new and special 'creations' which ultimately were works of God. Like Lyell he worried about the appearance of new species in past time but unlike Lyell gave an explicitly providential view of the phenomenon.[32]

On the explanation of change of climate in the past and on directionalism in the earth's history Phillips was at odds with Lyell. Phillips thought Lyell's theory of climate change was elegant, consistent, not improbable, and invoked real causes, but it was not sufficient to explain the magnitude of the effects. For decades Lyell felt that Phillips had slighted a key theory of the *Principles*.[33] Apropos the question of direction in the earth's history, Phillips advocated strongly the theory of the refrigeration of the earth which he saw as decidedly anti-Lyellian. It was incompatible with 'the uniform intensity of natural agencies taken separately, the continual compensation of their antagonist effects, and the production of equal effects in equal times'. By the late 1830s Phillips was a convinced believer in the theory of the progressive cooling and contraction of the earth, a physical theory advanced in mathematical form by Fourier and Hopkins. Phillips hoped that it would become a general mathematical theory, like that of gravity and the wave theory of light, from which 'characteristic phenomena in the real order of their succession' could be deduced. He even believed in the slow decrease of the heat of the earth, and its contraction, as the primary law of causation in geology because the resultant tension in the crust of the earth had led to a series of collapses upon the hot fluid nucleus of the earth. When that happened, elevation, subsidence, fracture, and displacement were the results. Phillips' obsession with progressive cooling of the earth led him to denounce an inconsistency in Lyell's theory that there was an equilibrium between the internal heat of the earth, which had melted sedimentary aggregate, and the external solidifying agents of the earth: 'this speculation is much too poetical to be examined according to the dry rules of the Baconian philosophy' because heat expended over Lyell's 'indefinite time' would have led to the gradual refrigeration of the earth, a consequence vehemently denied by him. In any event, Phillips rejected what he viewed as Lyell's postulate of 'unlimited duration'. Instead, Phillips quickly accepted the theory of cosmic evolution and progress popularised from 1837 by John Nichol who believed that the earth had been created when nebular expansion led to the condensation of planets: it had not existed always.[34]

Phillips welcomed Lyell's work on tertiary fossils for its peculiar power and interest but he criticised it for ignoring the diversity of local conditions of deposition in the tertiary period; and he rejected the assumption of the uniform rate of change in living forms as neither an inductive inference nor a mathematical principle. Phillips had first expressed his doubts about Lyell's percentage method in 1835 to the British Association: though in principle attractive, it was subject to considerable error in practice. This was a leading point in the assault launched by Edward Charlesworth in 1835, and continued in 1836 and 1837, on Lyell's view that the Suffolk crag was one formation which belonged to the pliocene period. In his attacks Charlesworth, egged on by Phillips, invoked him as an authority for this crag being composed of two formations, the coralline and red crag, probably of miocene age. For his part, Lyell saw Phillips as an accomplice of Charlesworth in the crag controversy. When Lyell at last ventured into Suffolk in 1838 to see the exposures, he soon accepted Charlesworth's view. Phillips noted the irony in Lyell's concession: Lyell's new position on the Suffolk crag was that of Charlesworth, who from detailed field work had strenuously maintained it against Lyell, 'who is now of course convinced by *his own researches*'.[35]

Phillips believed that geology should be an observational science which generated reliable facts from which limited generalisations could be cautiously inferred. He stressed the difference between positive geology and hypothesis: the paths of observation were hard and difficult whereas those of hypothesis were easy and inviting. Accordingly he saw geological theory as 'the summit of a cone whose base continually enlarges to include every known fact appertaining to the subject'. He was convinced that it was desirable to try to obtain by induction descriptive laws of phenomena before searching for laws of causation. In his advocacy of induction he was not a naïve empiricist because he invoked the active agency of the mind. In his view the mind operated on facts and transformed them into phenomena; it generalised phenomena into laws; and it interpreted the laws of phenomena to produce theory which in geology often involved duration and causation. Not surprisingly Phillips joined De la Beche in judging that too often Lyell offered premature and unsubstantiated hypotheses. Like Sedgwick Phillips was convinced that natural agents, acting with varying intensities in different combinations in varying conditions, had produced extremely diverse effects. It was especially these varying conditions which for him made Lyell's a priori doctrines as 'baseless as the fabric of a vision'. Lyell, who claimed to have established the principles of geology, was for Phillips a deviant from inductive geology and indeed a cosmogonist like William Whiston whose theory of the earth was a reverie. Presumably the sobriquet cosmogonist irked Lyell who in his *Principles* had argued that to conflate the aims of geology and cosmogony was a 'common and serious source of confusion'. Given such palpable hits, it is surprising that Lyell, who had a highly developed sense of *amour-propre*, was on good though guarded working and social terms with Phillips. Of course Lyell pumped Phillips for information and at times regarded him as a gauche but clever provincial of lowly origin; but he respected Phillips as a geologist and saw him as a formidable literary competitor. It is not entirely fanciful to see Lyell's *Elements* as his response to the success of Phillips' general works and of those of De la Beche whom Lyell saw in 1832 as a low-grade compiler of cheap books and a prospective plagiarist. For his part Phillips dismissed Lyell's *Elements*: it contained nothing which was not in the superior *Principles* so he did not buy it because it was of little service to him.[36]

In his general works of the 1830s Phillips covered the related questions of the achievement of Smith, fossil-labelling of strata, and definitions of geological systems and formations. Phillips remained loyal to his uncle, presenting him as a pioneering mapper and as the founder of inductive geology whose ideas had been illustrated by Greenough's map, Buckland's *Reliquiae*, and Conybeare and W. Phillips' *Outlines*. Phillips stressed that Smith had not only used fossils as markers to discrimate the strata and their succession; he had also perceived strata as the beds of past seas, thus stimulating work on the history of successive geological periods, and had introduced in England the notion that each stratum was a 'museum of that age of the world, containing a peculiar suite of organic exuviae, the remains of the creatures then in existence'. Phillips was here claiming that Smith, like Cuvier, was a historically minded palaeontologist interested in the succession of life on earth.[37]

Phillips also went out of his way to preserve and to refine Smith's notion of fossil-labelling. By 1836 Phillips appreciated that it was only locally true that strata of different ages contained 'distinct races' of fossils and he accepted that in a given stratum there were local differences in the character of the fossils. Thus the use of a few allegedly characteristic fossil species to identify strata was vitiated, so Phillips turned for help to the statistical tables he had compiled for Smith. Phillips drew up and published tables of genera of fossils present in five different geological systems, showing for example that terebratulae, a genus of lamp shell, existed in all five, but was fecund in the cretaceous system with fifty-seven species but less so in the saliferous system with only fourteen. It followed that strata could be identified by the plenty or paucity of coexisting genera expressed in numbers. Phillips' statistical palaeontology, he hoped, overcame the problem that some strata, widely separated geographically but apparently contemporaneous, contained few common fossil species. Phillips knew that the cretaceous systems of America and Europe had almost no characteristic fossil species in common, a contrast which he explained by invoking different local environmental conditions. By 1836 Phillips had used a Smithian technique to refine his uncle's notion of fossil labelling of strata. Having concluded that so-called characteristic fossil species and a few fossil species of one genus were unreliable, Phillips' answer was to use what he called in 1840 *'characteristic combinations of organic life'* embodied in statistical tables showing the preponderance of genera in various strata and geological systems.[38]

The notion of a system was crucial in geology in the early nineteenth century because it was a taxonomic term for a set of rock formations. It generated considerable controversy because ordering the rocks linguistically raised fundamental problems of method and offered many opportunities for geologists to disagree about intellectual procedures, property, and sometimes priority. The 1830s saw the creation of three important geological systems, the Cambrian announced in 1835 by Sedgwick, the Silurian also in 1835 by Murchison, who described it in detail in his huge monograph of 1839, and the Devonian redefined by Sedgwick and Murchison in 1839. This trio of systems classified the rocks immediately below the carboniferous strata as Devonian, Silurian, and Cambrian in descending order in the stratigraphical column. In 1841 Murchison introduced the notion of a Permian system immediately above the carboniferous strata. It is not therefore surprising that Phillips, a leading expert on carboniferous limestone, was particularly interested in the nomenclature of geological systems.

Just before the announcement of the Cambrian and Silurian systems Phillips had provided a general definition of a system: it contained certain recurrent rocks and similar types of fossils, it showed striking analogies of composition, and was apparently derived from convulsions of the same epoch. In the secondary strata there were therefore four systems, the cretaceous, oolitic, saliferous (containing new red sandstone and magnesian limestone), and carboniferous (containing coal, mountain limestone, and old red sandstone). His neologism of saliferous was contentious because there were interruptions in the English sequence of rocks which appeared to indicate great and sudden changes. Characteristically, Phillips argued that these interruptions were mainly local and arose from the absence of strata which were widely deposited in continental Europe. By 1838 Phillips had concluded that classifications of secondary strata into systems were at best only locally true so it was hazardous to argue, as Murchison had, that his Silurian rocks as they occurred on the Welsh border constituted a general system: far more research on other districts was needed to give to a local and geographically limited classification those comparative and comprehensive elements which would transform it into a system. Thus before Murchison's magisterial *Silurian system* was published, Phillips had two doubts: it was only locally applicable and, in using fossil species as the best way of characterising a system, Murchison underestimated the limitations of the procedure.[39]

5.5   Genesis and geology

Phillips was a convinced Anglican so he had to consider privately the question of the relation between geology and scripture. It was, however, impossible for him to be silent on this matter because the various posts he occupied made it incumbent on him to overcome his reluctance to declare his position in public. In York, an archiepiscopal city, his chief patron was Harcourt, an Anglican minister, and his chief irritant was William Cockburn, dean of York Minster and a scriptural geologist. He was a member of the Yorkshire Philosophical Society of which Phillips was secretary, a generous contributor (£50) to the building fund of the Yorkshire Museum of which Phillips was keeper, and a facilitator of Phillips' rainfall measurements made at the top of the Minster. As a believer in a young earth only 6,000 years old, Cockburn taunted Phillips in the York press and elsewhere about his belief in the great antiquity of the earth. Phillips was nationally prominent as assistant secretary of the British Association, in which geologists and their section (C) were popular. From its early days the Association was attacked by scriptural geologists, including Cockburn, as a mouthpiece for irreligious geology which undermined revelation and encouraged infidelity. Finally, as professor of geology in the avowedly Anglican King's College, London, Phillips had to consider how its motto, sancte et sapienter, illuminated the genesis and geology question.

After 1831 Phillips divorced himself entirely from scriptural geology and adopted a liberal Christian position, an inadequate term which avoids some of the difficulties of the notion of liberal Anglicanism.[40] Phillips was a leading British geologist and practising Anglican who had learned from experience that Catholics and dissenters, especially Unitarians, could be as good Christians as Anglicans. He was saddened by doctrinal sectarianism between Christians and within the Church of England. He wanted an irenic, accommodating, and intellectually tolerant church, so he was not a member of any church party. He therefore supported the way in which liberal Christians asserted the mutual autonomy of science and theology yet hoped for, accepted, or insisted on the ultimate congruence of natural with revealed religion. They believed that the books of nature and scripture were in ultimate accord, because their Author was the same, but that it was futile to try to reconcile geology with scripture or scripture with geology. This view gave geologists the liberty to pursue their science with total independence while encouraging them to reveal evidence of design and a designer in the natural world. It was hoped that geology, via natural theology, would corroborate or strengthen Christian belief. These views were espoused, wholly or partly and with varying degrees of enthusiasm, by progressive dissenters and middling Anglicans, including those with evangelical leanings; and they were denounced by scriptural geologists who enjoyed a vigorous phase in the 1830s and 1840s. Generally they believed that the Mosaic record irrefutably affirmed the creation of the universe in six days about 6,000 years ago; and they viewed any separation of science from religion as heresy.

Unlike his friends such as Baden Powell, Sedgwick, Buckland, and Conybeare, all Anglican clergymen, and his lay chum Daubeny, Phillips did not publish polemics about the relation between geology and scripture. Unlike his Unitarian clerical friends Kenrick and Hincks, he neither preached nor wrote about Biblical authority and scriptural Christianity. Unlike his Anglican clerical friends Whewell and Buckland, he did not publish Bridgewater treatises on astronomy and geology considered with reference to natural theology.[41] But from remarks in his general texts of the 1830s and his correspondence his position as a liberal Christian may be inferred.

Phillips found no antagonism between science and his religion and avoided collision between them. He thought it dangerous to confound the independent bases of religious and natural truth: the former rested on 'moral evidence and the nature of man; the latter on physical facts and the sure laws of nature. Both are true and cannot disagree, but we must know them both well before we attempt the serious task of determining the manner of their union'. Though Phillips was usually placid on such matters, occasionally he was vehement. He judged it 'a dangerous theological error to put in unequal conflict a few ill-understood words of the Pentateuch, and the thousands of facts which the finger of God has plainly written in the book of Nature'. Again occasionally he scorned in public the scriptural geologists for their blind and ignorant opposition to the progress of inductive geology, based on an erroneous view of the true meaning of the scriptures. Clearly he agreed with his friend Sedgwick that 'to confound the ground-works of philosophy and religion is to ruin the superstructure of both'. Unlike Sedgwick, Phillips did not scourge the scriptural geologists in public, but privately he deplored Cockburn's attack on Buckland's popular Bridgewater treatise.

> Christianity was not based on geology, will not be overthrown by it; the Mosaic cosmogony is one thing, and right or wrong, Christ's sermon on the mount never alludes to it; geology is another thing equally independent of Christian faith, practice and evidence ... Geology is a science resting on sure physical evidence; Christianity is based on moral evidence; the Mosaic cosmogony will not for an hour arrest geology: geology will not for a moment lay a real impediment in the path of religion. My opinion is that a few such essays as the Dean's will do more harm to the church than an army of infidels.

Thus Phillips not only agreed with Sedgwick's attack in his *Discourse* on the 'pseudo-Mosaic school of geology', which Sedgwick had advised Phillips to read; Phillips also regarded it as '*the height of impiety*' to denounce as dangerous the study of God's plan of creation via geology, which for him gave results in harmony with high and true views of God and religion.[42]

It is not surprising that in early 1839 Pye Smith, a Congregationalist minister who was about to give lectures on the relation between scripture and geology, turned for aid to Phillips, who helped him. Also drawing on the works of Phillips, 'an eminently accomplished geologist', Pye Smith's aim was to defend geology from those who believed in a literal interpretation of Mosaic history. Though Pye Smith made some gestures towards reconciling geology and scripture, in general Phillips approved his attempt to curb 'the balaam bellowers' so he not only answered Pye Smith's queries but he also attended some of the lectures. He thought Pye Smith's published lectures were doing good because they showed the futility of misguided divines going outside their province and trying to frame geology out of the Pentateuch of which they took a false view.[43] Phillips advised another clergyman in 1839 about the relation between science and religion. He urged Harcourt, president of the British Association in 1839, to concentrate in his address on the pre-Adamitic phases of the earth and organic life. Presumably to Phillips' disappointment Harcourt ignored his advice and did not entirely reject a reconciliatory position. Harcourt contended that there were 'certain common points in which reason and revelation mutually illustrate each other; but in order that they may ever be capable of doing so, let us keep their *paths* distinct, and observe their *accordances* alone; otherwise our reasonings will run round in a circle, while we endeavour to accommodate physical truth to scripture, and scripture to physical truth'.[44]

Following Sedgwick's approach Phillips believed that humans were geologically recent and that geology had to itself the pre-human history of the earth because on that subject scripture was silent. Thus the pre-human period of geology came neither within the spirit nor letter of revelation. It followed for Phillips that there was nothing in Genesis which limited the inferences or even speculations of geology about the vast pre-human periods which could be studied only by inductive geology. He therefore rejected the attempts by scriptural geologists to date the creation of the earth which they assumed to be about 4000 BC. One of them, George Young, estimated in 1838 that 1656 years had elapsed between creation and the flood. For Phillips the fixing of the earth's chronology was a vain hope but he had no doubts about the vast antiquity of the earth. The total thickness of sedimentary strata (about ten miles) and the nature of conglomerate rocks convinced him that it was folly to contract 'the long periods of geology into the compass of a few thousand years'. It was Phillips' insistence on the 'long succession of time elapsed during the construction of the visible crust of the globe' which in 1838 provoked Cockburn to initiate a correspondence with Phillips in a York newspaper about the antiquity of the earth, its state at its creation, and its condition 10,000 years before the birth of Adam. In 1840 Phillips was forced by Cockburn into a second newspaper war, this time about what Phillips had said at the 1840 meeting of the British Association about the vegetable origin of coal and whether it was formed *in situ*. In a pamphlet of 1840 Cockburn complained that 'the great professors Buckland and Phillips' were snobs who would not deign to reply to him because they, like French materialists in the Enlightenment, were chronicling the infinite time of the pre-Adamite world. He wittily asked 'How can we expect them to attend to an humble mortal of a pliocene period?' Phillips' reaction was generally to ignore Cockburn whom he thought a booby and an 'absurdity in canonicals' who produced slanderous trash.[45]

On the question of the Deluge Phillips followed the views of two clerical friends, Adam Sedgwick and John Kenrick. In his presidential address to the Geological Society in 1831 Sedgwick had recanted what he called his philosophic heresy when he renounced his former belief that gravel deposits had been produced by one violent deluge which he had equated with the Mosaic flood. His new position was that diluvial gravel resulted from many successive periods of violent aqueous action. This view was also held by Kenrick, a Unitarian minister and classical scholar from whom Phillips derived an optimistic view of the after life which was far removed from the evangelical obsession with death, judgement, and hell. He accepted Kenrick's belief that mental faculties, and the limited knowledge they had acquired from sense experience in a short terrestrial life, would be brought to maturity in the afterlife to produce a state of transcendental geological beatitude until the soul 'be reabsorbed into the great first cause of all'. In the 1830s Kenrick investigated Greek traditions of the deluge and was developing his view of Genesis as a mythic fiction which in popular belief was deemed literally true. For Kenrick it was futile and dishonest to try to make geology conform to scripture: for him diluvial phenomena were explicable only through the prolonged, repeated, and multiform operation of diluvial currents.[46]

Phillips' own field work led him to a similar conclusion. His further research on the gravel and erratic blocks of east Yorkshire indicated that there had been several inundations of which the latest was anterior to the agent responsible for erratics. In his general works Phillips urged that there had been several diluvial disturbances in a relatively short period of convulsion of land which had produced an agitated and elevated sea which flowed over dry and high land with great force and volume. These disturbances occurred after all marine strata had been deposited. By the late 1830s Phillips thought that sudden and energetic changes in the levels of land and sea had generated powerful oceanic currents capable of producing diluvium, but that this sort of explanation could be stretched only with difficulty to explain erratic blocks. In the case of the Alps, erratics had been transported, he thought, by floating icebergs formed by the breaking up of glaciers by violent disturbances in the Alps themselves. For erratics elsewhere, as in the north of England, Phillips was puzzled by the paradox that they had been moved considerable distances along lines which respected the present-day levels of the country apropos height and direction but that they had also sometimes crossed great vales and seas and ascended ridges, contrary to the course of existing drainage. It was a paradox that troubled him for years after the announcement in 1840 by Agassiz of his ice-sheet explanation of erratics.[47]

In his general works Phillips used suitable opportunities to present a theistic view of geology about which he had no doubts. For him the natural world was an expression of the supreme creator and lawgiver of the universe. Hence the study of geology, using the intellect conferred on humans by God, had important religious effects on the geologist: it produced awe of the glorious and beneficent author of nature and led to sublime communion with the creator of the universe. The fossil record showed the superintending care and providence of God and generated reverential thoughts of the divine lawgiver of nature. The earth's history revealed the power and wisdom of the creator and induced veneration of the universal cause. The convulsions which strata had undergone showed the creator's plan of unceasing care and comprehensive benevolence. The cooling earth was uninhabited by any form of life for a very long time until 'the unchangeable laws of nature relating to organic life, which are but the expression of the will of God, began to operate'. It seems that Phillips believed God had operated in the natural world through law-like secondary causes. In that sense nature was for him a mirror in which 'the Almighty and the Infinite is faintly typified in the vast and the diversified' works of nature.[48]

Phillips' public view of scripture was like that of Sedgwick and Conybeare: the Bible revealed the history of the dealings of God towards men. It seems that in private he agreed with Harcourt that the language of scripture was highly figurative in order best to convey transcendental truth. Hence the biblical account of the creation of the world was a fine poem rather than history. It is significant that in 1838 Phillips publicly aligned himself with Baden Powell, who argued in his *Connexion of natural and divine truth* that Genesis was not a historical narrative: it was written in the language of figure and poetry to convey the greatness and majesty of divine power. At Powell's request and to his 'peculiar satisfaction', Phillips provided an appendix on the geological evidence of former conditions of organised life and its unbroken succession. In this he said nothing about his view of scripture, confining himself to asserting that through geology one could rise to the highest human contemplation, i.e. the relation of humans to the unseen Author of all the visible material world. Even so his willingness to oblige Powell, the most outspoken critic of scriptural geologists and an 'advanced' Anglican, and to appear publicly in harness with him, confirms that Phillips accepted Powell's argument that ancient scripture was a parable for the Jews employing poetry and dramatic action. That was why in 1841 Phillips asked Kenrick whether his method of mythic explanation was applicable to Jewish history.

By 1838 Phillips had become a fellow traveller of Powell and Kenrick, who both thought reconciliation of scripture and geology was futile and interpreted ancient scriptures as poetry and myth. Perhaps like them he thought that the reconcilers too often equated Christianity with the cosmogony of Jews in the remote past, an identification that Powell exposed as absurd. The ancient scriptures had indeed delivered religion to the Israelites, yet the scriptural geologists were trying to make them supply astronomical and geological instruction to Christians. Unlike Powell, Phillips did not go out of his way to denounce Tractarians such as Edward Pusey or holders of biblico-geological views such as Pye Smith. His approach to Christianity was irenic and had much in common with Kenrick's view that a Christian should try to do what the gospel bids, to revere Christ, and to commune with as well as obey Christ and God.[49]

Notes

1. Phillips to Lubbock, 18 Sept, 8 Oct 1833, Lu P; Phillips to Sedgwick, 15 Nov 1833, Se P; Phillips to Murchison, 16 Nov 1833, MP; Certificates 1830-40, Royal Society London; John Edward Gray (1800-75), *DNB*, assistant zoological keeper, British Museum; William Clift (1775-1849), *DNB*, osteologist and curator of Hunterian Museum, London; John Taylor (1779-1863), *DNB*, mining entrepreneur, treasurer of BAAS 1832-61 and Geological Society 1823-43.
2. On early KCL see Hearnshaw, 1929, pp. 35-159; George D'Oyly (1778-1846), *DNB;* Robert Peel (1788-1850), *DNB,* home secretary, 1828-30; Charles Manners-Sutton (1755-1828), *DNB,* archbishop of Canterbury 1805-28; Arthur Wellesley, first Duke of Wellington (1769-1852), *DNB,* prime minister 1828-30.

3. Blomfield, 1831, pp. 14, 18; Charles James Blomfield (1786-1857), *DNB*, bishop of London 1828-56; Edward Copleston (1776-1849), *DNB*, dean of St Paul's 1828-49.
4. Rudwick, 1975a; Wilson, 1972, pp. 308. 340, 353-60, 376; Porter, 1982a, pp. 38-40; Scrope to Lyell, 12 April 1831, American Philosophical Society.
5. Lyell to Murchison, 22 Dec 1831, MP; Blomfield to Sedgwick, 7 Nov 1833, Se P; Buckland to Phillips, 21 Dec 1833, PP; KCL Council minutes, 17 Jan 1834; Phillips to Henry Smith, 26 Feb, 6 March 1834, KCL incoming correspondence; Phillips to sister, 1 June 1833, PP.
6. Phillips to Buckland, 23 April 1834, Bu P; Phillips to Harcourt, 26 April 1834, HP; *KCL calendar 1833-4*, pp. 128-30 for 1834 syllabus.
7. Phillips to Henry Smith, 13 and 23 April 1835, KCL incoming correspondence; Smith to Phillips, 25 April 1835, KCL letter book, 1834-43; *KCL calendar 1834-5*, pp. 167-9, for 1835 syllabus; Phillips to sister, 11 and 20 May 1835, PP; Henry Hallam (1777-1859), *DNB*, served on Council of Geological Society 1835-8.
8. Phillips to H. Smith, 16 and 28 April 1836, KCL incoming correspondence; Phillips to Forbes, 20 May 1836, FP.
9. Phillips to H. Smith, 15 and 22 April, 1 June 1837, KCL incoming correspondence; Phillips to sister, 11, 14, 21 May 1837, Monday [May 1837], PP; *KCL calender 1836-7*, pp. 125-6, for 1837 syllabus; William Otter (1768-1840) *DNB*, principal of King's College London 1830-6, bishop of Chichester 1836-40; Samuel Peace Pratt (1789-1863), Bath fossil collector.
10. Phillips to H. Smith, nd [fr 9 May 1838], KCL incoming correspondence; Phillips, lecture syllabuses, folder 12, PP, for 1838 syllabus.
11. Daniell to Phillips, 13 Dec 1837, Bell to Phillips, 11 April 1838, PP; John Forbes Royle (1799-1858), *DNB*, professor of materia medica 1837-58; Thomas Bell (1792-1880), *DNB*, professor of zoology 1836-80; David Don (1800-41) *DNB*, professor of botany 1836-41; Charles Wheatstone (1802-75), *DNB*, professor of experimental philosophy 1834-75; Henry Moseley (1801-72), *DNB*, professor of natural philosophy and astronomy 1831-44.
12. *KCL calendar 1838-9;* Phillips to H.J. Rose, 15 Oct 1838, Phillips to T.G. Hall, 24 Nov 1838 (q), Phillips to H. Smith, nd, annotated 27 Feb 1839, KCL incoming correspondence; Phillips to sister, 1 Mar 1839, PP; KCL Council minutes, 16 Nov 1838; Hugh James Rose (1795-1838), *DNB*, principal of King's 1836-8; James Tennant (1808-81), *DNB*, London mineral dealer.
13. Phillips to sister, Thursday (fr 12 Feb 1839), Monday (18 Feb 1839), 21/2, 23 Feb, 1, 6/7, 28 March 1839, PP; Phillips to Lonsdale, 6 Aug 1839, KCL incoming correspondence; Glennie to Phillips, 2 Aug 1839, Lonsdale to Phillips, 27 Aug 1839, PP; KCL Council minutes, 8 Nov 1839; reverend Thomas Grainger Hall (1803-81) professor of mathematics 1830-69; John Lonsdale (1788-1867), *DNB*, principal of King's 1839-43; for David Thomas Ansted (1814-80), *DNB*, professor of geology, 1840-53, and pedagogic workhorse, *KCL calendar 1840-1*.
14. Royal Institution managers' minutes, 2 July 1838, 1 April 1839; Phillips, lecture syllabuses, folder 12, PP; E.R. Daniell to Phillips, 23 Nov, 6 Dec 1837, PP; Brayley to Phillips, 12 June 1838, PP; Lecture handbills, London Institution, 1837-8, 1838-9, Corporation of London Guildhall Library; Phillips to sister, 21/2 Feb, 1, 6/7 March 1839, PP; James, 1993, pp. 144, 146, 149; Darwin to Phillips, 18 March 1839, in Burkhardt and Smith, 1986, p. 177; on the Royal Institution, Berman 1978; on the London Institution, Hays, 1974; on Faraday's character, Cantor, 1991.
15. Lyell to Horner, 12 March 1838, in Lyell, 1881, vol. 2, pp. 39-41; Phillips, 1837, 1839, vol. 1, pp. 310-13; Phillips to sister, 11, 19 May 1835, 23 Feb, 14 May 1837, 18 and 21/2 Feb, 1, 6, and 12 March, 2 April 1839, PP; William Devonshire Saull (1784-1855), *DNB*, eccentric London business man and geologist; Francis Baily (1774-1844), *DNB*, retired London stockbroker, astronomer, a general secretary of BAAS 1835-6; Thomas Weaver (1773-1855), *DNB*, consultant mining geologist; Woodbine Parish (1796-1882), *DNB*,

diplomat and geological facilitator; Francis Legatt Chantrey (1781-1842), *DNB,* drew geological specimens; Charles Stokes (1783-1853), London stockbroker; James Dunbar Douglas, sixth Earl of Selkirk (1809-85).

16. Phillips to sister, 14 May 1837, 21/2 Feb, 18 March 1839, PP; Augustus Frederick, Duke of Sussex (1773-1843), *DNB,* president of Royal Society 1830-8; Spencer Joshua Alwyne Compton, Marquis of Northampton (1790-1851), *DNB,* president 1838-48; William Henry Fox Talbot (1800-77), *DNB,* a pioneer in photography.
17. Phillips to sister, 14 May 1837, 12 March 1839, PP; John Pye Smith (1774-1851), *DNB.*
18. Darwin to Lyell, 9 Aug 1838, in Burkhardt and Smith, 1986, pp. 95-8; Topham, 2000; Secord, 2000, pp. 41-76.
19. Phillips, 1838b; *Athenaeum,* 1838, p. 872; Woodward, 1907, p. 131; Joseph Wilson Lowry (1803-79), *DNB;* I agree with Rudwick, 1985, p. 257, that the so-called first edition of the map described by Douglas and Edmonds, 1950, was a proof copy of an unpublished version.
20. Lyell, 1830,1832,1833; Conybeare and Phillips, 1822; Bakewell, 1833; De la Beche, 1831; Phillips, 1834a; W. Henry to Phillips, 18 Nov 1832, PP; *Athenaeum,* 1835, pp. 142-3 (q); Kenrick to Phillips, nd [1834], PP; Phillips to Murchison, 22 Feb 1835, MP; Longman to Phillips, 3 Feb 1835, PP; Phillips to sister, 27 Oct, 21 Nov 1834, 8 May 1835, nd, np [late May 1835], Torrie to Phillips, 18 May 1836, PP; Phillips to Murray, 27 Aug 1834, Mu P; obituaries of Lowry in *Nature,* 1879, vol. 20, p. 197 and *Athenaeum,* 1879, p. 796.
21. Challinor, 1963, pp. 290-3; Daubeny to Phillips, 24 June 1829, 6 Feb 1830, 6 March 1831, PP; Phillips to Daubeny, 3 June 1829, PP; Phillips, 1835c, p. 530; Buckland to Phillips, 12 April 1829, PP; Edward Smedley (1788-1836), *DNB;* Samuel Taylor Coleridge (1772-1834), *DNB.*
22. Phillips, 1837, 1839; Peckham, 1951; Hays, 1981; Prospectus for *Monthly Chronicle;* Morrell and Thackray, 1981, pp. 180-1, 258-62, 472-3, 497-8; Yeo, 1993, pp. 86-7; Dionysius Lardner (1793-1859), *DNB;* Edward George Earle Lytton Bulwer-Lytton (1803-73), *DNB.*
23. Phillips, 1838a; Yeo, 1991, p. 45; Black to Phillips, 30 Nov 1836, PP; Adam Black (1784-1874), *DNB.*
24. Fitton to L. Horner, 24 Dec 1827, Fitton to Coates, 20 Nov 1834, De la Beche to Coates, 22 Dec 1834, 10 July 1837, 26 July 1838, Kenrick to Coates, 20 Nov 1834, SDUK incoming correspondence; Coates to De la Beche, 17 Dec 1834, DLB P; *Penny cyclopaedia,* 1833, vol. 1, p. iv (q); Knight, 1864-5, vol. 2, pp. 203, 230; Phillips, 1841a, p. 160. For Phillips' articles, *Penny cyclopaedia,* 1838, vol. 11, pp. 127-51 (geology); 1840, vol. 16, pp. 487-91 (organic remains); 1840, vol. 17, pp. 153-4 (palaeozoic series), pp. 231-5 (parallel roads of Glen Roy); 1841, vol. 20, pp. 55-7 (rocks), pp. 354-5 (saliferous system); 1842, vol. 22, pp. 13-15 (Silurian system); 1842, vol. 23, pp. 106-8 (stratification); 1843, vol. 27, pp. 1-3 (Wales, geology of); 1838, vol. 11, pp. 297-9 (goniatites); 1840, vol. 18, pp. 375-7 (polythalamacea); 1843, vol. 25, pp. 434-5 (turrilites). Henry Peter Brougham (1778-1868), *DNB;* George Long (1800-79), *DNB;* Charles Knight (1791-1873) *DNB.*
25. Bartholomew, 1979.
26. Lyell, 1830,1832,1833; Sedgwick, 1831, p. 304; Secord, 1997.
27. Lyell, 1830, pp. 144-66; on Lyell's general strategy, Rudwick, 1990-1.
28. Lyell, 1830, pp. 92-124 (113 q).
29. Lyell, 1833, pp. 45-61; Rudwick, 1978; Gérard Paul Deshayes (1797-1875), *DSB.*
30. Sedgwick, 1831, pp. 306-7; Conybeare, 1830-1; Phillips, 1835c, pp. 570-1, 781-2; Phillips, 1837, 1839, vol. 1, p. 43, vol. 2. pp. 253, 261-6; Phillips, 1838a, pp. 260-3; Phillips, 1838c, pp. 140-2; on Élie's importance, see Greene, 1982, pp. 93-121.
31. Phillips, 1834a, pp. 61-4; Phillips, 1840b, pp. 490-1; Phillips, 1837, 1839, vol. 1, pp. 20, 96-8; Phillips, 1838a, pp. 286-90 (290 q).
32. Phillips, 1837, 1839, vol. 1, pp. 94-5, 100-1; Phillips, 1838a, p. 291; Phillips, 1838c, p. 147; on the French threat, Brooke, 1989.

33. Phillips, 1837, 1839, vol. 2, pp. 273-7; Lyell to Herschel, 21 Feb 1865, in Lyell, 1881, vol. 2, pp. 390-1.
34. Phillips, 1838c, p . 132-3 (q); Phillips, 1837, 1839, vol. 2, pp. 274-8 (278 q); Phillips, 1838a, pp. 22, 63, 282-3; Phillips, 1835c, pp. 798 (q), 800 (q); Nichol, 1837; Schaffer, 1989; John Pringle Nichol (1804-59), *DNB*, appointed professor of astronomy, University of Glasgow, in 1836.
35. Phillips, 1838a, pp. 178-80; Phillips, 1837, 1839, vol. 1, pp. 249-52; *BAAS 1835 report*, p. 62; Phillips, 1835d; Sedgwick to Lyell, 20 Sept 1835, in Clark and Hughes, 1890, vol. 1, pp. 446-8; Phillips to Charlesworth, 7 Dec 1835, British Library Add Ms 37951, f. 38; Charlesworth, 1835, 1836, 1837 pp. 6-9; Wilson, 1972, pp. 461-95; Phillips to Harcourt, 26 July 1839, PP (q); Edward Charlesworth (1813-98).
36. Phillips, 1835c,.p. 534; Phillips, 1837, 1839, vol. 2, p. 240 (q); Phillips, Dublin courses 1844-5, lecture 1, notebook 50, PP; Rudwick, 1975b; Phillips, 1838c, p. 135 (q); Phillips, 1838a, pp. 251-3; Lyell, 1830, vol. 1 p. 4 (q); Wilson, 1972, p. 343; Lyell, 1838; Phillips to Harcourt, 26 July 1839, HP.
37. Phillips, 1835c, pp. 533-4; Phillips, 1838c, p. 131 (q).
38. Phillips, 1836b, pp. 71-6; Phillips, 1838a, pp. 156-7; Phillips, 1840b, p. 491 (q).
39. Phillips, 1835c, pp. 539, 609; Phillips, 1834a, pp. 16-19; Phillips, 1838a, pp. 37, 85-7, 124.
40. For criticism of 'liberal Anglicanism' as a category, Hilton, 1988, pp. 28-31; Brooke, 1999; for the uses, limitations, and language of natural theology, Brooke, 1979, Brooke, 1991, and Brooke and Cantor, 1998, pp. 176-206.
41. Corsi, 1988; Morrell and Thackray, 1981, pp. 224-45; Topham, 1992, 1998.
42. Phillips, 1837, 1839, vol. 1, pp. 266-7 (q), 232 (q); Phillips, 1838a, p. 146; Sedgwick, 1833, pp. 102-9 (104 q); Buckland, 1836; Cockburn, 1838; Phillips to Henry Robinson, nd [1838], Robinson papers, DDX/65/27, Humberside County archives (q); Sedgwick to Phillips, 18 Dec 1833, PP; Phillips to Powell, 15 March 1837, private; William Cockburn (1773-1858), dean of York Minister 1822-58.
43. J.P. Smith to Phillips, 6 Feb, 25 Mar, 19 Aug 1839, 11 Apr 1840, PP; Smith, J.P., 1839, pp. 41, 52, 83-6(83q), 372-4, 392, 396, 412; Phillips to sister, 12-14 and 28 March 1839, PP; Phillips to Murchison, 7 Dec 1839, MP.
44. Phillips to Harcourt, 12 June, 22 July 1839, HP; Harcourt, 1839, pp. 17-22 (17 q).
45. Sedgwick, 1833, p. 105; Phillips, 1834a, pp. 59, 47(q); Young, 1838, p. 42; Phillips, 1838a, pp. 291-5 (293 q); Cockburn to Phillips, 12 Sept 1838, PP; Cockburn, 1840, p. 38; Phillips to sister, 27 Oct 1840 (q), PP.
46. Sedgwick, 1831, p. 313; Phillips to sister, 20 May 1835, Fri evening [Nov 1836] (q), PP; Kenrick, 1834, 1835, 1846; Kenrick to Phillips, 17 April 1841, PP.
47. Phillips, 1835b, pp. 688-90, 697; Phillips, 1838a, pp. 207, 210-15; Phillips, 1837, 1839, vol. 1, pp. 277, 297-8.
48. Phillips, 1837, 1839, vol. 1, p. 5; 1834a, pp. 32, 66-7; Phillips 1835b, pp. 542-3 (542q); Phillips, 1838a, pp. 96-7 (q). For the importance of theistic science see Brooke, Osler, and Van der Meer, 2001.
49. Sedgwick, 1833, pp. 104-8; Conybeare, 1834, p. 309; Harcourt, 1839, p. 20; Phillips to Robinson, nd [1838], Robinson papers, Humberside County archives; Powell, 1838, pp. 227-8, 247, 256-60, 294-5; Powell to Phillips, 15 Sept 1837, PP; Phillips, 1838d; Kenrick to Phillips, 17 April 1841, PP; Martineau, 1878, pp. 28-9; Edward Bouverie Pusey (1800-82), *DNB*.

Chapter 6

The Geological Survey 1836-1841

6.1 The Survey under attack

The Ordnance Geological Survey was established in 1835 with De la Beche as its director (Fig. 6.1). His enterprise was not the first national geological survey to be patronised by a government: in France and the USA surveys were begun in 1825 and 1823 and employed geologists temporarily. De la Beche's Survey was unusual in two ways. It endured and expanded even though government regarded it as temporary; and it survived whereas other geological ventures, pursued under the auspices of the Ordnance Survey, became unpopular with government. John MacCulloch's Scottish work, begun in 1814 and completed in 1836, was for government a painful example of jobbery and uncontrolled expenditure. From 1832 Joseph Portlock worked officially in the Irish Ordnance Survey on geology until in 1840 the project had become so expensive that government stopped it.[1]

The founding of the Geological Survey was the government's response to the surveying work already done by De la Beche and to his careful lobbying in which he stressed the utility and cheapness of his schemes. In 1831 he lost an independent income of about £3,000 per year as a slave owner with estates in Jamaica so he found himself having to earn his livelihood with his geological skills. Supported by Murchison and Sedgwick, who lobbied Lord Lansdowne, a member of Grey's cabinet as Lord president of the council and Lord Kerry, his eldest son, De la Beche persuaded government in 1832 to permit him to establish temporary links with the Ordnance Survey by colouring geologically eight sheets covering Devon and adjacent areas. Presenting himself as an effective man of despatch and business, allegedly unlike some of the gentlemanly worthies of the Geological Society, De la Beche received a fee of £300 but no salary and no expenses. Thomas Colby, the superintendent of the Ordnance Survey, supported this arrangement because it was in harmony with the way he encouraged a few of his surveyors, especially Henry McLauchlan and Henry Still, to colour Ordnance maps geologically in the early 1830s.[2]

By May 1835 De la Beche had completed his colouring work for the Ordnance which was so pleased with it that his fee was increased by 50%. He immediately proposed to the Master General and Board of Ordnance that such work be extended to other areas under his direction. Supported by the Master General and the Board, by Sedgwick and Buckland as Oxbridge professors of geology, by Lyell as president of the Geological Society, and by Lansdowne and Spring-Rice, two key members of Melbourne's second cabinet formed in April 1835, the whig De la Beche was appointed in summer 1835 to direct a geological examination of English counties, particularly by colouring geologically the Ordnance county maps. As a government scientist he was to be paid a salary of £500 pa and estimated expenses of £1,000 a year. These were negligible sums compared with the cost of the Ordnance which was £1,273,000 in 1835.[3]

Fig. 6.1   Sir Henry Thomas De la Beche, an engraving of 1848 by W. Walker
of a painting in enamel by H.P. Bone. His skills as a jobber and
intriguer helped him to promote the governmentally supported
Geological Survey.

From 1835 De la Beche became an even more accomplished lobbier of government. He exploited the sly fait accompli, covertly changed the terms of an agreement, hoisted his critics with their own petard, inserted the thin edge of the wedge, appealed to precedent, and built up a state-supported geological empire bit by bit. He successfully lobbied Spring-Rice for a Museum of Economic Geology which was effectively established in 1837 in Craig's Court, Whitehall, London, under his direction and the control of the Department of Woods, Forests, Land Revenues, Works, and Buildings. Two years later, as a result of pressure exerted by Sopwith and the British Association, De la Beche was put in charge of the adjacent Mining Record Office which opened in 1840. In 1845 he transformed the Survey from a military to a civilian enterprise when it was transferred from the Board of Ordnance to the Department of Woods and Forests. His final coup occurred in 1851 when the new Museum of Economic Geology, in Jermyn Street, London, was opened and the government School of Mines was established there. By that time he superintended four different geological ventures, the Survey, the Museum, the School, and the Mining Records Office, all of which were government institutions.

De la Beche's equally successful attempts to expand his staff were based on the same incremental approach, which had the distinct virtue of not alienating either the Board of Ordnance or the Treasury. Until the end of 1836 De la Beche had as assistants only two experienced Ordnance surveyors, McLauchlan and Still, who made topographic corrections to outdated maps and helped geologically, plus a hired miner who examined mines and excavated geologically debatable ground. Subsequently several geologists, such as William Logan, Harvey Holl, and William Sanders, worked voluntarily for the Survey which did not pay them a salary.[4] Colby's preference was to train existing Ordnance men for work with the Survey but De la Beche's first appointment to it was that of Phillips who was connected with it in four ways. From November 1836 to summer 1838 he worked for the Survey at De la Beche's invitation voluntarily and without pay. From July 1838 to early 1841 he was paid to draw and describe the fossils of Cornwall, Devon, and west Somerset. During the financial year 1841-2 he was employed for part of it as a full-time salaried palaeontologist. Starting in 1842 he was paid by the Survey as its palaeontologist on a tacitly permanent basis in that his post was regular but annually renewable.

De la Beche approached Phillips to help the Survey not from altruism but in order to serve certain interests dear to himself. One of these was the Devonian controversy which had erupted in 1834 when De la Beche reported a paradox: Devon culm, apparently an integral part of the folded greywacke rock, which was widely assumed to be the oldest secondary stratum, contained fossil plants like those of the much newer coal measures. As Rudwick has shown, this controversy was not just about the anomaly of fossil plants in such ancient rocks: it involved heated and protracted discussion of such central issues as how to use fossils in determining the order of strata. By autumn 1836 the battle lines were clearly drawn with Murchison, Lyell, and Sedgwick lined up against De la Beche. At the 1836 meeting of the British Association, his work on Devon was severely attacked by Murchison and Sedgwick in the presence of Spring-Rice who had authorised the Survey the previous year. They asserted that, contrary to De la Beche's proposed structure, there was a coal basin in central Devon, a claim which he conceded immediately. They also proposed that the order of strata going upwards was Cambrian, greywacke, which they thought belonged to a newly invented Devonian system, the Silurian system, and carboniferous strata containing the fossil plants. They dated the culm as carboniferous, an interpretation which De la Beche did not accept: for him it was an integral part of the greywacke and much older than carboniferous. Having confessed his inadequacy as a structural geologist and having been castigated as a palaeontologist, De la Beche feared that Murchison would use his undoubted influence to destroy the state-supported Survey and thereby preserve his own intellectual dominance and geological territory. He suspected that Lyell would make sly cuts at the Survey whenever possible. Moreover, he knew that Sedgwick had put himself in a contradictory position: in 1832 and 1835 Sedgwick had assured government about De la Beche's geological ability, but in 1836 he had publicly questioned it. De la Beche feared that government and especially Spring-Rice would regard specialist scientific advice, such as Sedgwick had given, as deceitful or incompetent. By late November 1836 De la Beche knew that Sedgwick had gone out of his way to protect the Survey, as well as to justify his own integrity as a scientific consultant to government. Sedgwick had told Spring-Rice that De la Beche had mapped Devon well according to the notions then in vogue and was therefore not to be blamed. He explained that Murchison and he were not attacking De la Beche's competence but were proposing a new classification of what everyone agreed was a very difficult area. Sedgwick concluded by averring his continued support for the Survey.[5]

6.2   The unpaid helper

De la Beche was in an unstable position and defensive mood apropos the Survey when in November 1836 he asked Phillips to identify Cornish fossil specimens collected by it. His reasons were clear. At heart De la Beche was not a palaeontologist, yet the 1836 fracas at Bristol had shown him that stratigraphical geology was becoming more dependent on palaeontology and less dependent on mineral composition and on the observed order of superposition of rocks. He therefore needed help in the task of identifying strata from their fossil contents. Phillips was ideally qualified to do this. He was an experienced field surveyor, mapper, and geologist as well as a palaeontologist who illustrated his works with his own drawings of fossils. In late 1836 he was preparing for publication two palaeontological works, one on fossils in general and the other on belemnites. As the leading British expert on the carboniferous limestone, Phillips was prospectively useful to De la Beche whose 1835 classification of strata put this limestone immediately above the old red sandstone and the problematic greywacke rocks of Devon.[6] De la Beche knew that small quantities of limestones occurred in greywacke rocks. In the event of these turning out to be crucial in the Devonian controversy, De la Beche would have a recognised expert to call on, at best to secure victory, or as a pis aller to muddy the water.

Furthermore De la Beche and Phillips shared several attitudes. As Secord has stressed, both were interested in the reconstruction of ancient environments and past localities. Both were deeply aware that varying local circumstances of deposition of a stratum modified its fossil contents and mineral composition so that it was erroneous to assume that any stratum, which occurred over a large area, would contain the same fossil species. It was therefore hazardous in their view to make correlations of strata over large distances on the basis of a few and allegedly unique fossil species. It was equally rash to try to establish general geological systems, such as Murchison's Silurian, on the basis of merely local investigations. For these reasons De la Beche had proclaimed in 1835 that, in identifying strata, fossil content and mineral composition were subordinate to observed superposition. But Phillips, as De la Beche well knew, had revealed by spring 1836 that strata could be identified using a statistical palaeontology based on the total number of fossil species in each stratum and on the distribution of those species, arranged by zoological classes, among the strata (see section 4.2).[7]

De la Beche and Phillips were also united in their opposition to what they regarded as the premature and speculative theorising of Lyell (see section 5.4). In De la Beche's case this intellectual hostility involved personal antagonism between Lyell and himself in the 1830s and 1840s. As Secord has indicated, De la Beche's sustained opposition to Lyell's preconceived opinions helped to give to the Survey coherence as a collective research enterprise. But De la Beche and Phillips also shared positive beliefs, such as that geology needed the help of physical sciences. In 1835 Phillips revealed to De la Beche, whom he hardly knew, that on many occasions when reading De la Beche's works he had been 'arrested at many points by a very unexpected coincidence of thought and feeling ...'.[8]

Phillips was therefore ideally qualified to help De la Beche but there were two drawbacks. Firstly, Phillips' leading London patron was Murchison who was arrayed from 1836 against De la Beche on the Devonian question. If Phillips were to work for the Survey he would face difficulties in balancing his loyalties and any increased allegiance to De la Beche could be achieved without discord only if Phillips became more involved in the Survey gradually and not suddenly. Secondly, De la Beche knew that Phillips might qualify or reject his own solutions to palaeontological problems. This had happened at the 1836 meeting of the British Association when, in response to Murchison and Sedgwick's interpretation, Phillips had argued that the Devon culm strata were carboniferous and not pre-carboniferous, as De la Beche still claimed and as Phillips had previously believed.[9]

From November 1836 to summer 1838 Phillips worked for De la Beche without pay and when he could. In return for preferential access to the Survey's fossils and freedom to exploit them in any publication, he agreed to identify them and to permit his identifications to be used in Survey reports. By May 1837 he was inspecting Cornish fossils for De la Beche, who had become a close associate, and also helping Murchison with his *Silurian system*. Just at this time Lyell and Murchison were apparently spreading rumours about De la Beche's incompetence. Indeed, Murchison told Harcourt that their friend Phillips was the only fit person to run the Survey: 'he would have done the whole thing *on a great scheme* and on sound principle and would not have started in *chaos*'. De la Beche's Survey survived in spring 1837 because it was backed by Colby, Greenough, and several members of Parliament, one of whom primed the Master General of Ordnance, and because government, wishing to extinguish rumours about De la Beche's incompetence, had agreed to publish his memoir on Devon and Cornwall at its expense and not his. Consequently from spring 1837 De la Beche gradually lured Phillips into taking on more palaeontological work, by sending to him Survey specimens from Cornwall and then fossils from several private collections including some important Devon ones. Simultaneously, De la Beche stressed to Phillips that the unfraternal and selfish conduct of Murchison and Sedgwick deviated from the invincible liberality of sentiment allegedly characteristic of the gentlemanly brethren of the hammer.[10]

By February 1838 De la Beche, ready to go to press with his *Report* on Devon and Cornwall, was desperate for Phillips' account of the fossils of the area, but was in no position to compel compliance from his voluntary helper. Phillips had completed his identifications of the fossils from no more than three private local collections, those of William Harding, Samuel Pattison and Charles Peach. De la Beche wanted quicker service from Phillips; and in order to show fellow geologists that the Survey was up to date in its approach and to convince government that his Survey was efficient, De la Beche wanted a palaeontological companion volume to his 1839 *Report*. By spring 1838 Phillips had done enough good work on the fossils from Devon and Cornwall for De la Beche to be convinced that Phillips could provide just that. In general De la Beche wished to patronise a new sort of palaeontology which would show that the days of the individual gentleman geologist such as Murchison were numbered: in late 1837 he had welcomed Phillips' approach to the study of fossils as 'a thousand-fold richer than the mere bald and trifling notion of the identification of strata by their organic contents'. Moreover, in late 1837 De la Beche had begun the Survey's work in south Wales, which the Treasury and Ordnance had authorised in January. Here he soon received enthusiastic assistance in mapping from William Logan whose minute accuracy, in his traverse sections on a scale of six inches to the mile, was then unparalleled in the United Kingdom; but Logan was no palaeontologist.[11]

6.3   The Survey employee

In 1838 De la Beche transformed Phillips' voluntary and unpaid connection with the Survey into a contractual, paid, yet temporary *ad hoc* arrangement. After almost five months of negotiations, on 27 July 1838 the Treasury issued an order for Phillips to draw and describe the fossils of Cornwall, Devon and west Somerset for a fee of £250 plus expenses. This second type of connection with the Survey occupied Phillips until early 1841 and it culminated in the publication in that summer of his *Palaeozoic fossils*. It seems that in early March 1838 both De la Beche and Phillips envisaged a fixed and permanent appointment for Phillips as palaeontologist to the Survey, but they settled for the *ad hoc* fee arrangement as easier to secure from government. Each man was adopting the tactic of gaining a toe in the door. In 1835 De la Beche had persuaded government to transform an *ad hoc* fee arrangement for himself into a regular salaried position; and thus he saw no great difficulty in eventually doing the same for Phillips, for whom the *ad hoc* arrangement was a prelude to a fixed appointment with the Survey. It was Phillips who on 6 March 1838 suggested to De la Beche that government might give a special grant for the Devon and Cornwall fossil work, which Phillips rashly estimated would take no more than a year. He proposed a fee of £250 plus expenses, an arrangement which De la Beche quickly put to the Treasury. By 12 March he had obtained a gentleman's agreement from Francis Baring, secretary to the Treasury. De la Beche then applied formally to the Treasury, supporting his application with testimonials from Sedgwick, Buckland, and Whewell as president of the Geological Society about Phillips' prowess and suitability. He did so partly to disable any intriguer opposed to the arrangement and partly to justify it to Parliament.[12]

From summer 1838 De la Beche leaned hard on Phillips who knew that he would be paid only on completion of his task. De la Beche stressed that cabinet work would not be enough: it would be necessary to inspect the sites from which specimens had been taken. Simultaneously he made a confidential offer to Phillips to join Charles Barry and himself on a royal commission appointed to discover the most suitable stone for the building of the new Houses of Parliament, of which Barry was the architect. The commission, De la Beche revealed, was partly his own brainchild: it had originated in a conversation between Barry and himself at the Museum of Economic Geology. Though De la Beche later complained to Colby that the Parliament stone commission was distracting him from his south Wales field work and delaying publication of his 1839 *Report*, he welcomed the opportunity of demonstrating to government the economic utility of the Survey, and of bringing Phillips into advantageous contact with what he called 'the higher powers'. Phillips' response was decisive: though he was aware of the patronage that membership of the commission might generate, and that his time and expenses would be paid for, he simply could not squeeze in two months' extra travelling. In addition, his old uncle, William Smith, was pained by the notion of Phillips doing what he himself could do. Phillips soon decided that Smith was the man for the job: by the end of August 1838 he had persuaded Barry and De la Beche to put Smith on the commission which would pay him a fee of £150. In his busiest ever year as a lecturer Phillips wished to release himself from the 'plague of going to see stones and buildings', though in principle he approved the appointment of the commission. Phillips was also happy to help his uncle financially because government had refused from 1837 to increase his pension of £100 pa on which he was totally dependent.[13]

For the remaining part of 1838 Phillips' pressing engagements prevented him from making much progress with the Devon fossil work. In early 1839 the appearance of books by Murchison and De la Beche made that fossil work all the more important. In January Murchison published his *Silurian system*, which elevated the old red sandstone into a geological system, while referring occasionally to his views on Devon. The following month De la Beche produced his 1839 *Report* on the geology of south-west England, in which he accepted Murchison and Sedgwick's interpretation of the structure of Devon, without giving more than the scantiest indication that they had first proposed it in 1836. On the vexed question of the age of the Devon culm, he referred to it with deliberate vagueness as carbonaceous, instead of conceding to Murchison and Sedgwick by calling it carboniferous. Needless to say Murchison was outraged by this mixture of plagiarism and obfuscation, deploring to the president of the Geological Society 'the ordnance jockeyship of riding home with false weights'. As part of his obfuscation De la Beche asserted that Phillips' work on the north Devon culm, which Phillips had already publicly dated as carboniferous, could be interpreted another way: it was rash to assume that these fossils were limited to the carboniferous epoch. On the related question of the lack of resemblance between the greywacke in north and south Devon, De la Beche offered two conjectures: first, that there was a difference in age, the south being younger; second, that they were of the same age, but the conditions of fossilisation varied between north and south. He then pointedly announced that Phillips would give the problem proper attention. Meanwhile Murchison was fretting about whether he and Sedgwick should let Phillips see all their fossils: he feared parting with his palaeontological weapons and he suspected that De la Beche would induce Phillips to do field work in Devon to supplement De la Beche's patent deficiencies. Simultaneously Murchison acknowledged that Phillips' future work for the Survey 'would be well for the country'.[14]

His suspicion that De la Beche needed Phillips was well founded. On 28 February 1839, with his *Report* published, De la Beche launched the negotiations which eventually culminated in Phillips being elevated yet another rung on the Survey ladder, this time as a salaried Survey geologist. Initially De la Beche offered tempting

bait: as part of a projected expansion of the small Survey empire, Phillips was asked
to be a sub-director responsible for half of the English work, at a salary of £400 pa
(£100 less than De la Beche) with £100 pa for expenses (the same as De la Beche).
Phillips agreed to let his name be used in De la Beche's negotiations because the
proposed renumeration would permit him to relinquish all lecturing and avoid its
concomitant anxiety.[15]

Phillips had less than a month to mull over the possible consequences of accepting
De la Beche's offer before yet another fracas occurred between Murchison and De
la Beche, this time in connection with the April 1839 paper by Murchison and
Sedgwick on Devon and Cornwall, which contained a startling redefinition of the
Devonian system compared with their views of 1836. This redefinition owed much to
Robert Austen and William Lonsdale. In late 1837 Austen had published his view that
the fossils he had found in the south-east Devon limestone bands of the greywacke
were disguised mountain limestone fossils and Lonsdale had identified these fossils
as not much older than those of the mountain limestone. Drawing on this research,
Murchison and Sedgwick argued that the older rocks of Devon were representatives
of a previously unrecognised geological system, called the Devonian, which was
intermediate in its age and fossils between the Silurian and carboniferous systems.
Thus Murchison, the prime author, had renamed as the Devonian what he had called
the old red sandstone system in his 1839 book. This redefinition was audacious.
Having postulated in 1836 that the greywacke was older than Silurian strata, in 1839
he claimed it was younger. It also looked different from the old red sandstone to
which it was allegedly equivalent; and its fossils, too, were different in that fossil fish,
found in the sandstone, had not then been found in it. Another purpose of Murchison's
paper was to expose De la Beche's alleged plagiarism: De la Beche's 1839 map, an
integral part of his *Report*, drew on Murchison's and Sedgwick's 1836 account of
Devon, but failed to acknowledge that debt. In Phillips' view De la Beche had laid
himself open to such severe animadversions that his influence was likely to end soon.
Phillips, too, was attacked by Murchison for alleged ungentlemanly plagiarism: in his
Index geological map of Britain, published in summer 1838, Phillips had coloured
the Devon culm as carboniferous but not as a productive coalfield and had cited De
la Beche as his authority. Phillips also found that he had 'promised' to correct this
misacknowledgement and that De la Beche wondered whether this was so. Though
Phillips feared acrimonious controversy as personally disagreeable and a deviation
from the true ethos of geology, he decided to assert his independence by publishing
an explanation. Murchison expected Phillips to knuckle under and to withdraw his
protest, but was mistaken: in May 1839 Phillips objected to the pledge which had
been unexpectedly given for him and defended his own and De la Beche's scientific
reputations. He made it clear that in early 1838 he *had* taken his colouring of Devon
from De la Beche, but that it was unnecessary to give specific acknowledgement for
the dating of the culm as carboniferous because that was common knowledge. These
defiant remarks were printed immediately before a further statement by Murchison
and Sedgwick, who expressed their deep regret about their unnecessarily sharp
words about De la Beche's behaviour, but ignored Phillips' rejoinder because it
was unanswerable. From May 1839 onwards Murchison, previously Phillips' chief
London patron and his superior in the secretariat of the British Association, knew
that Phillips would not kowtow to him. By refusing to be silent or silenced, he had
asserted his independence, demonstrated his moral allegiance to De la Beche, and
might well repeat the performance.[16]

Warmed by Phillips' public loyalty to the Survey, De la Beche began to exert what he called sincere and sometimes unpleasant pressure on Phillips to continue the Devon fossils work and not defer it. De la Beche feared that the Treasury might notice Phillips' inaction and become more obstructive in future, irrespective of which party happened to be in power. Moreover, in June Austen had left Devon for Surrey, taking his fossils with him, so that he was no longer available to give help locally in Devon. For his part, Phillips pointed to some progress he had made in studying specimens and to practical difficulties, such as Austen's collection being split between three places. Phillips reaffirmed his dedication to producing 'a true history' of the Devon fossils, but confessed he was intolerably busy. From August 1839 Phillips was deflected even more from this task because William Smith died in Northampton, leaving him with the taxing responsibilities of clearing debts and of looking after an insane widow. These were presumably responsible for Phillips applying to De la Beche for a financial advance for the Devon fossils work, a request the latter had to refuse as contrary to the Treasury minute authorising it.[17]

In 1840 Phillips was able at last to spend two months (March and April) on field work in Devon accompanied by his friend William Sanders. He wished to discover the bias in each of the cabinet collectors whose specimens, often incomplete, had been sent to him. He wanted to inspect for himself the fragmented fossils found in quarries, often ignored by collectors but valuable for purposes of identification. Once in Devon he soon found many new species, thus showing that De la Beche's reliance on local civilian collectors and Ordnance men was flawed. When necessary Phillips drew fossil specimens in situ; but preferably he collected fossils, weighing 336 pounds, and took them home where he made casts of the cavities left by shells, and split some specimens parallel to their stratification, in order to make the most of his materials.[18]

Phillips was just beginning his field work in Devon when De la Beche broached yet another proposition to him. Nothing having come of the February 1839 scheme for Phillips to be a full-time sub-director of the English Survey, De la Beche fell back on a less ambitious idea, which involved Phillips working some time each year as palaeontologist to it. This proposal, launched in March 1840 and fully authorised by government in February 1841, inaugurated the third type of connection Phillips had with the Survey, namely, as a salaried palaeontologist, full time for part of the financial year 1841-2, his tenure being officially limited to that year. De la Beche certainly went out of his way to secure Phillips on this basis. He thought hard about how to stop his proposal being referred back by the appropriate ministers who he feared might use this device to ignore it. He accommodated Phillips about the time of the year and the salary Phillips wanted. He dangled the visions of Phillips lecturing at the Museum of Economic Geology and of Phillips arranging the fossils in some large national museum. He arranged for Phillips' work on the Devonian fossils to be printed by order of the Treasury and secured helpful co-operation from the Stationary Office. He consulted Phillips' convenience in arranging for the lithographing of the plates to be done privately in York against the Office's rules and accepted the awkwardness of having the plates lithographed in York and worked off in London. Finally, in December 1840 he used his skill as an experienced lobbyist to secure unwritten agreement from government about Phillips being employed for one year as palaeontologist to the Survey. He knew that five minutes' talk with Francis Baring, the Chancellor of the Exchequer, was worth six months of scribbling memoranda, not least because the private secretaries made sad havoc with scientific matters. He took action in early December 1840 when ministers were in London and Parliament not sitting, having spotted that part of the previous years' Ordnance estimates were unexpended and therefore ripe for appropriation by the Geological Survey. Before Christmas he had secured the Chancellor's agreement. The following month he outlined his scheme to the Master General of Ordnance: as there would be by March a balance of £570 in hand, Phillips could be employed for eight months on field work in Wales in the financial year 1841-2 at a salary of £300 pa plus expenses. Moreover, as De la Beche stressed, Phillips could be relied on: for his fee of £250 Phillips had just finished his important book on the Devon fossils. The Ordnance authorities agreed to De la Beche's proposal, Colby arguing that Phillips would be a faster and cheaper worker than anyone De la Beche could find locally in Wales.[19]

Phillips had good reasons for accepting his third type of connection with the Survey. He would work in the field for the warm part of the year, while retaining as his York base the refurbished St Mary's Lodge. He would be well paid for doing what he enjoyed, that is, field work and studying fossils. He could anticipate more time for research and more opportunities for government-sponsored publication, while being able to relinquish most of the types of employment he had been forced to take in the past. He hoped that the one-year position would be a prelude to an appointment as permanent fossilist. With respect to income, the prospect of £300 for eight months' field work plus expenses meant that he would be free from taking various hack jobs to make ends meet. His own estimate in October 1840 of his total income in the period 1841-2 was £600. In January 1841 he augmented this to £700 when the British Association recognised his importance to its machinery by raising his salary as assistant secretary from £200 to £300 per annum. He could look forward to being well paid for one year for contributing to what he regarded as an important national undertaking. That was why he not only accepted but welcomed De la Beche's insistence that the conclusions of his forthcoming Devon fossil book be kept secret: they agreed that government had the right to be the first announcer of the results of the work for which it had paid. In several ways, then, the one-year Survey post was ideal, as Phillips jubilantly told De la Beche:

> I cannot perceive in my country any other mode of occupation more consonant to my habits and feelings, more beneficial to my health, or more fertile of opportunities of performing useful labours in science, and so of founding or strengthening a claim to be remembered among the geologists of this age.[20]

6.4   Palaeozoic fossils

This last wish was realised in July 1841 when Phillips' *Palaeozoic fossils* was published by Longman by order of the Treasury. Costing 6 shillings it was heavily subsidised by government. If published commercially it would have sold for about £1. Much of the book was devoted to sixty plates containing 750 drawings, mostly done by Phillips and lithographed in York by William Monkhouse, and accompanying descriptions of 277 species of fossils from Cornwall and Devon. As such it was a parallel palaeontological volume to De la Beche's 1839 *Report* and the first palaeontological monograph to be produced by the Geological Survey. As the full title of the book proclaimed, the figures were important. Phillips numbered them but did not add names of species to the plates because names were subject to the vagaries of nomenclature. He hoped that his figures would be enduringly reliable for identifying fossil species so he offered multiple perspectives, magnified views, and differentiation between young and mature forms, in order to highlight salient characteristics. For the same reason he took great care about the lithographic fidelity of his drawings. He also hoped that his descriptions would be useful for identifications but thought it presumptuous to claim that they were 'completely diagnostic ... because in a fossil state we seldom meet with specimens in such abundance, in such perfection, in such various states of growth, or conditions indicative of past existence, as to render this possible'.[21] Thus in both his figures and descriptions Phillips was a cautious phenomenological palaeontologist.

Though *Palaeozoic fossils* was ostensibly an illustrated catalogue of fossils, it also vindicated Murchison's Devonian interpretation of Devon but it opposed the notion that there were well defined geological systems such as the Devonian. As De la Beche was still sceptical about the existence of a Devonian system, Phillips' book gave face-saving and temporising protection to his boss. But Phillips did more than sit on the fence and try to maintain the geological peace. With cautious modesty he claimed that his book offered merely the 'natural results of a simple arrangement and contemplation of the phenomena' and that he had been unable to remove completely 'the veil which has long obscured the age and affinities of the strata of Devon and Cornwall'. In fact he revealed a statistical palaeontology, a new terminology for sedimentary strata, a suspicion of sharply separated geological systems, and an obsession with local conditions of deposition of organisms which became fossilised, all of which were anathema to Murchison.[22]

Phillips had been aware for years that, to minimise the difficulties of labelling strata by their fossils, it was advisable to study whole assemblages of fossils in strata thought to be analogous, preferably focussing on strata not widely separate geographically. Like De la Beche he realised that the occurrence of identical fossil species in different strata was not proof of their contemporaneity and that the occurrence of different fossil species in different strata did not necessarily indicate their different age. He sympathised with Sedgwick's presidential dictum of 1830 that, in correlating strata using fossils, only large zoological groups of fossils were reliable, particular genera and species being useless. Lyell, too, was suspicious about using particular species but felt that certain 'sets of fossils' could serve as reliable indicators of the chronology of strata. Wishing to avoid the sort of nihilism epitomised by Thomas Huxley in 1862 that all that geology could prove was local order of succession, Phillips employed two modes of numerical tabulation, based on assemblages of fossils, to reveal the degree of similarity between different regions. The first method compared the number of species in several zoological classes in various regions; the second gave the number of identical or closely allied species, arranged by zoological class, in various regions. In both cases he used as reference data Murchison's published catalogue of 336 Silurian fossils and his own of 420 from the mountain limestone. Using the first method, he drew up tables showing the absolute and proportionate numbers of fossil species, divided into several zoological classes, which occurred in the Silurian strata (which he called lower palaeozoic), Devon and Cornwall, and the mountain limestone (which he called upper palaeozoic). Significantly he used virtually the same zoological classes as in his 1836 monograph. The result of the triple comparison was that in general character the Devon and Cornwall fossils were intermediate between those of the lower and upper palaeozoic periods, a conclusion in harmony with Murchison's claim that the Devon greywacke belonged to a Devonian system which was younger than the Silurian and older than the mountain limestone. Again applying the first method, Phillips compared the fossils of three areas in Devon and Cornwall with those of the lower and upper palaeozoic eras. His tables revealed that north Cornwall and north Devon fossils had affinity with those from upper palaeozoic whereas those from south Devon were closer to the lower palaeozoic forms. This more sensitive application of the first method showed that there were no hard and fast distinctions between the Silurian, Devonian and carboniferous systems.

The second method adopted by Phillips measured the extent to which Devon and Cornwall fossils, categorised by zoological class, were found in the Silurian and mountain limestone strata in Britain and in the rocks of the Eifel and Bensberg, near Cologne, fossils from which had been shown to Phillips by Murchison and Sedgwick. Phillips concluded that the fossils of south Devon had a strong affinity with those of the Eifel and Bensberg and that all of them belonged to the middle palaeozoic era. Otherwise the results of the second method confirmed those of the first, though both made several hazardous assumptions to which Phillips drew attention.[23]

In order to avoid the difficulties inherent in conceptualising geological systems as discrete units and in order to embody his view that in the past the totality of living forms had slowly changed, Phillips devised a terminology which recognised three great periods of life, which nevertheless showed transitions from one to another. These periods he called palaeozoic to describe ancient life, mesozoic for middle life, and cainozoic for recent life. Each of these three periods was subdivided by Phillips into upper, middle, and lower. His nomenclature had the advantage of being broadly compatible with existing terminology. For example, Lyell's division of tertiary strata into pliocene, miocene, and eocene, corresponded to the upper, middle, and lower cainozoic periods. But Phillips' use of the term palaeozoic was controversial; it covered all forms of life before the epoch of the new red sandstone and subsumed Sedgwick's Cambrian and Murchison's Silurian and Devonian systems. Phillips was careful to restrict the term Devonian to its topographical sense, which gratified De la Beche. Phillips also repeatedly used his term 'lower palaeozoic' instead of Murchison's Silurian system because he wished to avoid any reference to particular localities and to refer only to periods of time elapsed. All Phillips' conclusions were explicitly subject to qualifications: his data were incomplete; it was dangerous to generalise from local truths; and differences in fossil populations might depend on merely local physical and ecological conditions and not on sequence of time which, in his proportionate tables, was assumed to be all important.[24]

Phillips' statistical palaeontology and his emphasis on using characteristic combinations of fossils were refinements of his uncle's concerns. The tables of fossil distribution had their origin in the tabular synopses, of the distribution of several groups of fossils in various strata, which he drew up by 1817 under Smith (see section 1.2). Phillips' interest in statistical thinking about fossils was stimulated in the late 1820s by the botanical arithmetic of his friend, Adolphe Brongniart, whom Phillips admired as a statistical pioneer. For instance in 1828 Brongniart published a table which showed the number of fossil species in each of six classes of plants for each of five geological epochs, the last being the present. Perhaps inspired by Brongniart, in his general texts of the 1830s Phillips published several distributional tables of fossils. Some compared recent and fossil species whereas others showed the distribution of fossils, categorised zoologically, among geological systems and topographical areas. One of these tables (of 1836) used almost the same classes as were employed in 1841; another (of 1837) used the technique of proportionate numbers, also exploited in 1841. Phillips' 1836 mountain limestone book contained an important table showing the distribution of fossils among different strata and a list giving the numbers of various fossils common to them. Thus Phillips' statistical palaeontology of 1841, as revealed in distribution tables, had its roots in Smith's interest in them and was publicly matured by Phillips in the 1830s. His strong interest in censuses of fossils was noticed abroad: in 1839 he was elected a member of the Société française de statistique universelle.[25]

Phillips' concern to identify strata, by the preponderance of fossils belonging to various zoological classes and families, was expounded in his general texts of the 1830s and his 1836 monograph (see sections 4.2 and 5.4). It was also manifest in the discussion of cornbrash fossils in his 1829 book (see section 2.6). It was probably derived from working with Smith's collection before it was sold to the British Museum. Knell has stressed that Smith liked his most characteristic fossils to be arranged on shelves in order of the strata with the other fossils in drawers but cross-referenced to each stratum. This difference between the most characteristic fossils and all the fossils of a stratum was surely not lost on the young Phillips who saw Smith's collection daily and catalogued it (see section 1.2).[26]

There was nothing new in using a triple division and triple subdivision of the strata. The terminology of primary, secondary, and tertiary, established before 1800, was used in the late 1830s by Phillips in his two *Treatises* and in his article on geology for the *Penny cyclopaedia*. In his monograph of 1829 Phillips claimed there were three divisions of secondary strata, characterised by plant and mollusc fossils. And, of course, he was familiar with Lyell's triple division of the tertiary strata into pliocene, miocene, and eocene. There was also nothing new in using terms which alluded primarily to periods of past time: in the late 1820s Brongniart had promulgated the view that there were four distinct periods of plant life in the past; and Lyell's three terms referred explicitly to geological epochs. Phillips thought that proper general terms for strata should reflect the sequence of geological time (like the old primary, secondary, and tertiary nomenclature), be elastic enough to be compatible with future research, and should be based on systems of organic life characterised by the preponderance of certain fossils. He therefore invented two neologisms, mesozoic and kainozoic, first published without explanation and without subdivision, in his 1840 article on the palaeozoic series for the *Penny cyclopaedia*.[27]

Phillips' third term, palaeozoic, was not new: Sedgwick had used it in 1838 to describe the Cambrian and Silurian systems. In 1840 and 1841 Phillips twice extended Sedgwick's concept of palaeozoic (Fig. 6.2). In the article on the palaeozoic series Phillips used the term, without triple subdivision, to include all strata up to and including the old red sandstone; and he suggested it might need to be extended to include carboniferous rocks and even the magnesian limestone. Early next year, in his article on the saliferous system for the *Penny cyclopaedia*, he had done precisely that and also created the three sub-divisions of each of the palaeozoic, mesozoic, and cainozoic eras. He then put into the upper palaeozoic the mountain limestone, millstone grit, coal measures, and magnesian limestone, all of which he had classed as mesozoic in 1840. He did so to accommodate his conclusion that the fossils of Murchison's Devonian system, alleged to be equivalent to the old red sandstone, were intermediate between those of the Silurian and Carboniferous eras. If the Silurian and Devonian systems were classed as lower and middle palaeozoic, it made sense to classify carboniferous strata as upper palaeozoic and to include in this subdivision the magnesian limestone because its fossils were analogous to or identical with those of the mountain limestone. In his 1841 book Phillips reproduced exactly the ninefold nomenclature first published in early 1841 in the *Penny cyclopaedia*. Phillips' nomenclature disturbed the widely accepted classification of primary, secondary, and tertiary strata because he put into the palaeozoic era formations usually designated as secondary, namely, the old red sandstone, mountain limestone, millstone grit, and coal measures, and the magnesian limestone of the new red sandstone system.[28]

Main formations and systems	Sedgwick 1838	Phillips 1840	Phillips 1841
New red sandstone			
Magnesian limestone			XU
Coal measures			XU
Millstone grit			XU
Mountain limestone			XU
Old red sandstone = Devonian		X	XM
Silurian	X	X	XL
Cambrian	X	X	XL

Fig. 6.2   Three definitions of palaeozoic rocks (U = upper, M = middle, L = lower).

Phillips had rejected the term protozoic which Murchison had suggested orally in spring 1838 and then in his *Silurian system* in 1839. By protozoic Murchison meant all fossiliferous strata below the old red sandstone, ie. the Silurian and Cambrian systems. Though Murchison claimed that the term was atheoretical, it proclaimed his belief that no life had existed before the Cambrian epoch. Phillips thought that protozoic was an inadequate term: it was ambiguous in relation to zoological protozoa; and it asserted more than was either necessary or known. Rather mischievously Phillips' category of lower palaeozoic was synonymous with Murchison's protozoic terminology. Sedgwick, too, was uneasy about Murchison's term. In a paper published in 1838 he used palaeozoic to describe the Cambrian and Silurian systems, and reserved protozoic for strata below the Cambrian if fossils were discovered in them. Phillips took the term palaeozoic from Sedgwick, gave to it a temporal dimension, and extended its meaning twice.[29]

6.5   Palaeozoic converts

Phillips was well aware that the novelties of his *Palaeozoic fossils* would not please everyone. He expected it to be useful in the progress of the application of zoology to geology but he feared that some geologists would not like his statistical approach and he knew that some of the Devonian controversialists would not be gratified because he had not given them a handsome notice. He felt, however, that his new scheme of palaeozoic nomenclature was both secure and flexible. He told Andrew Ramsay that before 1841 schemes of palaeozoic classification were like soap bubbles and iridescent phantoms. Phillips' scheme was widely accepted within about two years, the most reluctant convert being Murchison who feared that Phillips' renaming of the Silurian system as lower palaeozoic would sully his renown.[30]

Even before *Palaeozoic fossils* was published, Murchison worried that at the forthcoming meeting of the British Association in Plymouth, Devon, Phillips and De la Beche, as president of the geology section, would try to blur what Murchison regarded as clear and distinct differences between the geological systems he and Sedgwick had created. He thought that Phillips had rejected the Devonian system on account of his great obligations to De la Beche. At the Association's meeting, held just after the publication of his book, Phillips took advantage of the absence of Murchison and Lyell from section C by dominating it with a bravura performance in which he pressed it to adopt his three classes of past life which were based on 'prevalent forms of life characteristic of successive periods'.[31]

When Murchison returned from Russia in autumn 1841 Phillips wrote to him to point out the necessity of regarding his cherished systems as merely locally applicable because they were too dependent on a few characteristic fossil species whose use was unreliable because of the 'physical peculiarities of land and sea' when they were deposited. This lesson was not to Murchison's taste, not least because in late 1841 he revealed in print his latest geological system, the Permian, based on his recent research in Russia. According to Murchison the Permian was composed of strata which were represented in Britain by the magnesian limestone and the supra adjacent lower new red sandstone, which together constituted the lower part of what Phillips had called for several years the saliferous system. Murchison claimed that the affinities of the Permian, as revealed by saurian fossils, were with younger rocks yet Phillips had just bracketed the magnesian limestone with the older carboniferous rocks on the basis of their populations of fossil molluscs. In his presidential address to the Geological Society delivered in February 1842, Murchison attacked Phillips' palaeozoic nomenclature because with wounded vanity he feared that it would obliterate his own zealously promoted geological systems on which he had built his reputation. Murchison deplored Phillips' cribbing of the term palaeozoic from Sedgwick, his intellectual colonialism in extending the meaning of the term, and his intolerable coxcombry in substituting lower palaeozoic for Silurian. Murchison also reiterated his belief that the Permian system had little affinity with the mountain limestone. He felt so angry for six months that he broke off correspondence with Phillips who objected in private but not in public to Murchison's verdict which he had not heard in person.[32]

They met in June 1842 at the British Association meeting. With Murchison in the chair of the geology section, Phillips reiterated his opposition to the simplistic use of fossils in correlating strata and forcibly argued for the use of a large number of forms; but De la Beche accepted the Devonian and Silurian systems. In his address of early spring 1843 to the Geological Society, Murchison recanted his view of the previous year about the magnesian limestone's position: he conceded that, because of the affinity between magnesian and carboniferous limestone fossils, the Permian system had to be classed as palaeozoic. He was converted to this view by Philippe de Verneuil and count Alexandr Keyserling, his geological companions in Russia, and Étienne d'Archiac, Verneuil's collaborator. At the 1843 meeting of the British Association Murchison and Phillips, 'his distinguished friend', at last reached a consensus. Murchison displayed there his recently published geological map of England and Wales, the key to which used palaeozoic in Phillips' sense to embrace the Silurian, Devonian, carboniferous, and Permian systems. In a paper on the Permian system, Murchison again recanted his former view about the fossil affinities of magnesian limestone, being this time guided not by misleading fossil saurians but by fossil plants and fish. For his part Phillips accepted in public in 1843 that there was sufficient new fossil evidence from the Rhineland, Russia, Cornwall, and Devon to persuade him that there was a Devonian system.[33]

Sedgwick was much easier to convert than Murchison. At the 1841 meeting of the British Association Sedgwick accepted that the boundaries between geological systems were arbitrary if they were based on allegedly typical fossils. Then in November 1841 he committed intellectual suicide by announcing that no distinctive set of fossils occurred below the lower Silurian strata and thus conceded that the Cambrian system had no palaeontological base. He was thus sympathetic to Phillips' view that geologists needed to unsystematise their minds and to accept terminology better suited to 'the soft *shades* of mother nature'. By June 1843 Sedgwick was so aware of the way in which geological formations and systems seemed to melt into each other, both stratigraphically and palaeontologically, that he adopted Phillips' notion of the palaeozoic as a flexible term capable of incorporating Murchison's Permian system.[34]

An early admirer of *Palaeozoic fossils* was Lyell who recommended it to Henry Rogers, a leading geologist in the USA who put it in the same bracket as Murchison's *Silurian system*. Another was De la Beche who was so pleased with Phillips' book and field work that in October 1841 he began the lobbying which led the following month to the arrangement that for the financial year 1842-3 Phillips would be paid from the regular expenditure by government on the Geological Survey. This tacitly permanent position was not secured by going through the bureaucratic mills of the Ordnance and the Treasury. De la Beche was so keen to secure Phillips on a regular basis that initially he approached indirectly no less than Robert Peel, who became prime minister in September 1841 and was generally thought to be better disposed towards science than Melbourne, his predecessor. De la Beche used two intermediaries, Harcourt and Buckland, Peel's scientific adviser. Both were deputed to argue that little was being asked for, namely, a regularisation of the way in which Phillips was paid, without any increase in his salary of £300 pa for eight months' work. Buckland and Harcourt wrote to Peel within three days of each other, Buckland stressing that if Phillips were to leave the Survey it would be a great public loss and a calamity for geology. By 8 November 1841 Peel had resolved that the Survey estimates for 1842-3 should not be curtailed, thus making the posts of Phillips and young Andrew Ramsay tacitly permanent though formally renewable each year. The prime minister had taken the matter into his own hands, given his imprimatur to the Survey, and secured Phillips' future with it as a quasi-permanent fossilist. De la Beche celebrated this coup by organising Phillips' election to the Athenaeum, London's leading club for the intelligentsia, in February 1842. Two months later De la Beche was knighted, in preference to Murchison who had supplicated Peel for this honour in November 1841. Murchison had to wait until 1846 and Lyell until 1848. Phillips was overjoyed, assuring De la Beche that 'we of the Ordnance will fire a *feu de joie* in your absence, for in your presence I doubt if any of us can feel more than our wonted and affectionate regard'.[35]

Notes

1. Hendrickson, 1961; Eyles, V.A., 1950, 1937; 'Mineralogical survey of Scotland. Account showing the several payments ... of ... expenses ...', *Parliamentary papers*, 1830-1, vol. 14, pp. 53-83; Davies, Herries, 1983; Andrews, 1975; John MacCulloch (1773-1835), *DNB*.
2. The standard sources on the Survey are Flett, 1937; Bailey, 1952; Geikie, 1895; McCartney, 1977; North, 1934; Secord, 1986b. Colby to De la Beche, 9 April 1832, Byham, Secretary to Ordnance Board, to De la Beche, 2 May 1832, DLB P; De la Beche to Sedgwick, 26 March 1832, 22 Nov 1834, Se P; De la Beche to Master General and Board of Ordnance, 29 March 1832, Colby to Bryce, 9 April 1832, De la Beche to Byham, 9 May 1832, De la Beche to Colby, May 1832, all in Ordnance Survey letters relating to Geological Survey, GSM 1/68, British Geological Survey, Keyworth; Lansdowne, Lord president of council 1830-4, had been a vice-president of the Geological Society and would have been president of the British Association in 1836 had not Lord Kerry (1811-36), MP for Calne 1832-6, been fatally ill; Charles Grey, second earl Grey (1764-1845), *DNB*, prime minister 1830-4; Thomas Frederick Colby (1784-1852), *DNB*, had served for seven years on the council of the Geological Society.
3. De la Beche to Byham, 25 May 1835, report by Buckland, Sedgwick, and Lyell, 12 June 1835, Colby to Master General, 13 June 1835, all in GSM 1/68; Byham to De la Beche, 18 Nov 1834, Colby to Byham, 14 July 1835, DLB P; Spring-Rice, 1836, p. 7; the Master General of Ordnance was Richard Hussey Vivian (1775-1842), *DNB*; Lansdowne was Lord president of council, April 1835-41; Thomas Spring-Rice (1790-1866), *DNB*, whig MP for Cambridge 1832-9 and political ally of Sedgwick, was chancellor of the exchequer April 1835-9.
4. William Edmond Logan (1798-1875), *DNB*, collaborated with the Survey from 1838 and became first director of the Geological Survey of Canada 1842-69; Harvey Buchanan Holl (1820-86) spent six months with De la Beche in the late 1830s; William Sanders (1799-1875), *DNB*, Bristol corn merchant who worked for the Survey mainly 1841-2.
5. Rudwick, 1985, pp. 93-107, 159-70, 173-81; Sedgwick to Spring-Rice, 28 Oct 1836, Monteagle papers 13382, National Library of Ireland.
6. De la Beche, 1835, pp. 13-16.
7. Secord, 1986b, pp. 241-51; De la Beche, 1835, pp. 16-19, 23-5, 220-1, 238-9, 306-7.
8. Secord, 1986b, pp. 244-5; Rudwick, 1975b; De la Beche, 1834, used astronomy, physics, and chemistry; Phillips to De la Beche, 1 April 1835, DLB P.
9. Rudwick, 1985, pp. 121-3, 167-8.
10. De la Beche to Phillips, 6 Nov 1836, 21 and 31 May, 1837, PP; Murchison to Phillips, 3 April 1837, PP; Murchison to Harcourt, 1837 [Feb], HP; De la Beche to Greenough, 5 Feb, 2 March, 8 April 1837, GP; Colby to De la Beche, 3, 17, 18 April 1837, Greenough to De la Beche, 14 and 21 April 1837, Kerr to De la Beche, 12 and 14 April, 17 May 1837, all DLB P.
11. De la Beche to Phillips, 29 Oct, 13 Nov (q), 12 Dec 1837, 5 and 19 Feb 1838, PP; Phillips to De la Beche, 4 Nov 1837, DLB P; on Logan's voluntary help see Harrington, 1883, pp. 52-5, 59-71, 127-31, and Torrens, 1999; William Harding (1792-1886), *DNB*, Tiverton soldier; Samuel Rowles Pattison (1809-1901) Launceston solicitor; Charles William Peach (1800-86), *DNB*, revenue coastguard at Gorran Haven, Cornwall.
12. Phillips to De la Beche, 6 March 1838, DLB P; De la Beche to Sedgwick, 12 March 1838, Se P; De la Beche to Whewell, 12 March 1838, WP; De la Beche to Baring, 17

March 1838, DLB P; De la Beche to Phillips, 16 April, 26 and 30 July 1838, PP; Francis Thornhill Baring (1796-1866), *DNB*.

13. De la Beche to Phillips, 5 August 1838, PP; De la Beche to Colby, 27 August 1838, Ordnance letters, GSM 1/68; Phillips to sister, 9, 21 (q), 28 August 1838, PP; Phillips, 1844a, pp. 121-2; Phillips to Sedgwick, 2 Feb 1837, Se P; Smith to Phillips, 24 Jan 1839, SP; Charles Barry (1795-1860), *DNB*.

14. Murchison, 1839, proposed eight systems, namely, tertiary, cretaceous, oolitic, new red, carboniferous, old red, Silurian, and Cambrian, of which only the old red was novel; Murchison to Whewell, 7 Feb 1839, WP; De la Beche, 1839, pp. 136-8, 153-4; Murchison to Sedgwick, 22 Feb 1839, Se P; Rudwick, 1985, pp. 259-72, gives a telling account of this 1839 battle of the books.

15. Phillips to sister, 1 March 1839, PP.

16. Sedgwick and Murchison, 1839a and b; Phillips, 1838b, 1839b; Rudwick, 1985, pp. 257-9, 280-7; Phillips to sister, 28 March 1839, PP; De la Beche to Phillips, 7 April 1838, 9 April 1839, PP; Phillips to Sedgwick, 16 April 1839, Murchison to Sedgwick, 16 April 1839, Se P; Robert Alfred Cloyne Austen, later Godwin-Austen (1808-84), *DNB*, of Newton Abbot and Guildford.

17. De la Beche to Phillips, 26 May, 23 June, 18 Sept 1839, PP; Phillips to De la Beche, 26 June 1839, DLB P; Phillips, 1844a, pp. 123-4.

18. Knell, 2000, pp. 225-37; Phillips, 1841a, pp. v-vii.

19. De la Beche to Phillips, 25 March 1840, 10 April, 3 May, 4 and 15 Oct, 22 Nov, 10, 23 and 24 Dec 1840, PP; MacCulloch, Stationery Office, to Phillips, 1 Dec 1840, PP; De la Beche to Vivian, 27 Jan 1841, Colby to Mulcaster, 6 Feb 1841, Byham to Mulcaster, 17 Feb 1841, Colby to De la Beche, 23 Feb 1841, all in Ordnance letters, GSM 1/68; Phillips to De la Beche, 19 Feb 1841, DLB P.

20. Phillips to De la Beche, 6, 20 and 27 April, 16 June, 20 Oct 1840, 30 Jan 1841 (q), DLB P; Phillips to sister, 17 Sept, 20 Oct 1840, PP.

21. The cost of printing was £235. By late 1843, 207 copies had been sold. See 'Report of the commissioners appointed to enquire into the facts relating to the Ordnance memoir of Ireland', *Parliamentary papers*, 1844, vol. 30, pp. 259-385 (266). Phillips, 1841a, pp. viii-ix (q); Phillips to De la Beche, 8 Dec 1840, 31 Jan 1841, DLB P; Knell, 2000, pp. 237-40.

22. Rudwick, 1985, pp.371-5; Secord, 1986a, pp. 132-4, 140-2; Phillips, 1841a, pp. xi, xii.

23. Phillips, 1841a, pp. 158-60, 164-92; De la Beche, 1834, pp. 223-61; Sedgwick, 1830, p. 204; Lyell, 1837, pp. 512-3; Huxley, 1862, p. xlvi; Thomas Henry Huxley (1825-95), *DNB*.

24. Phillips, 1841a, pp. 160, 162-3, 171, 178, 187.

25. Phillips, 1835c, pp. 548-9; Phillips, 1840b, p. 487; Brongniart, 1828b, p. 219; for tables see Phillips, 1836b, pp. 75-7, Phillips, 1837, 1839 vol. 1, pp. 80-5, Phillips, 1836a, p. 244; C. Moreau to Phillips, 16 Aug 1839, PP.

26. Knell, 2000, pp. 23-7.

27. Phillips, 1829b, pp. 154-5; Phillips, 1840c.

28. Phillips, 1840c, 1841c, 1841a, p. 165.

29. Murchison, 1839, p. 11; Murchison to De la Beche, 6 April 1838, DLB P; Phillips, 1840c, p. 153; Sedgwick, 1838, pp. 684-5.

30. Phillips to Forbes, 17 Oct 1841, FP; Phillips to De la Beche, 13 Jan 1844, DLB P; Phillips to Owen, 25 Dec 1841-1 Jan 1842, OP; Phillips to Ramsay, 10 July 1842, RP; Sanders to Phillips, 19 July 1842, PP.

31. Murchison to Sedgwick, 8 April 1841, Se P; *Athenaeum*, 1841, p. 675 (q); Rudwick, 1985, p. 375.

32. Phillips to De la Beche, 14 Dec 1841 (q), 15 Feb, 31 March, 27 May, 2 Sept 1842, DLB P; Murchison, 1841; Murchison to Sedgwick, 26 Feb 1842, Se P; Murchison, 1842, pp. 645-9; Rudwick, 1985, pp. 381-6.

33. Rudwick, 1985, p.389-93; Murchison, 1843a, pp. 83, 106; d'Archiac and Verneuil, 1842, pp. 335; Murchison, 1843b; Étienne Jules Adolphe Desmier de St-Simon d'Archiac (1802-68), *DSB*, affluent French palaeontologist; Philippe Édouard Poulletier de Verneuil (1805-73), *DSB*, Parisian lawyer; Alexandr Andreevich von Keyserling (1815-91), *DSB*.

34. Rudwick, 1985, pp. 375, 390; Secord, 1986a, pp. 130-2, 140-2; Phillips to Sedgwick, 24 Jan 1842, Se P (q); Sedgwick, 1843, p.223.

35. Gerstner, 1994, p. 122; De la Beche to Phillips, 25 Oct 1841, Phillips to Harcourt, 26 Oct 1841, HP; Phillips to De la Beche, 26 Oct, 8 Nov 1841, 15 and 23 Feb 1842, 18 (q) and 19 April 1842, DLB P; De la Beche to Buckland, 1 Nov 1841, BL Add Mss 40494, ff.9-12; Buckland to Peel, 4 Nov 1841, BL 40494, f. 7; Harcourt to Peel, 6 Nov 1841, BL 40494, ff.85-8; Peel to Buckland, 8 Nov 1841, BL 40494, f. 13; Peel to Harcourt, 10 Nov 1841, BL 40494, f89; Harcourt to Phillips, 12 Nov 1841, PP; Murchison to Peel, 12 Nov 1841, BL 40494, ff.331-3; Peel to Murchison, 13 Nov 1841, ff. 334-5; Henry Darwin Rogers (1808-66), *DSB*, expert on geology of Pennsylvania; Andrew Crombie Ramsay (1814-91), *DNB*, director-general of Geological Survey 1871-81.

Chapter 7

The Geological Survey 1841-1849

7.1   Maps, sections, and obsessions

Late in 1837 De la Beche transferred the work of the Geological Survey to south Wales partly to pursue economic geology by surveying its coalfield and partly to revive his interest in the complicated geology of Pembrokeshire on which he had published a memoir and map in 1823. A basic responsibility of the miniscule Survey was to colour geologically the one-inch-to-the-mile Ordnance Survey maps so that the main areas and boundaries of different strata were shown. Having begun near Swansea in Glamorgan, De la Beche expanded his Survey westwards through Carmarthenshire into Pembrokeshire and its coalfield, and eastwards to the coal measures of the eastern end of the Mendip Hills, Bristol, and the Forest of Dean. In 1842 the Survey pushed north of the Forest to the Malvern Hills, which lay south of the Bewdley coalfield near Kidderminster. By 1846 the Survey had issued geological maps which covered not only Cornwall and Devon but also five counties of south Wales (Pembrokeshire, Carmarthenshire, Glamorgan, Brecknock, Monmouthshire) and parts of Herefordshire, Gloucestershire, and Somerset. In the rubrics of the maps De la Beche stressed that his Survey was mapping areas containing or bordering coal fields.

This emphasis on economic geology was also revealed in two sorts of geological sections published by the Survey to supplement its maps. By 1846 seventeen sheets of traverse sections and fifteen of columnar ones had appeared. The former showed the sequence and changes of strata along particularly interesting traverses. They were novel in three ways. They used a big scale of six inches to the mile to show detail, they avoided distortion because they used the same vertical and horizontal scales, and they deployed a new way of depicting the lithologies of the strata. Using a scale of 132 inches to the mile, the columnar sections were mostly devoted to coal measures and gave stratigraphic details which were too fine to be included in the traverse sections. The steady publication of maps and sections, each of which was separately two-dimensional, gave when they were combined a three-dimensional representation of strata which had been produced by group endeavour and the regular use of measuring instruments. Such publication enhanced the credibility of the Survey with the public and government, gave it authority with fellow geologists, and revealed that it was capable of work beyond the reach of private individuals.[1]

Secord has also stressed that in the 1840s De la Beche's Survey did more than produce maps and sections. De la Beche wished it to promote general views in geology, especially those concerned with past environments, erosion, deposition of strata, and the distribution of past and present life on the earth. In 1846 the Survey's wide interests and its communal research programme were revealed in the first volume of Survey *Memoirs*, to which De la Beche contributed a huge essay on the ways in which the rocks examined by the Survey had been produced. The Survey's commitments were apparent, too, in part one of the second volume of *Memoirs* published in 1848, the 400 pages of which were devoted to Phillips' research on the Malvern Hills and area. Publicly, Phillips happily acknowledged that De la Beche and he had enjoyed 'innumerable conferences and discussions on the philosophical bearings of the facts' discovered by the Survey and that generally De la Beche had excited and encouraged a spirit of philosophical enquiry which pervaded it. Privately, Phillips went so far as to tell De la Beche that he had been gratified in the field by 'unrestrained communication of opinion and candid discussions of phenomena'. This emphasis on the theoretical bearings of the Survey's work was apparent in its approach to maps and sections. De la Beche and Phillips worried about the colouring of formations on the Survey maps because the representation of a complicated area in one particular colour ignored its complexity. As Phillips told De la Beche, 'what we esteem to be truth often looks best *without colour* ... '. Both were only too well aware that a geological map looked assertively positive and reliable but in fact it fixed on a chart unstable doubts and hazardous inferences. The traverse and columnar sections were motivated by De la Beche's theoretical obsessions. From the former he hoped to infer the manner in which the depicted deposits had been formed. The latter, which recorded the detailed stratigraphy of dozens of beds of different rocks, indicated to him the abstraction and distortion involved in categorising them as a single geological formation and confirmed his suspicion of geological systems.[2]

7.2   Halcyon years

From spring 1841 until spring 1844 Phillips was employed as the Survey's palaeontologist on an annually renewed basis. Then for six months he was paid by special arrangement because it was unclear how he could combine his Survey responsibilities with those of his chair at Trinity College, Dublin. On 30 September 1844 his first period of employment with the Survey ended. In these fruitful years Phillips occupied a post which he saw as central to its work. He thought that it was particularly important for him to examine *in situ* the distribution of fossils in stratified rocks, to arrange on the spot or at home the data collected in the field, and to consider the important general questions which arose. Simultaneously Phillips was proud of the Survey's novel approach which involved scrupulous and exact delineations of the areas occupied by different rocks, the boundaries between them, and the dips, thicknesses, and flexing of the different beds. This sort of geological surveying had never been attempted, except in mining districts, never mind regularly carried out. After his anxieties of the late 1830s Phillips found that the exercise and fresh air of field work improved his health greatly. He enjoyed the camaraderie in the Survey and welcomed the opportunities of teaching his young comrades the skills of field work, from fossil hunting to measurement techniques. Initially, all the recruits were not salaried but waged and some found that the Survey could involve drudgery and disappointment. But for Phillips, a salaried employee, the Survey gave him the opportunity of seeing many new phenomena and increased his power of generalising on geological questions. Until the disappointment of the Irish Geological Survey affair in 1844, Phillips enjoyed a happy, easy, and bantering relationship with De la Beche who treated him as an equal. Generally Phillips liked the Survey life so much that he did not repine for his days of lecturing and writing general geological works. In only one respect was he uneasy. As palaeontologist to the Survey he loved being in the field but procrastinated about cabinet work, especially in London. De la Beche became so frustrated with Phillips' evasiveness that he used the imminent reorganisation of the Survey to appoint Edward Forbes as its palaeontologist in November 1844.[3] Having left the Survey in autumn 1844 Phillips rejoined it on 1 April 1845 on a special commission to survey the north of England. Though still on a salary of £300 a year, the same as that of Forbes and Ramsay, who had been promoted to be local director for England and Wales, his second period with the Survey lacked the heady excitement of his first: there was no equivalent to the *Malvern memoir* and the roll-call of pupils was much smaller. When Phillips agreed to be a government commissioner reporting on ventilation in coal mines, he was required to resign from the Survey and did so effectively from 30 June 1849.

Phillips' first stint in the field lasted from April 1841 until the end of the year, with an interruption of about six weeks to prepare and attend the British Association meeting at Plymouth (Fig. 7.1). He began at Usk, continued in late spring at Tenby for two months, resumed in August at Bristol, and spent much of the rest of the year in Pembrokeshire and Carmarthenshire, St Clears being his main centre. His chief concerns were mapping boundaries, making coastal and inland sections, collecting fossils, exchanging queries and results with De la Beche by letter, and later in the year correlating the strata around Llandeilo with those further west. He was greatly helped by his sister who brought her maid and their dog to Tenby and Bristol, where they rented property which functioned as his temporary headquarters. She looked after his instruments and small portable library as well as helping to sort and catalogue the fossils and minerals he had collected. She accompanied him to

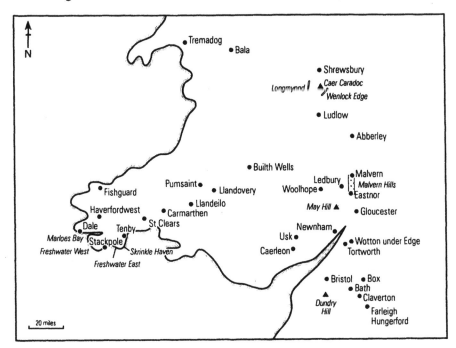

Fig. 7.1   Map showing important geological sites in Wales, the Welsh borders, and south-west England, for Phillips' work for the Survey.

the British Association at which her 'merry laugh' was appreciated. She would have organised a third base in autumn had not a temporary tenant of their home left it in such confusion that she refused to let it again that year. At Bristol Phillips was mainly working on the fossils he and his aides had collected, but during the Tenby period he often spent several days away from his headquarters, staying locally near his research site, and then returning to base. In the autumn he moved around from site to site, again staying locally, and he began to use a horse regularly to reach them, thus saving time and energy for field work on foot.[4]

The autumn was often wet but only rarely did Phillips retreat indoors. Even after five successive days of rain, he reported that 'we have each day confronted the storms, getting wet as Neptunian geologists ought ever to be'. On occasion he was soaked to '*the bone*', and endured 'rain, of the heaviest and wettest sort, darkness palpable, and cold excruciating'. He did section work on a wet Christmas Day, having left Llandovery the previous day for a 'voyage with oars and fins to Pumsaint' which 'was all one splash, much water shipped'. It was only in November that he devised a combination of hat, oiled silk coat, and double proof shoes, to keep himself reasonably dry. In compensation for his aquatic labours Phillips was not only buoyed up by the prospect of ultimately succeeding De la Beche but also enjoyed idyllic days in June, especially on the Marloes peninsula in Pembrokeshire. He felt satisfied and energised by the Survey's 'merry fellows (some of them joyous), hammering, measuring, reasoning, running, swimming – without a shade of any thing but good humour'. He felt privileged to have 'this *merry* company', which in his view rarely befell labourers in science. The Survey work provided 'a scene of unclouded enjoyment to many who never met in so much friendship before'.[5]

Phillips' second year of field work was spent mainly around Malvern where he and his sister, having let their York home for nine months, rented two headquarters, one in April and May 1842 and the other from July to mid-December. As before, Phillips had to forsake his Survey work to attend in Manchester the British Association meeting, which in prospect he regarded as an intrusive bore. In contrast with 1841 his sister did not go with him to the BAAS: she stayed in Malvern overseeing their removal. They enjoyed the pure air and water of the spa town. Malvern turned out to be a better centre for transport than Phillips anticipated: he used regular coaches, occasionally hired a phaeton, and late in the year at Gloucester took the train. At his research sites he worked mainly on foot until July when he began to use a pony as a useful aide de camp in distant rambles. Geologically Phillips and his sister enjoyed 1842. He revelled in the structural problems posed by the Malvern Hills and by the end of the year had done much of the basic research which confirmed his provisional answers to them. He was delighted with his sister's crucial discovery of a conglomerate which is still known as Miss Phillips' conglomerate. As De la Beche's lieutenant general in the field, he continued to refine the Survey's instrumental skills and to teach them to the noviciates. He remained proud of the Survey's minuteness and accuracy which in his view were unattainable by what he regarded as the necessarily hurried way in which 'gentleman geologists' worked.[6] Indeed when distinguished geologists, from home and abroad, wanted to see for themselves in 1842 the Survey's novel techniques, they visited Phillips' corps in action and not De la Beche's.

Under what he felt was the gentle and pleasant rule of De la Beche, whom he addressed as your worship, your reverence, your excellency, or good masthur, Phillips was a free knight of the hammer. In a light-hearted correspondence, he continued as De la Beche's principal adviser on mapping matters, making for example an important input in early summer 1842 to the colouring of the Carmarthenshire sheets. Phillips appreciated that banter sustained the Survey's *esprit de corps* so, knowing De la Beche's penchant for caricature, he flavoured his letters with parodies which often made a serious point. To show his suspicion of rigid geological systems he quoted to De la Beche a triad, allegedly by Pope:

> "Siluria to old red is near allied
> And thin partitions do their bounds divide
> In the same knot is crooked Cambria tied."[7]

Phillips' third year of field work was spent mainly at Wotton-under-Edge, on the edge of the Cotswold Hills, and at Claverton, just east of Bath, an area with which he was familiar from his boyhood. He reached the former in late April 1843, having reviewed perplexing geological points at Malvern and Newnham in the Forest of Dean. For three months at Wotton he 'prosecuted steadily our trade of oolitising', as he told his *chef de battaile,* De la Beche. Though his sister, it seems, stayed in York, Phillips enjoyed Wotton where the only drawbacks were capital dinners, lovely weather, and easy work. He used his pony to ride from Wotton to areas shown on the edges of the maps he was colouring and was thus saved from taking several temporary lodgings. He was happy to receive the occasional visitor such as Lyon Playfair, an ambitious young chemist, and Alphonse Favre, later director of the geological survey of Switzerland. After a short trip to London, Phillips went to Ireland for the British Association meeting held in Cork in August. While there he examined the promontories of counties Cork and Kerry, the latter with Murchison and Griffith, because it was important for the Survey's Welsh work to know about the peculiarities of the alleged Cambrian, Silurian, and old red sandstone strata in southern Ireland. Contrary to the representation on his *Index* map and Griffith's view, Phillips concluded that in Kerry the dominant rock was old red sandstone. Having visited Birr Castle and looked at the moon through the Earl of Rosse's telescope, he then made a quick survey in mid-September of Bala in north Wales where he concluded that, if the Berwyn Hills were sub-Silurian, there was a paradox: the fossils at Bala were Silurian yet the traverse sections seemed to indicate that the Bala rocks were lower and older than Silurian and therefore Cambrian. After a short period at York, Phillips left his home and sister in mid October, met De la Beche at Frome in Somerset, and then spent about six weeks at Claverton where he found that tracing the boundaries of such familiar strata as the forest marble and cornbrash was so difficult that he was more plagued than Job. In mid-December he spent a few days at Frome, again meeting De la Beche, before going home.[8]

Phillips' fourth stint of field work was interrupted not so much by the British Association, which met in York beginning 26 September 1844, but by his new position as the first professor of geology and mineralogy at Trinity College Dublin. He was elected in December 1843, effective March 1844. In January Phillips had formally accepted this offer and told De la Beche that he was required to give at least twelve lectures in each of the three terms, beginning in November, February, and May. If delivered at the rate of two a week, he would still have thirty-three to thirty-four weeks left for Survey work. He proposed to De la Beche to spend the winter and Easter vacations, each of six weeks, in Ireland and the long one of twenty weeks in England. He called this 'a pretty fair division of the eight months'. He also said that he had no intention of resigning his post with the British Association but failed to state that usually its annual meeting consumed about a month of his time. Even so Phillips claimed that he could combine his Dublin chair with his Survey responsibilities, whether as its palaeontologist in England and Wales or more easily as the first director of a new Irish branch of the Ordnance Geological Survey.[9]

Given these new circumstances De la Beche failed to persuade the Board of Ordnance to pay Phillips for eight month's work in the financial year 1844-5. Not surprisingly it took the view that his chair precluded his being available for employment in England and Wales for that length of time. As a *pis aller* De la Beche managed to make special arrangements for Phillips to be paid for six months from 1 April 1844. At the end of May 1844 at the latest Phillips was aware of the Board's view and his imminent departure from the Survey. In April 1844 Phillips returned to Malvern where he revised and extended his previous work there. Then at the beginning of May he left for Dublin to begin his first course of lectures on 10 May. Later that month, in a gap between them, he returned in twenty hours to Malvern for a week's research on the Abberley Hills. He was desperate to show that his Dublin chair did not entirely preclude his working in England for the Survey. The week was rewarding: his sister discovered more conglomerate which confirmed further his interpretation of the structure of the Malvern Hills. Less happily, at the end of May Phillips found his competence as a mapper challenged for the first time by De la Beche and implicitly by the Survey's two most promising field workers, Andrew Ramsay and Henry Bristow. Phillips was sure that the marlstone, on the south side of the valley at Box, was thin and discontinuous whereas Ramsay and Bristow were convinced that it was continuous. To De la Beche's irritation Phillips was unable to examine the disputed boundaries near Box because for most of June he was professing in Dublin, having left his sister in Malvern. Even on his return to Malvern in late June, Phillips jibbed at visiting the Box valley, partly because De la Beche had already revised the boundary lines. Phillips preferred to finish his Malvern work but while doing so received in mid-August from Henry James (not De la Beche) the disappointing news that James was to direct the new Irish Survey. Probably in late August Phillips went home from Malvern, to prepare for the approaching York meeting of the British Association and to face a future in which he would no longer be the Survey's palaeontologist and De la Beche's most experienced field worker.[10]

7.3   Surveying and pupils

As De la Beche's field lieutenant from 1841 to 1843, Phillips deployed techniques of geological surveying which drew on his long experience of measuring devices which he handled far more adeptly than De la Beche. These included theodolites and chains which he had learned to use in the early 1820s when he was an apprentice surveyor. Phillips thought that the Survey was unique in using exact measurement and not conjecture to generate accurate delineation of the boundaries of strata and detailed traverse sections which were true to nature because their horizontal and vertical scales were the same. The Survey's techniques were different from those of such an accomplished gentleman geologist as Fitton who at that very time acknowledged that his published section of greensand on the Isle of Wight was not metrically exact: he had measured distances with a rope 300 feet long, and heights and thicknesses with a light rod twenty feet long. Even Murchison, who in 1842 saw for the first time the Survey corps in action under Phillips, applauded its techniques of 'trigonometrical mensuration' made by Phillips' theodolite men. By 1845 an established member of the Geological Survey, such as Bristow, had at his disposal no fewer than eight sorts of surveying instrument, namely, a five-inch theodolite, a prismatic compass, a clinometer for measuring the dip of strata, an ivory protractor, an ivory plotting scale, three ranging poles, a chain (Gunter's) with arrows, and a tape fifty feet long.[11]

Phillips' obsession with measuring instruments showed itself in the new words he coined in connection with the Survey's field work. By autumn 1841 he was using such verbs as theodolize, clinize, and barometrize. Next year his fruitful facility with measuring instruments was made abundantly clear. For making approximate short sections he used a mounted small spirit level which could be moved over ground nearly at walking rate. He thought a theodolite far superior to a chain on sloping or rough ground so he was concerned to increase its accuracy to 0.1% by inserting measuring hairs in it. In his surveying of the Malvern Hills, Phillips measured the heights of Worcester Beacon and North Hill in two independent ways, ie trigonometrically using a five-inch theodolite, the largest that could be carried by hand, and barometrically. By both means he obtained 1,444 feet as the height of the Beacon, which agreed with the Ordnance Survey's figure which Murchison had adopted for his Silurian work. For North Hill Phillips obtained 1,359 and 1,361 feet by his two methods, showing the accuracy of each and casting doubt on Murchison's figure of 1,335 feet. Henceforth Phillips could rely on elevations determined with a mountain barometer when it was impossible to use a theodolite. He also graduated a levelling staff for reading distances 'to dispense with chaining and thus to allow of an order being given to *measure everything*'. Provided it incorporated a spirit level, rested on level pegs, and was used forwards and backwards, it attained an accuracy of 0.2%. He doctored and set up various measuring instruments for his colleagues. Even De la Beche relied on Phillips for detailed advice about levelling staffs and theodolites so when Murchison and Keyserling wished to inspect the Survey in action they visited Malvern in late July 1842 and returned in early September expressly to see how Phillips made a traverse section westwards from Worcester Beacon. It was this very section, which recorded reverse dips and strikes of Silurian strata to the nearest degree, which was avowedly central to Phillips' anti-Murchisonian conclusions about the structure of the Malvern Hills (Fig. 7.2). Murchison and Keyserling, who had been sent by his government to learn about the Survey's way of making sections, found Phillips busy with his theodolite and were impressed by his approach which was so different from their own hop-step-and-jump kind of field work in Russia: he was elevating geology by making it in part a physical science. The key to Phillips' success, it seems, was that when possible he avoided the use of chaining in favour of a staff and a theodolite, the telescope of which contained either measuring hairs or two vertical wires on each side of a central one. Indeed, in 1850 Beete Jukes, who became Director of the Irish Geological Survey that year, had his own theodolite altered so that he could make sections as Phillips did. No wonder that in his *Malvern memoir* Phillips included a drawing of a Survey geologist at work, not with a hammer, but with a theodolite on a tripod (Fig. 7.3).[12]

From spring 1841 to the end of 1843 Phillips and De la Beche sometimes joined forces in the field but more often Phillips either led his own corps of surveyors or was the senior man left in the field when De la Beche was engaged in administration in London. As the Survey's director who gave social and intellectual cohesion to it, De la Beche was known to its young members as Daddy. It is unclear whether he routinely taught field techniques but it is certain that every field geologist who

The best and most interesting of all the sections, that beneath the Worcestershire Beacon, may now be described at greater length, commencing on the east, and proceeding toward the west, where the Mathon Park road crosses the little valley, in the lower part of the Wenlock shale, of which a thickness of nearly 50 feet is here exposed.

The following diagram may be consulted.

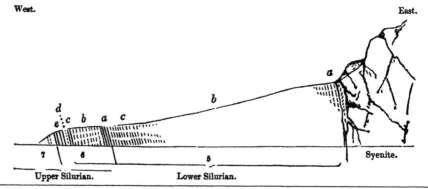

Fig. 7.2   Phillips' traverse section west of Worcester Beacon, Malvern Hills, was interesting because it showed two phenomena which Murchison's interpretation could not explain: the conglomerate between the syenite and the Caradoc sandstone (5); and the reversed dip of 63° of the Caradoc, the Wenlock limestone alternations (6), and the Wenlock shales (7). Using this section Phillips reached the anti-Murchison conclusion that the distinction, between upper and lower Silurian strata and periods, was not firm but arbitrary.

worked for the Survey from 1841 to 1843 at some time learned valuable skills from Phillips. Many of these noviciates were fifteen to twenty years younger than he, so to them he was 'old Phillips' because he was only three years younger than De la Beche and belonged to his generation. They respected but did not share Phillips' sense of propriety. In 1847 Bristow wished to be remembered via Ramsay to Phillips but advised that he should avoid Cerne Abbas in Dorset because he would be upset by the famous giant, an enormous effigy cut into chalk hills 'in a most outrageous state of excitement'. To Phillips the Survey youngsters were sympathetic friends who made up a happy family party. He was particularly fond of the Malvern area because it was excellent ground on which to run 'a capital bit of a school for the younger members' of the Survey.[13]

Four of his pupils stayed with the Survey, three of them for forty or more years, and rose to different degrees of eminence. Ramsay became its third director general, being knighted on his retirement. Bristow, who was handicapped by inveterate deafness, succeeded Ramsay as director for England and Wales. Talbot Aveline ended his career in the next most senior post, that of district surveyor (1867-82) and was awarded the Murchison medal of the Geological Society in 1894. It is surely not accidental that the three top men in the English and Welsh Survey in the 1870s all

Fig. 7.3   **Phillips' drawing of the Geological Survey in action under his direction in the Malvern area in the 1840s. The geologist is unusual: he is neither hammering rocks nor collecting specimens in a bag; he is using a measuring instrument.**

learned from Phillips in the early 1840s. A fourth pupil, Richard Gibbs, worked as a highly effective fossil collector from 1843 to 1872. Though William Baily cannot be considered as a pupil of Phillips, he was keen to work for the Survey drawing fossils under Phillips' superintendence and, having been vetted by Phillips, was appointed by him as draughtsman effective January 1844. Phillips then urged De la Beche that Baily be employed as a Survey geologist, a move accomplished in 1845. Phillips and Baily remained on good terms: after the deaths of three of Baily's offspring, Phillips sent him a cheque as a token of sympathy and regard.[14]

Ramsay was an experienced field geologist when he began work with the Survey in spring 1841. At the meeting of the British Association held in 1840 at Glasgow he had made such a favourable impression with his map, sections, and model of the Isle of Arran, and his accompanying paper, that Murchison successfully urged De la Beche to employ him. When Ramsay began in Pembrokeshire his coast sections were vetted by Phillips whose mapping he admired. Ramsay and Phillips soon became good friends and when working in different areas kept in touch by correspondence in which Phillips acted as a commentator on Ramsay's discoveries and gave advice on fossil collecting and locations. Their camaraderie was cemented by an addiction to such bad puns as hand right in for handwriting and litterarrear for literary. In 1843 Ramsay learned how to map oolitic strata, with which he was unfamiliar, between Wotton and Bath under what he regarded as Phillips' kind supervision and was shown how to distinguish between adjacent strata such as forest marble and Bath oolite. Even after 1845, when Ramsay became local director for England and Wales, he still benefited from Phillips' expertise on such matters as the pros and cons of different makes of mountain barometer. Though they inevitably lost their previous closeness,

Ramsay remained a warm admirer of Phillips as an 'ancient master' or 'old master' of both geology and himself.[15]

Bristow joined the Survey in April 1842, having studied from 1838 to 1841 at King's College London where he could have attended Phillips' last course of lectures. Bristow registered at King's as a student of civil engineering and mining, became an associate of King's (equivalent to a degree), and compiled a descriptive catalogue of the minerals in King's Museum. In April 1843 Bristow moved from Wales to the Cotswolds where he soon impressed Phillips who acted as his schoolmaster, teaching him to draw fossils, to make sections, and to map boundaries. Phillips introduced Bristow to oolitic strata at Wotton and encouraged him by giving him a copy of his *Guide* and by supporting his successful candidature as FGS in 1843. After they had gone their separate ways, Bristow still consulted Phillips on such a technical surveying matter as converting a true dip to the dip on the line of a section. When Bristow was promoted to the directorship for England and Wales, he went out of his way to tell Phillips that, as an old pupil, he remembered their pleasant days together and the advantages he had derived from Phillips' teachings.[16]

Aveline and Gibbs were not as close to Phillips but he helped each of them to become good all-round field workers. Aveline, who had joined the Survey in August 1840, was Phillips' assistant for much of the period 1841-5: Phillips gave him a copy of his *Guide* and instructed him in fossilising, mapping, and sectioning. Having learned useful techniques from Phillips, Aveline's silent demeanour precluded his becoming a disciple. His singular taciturnity became a legend in the Survey. On one long day in the field, he uttered only three words in response to his hammering: in the early morning 'grits' and in the late afternoon 'more grits'. Phillips' verdict of 1845 about his 'good natured young friend' was prescient: Aveline was 'very steady and sure of hand and head; and is worthy of being taught and brought out a bit'.[17] Gibbs was appointed to the lowly position of fossil collector in May 1843 and helped Phillips until 1845. Phillips liked Gibbs' work, successfully pressed De la Beche to increase his wage, and extended his skills by teaching him to map boundaries and to use a clinometer when producing sections. Gibbs sacrificed much for the Survey: he disliked winters and sometimes went almost a year without seeing his children.[18]

Trevor Evans James was also loyal to the Survey but suffered a different fate. Having joined it in 1840, he spent much time from 1841 to 1844 with Phillips in the field where generally he was a good aid. Phillips drilled him on the lithology and mapping of problematical strata, set him to work on sections, guided him on trigonometrical mensuration, and identified fossils he had collected. Known as Jacobus 2 he worked for a short time under Henry James, Jacobus 1, on the Irish Survey but their relations were so strained that in autumn 1845 De la Beche sent him to Derbyshire to be Phillips' assistant once again. Next year De la Beche, who had renewed his suspicions of James' mapping, told him in April that he had to leave in September owing to reorganisation of the Survey. Though James always like Phillips, James so resented De la Beche's shameful deceit that he told Ramsay that 'it would have been much better for me if I had never been on the Survey for now my prospects are blasted forever'. Phillips had been happy to tutor in Derbyshire an old friend with whom he had 'so long wrought together': when he heard about James' unhappy fate he gave him advice and lent him 22 pounds to keep him going.[19]

Three of Phillips' colleagues in the field resigned from the Survey. Josiah Rees, who had joined in 1839, was supervised by Philips in late 1841 and quickly impressed him. Rees helped Phillips, whose advice he valued, in the Malvern work next year.

To Phillips' regret, Rees resigned in late 1842 and eventually became a judge in Bermuda.[20] Alexander Murray and David Hiram Williams were also supervised by Phillips for two months in summer 1842 in the Malverns, to help with the work that Aveline, James, and Rees were doing under his direction. Murray joined the Survey in 1842 to prepare for his subsequent work with the Canadian Survey and resigned next year. He impressed Phillips who deemed him a useful hand, a good observer, and a promising man. Phillips' prescience was shown by Murray's career as founder and first director of the Geological Survey of Newfoundland.[21] Williams, who had joined the Survey in 1841, suffered a tragic end. He resigned in 1845 to survey India for coal for the East India Company for £800 a year. Having fallen from an elephant, he died in 1848 of jungle fever, leaving a widow and four young children.[22]

Another pupil of Phillips was Henry James who ironically was appointed in 1844 director of the Irish Geological Survey, a post which Phillips coveted. In 1842 the Board of Ordnance wanted any Irish geological work, if resumed, to conform to De la Beche's English and Welsh Survey so it sent James, a geologically minded engineer who had been promoted to captain in June, to learn the Survey's field procedures. He spent much of autumn 1842 as an apprentice field geologist under Phillips, acquiring valuable techniques which he realised he could gain in no other way. He saw how to run a section at Malvern, he spent agreeable days at Wotton learning about the liasic and oolitic strata, and he was taught to draw fossils. Phillips also supported his candidature as FGS. Even when director of the Irish Survey, James invited Phillips to visit it and wanted Phillips with him in the field at Wexford to guide him.[23]

The most important of the volunteers who learned from Phillips was his close friend William Sanders, who joined Phillips as a pupil in south Pembrokeshire in late spring 1841 and brought with him from Bristol another volunteer, Samuel Stutchbury, curator of the Institution there. From June Sanders worked solo for the Survey, which had agreed to pay his expenses, on the geology of the Bristol area. Hoping to insinuate himself into the Survey, he put in 152 days in 1841; but De la Beche, who was happy to have him as an unsalaried surveyor, did not reply to his claim for expenses incurred. Sanders felt slighted and peeved, especially as he was out of pocket in 1841 to the tune of £114. By February 1842 Sanders was so disappointed and hurt by De la Beche's behaviour that he decided to reduce his work for the Survey, a move that Phillips felt was a loss. The trusting Sanders had not fully appreciated De la Beche's weasel words that the Survey would find money which 'might tend to cover, if it didn't not altogether cover, which probably it would' Sanders' expenses. By 1844 he had lost all hope of being permanently employed on the Survey: regarding himself henceforth as an amateur, he produced in 1862 at his own expense his map of the coalfield of Gloucestershire and Somerset on the generous and revealing scale of four inches to the mile, an achievement which led to his FRS in 1864.[24] Though local clergy and landowners sometimes facilitated Phillips' field work, only one aristocratic family joined his corps temporarily. The Earl of Cawdor and his son, Lord Emlyn, not only offered hospitality at Stackpole Court, in south Pembrokeshire, and at Golden Grove, Llandeilo, but the young laird joined Phillips for geological walks, hammers in hand.[25] Another temporary helper of Phillips in 1841 was Tremenheere, described by Phillips as a 'gentleman amateur'.[26]

Phillips had a distant pupil in the USA in Henry Darwin Rogers, who from 1835 was in charge of the geological survey of New Jersey and from 1836 that of Pennsylvania. In 1833, while on an extended visit to England, Rogers not only enjoyed the privilege of learning surveying techniques from De la Beche in Devonshire but also attended the British Association meeting at which Phillips became a mentor. Next year Rogers' commissioned report on the geology of north America was presented to the Association by Phillips and Murchison and took pride of place in its *1834 report*. By 1840 Rogers saw himself as a distant disciple of Phillips via correspondence. Rogers admired the 'just and philosophical cast' of Phillips' works and anticipated that the proudest privilege of his next visit to England would be to sit as a pupil at Phillips' feet. In May 1842 he wanted Phillips to visit the USA to give to its geologists a beneficial impulse, as Lyell had just done with his tour. Rogers was too busy to travel to the meeting of the British Association in 1842; but a condensed version of his paper on the structure of the Appalachian mountain chains, written with his brother William, was presented to it through Phillips' good offices. Their surveying was like that of the British Survey in that in their traverse sections the vertical and horizontal scales were the same, thus avoiding vertical exaggeration. Their mapping showed that the strata of the Appalachians, 1,500 miles long and 150 miles wide, were regularly and sinuously flexed and sometimes overfolded. They thought these phenomena indicated that lateral and vertical compression had occurred and they explained this combination of forces by assuming the existence of undulatory waves in the earth's crust. At the 1842 meeting De la Beche welcomed the Rogers brothers' data but rejected their explanation. Phillips was too busy on secretarial business to hear the paper but he liked its mechanical reasoning. After 1842 the brothers continued to view Phillips as their geological father to whom they were indebted for valuable philosophic methods, views on the elevation of mountain chains, palaeozoic classification, and advice about conducting geological surveys. Eventually they visited England separately. In 1848 Henry regarded Phillips as a zealous and most useful friend and looked forward to a glorious time with him in the field in Derbyshire. Next year William met Phillips at the British Association, continued to bracket Phillips, Sedgwick, Murchison, and De la Beche as leading European geologists, and was gratified by the way in which his new wife was looked after by Phillips' sister. It is therefore not fanciful to see the pioneering work of the Rogers brothers on folded mountain structures as encouraged by Phillips, who continued to regard Henry as a capital geologist.[27]

7.4   Sparring with Murchison

In the early 1840s the Geological Survey broadly confirmed and extended Murchison's Silurian domains, but Phillips' detailed investigations showed Murchison's inadequacies as a field geologist on the outskirts of Siluria as defined in his *Silurian system*. In 1841 in south-west Wales Phillips found deficiencies in Murchison's coast section at the key location of Marloes Bay and mistakes in Murchison's map engraved in 1837 for his 1839 book. Murchison was utterly wrong on the carboniferous limestone near Laugharne and had missed a conspicuous igneous intrusion near Caermarthen. Phillips gleefully told De la Beche that what Murchison depicted as lower Silurian was a post- 1837 volcanic hill containing felspar. In 1842 even Phillips' field pupils, Rees and Trevor James, found errors in Murchison's map as did Phillips himself who surmised that Murchison was in a 'mesmeric dream' when he made the 'sad blunder' of confusing lias with new red sandstone over an area of five square miles near Gloucester.[28]

These corrections by Phillips of Murchison's map constituted continuing differences between them. Though the controversy between the Survey and Murchison about the Devonian question ended in 1843, Phillips' research on the Malverns, carried out from 1842 and fully published in 1848, showed that in his hands the Survey was still at odds with Murchison, the leading British gentleman stratigrapher. It is significant that in response Murchison neither defended himself nor attacked Phillips, as he had done in 1842 apropos *Palaeozoic fossils*. He realised that Phillips' elaborate and precise approach, and the results to which it led, were beyond the capacity of the gentleman devotee working solo. Indeed, in 1854 in his *Siluria*, which revised and enlarged his *Silurian system*, Murchison praised Phillips' *Malvern memoir* as reflecting great credit on the Survey. Perhaps he was flattered by Phillips' assertion that most of his facts and inferences about the Malvern area were 'remarkably in harmony with the earlier statements and original opinions of the author of the SILURIAN SYSTEM'. Phillips gave the impression that his *Memoir* corroborated Murchison's views and classifications, except on minor points, so that in a recent authoritative account by Oldroyd it has been depicted as mainly Murchisonian; but Phillips' modesty was deceptive. His *Malvern memoir* not only continued and extended the critique in *Palaeozoic fossils* of Murchison's simplistic use of fossils: it also offered an anti-Murchisonian interpretation of the geological history of the Malverns, which Murchison quickly accepted and subsequently incorporated into his *Siluria*.[29]

In his *Silurian system* Murchison's outline description of the Malvern Hills revealed that his interpretation was basically Huttonian. Like others before him, Murchison was impressed by the peculiar character of the ridge of the Malverns. Made mainly of syenite, an igneous rock, it was straight, long (eight miles), and narrow (half a mile). Immediately on its west were distorted Silurian strata and to its east undisturbed new red sandstone. Using his knowledge of Scottish intrusions, such as the famous one at Glen Tilt cited by Hutton, and the well-known dykes on the Isle of Arran, Murchison postulated that molten syenite had forced its way upwards through a straight, long, and narrow fissure to form a ridge. Simultaneously the earth movements altered and bent the surrounding Silurian rocks. For Murchison the syenite ridge was therefore younger than the Silurian strata. Subsequently new red sandstone strata were deposited peaceably to the east of this ridge. This Murchisonian interpretation was espoused by Phillips before he began serious field work in the Malverns, an area which Phillips regarded as Murchison's own kingdom.[30]

Phillips' detailed surveying soon revealed that the igneous rock of the Malvern ridge was endlessly varied: though mainly composed of syenite, it also contained granitic and schistosic materials. Parts of the ridge rock were laminated and sometimes metamorphosed, which suggested to Phillips that they existed before the syenite and had been partly changed on its eruption. Phillips also found mistakes in Murchison's mapping of igneous rocks. West of the Malverns near Eastnor Murchison had missed totally the igneous rocks which occurred as hills and bosses and were different from those of the Malvern ridge. In the Abberley district Phillips subsequently discovered that three hills depicted by Murchison as syenite were composed of new red sandstone conglomerate containing syenite fragments.[31]

Once in the field Phillips soon began to suspect Murchison's interpretation of the Malvern Hills. Concentrating on their 'whole *physique*' and 'physical history', and drawing on his knowledge of contact metamorphism derived in 1826 from his Scottish tour, he discovered within two weeks that the Caradoc sandstone, a lower

The next thing to determine was the position of this conglomerate in relation to the ridge of syenitic rocks amongst the detritus of which its fragments lay. This was difficult. The abundance of detritus on all the slopes is so great as to conceal for the most part the junction of the stratified and unstratified rocks. The loose shelly pieces we found abundantly for fully one third of a mile along the mountain side, and at length the conglomerate rock itself was plainly,seen adhering to the extreme western nearly vertical face of the trap mass, west of the Worcestershire Beacon, in a situation laid open by a large excavation close to the road, and north of the little stream.

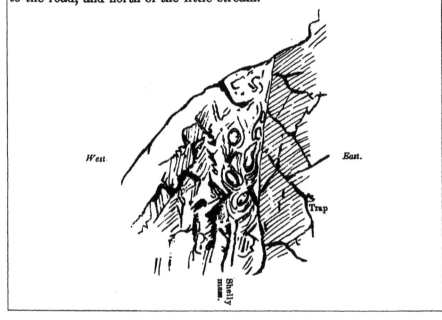

Fig. 7.4　**Phillips' drawing of the Miss Phillips' conglomerate, which she first found in detached blocks west of North Hill, Malvern Hills. She and he then found it, the 'shelly mass', adhering to the syenite, the 'trap', west of Worcester beacon.**

Silurian rock in contact with the syenite ridge and sometimes enclosed in it, had not been altered or baked. This lack of contact metamorphism indicated to Phillips that some parts of the syenite had solidified before the deposition of the sandstone; and it contrasted with Eastnor, where local eruptions of different igneous rocks had baked the lowest Silurian strata, from which Phillips inferred that these particular igneous rocks solidified after the early Silurian period. By late May Phillips had produced what he regarded as a good east-west traverse section of the Malverns, showing the general relation of the Malvern ridge to the flexed palaeozoic strata to its west and the unflexed mesozoic ones to its east. About this time Phillips matured a hypothesis about the relation between the syenite and the Silurian strata. If the solid syenite ridge had been elevated above the sea in the period of the Caradoc sandstone, then bits of the syenite could have broken off, fallen into the sea, been shaped into pebbles by it, and could have become embedded with the molluscs and corals of the Caradoc sea to form a conglomerate adjacent to or even attached to the syenite. On the basis of these suppositions about an ancient local environment, Phillips asked his sister to look for this conglomerate when he was busy in late July and early August 1842 showing his discovery of unaltered Caradoc sandstone to Murchison, Keyserling, and Sedgwick. To his delight on 1 August she found in heaps of fallen stones on the western flanks of the ridge what became known as the Miss Phillips' conglomerate, a discovery which confirmed his view that part of the Malvern ridge was in 'quiet slumber and coolness when the Silurians were formed' and refuted the 'dogma of *elevation in a fluid state*' of the syenite among them. Within a fortnight she and Phillips, who had invited De la Beche to join him, found the vertical contact between the conglomerate and the syenite of the ridge and showed it to reliable witnesses (Fig. 7.4).[32]

Phillips regarded his sister's discovery as a nice hit which should be published immediately to establish the Survey's priority. De la Beche agreed, the paper appearing in the October issue of *Philosophical magazine*. In it Phillips set out briefly the interpretation given at length in the 1848 *Memoir*. Drawing heavily on his own research on large scale faulting and elevation in the north of England after the carboniferous period, he proposed that the syenite of the Malvern ridge was accumulated and indurated before the aggregation of the Caradoc sandstone, and that after the carboniferous period it was raised from the sea, along with vast piles of palaeozoic strata deposited above and around it, along a great fault line which he thought passed through its eastern side. During this violent elevation of two to three miles, the palaeozoic strata were crumpled and even overturned. After the carboniferous period, mesozoic strata were deposited against the fluted and broken edges of the syenite on its eastern flanks (Fig. 7.5). In the *Memoir* Phillips extended his Malvern interpretation to explain the elevation of much of Wales. He postulated that along a great line of dislocation, 120 miles long from Flintshire to Somerset, there had occurred after the carboniferous period great elevatory earth movements which produced a long, winding, and broken cliff, against which the new red sandstone was eventually deposited. Hugh Strickland, who was an expert on this new red sandstone, regarded Phillips' demonstration of this set of operations, very intense and on a large scale but limited in duration, as 'one of the grandest generalisations at which British geologists have arrived'. It was a tour de force of structural geology which owed everything to Phillips' research on faulting and elevation in Yorkshire, carried out before 1829 and published that year, and nothing to Élie de Beaumont's theories of mountain elevation, promulgated from 1829, and Hopkins' theories of elevation of the mid-1830s.[33]

Phillips' skill with instruments was used to produce traverse sections which revealed the subterranean geometry of the strata in the Malvern area by showing their thickness, succession, dip, flexing, and faulting. These sections, which were printed in 1845 independently of the 1848 *Memoir*, were so central to the Survey's reputation that they were not shown to interested individuals until their publication was ordered. It was indeed Phillips' mode of producing measured and not guesstimated traverse sections which induced Murchison to acknowledge the Survey's superiority in this aspect of field geology. Phillips was by no means the first to notice the flexing and folding back of strata west of the Malvern Hills; these contortions had been

Murchison's interpretation of the Malvern ridge (simplified)

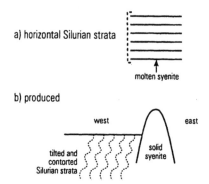

a) horizontal Silurian strata

molten syenite

b) produced

west   east

tilted and
contorted
Silurian strata

solid
syenite

Phillips' interpretation of the Malvern ridge (simplified)

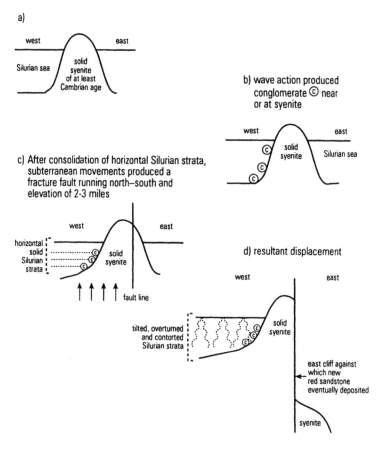

a)

west   east

Silurian sea

solid
syenite
of at least
Cambrian age

b) wave action produced
conglomerate © near
or at syenite

west   east

solid
syenite   Silurian sea

c) After consolidation of horizontal Silurian strata,
subterranean movements produced a
fracture fault running north–south and
elevation of 2-3 miles

west   east

horizontal
solid
Silurian
strata

solid
syenite

fault line

d) resultant displacement

west   east

tilted, overturned
and contorted
Silurian strata

solid
syenite

east cliff against
which new
red sandstone
eventually deposited

syenite

Fig. 7.5   **Conflicting interpretations: Murchisonian intrusion versus
Phillipsian faulting.**

described in 1811 by Horner in a memoir that Phillips admired. Phillips' novelty was to delineate these features with a precision hitherto not attained so that on the basis of measurements of the dip and the thickness of strata he inferred the geometry of various synclines (trough or basin structures) and anticlines (dome structures), parts of which had been denuded. Such flexing included the reversed dip of unfaulted strata in the Malverns, explained by Phillips as the result of anticlinal curvature caused mainly by lateral pressure. He was particularly fascinated by the Abberley Hills, where flexing and faulting, caused by vertical pressure, inverted the normal order of strata by putting the upper Ludlow shales, which were Silurian, above the old red sandstone. Above all, sections helped Phillips to compare the Malvern Hills with other Silurian districts of Abberley, Woolhope, May Hill, Tortworth, and Usk. Sections revealed that the succession and thicknesses of the strata of all these areas were the same, from which Phillips inferred that they were co-eval and had been deposited in one peculiar part of the palaeozoic sea. Sections also revealed in detail 'the mutual adjustment and geometrical relations of the uplifted and depressed strata', from which Phillips concluded that all the Silurian districts, having been part of a sea bed, were disturbed and elevated by a great system of mechanical forces, both lateral and vertical, which operated through one period of time. Thus Phillips' traverse sections helped him to produce a physical history of the Malverns and cognate areas which differed not only from Murchison's, in that it stressed faulting and elevation, but also from Lyell's because it stressed the violent nature of the upheaval of the Malvern area. Phillips' conclusions were no doubt agreeable to De la Beche in two ways. He was always happy to see Murchison and Lyell taken down a peg or two, and for almost twenty years he had been a leading supporter in Britain of Élie de Beaumont's theory of elevation of mountain chains.[34]

The approach to palaeontology taken by Phillips in his Malvern research and *Memoir* continued and developed that adopted in his *Palaeozoic fossils*. It was therefore anti-Murchisonian and it led to Murchison making significant revisions in his *Siluria* of the views expressed in his *Silurian system*. Phillips was at odds with Murchison about the identification of a significant stratum. Whereas Murchison thought that the black shale on the west flanks of the Malverns, which was important as a marker stratum, was of Llandeilo age, Phillips found two types of fossil trilobite in it and concluded that it was older, a claim subsequently acknowledged by Murchison. Phillips thought this discovery so significant that he published it rapidly in 1843 in order to show the superiority of the Survey's field work. More generally his research on Silurian strata in the Malvern area confirmed his suspicion of hard and fast distinctions within geological systems. In this locality he found that the differences between upper and lower Silurian strata were blurred because peculiar local physical conditions had broken or mingled past life, sometimes giving older fossils in later deposits and later fossils in older deposits. Such palaeontological paradoxes had their human equivalent for Phillips: 'in the history of nations the germs of new social systems may be traced ... before older systems have died away; and ... a banished race may return and possess for a time its old domains'. The consequence of this view was that fossil-labelling of strata was only reliable if comparisons were made over very wide areas and over 'geological periods long enough to neutralise the influence of peculiar physical conditions', a point and a phrase Murchison was compelled to accept and cite. In Phillips' discussion of the Usk district, he rejected Murchison's distinction between Caradoc and Wenlock strata within the Silurian system because sandstones, which were mineralogically Caradoc, had been deposited in an ocean full of creatures of the Wenlock period.[35]

Phillips believed that it was possible to distinguish strata by their peculiar fossil species only at a few localities. Otherwise it was necessary to use all the groups of living things of the past. This belief informed Phillips' mode of fossil hunting, which ignored novelty and beauty. Instead he collected as comprehensively as possible in order to secure 'good data for numerical estimates and comparative tables of the distribution of life in the ancient sea, during successive palaeozoic periods'. Such palaeo-ecological results, claimed Phillips, were as reliable and complete as those about the distribution of life in the present-day sea obtained by dredging. Phillips' aim was indeed to unite palaeontology, the science of ancient life, with biology, the science of life. He therefore compiled two sorts of tables for eight zoological classes of fossils. The first sort showed the geographical distribution of the fossil species of a particular class, such as gastropods, between six Silurian districts of west Wales (Marloes, Freshwater, Haverfordwest, Caermarthen, Llandeilo, Builth) and six eastern ones (Usk, Tortworth, May Hill, Woolhope, Abberley, Malvern). The second sort revealed the geological distribution of fossil species of a particular class between eleven Silurian strata for the eastern region as a whole, with extra tables for Marloes (one), the river Towy region (one), Llandeilo (three) in the western region. From these twenty-one distribution tables Phillips concluded, in line with his existing views, that strata were better characterised by the predominance of fossil species than by their mere occurrence. But some of his conclusions were novel. For him the plenitude of Silurian life varied between strata: it had peaks and troughs but showed no grand epochal creations and destructions. Rather there had been in the Silurian period gradual introductions and withdrawals of life; and sometimes older races occasionally returned and new species ceased for a time. Phillips' explanation of these phenomena was that there had been local centres of life which were spread by sea currents and that extinction was caused by a change of physical conditions 'beyond the range of possible adaptation'. Phillips' conclusions were not lost on Murchison who in his *Siluria* acknowledged that his Silurian system was characterised by the profusion of certain fossils.[36]

The palaeontological appendix of the *Malvern memoir* was a collaborative Survey effort which updated Murchison's *Silurian system*. In compiling this catalogue of Silurian fossils, Phillips enjoyed from 1846 the novel experience of being helped by two Survey experts, John Salter, a specialist on fossil crustacea, and Edward Forbes, both of whom revised the catalogue and wrote parts of it. Phillips himself successfully urged that the collected descriptions of fossils be called a palaeontological appendix because, as he confessed to De la Beche, this device delivered field men from a task which was beyond them. It seems that Phillips, the Survey's palaeontologist until 1844, welcomed the assistance of Salter and Forbes: the former had detailed and up-to-date knowledge of trilobites and the latter an approach to description of fossils which benefited from his training in zoology. Consequently the descriptive taxonomy in the Malvern catalogue of fossils was fuller than in Phillips' *Palaeozoic fossils*. The twenty-six plates of Silurian fossils were also a joint stock affair which revised those in Murchison's *Silurian system*. Though Phillips had organised the collecting of the fossils, they were not drawn by him but by Charles Bone, the Survey's artist, and Baily.[37]

7.5   Cabinet palaeontologist

As the formally appointed palaeontologist to the Survey from 1841 to 1844 Phillips liked the field work but persistently procrastinated about preparing for publication descriptions and drawings of fossils collected by the Survey, especially if the required cabinet work had to be done in London. This tardiness was in part caused by the big harvest of fossils collected by him and his assistants during eight months in the field. In the remaining part of the year he was expected to examine these fossils, usually at his home. It became clear to De la Beche within a year that Phillips was making slow progress in his cabinet work. Phillips thought that ideally drawings and catalogues of fossils should be done before the appropriate Ordnance sheets were coloured, but by spring 1842 he confessed to De la Beche that he had fallen behind in describing and drawing the ton of Pembrokeshire fossils that had been sent to York in 300 barrels and that he would need the opening three months of 1843 to finish this task. He also revealed to De la Beche that he wanted the haul of fossils from the Malvern area to be less than that of 1841 and that he supposed publication 'is not only not urgently wanted, but rather urgently *not* wanted'.[38]

Once in the field in 1842 Phillips realised he was postponing too long the necessary work of describing fossils of the Malvern area. By December there were twelve casks of fossils collected in 1842 which were left untouched in Malvern over the winter, even though by this time Phillips appreciated that plates of fossils not only gave the Survey useful publicity but also could become its enduring monuments. De la Beche presumably realised that it was unrealistic to expect Phillips to combine cabinet and field work so in December 1842 he devised his scheme of keeping the fossils collected by the Survey in a central depot where they would be classified, described, drawn, engraved, and exhibited by the Survey's employees under Phillips' supervision. De la Beche saw this proposal as consolidating the Survey, as a big general scheme of publication which would be cheaper than monographs on districts, and as a way of securing more cabinet work from Phillips. By March 1843 the British organic fossils scheme was launched with Phillips expressing strong support for the project, sending drawings to De la Beche to whom he gave detailed advice about drawers and trays for about three tons of fossils which would end up in London, and proclaiming that he was prepared to move to London for many months of the year. Though Phillips was reluctant to leave the facilities and comforts of his York home, he told De la Beche that he would live in a village near London to which he would walk in order 'to keep the life in me'. Though De la Beche had appointed Lowry as the scheme's engraver and found accommodation for it in Whitehall Yard, London, he had not secured authorisation from government to publish, so he sent Phillips to Wotton in spring 1843 and then for all of July to London to organise and draw fossils collected by the Survey and stored in Whitehall Yard. Phillips tackled this work with gusto because he supported the plan for an Ordnance Geological Museum, but he was clearly worried about his residence, pay, and responsibility if the scheme were to be realised. He was also very unhappy in London, a 'roystering abode of anxiety' full of 'multitudinous cries of distress', and yearned for a healthy and quiet outdoor life.[39]

By late September 1843 Phillips knew he was trying De la Beche's patience because he had produced so few plates of fossils. Moreover Phillips did not conceal his reluctance to be engaged in cabinet work. He knew that descriptions of fossils had to be full and told De la Beche that 'the preparation of them is a most serious labour of *the cabinet*! It can never be done in the field! (Alas!) and what is worse it can not be delegated ....'. In early January 1844 Phillips accepted the chair of geology at Trinity College, Dublin, not only because he wished to avoid cabinet work in London but also because he was unhappy with the uncertainties in his Survey career. He had devoted time and thought to the British fossils and the Survey Museum projects but they were still embryonic and could well remain so. In September 1843 he received only a fortnight's notice from De la Beche about whether he was to do plates or work in the field.[40] Phillips' position with the Survey was no more than annually renewable and, though he still trusted De la Beche, he was aware that he was peculiarly dependent on De la Beche's ability as a fixer. In early February 1844 Phillips told De la Beche that his Dublin chair precluded his continuing as palaeontologist in charge of the Survey's fossil collection in London, arguing that its distributive aspect could be covered only by a field geologist, i.e. himself, and the descriptive aspect be superintended by a naturalist. Phillips wished to continue with the Survey, not as a London-based cabinet-worker but in the field as an all round geologist. With intellectual honesty but political rashness, he suggested that Edward Forbes, whom he regarded as an eminent naturalist interested in geology, be put in charge of the London collection. De la Beche took Phillips' advice about his successor, sounded out Forbes for this post later that month, and with Ramsay's warm support offered to Forbes the star position in British palaeontology in late October 1844, effective 1 November, at £300 pa. Forbes accepted: unlike Phillips he was prepared to carry out sustained cabinet work in London. Meanwhile Phillips' post as Survey palaeontologist formally ended in late March 1844 because the Ordnance Board decided that it was incompatible with his Dublin chair.[41]

For the next six months Phillips was paid by special arrangement to continue his Survey work and became increasingly unresponsive to De la Beche's proposals. Though in May Phillips was doing only a few drawings of fossils for Lowry to engrave, he told De la Beche that he could run the British fossils scheme from Dublin and York and claimed erroneously that the notion of a London base was a new and false doctrine. When it became clear that De la Beche rejected this plan, Phillips refused to undertake 'even one cahier of the organics': he was not prepared to shut himself in London for his last three months with the Survey to originate a difficult project which he would soon leave. Next month, when Phillips knew that Forbes was in line to succeed him, he wildly accused De la Beche of taking the post of palaeontologist from him; and he was prepared to visit London to instruct Forbes in the 'generalities touching the collection' only when Forbes was officially in post. It seems that Phillips and Forbes did not meet in London because the day after Forbes assumed his Survey post Phillips sailed to Dublin, having spent the previous three weeks examining again the puzzling Berwyn Hills. Subsequently, Phillips sent to De la Beche a specimen catalogue of fossils and a short account of how the British fossils project should proceed, but it was not until January 1845 that Phillips communicated directly with Forbes who thought he was sulking. Forbes told Phillips that the catalogue was unimprovable and would be followed; but in drawing and describing fossils Forbes proposed to compare them with their living congeners or family allies.[42]

To De la Beche, however, Forbes revealed a much more critical attitude to Phillips' descriptive palaeontology which he thought 'notoriously defective' and replete with 'sad deficiencies'. For the rest of the decade Forbes did not change his mind: he confided to his friend Ramsay that Phillips' palaeontology was disgraceful, blind, blundering, and villainous, which proved that Phillips '*cannot possibly have* judgement or sound observing powers'. Though Forbes was a braggart and too ready to criticise other geologists, his denunciations were driven by a very different approach to palaeontology. Phillips used fossils as an important resource for regional stratigraphical geology and eliciting the succession of life on the earth but was not an expert on their detailed morphology. In contrast Forbes was a zoologist, trained in Paris by Henri Blainville and Geoffroy St Hilaire, who were experts in descriptive and taxonomic zoology. Unlike Phillips, Forbes was capable of producing a standard work in four volumes on British molluscs. He therefore approached geology through biology, especially that of invertebrates. In line with his view that palaeontology should be pursued in accordance with the existing state of zoology and botany, Forbes focussed on the zoological characteristics and relations of fossil creatures. His fastidiousness contributed to the delayed appearance in 1849 of the first Survey publication devoted to figures and descriptions of British fossils, and helped to ensure that in this genre it was forestalled by the Palaeontographical Society, founded in 1847, which published monographs containing plates and descriptions of fossils.[43]

7.6   Irish stews

The notion of establishing a chair in geology at Trinity College, Dublin, to be held by Phillips, was created by the reverend Humphrey Lloyd, a key figure in the college and a leader of the Dublin mathematical physicists, who first mooted it to Phillips in 1837. As part of his scheme for an engineering school, Lloyd wished to have Phillips professing at Trinity and 'training up a young geological brood'. By that time Phillips had consolidated his high reputation in Dublin through his performance at the meeting of the British Association held there in 1835 when he was the guest of Lloyd with whom he rapidly developed a close friendship. Lloyd assured Phillips that as a leading light of the Association he had made a very favourable impression; James Apjohn told him he had made many friends in Dublin; George MacKenzie congratulated him on his public speaking which was so lucid because of its 'strict and plain consecutiveness'. No wonder that Phillips felt that his stay in Dublin in 1835 had been 'all one long triumph'.[44]

In 1841 the civil engineering school at Trinity was established by Lloyd who was supported by James McCullagh and Thomas Luby. In his proposal of April Lloyd stressed that training via apprenticeship was inadequate because its products were deficient in the theoretical knowledge necessary for dealing with new situations. He argued that the new school should teach not only practical engineering, but also the principles of mathematics, mechanics, chemistry, and geology, and their application to the arts of construction. Only two new chairs, one in chemistry and geology and the other in practical engineering, would be needed because the remaining teaching could be done by existing staff. Stimulated by the arrival of the railway in Ireland and worried by the threat of a polytechnic in Dublin to be founded by Robert Kane, the college accepted Lloyd's proposal in June 1841. The school of engineering opened in November 1841 with a luminous address by Lloyd on teaching applied sciences.[45]

It was launched by a powerful quartet of teachers. Luby covered mathematics, Andrew Hart mechanics, and Lloyd practical mechanics, heat, and the steam engine. The chair of chemistry and geology was occupied by Apjohn, a chemist who was au fait with mineralogy and Irish economic geology but not with geology as a science. Late in 1842 the staff was augmented when John MacNeill was elected professor of engineering. It soon became clear that Apjohn was giving insufficient attention to geology so in 1843 Lloyd used Sabine, a fellow magnetician, as an intermediary to negotiate with Phillips about his holding a new chair of geology which Lloyd wished the College to establish. On 30 December 1843 Phillips was elected professor of geology and mineralogy for five years from 25 March 1844 at a salary of £200 pa until the death of Whitley Stokes, recently retired Regius professor of medicine and sinecurist lecturer in natural history who received £800 pa; thenceforth Phillips would be paid £400 pa. It was expected that he would also look after the fossils and minerals in the College museum. He accepted Trinity's offer with alacrity and committed himself to delivering at least twelve lectures at the rate of two a week in each of the three terms. This appointment, he told Sedgwick, was a '*great fact*' in his life. It made Phillips the equivalent in Dublin of Sedgwick and Buckland, who professed geology at the two ancient Anglican universities in England. Given that Stokes was an eighty-year old invalid, there was a good chance of Phillips' salary being soon increased from £200 to £400 pa which, with his British Association salary of £300 pa, would render him comfortable financially. He anticipated being a welcomed star at the Geological Society of Dublin. He had many friends in Dublin, especially the physical scientists at Trinity, and he knew that the College was about to increase its commitment to science by appointing a director of its museum. Before accepting the chair he had been advised by De la Beche that it could be compatible with his holding the new post of director of the Irish branch of the Geological Survey which might be soon created. Such prospective cumul was alluring to Phillips because it would make him king of geology in the Emerald Isle and enable him to fruitfully combine field work and teaching.[46]

Phillips was aware that the Irish Survey post was not a foregone conclusion. He knew that Portlock's geological surveying of Ireland under the auspices of the Ordnance Survey had been contentious. In 1837 the Survey published its memoir about the parish of Templemore, which embraced most of the city of Londonderry; it included a geological contribution by Portlock. In order to attack further parishes, he ran a geological office-museum-bureau in Belfast employing thirty-five people in the late 1830s. In 1840 retrenchment occurred: his staff was reduced to three, the office moved to Dublin, and Henry James joined him to hasten the completion of the geological survey of county Londonderry. Next year James was put in charge of the Dublin geological office, the collections of which he rearranged until May 1843. Early in 1843 Portlock's huge report on the geology of county Londonderry was published just after he had been moved from the Ordnance Survey and posted to Corfu.[47]

In June 1843 the prime minister set up a three-man commission to report on the future of the scientific work of the Irish Ordnance Survey and on the form of any resumed geological survey of Ireland. In July 1843 De la Beche and Phillips gave evidence on the same day to this commission. The latter argued that a new Irish geological survey, paid for by government, should be separate from any general memoir scheme and that the fossils collected should be described and figured under one person, on the plan adopted in the English and Welsh Survey. The former recommended that Phillips could easily extend his palaeontological responsibilities to include Ireland. Next month Phillips was in Ireland at the Cork meeting of the British Association, which was attended by Thomas Robinson and Thomas Larcom, of the Irish Ordnance Survey, both of whom had given evidence to the commission, and by Portlock and Griffith, the two leading geological mappers of Ireland. It is inconceivable that Phillips did not learn from them and others about the various difficulties facing geological surveying in Ireland. The commission's report, dated November 1843, recommended that there be a geological survey of Ireland, separate from any memoir scheme but still under the aegis of the Ordnance, to be run overall by De la Beche with James as local director and Phillips in charge of naming and drawing the fossils collected. When Phillips accepted the Trinity chair in January 1844 he did so in the hope that he would continue his palaeontological work for the English Survey and, if an Irish Survey were to be established, extend it to Ireland. Phillips also hoped that if De la Beche could implement his cherished plan of civilianising the Geological Survey and the putative Irish branch then he might gain the power to replace James by Phillips. Though this hope rested on suppositions, it was not entirely forlorn: in lobbying government about Phillips, De la Beche had never failed; and by late 1844 he did indeed convince no less than the prime minister that the Survey should become a civilian enterprise.[48]

Before Phillips arrived in Dublin in May 1844 to give his first course of lectures, he began to be carried away by his own detailed plans for the non-existent Irish survey, even though he thought that Portlock had been treated unfairly. Phillips felt his appointment would be welcomed by Irish men of science, that Griffith would cooperate by lending valuable maps and documents, and that being both professor and director was not only feasible but desirable: the theoretical knowledge required for the former was essential for guiding the Survey and field knowledge was essential for teaching. But De la Beche's scheme for Phillips to direct a future Irish Survey was opposed by Colby, Larcom, and two of the three commissioners (John Young, an important Treasury official, and Lord Adare) who preferred it to be totally military and the directorship to be given to James. Yet even Larcom admitted that Phillips was a much better geologist than James and would direct a survey well. Capitalising on such feelings and supported by the other commissioner, Henry Boldero, De la Beche pushed from February 1844 for Phillips as Irish director, with Forbes as palaeontologist to the expanded Survey, and spread rumours that the matter was settled. In April and May De la Beche changed his tune in private. He told James that he had consented to the commission's recommendations. He floated to Colby the idea of having James as Irish director subject to a proviso. He informed Adare that either James or Phillips would be acceptable as director. No wonder that James regarded De la Beche as an old dodger and fox. In late May De la Beche told Phillips that it was likely that the struggle about the directorship would go straight for Phillips but it needed watching and care. For his part Phillips believed by mid-June that the affair had become tortuous even though De la Beche was pertinacious. He revealed to Ramsay that he felt to be a tactical pawn in a strategic struggle between Colby and De la Beche: 'had *I* known all the nature of the game, *on either side*, it might have been in my power to throw in a grenade, but as I told you in London we have been merely played upon and played with'. Meanwhile Phillips delivered his first course of twelve lectures at Trinity in May and June 1844. As at King's College, London, he was a visiting lecturer and not a resident professor. During the Whitsuntide break he returned to Malvern. After his last lecture on the afternoon of 25 June, he caught the packet boat to England at 6 pm that day and resumed Survey work in Malvern two days later. Though he judged his lectures a success, he left Dublin like a shot and did not return until November.[49]

In July 1844 Phillips made it clear to De la Beche that he was losing enthusiasm for the Irish directorship because even if secured he would lack 'sufficient directive power and means of working it'. He was also hurt by De la Beche's refusal to reveal the details of the arrangements for the Irish Survey that he, De la Beche, had recently proposed to the Ordnance Board, even though De la Beche reassured Phillips: '*You* ought to be in charge of the Irish Survey: it does strike me most forcibly that that is the *real honest useful* public thing to do.' Deep in Wales at Builth, De la Beche it seems did not know that a Treasury minute of 23 July 1844 had decreed that an Irish geological survey be revived as an Ordnance responsibility, with Colby in overall control and James as local director. The Treasury agreed with the commission that James occupy this position but disagreed about De la Beche's role because the prime minister thought De la Beche had enough to do in England and Wales and should therefore be excluded from Ireland. Though the commission, Peel, and James all wanted Phillips to be the fossilist for Ireland, Colby acted *ultra vires* in early August by proposing that Forbes and not Phillips look after Irish palaeontology. Meanwhile, De la Beche, who was surely too wily an operator not to have heard quickly about the Treasury minute, left Phillips to hear about it for himself. He did so on 13 August, his informant ironically being his rival, friend, and former pupil, Henry James, who also reported that Forbes was talked of as the palaeontologist. Phillips had lost the Irish directorship because plurality of office was unacceptable and Colby had defeated De la Beche on the question of whether the post should be a military or civilian one. Phillips felt himself to be a victim of 'conspiring forces' and decided immediately that he would have no connection in any capacity with the Irish Survey. He wished to avoid the intrigue and personal contests associated with it. He was not prepared to be treated by De la Beche and Colby 'as pawns are used in chess, to be given, exchanged, or destroyed at the pleasure of the players'. Yet in a remarkable display of magnanimity or guile, Phillips told De la Beche in late August that he was not irritated but felt decided relief; and only two days later invited De la Beche to stay with him in York in September during the British Association meeting '*as soon as you can make it* convenient, and for as long a time as you find agreeable'.[50]

De la Beche presumably did so and, having just begun to lobby the Department of Woods about the transfer of the Survey to it, suggested to Phillips in person in late September that in due course it might be possible for him to undertake Yorkshire for the Survey. Other intrigues were also taking place. In March 1844 Griffith had proposed to the chancellor of the exchequer that he himself could do a better and cheaper geological survey of Ireland than anybody, especially if aided by a good fossilist such as Phillips. The chancellor referred Griffith's scheme to the Master General of Ordnance who set up a four-man panel (Sedgwick, Buckland, Henry Warburton as president of the Geological Society of London, and Charles Hamilton as president of the Geological Society of Dublin) to advise about whether it should supersede the recently authorised geological survey of Ireland under the Ordnance. The panel met in York during the Association's meeting and rejected Griffith's proposal. Also at York Thomas Robinson, an increasingly important savant in Ireland, floated yet another scheme, namely that James should be replaced as local director of the Irish Survey by Phillips under Larcom and not Colby. Robinson wanted the Irish Survey to be run competently and to 'squelch' both Griffith and De la Beche, whom Robinson regarded as selfish and ungrateful. Robinson, James, De la Beche, Sedgwick, and Hamilton met in Phillips' home and it was probably then that he refused to be involved in Robinson's scheme. Undeterred Robinson lobbied Adare who in December urged to the Earl of Lincoln, the first commissioner of the Department of Woods, that Phillips be in charge of the Irish Survey instead of James, an idea that Lincoln liked. Meanwhile Phillips and his sister had taken a house in Kingstown, near Dublin, from early November and Phillips gave his second course of lectures before Christmas.[51]

After Christmas the vexed question of the Irish Geological Survey was settled rapidly. Acting on Buckland's advice against it continuing under the Ordnance and knowing that the Master General of Ordnance regarded geological work as alien to its topographic duties, Peel and the Treasury produced on 27 December 1844 a minute which recommended that the Ordnance should no longer do geology and that a geological survey of the United Kingdom be established under the control of the Department of Woods. Lincoln, who became a cabinet minister in January 1845, was enthusiastic: on 13 January 1845 he set out his plan for the existing Geological Survey of England and Wales and the new Irish one to be united, under the auspices of the Department of Woods, with De la Beche as general director, Ramsay as local director for England and Wales, and James as local director for Ireland. On 31 January the Treasury accepted these proposals which came into effect on 1 April 1845. De la Beche had secured his own aims except that the Irish Survey was being run not by Phillips but by James whose relations with De la Beche deteriorated so much and so quickly that he resigned as local director in 1846. He prospered subsequently as director general of the Ordnance Survey 1854-75, being knighted in 1860. As local Irish director he was replaced by Thomas Oldham who from 1845 was Phillips' successor at Trinity. In a civilian regime Oldham was able to practise the pluralism denied to Phillips.[52]

If De la Beche had been tortuous and if Phillips had felt misled and ill-used by him, he made amends in two ways early in 1845. Having failed in 1842 to secure for Phillips the Wollaston medal, the highest award of the Geological Society of London, he succeeded in February 1845. Phillips judged that the medal, given for his services to geology through published works and received in person by him that month, was entirely the result of the good opinion and persevering advocacy of his dear old friend De la Beche.[53] Also, in January 1845 De la Beche, then confident of freeing the Survey from the Ordnance, revived the proposal made in September 1844 that, during his long vacation from Trinity, Phillips should work for the Survey on Yorkshire and adjacent districts. Phillips was in principle keen; he knew more about Yorkshire geology than anyone and could use his York home as the Survey headquarters from which he would travel by rail for field work. By late January, before he had begun his third course of lectures at Trinity, he decided to resign from his chair. In March 1845 he did so, thus releasing himself for Survey work from April, and went home in mid-month as soon as his lectures were completed. Lloyd and McCullagh deplored his loss; and, after Stokes' death on 13 April 1845, Phillips was asked to withdraw his resignation because his salary would be doubled to £400 pa but he declined to do so. He was tired of Irish enigmas and intrigues, and he disliked the languor induced by professorial desk and head work. He had not been a central professor at Trinity: while giving three courses of twelve lectures, he spent only six months in Dublin between May 1844 and March 1845; and he showed little interest in the flourishing engineering school which was so dear to his friend Lloyd.[54]

7.7   Anti-climax

From 1 April 1845 to 30 June 1849 Phillips was employed by the Survey on a special commission to work on the north of England, with his York home as his base. He received his previous salary of £300 pa, which put him on the same level as Ramsay and Forbes. He was welcomed back by his old friends in York and was happy working in an area he wanted, knew, and liked. But he was no longer De la Beche's number two and his lieutenant in the field. Except when Trevor James was his assistant for a year, Phillips worked solo and had no opportunity to train young geologists in the field. He reported quite regularly to De la Beche who thought he was doing beautiful work but was not as interested as before in the general questions raised by his surveying. Visitors were rare and only two colleagues in the Survey, Beete Jukes and Warington Smyth, showed that they related their own concerns to Phillips'. In 1845 he spent a month in the Malvern area completing his field work and in 1846 began to compose the *Malvern memoir*. Otherwise his field research was devoted to Derbyshire and areas surrounding it, where he and his sister had temporary local stations from which he rode on his pony or took coaches to field sites.[55] He was well qualified to cover this area, being familiar with carboniferous strata from his research on Yorkshire and with mesozoic strata from that on the Malverns.

Much of this field work involved the standard procedures of the Survey: he inferred the boundaries of strata, traced the lines of mineral veins, coloured Ordnance maps geologically, and made long traverse sections. He also investigated the curious Derbyshire toadstone, the irregular masses of which contained almond-shaped nodules and interlaminated the mountain limestone. As late as the 1870s the number of toadstones, an igneous rock, was still uncertain and had aroused acrimonious controversy. Early in the century Farey claimed it occurred in three unfaulted bands but in the 1830s Hopkins alleged it occurred in one greatly faulted band. Phillips concluded that in general there were two bands. But his research on the toadstone, which he thought was the key to Derbyshire's geology, was discouraged by De la Beche who preferred him to map more. To Phillips' regret his findings about the toadstone and the relation of Derbyshire's carboniferous strata to their fuller development in Yorkshire were not published in a Survey memoir or in any other form. Moreover Phillips left unfinished four sheets of the Ordnance map of Derbyshire. His long stints of field work in and around Derbyshire, recorded in seven notebooks, led to nothing more than the publication in the early 1850s of a few geological maps.[56]

During Phillips' second period with the Survey, it expanded considerably but new recruits were not trained in the field by Phillips as they had been before. He was apparently content with his special commission but this very term implies he was no longer seen by De la Beche as a regular and central member of the Survey. Until 1844 Phillips had been in some ways the local director for England and Wales, de facto if not *de jure,* but from 1845 this post was occupied by Ramsay; and as the Survey's palaeontologist he had been replaced by Forbes in 1844. Compared with the Malvern period, that spent in Derbyshire was an anti-climax. From 1845 Phillips was side-lined in the Survey from which he was required to resign in 1849 when he was appointed by government to report with John Blackwell on the important question of ventilation in coal mines.[57] When the report was completed in 1850, Phillips made no attempt to rejoin the Survey, De la Beche did not invite him to do so, and their previous regular correspondence lapsed for over three years. Thus the fruitful period from 1841 to 1844 was replaced by demotion and marginalisation from 1845 to 1849, except for the composition and publication in these years of the *Malvern memoir*.

Notes

1. Flett, 1937, pp. 38-45; Secord, 1986b, pp. 244-6.
2. Secord, 1986b, pp. 242-6, 254; De la Beche, 1846; Phillips, 1848a, pp. 2, 225(q); Phillips to De la Beche, 16 Jan 1848(q), 20 June 1842(q), 30 May 1842 (q), DLB P.
3. Phillips to Harcourt, 26 Oct 1841, HP; Geikie, 1895, pp. 43-7; Edward Forbes (1815-54), *DNB*, palaeontologist to Geological Survey 1844-54.
4. Phillips to sister, 30 March, 1, 8, 15, 19, 22, 23, 28, 30 April, 6, 8, 10, 15, 17 June, 21 and 24 Sept 1841, PP; Phillips to De la Beche, 49 letters, 1 April 1841-1 Jan 1842, DLB P; Woollcombe to Phillips, 24 Aug 1841 (q), PP; Knell, 2000, pp. 255-60.
5. Phillips to De la Beche, 7, 11 Oct 1841 (qs), 24, 25 Dec 1841 (qs), DLB P; Phillips to sister, 8, 15 June 1841 (qs), 9 Oct 1841 (q), 4, nd, fr 14, 20 Nov 1841, PP.
6. Phillips to De la Beche, 61 letters, 8 April-16 Dec 1842, esp 18 March, 8, 14 April, 7, 10, 21 May, 9 Dec, DLB P; Sanders to Phillips, 19 July 1842, PP; Phillips to Ramsay, 15 July 1842, RP; Phillips, talk on Survey, 6 March 1843 (q), YPS evening meetings minute book; Knell, 2000, pp. 261-7.
7. Phillips to De la Beche, 23 Feb, 12, 18, 21, 30 May, 11, 20 June 1842 (q), DLB P.
8. Phillips to De la Beche, 34 letters, 9 April- 23 Dec 1843, DLB P, esp 4, 11 (q), 14, 19, 21, 27 May, 9 June, 10, 31 Aug, 5, 13 Sept, 7 Oct, 12, 14 Nov; Phillips to Mantell, 11 Dec 1843, Ma P; Davies, 1983, pp. 73-4; Favre to Phillips, 8 June [1843], PP; Phillips, Geological notes, north Wales, Frome, Claverton, Sept-Dec 1843, notebook 46,and Geological notes, Ireland, summer 1843, notebook 47, PP; Lyon Playfair (1818-98), *DNB*, chemist at the Museum of Economic Geology, London, 1845-53 and lecturer at the School of Mines 1851-3; Jean Alphonse Favre (1815-90); William Parsons (1800-67), third Earl of Rosse, *DNB*.
9. Phillips to De la Beche, 2 Jan 1844, DLB P.
10. De la Beche to Phillips, 27 May 1844, Phillips to De la Beche, 1, 9, 25, 30, 31 May, 2, 17, 30 June, 2 July, 24 August 1844, DLB P; Phillips, 1848a, p. 69; Henry William Bristow (1817-89), *DNB*; Henry James (1803-77), *DNB*.
11. Phillips, 1843a; Fitton, 1847; Murchison, 1843a, p. 76 (q); Bristow to Ramsay, 15 April 1845, RP.
12. Phillips to De la Beche, 4 Oct 1841, 8 April, 27, 29 July, 28 Aug, 2, 11 (q), 29 Sept, 10, 22 Oct 1842, DLB P; Phillips 1848a, pp. 19-20, 65, 73, 181 (drawing); Geikie, 1875, vol. 1, pp. 369-72; De la Beche to Ramsay, 11 Sept 1842, RP; Phillips to Murchison, 5 Sept 1842, MP; Keyserling to Phillips, 27 Sept 1842, PP; Jukes to Phillips, 14 Feb 1850, PP; on Joseph Beete Jukes (1811-69), *DNB*, as director of Irish Survey 1850-69, see Davies, 1983, pp. 156-91. A five-inch theodolite had a horizontal graduated circle five inches in diameter. On theodolites see Turner, 1983, pp. 247-61.
13. Secord, 1986b, pp. 240-1; Ramsay, diary, 17 Jan 1848 (q), RP; Bristow to Ramsay, 7 March 1847 (q), RP; Phillips to De la Beche, 27 May (q), 2 Sept, 6 Dec 1842, DLB P.
14. Baily to Phillips, 18 Dec 1843, 25 March 1872, PP; Phillips to De la Beche, 23, 27 Dec 1843, 24 Aug 1844, DLB P; Richard Gibbs (d 1878); William Talbot Aveline (1822-1903); William Hellier Baily (1819-88) worked for the Survey 1844-88.
15. Phillips to De la Beche, 21, 22 May 1841, 5 May 1843, DLB P; Phillips to Ramsay, 14 May, 4, 11 June, 10, 20 July, 14, 21 Aug, 6 Sept, 20 Oct 1842, 26 Jan, 12 Nov 1843, 19 Aug 1845, 14 Aug 1846, RP; Ramsay to Phillips, 20 May 1867 (q), EUL Gen 784/1; Ramsay, 1894, p. 46 (q); Geikie, 1895, pp. 43, 50, 206.

16. Bristow entrance paper, King's College archives; *KCL calendar 1841-2*; Phillips to De la Beche, 29 April, 4, 8, May, 12 Nov 1843, DLB P; Bristow to Phillips, 30 July 1843, 26 June 1844, 14 June, 17 Nov 1845, 13 Jan 1864, 16 March 1872, PP.
17. Aveline to Phillips, 13 Jan 1842, PP; Phillips to De la Beche, 14 Dec 1841, 16 April 1845 (q), DLB P; obituary by Geikie, 1904.
18. Phillips to De la Beche, 18, 19 May 1843, 30 May 1844, 26 March 1845, DLB P; Gibbs to Phillips, 19 May 1844, PP; Knell, 2000, p. 269; Gibbs to Ramsay, 20 Nov 1848, RP.
19. Phillips to De la Beche, 1 April, 18, 21 June, 21 Oct, 8 Nov, 13, 14 Dec 1841, 28 Sept 1842, 26 May 1843, 19 Sept 1845 (q), 28 May, 25 Sept 1846, DLB P; James to Ramsay, 19 June, 13 Oct 1846 (q), RP; James to De la Beche, 28 April 1844, 28 June 1846, DLB P; James to Phillips, 17 Oct 1842, PP; De la Beche to James, 23 April, 10 July 1846, DLB P; Phillips to Ramsay, 20 June, 5, 14 Oct 1846, RP.
20. Phillips to De la Beche, 14, 29 Dec 1841, 1, 28 Jan, 21 Aug, 10 Oct 1842, DLB P; Rees to Phillips, 1 May 1842, PP; Josiah Rees (1821-99) became a barrister in 1851, chief justice of Bermuda in 1878, and was knighted in 1891.
21. Logan to De la Beche, 26 July 1842, DLB P; Phillips to De la Beche, 28 Aug, 1, 28 Sept 1842, DLB P; Alexander Murray (1810-84), *Dictionary of Canadian Biography*, director of Newfoundland Survey 1864-83.
22. Phillips to De la Beche, 10, 21 July, 5, 23 Aug 1842; Theobald to De la Beche, 7 Dec 1848; East India Company memo, enclosed with Dickinson to De la Beche, 2 March 1850; De la Beche to East India Company, 14 March 1849, all DLB P; David Hiram Williams (d 1848).
23. De la Beche to Ramsay, 11 Sept 1842, RP; Phillips to De la Beche, 11, 13 Sept, 13 Nov, 6 Dec 1842, DLB P; James to Phillips, 9 Dec 1842, 4 Feb, 16, 27 April 1843, 26 Dec 1844, 30 July 1845, PP; on James' short and dismal career as Irish local director see Davies, 1983, pp. 126-38.
24. Sanders to Phillips, 24 March, 25 June, 21 Aug, 19 Dec 1841, 23 March 1844, PP; Phillips to De la Beche, 18 June 1841, 23 Feb, 5 March 1842; Sanders to De la Beche, 13 Nov 1840, 21 Feb (q), 12 March 1842, all DLB P; Samuel Stutchbury (1798-1859).
25. Phillips to De la Beche, 27 Oct, 2 Nov 1841, DLB P; Phillips to sister, 27 Oct, 1 Nov 1841; Sanders to Phillips, 25 June 1841; Emlyn to Phillips, 26 Oct 1841, all PP; John Frederick Campbell, first Earl of Cawdor (1790-1860), FGS 1821; John Frederick Vaughan Campbell (1817-98), Lord Emlyn 1827-60.
26. Phillips to De la Beche, 7 Oct 1841, DLB P; Phillips to sister, 19 Sept 1841 (q), PP; probably Hugh Seymour Tremenheere (1804-93), *DNB*, barrister, bureaucrat, and amateur geologist.
27. H.D. Rogers to Phillips, nd [fr Aug 1834], 20 May 1835, 20 Aug 1840 (q), 30 May 1842, 27 March, 10 July 1843; W.B. Rogers to Phillips, 11 July 1843, PP; Phillips to H.D. Rogers, 1 Oct 1834, Massachusetts Institute of Technology Library; Rogers, 1896, vol. 1, pp. 106-7, 194-9, 218-19, 289-91, 304-6; Greene, 1982, pp. 122-5; Phillips to Grove, 20 July 1848, Gr P; Gerstner, 1994, pp.105-16; for Lyell's tours of USA from 1841, see Wilson, 1998; Henry Darwin Rogers (1808-66), William Barton Rogers (1804-82), *DAB*.
28. Phillips to sister, 6 June 1841, PP; Phillips to De la Beche, 4, 7, 16 Oct 1841, 28 Nov 1842 (q), DLB P; Phillips to Strickland, 4 Oct 1841, Strickland Papers; James to Phillips, 13 April 1842, Rees to Phillips, 1 May 1842, PP.
29. Murchison, 1854, pp. 92-7; Phillips, 1848a, pp. 3, 207 (q); Murchison, 1843a, p. 78; Oldroyd, 1992, p. 411.
30. Murchison, 1839, pp. 409-26, sections on plate 36, figs 7 and 8; Oldroyd, 1992, pp. 407-10; Phillips, 1837, 1839, vol. 1, p. 145; Phillips to Owen, 17 April 1842, OP.
31. Phillips, 1848a, pp. 32, 38-49, 55, 162; Phillips to De la Beche, 6, 7 May 1842, DLB P.
32. Oldroyd, 1992, pp. 410-15; Phillips to De la Beche, 14 (q), 17 April (q), 27 May, 29 July, 5 Aug 1842, DLB P; Phillips to Harcourt, 23 May 1842, HP; Phillips to Ramsay, 4 June, 21 Aug 1842 (q), RP; Phillips to Sedgwick, 21 Aug 1842, Se P; Phillips, 1848a, pp. 34, 64-9; Phillips, 1842a.

33. Phillips to De la Beche, 5 Aug 1842, DLB P; Phillips, 1842a; Phillips, 1848a, pp.5-8, 44, 49, 66-8, 144; Strickland, 1851, p. 359; Hugh Edwin Strickland (1811-53), *DNB*, deputy reader in geology, University of Oxford, 1850-3.

34. Phillips to Mantell, 8 Oct 1845, Ma P; Phillips to De la Beche, 28 March 1842, DLB P; Horner, 1811; Phillips, 1848a, pp. 2, 6, 49, 135, 151-4, 184-5, 206-7 (q); Greene, 1982, pp. 93-106.

35. Phillips, 1848a, pp. 54-5, 74-5 (qs), 200-2; Phillips to De la Beche, 12 April 1843, DLB P; Phillips, 1843c; Murchison, 1854, pp. 92-3, 109-10.

36. Phillips, 1848a, pp. 2 (q), 8, 208, 213 (q), 233-321; Murchison, 1854, p. 463.

37. Phillips, 1848a, pp. 225, 331-86; Phillips to De la Beche, 23 May, 1, 4, 20 Dec 1846, 23 May, 11 June, 3 Oct 1847, 16 Jan 1848, DLB P; for expertise and career of John William Salter (1820-69), *DNB*, Survey palaeontologist 1846-63, see Phillips, 1848a, p. 334 and Secord, 1985; Charles Richard Bone (1808-75), Survey artist 1845-75.

38. Phillips to Sedgwick, 16 Oct 1841, Se P; Phillips to sister, 5 Dec 1841, PP; Phillips to De la Beche, 24, 26 June, 24 Dec 1841, 16 Jan, 15 Feb (q), 15, 28 March 1842, DLB P.

39. Phillips to De la Beche, 7, 27 May, 10, 13 July, 12, 29 Dec 1842, 19, 23 Jan, 10 (q), 16, 22 Feb, 11 March, 8 May, 24, 26 July 1843, DLB P; Phillips to Ramsay, 4 July 1843 (q), RP; De la Beche to Buckland, 21 Jan 1843, Pl P.

40. Phillips to De la Beche, 19, 27 Sept (q), 12, 19, 30 Nov, 23, 27 Dec 1843, DLB P.

41. Phillips to De la Beche, 3, 4 Feb 1844, DLB P; Wilson and Geikie, 1861, pp. 358-9.

42. De la Beche to Phillips, 27 May 1844; Phillips to De la Beche, 7, 25 May, 19 June (q), 13 (q), 16, 24 July, 11 Nov, 16 Dec 1844, Forbes to De la Beche, 11 Nov, 17 Dec 1844, all DLB P; Forbes to Phillips, 4, 18 Jan 1845, PP.

43. Forbes to De la Beche, 17 Dec 1844, DLB P; Forbes to Ramsay, 11 May 1846, 7 July 1849, RP; Forbes to Ramsay, 1 June 1847 (q), EUL Gen 1999/1/7; Secord 1986b, p. 257; Wilson and Geikie, 1861, pp. 263, 314-20; Forbes and Hanley, 1853; Henri Marie Ducrotay de Blainville (1777-1850), *DSB*, specialised in invertebrate zoology, especially molluscs.

44. Spearman, 1981; Lloyd to Phillips, 8 Sept, 19 Nov 1835, 29 March 1837 (q); Phillips to sister, 19 Aug 1835 (q); Apjohn to Phillips, 23 Aug 1835; MacKenzie to Phillips, 28 Aug 1835, all PP; James Apjohn (1796-1886) professor of applied chemistry and geology, TCD, 1841-4; professor of applied chemistry, TCD, 1844-81; professor of mineralogy, TCD, 1845-81; professor of chemistry, TCD, 1850-75; Sir George Stuart MacKenzie (1780-1848), *DNB*.

45. Spearman, 1981, pp. 45-6; McDowell and Webb, 1982, pp. 180-4; TCD Board minutes, 8 May, 5, 26 June, 10 July 1841, TCD Library; James McCullagh (1809-47), *DNB*, professor of mathematics, TCD, 1836-43; Thomas Luby (1800-70), *DNB*, fellow of TCD and lecturer in mathematics; Robert John Kane (1809-90), *DNB*, professor of chemistry, Apothecaries' Hall, Dublin, 1831-45, professor of natural philosophy, Royal Dublin Society, 1834-47, did not establish a technical school but in 1845 created the Museum of Irish Industry, Dublin.

46. *TCD calendar 1841-2*; Lloyd to Sabine, 30 Dec 1843, Sa P, PRO, file 13; TCD Board minutes, 30 Dec 1843, 6 Jan 1844; Phillips to Sedgwick, 10 Jan 1844 (q), Se P; Phillips to De la Beche, 5, 18 May, 16 July 1844, DLB P; Andrew Searle Hart (1811-90), *DNB*, fellow of TCD, published textbooks on mechanics, hydrostatics, and hydrodynamics in the 1840s; John Benjamin MacNeill (1793-1880), *DNB*, professor of engineering 1842-52; Whitley Stokes (1763-13 April 1845), *DNB*, regius professor of medicine, 1830-43; Robert Ball (1802-57), *DNB*, made director of TCD Museum, April 1844.

47. Davies, 1983, pp. 95-106; Phillips to De la Beche, 14 May 1844, DLB P; Portlock to Greenough, 4, 9, June 1840, GP; Colby, 1837; Portlock, 1843.

48. Davies, 1983, pp.108-9; Report of the commissioners appointed to enquire into the facts relating to the Ordnance memoir of Ireland, *Parliamentary papers*, 1844, vol. 30, pp. 259-385 (271-2, 285-8, 290-3); Thomas Aiskew Larcom (1801-79), *DNB*, administrator in Dublin of Irish Ordnance Survey 1828-46, had suggested the Ordnance memoir scheme.

49. Davies, 1983, pp. 109-14; Phillips to De la Beche, 13 Jan, 3 Feb, 7, 9, 13, 14, 18 May, 2, 30 June 1844, DLB P; Larcom to Colby, 22 Jan 1844, La P, 7555; De la Beche to James, 21 Feb; Larcom to Adare, 24 Feb; Colby to Larcom, 23 March; James to Larcom, 23 March, 11, 19 April, 17, 23 May; Adare to Larcom, 22 May 1844, all La P, 7556; De la Beche to Phillips, 27 May 1844, DLB P; Phillips to Ramsay, 7, 13, 18 June 1844 (q), RP; Phillips, 'Philosophy of geology; twelve lectures, May, June 1844, TCD Library, Mun P/1/1771a; Phillips, Dublin lecture courses 1844-5, notebook 50, PP; Edwin Richard Windham Wyndham-Quin (1812-71), *DNB*, Lord Adare, conservative MP for Glamorgan 1837-51, had wide scientific interests, was a vice-president of BAAS 1843, and specialised in Irish antiquities; John Young (1807-76), *DNB*, tory MP for county Cavan 1831-55, secretary to Treasury, 21 May 1844-6; Henry George Boldero (1797-1873), conservative MP for Chippenham 1835-59 and clerk of Ordnance 1841-6.

50. Phillips to De la Beche, 13 (q), 16 July, 24 (q), 26 Aug 1844 (q), DLB P; De la Beche to Phillips, 29 July 1844 (q), PP; Robinson to Larcom, 23 July, 26 Aug; Robinson to Young, 13, 14 Aug; James to Larcom, 15, 23 Aug, 20 Sept 1844, all La P, 7556; Phillips to Sedgwick, 24, 29 Oct 1844 (q), Se P. My interpretation differs on some points from that of Davies, 1983, pp. 113-15, and 1995, pp. 14, 18. He depicts Phillips as a guileless pawn who was pushed onto the Irish stage by De la Beche, a devious jobber who misled Phillips who was trebly aggrieved.

51. T.W. Philipps to De la Beche, 1, 8, 15 Sept 1844; Phillips to De la Beche, 11 Nov 1844, 4 Jan 1845 (misdated 1844 by Phillips), DLB P; James to Larcom, 23 April; Robinson to Larcom, 25 (q), 27 Sept, 21, 24 Nov; Larcom to Robinson, 28 Sept; Adare to Larcom, 12 Dec 1844, all La P, 7556; on the Griffith 'diversion', Davies, 1983, pp. 115-18, and Byham to Sedgwick, 9 Sept 1844, Se P; Henry Warburton (1784-1858), *DNB*, president Geological Society of London 1843-5; Charles William Hamilton (?1802-80); Henry Pelham Fiennes Pelham Clinton (1811-64), *DNB*, Earl of Lincoln, first commissioner of Woods 1841-6.

52. Davies, 1983, pp. 120-2; Robinson to Young, 22 Nov 1844, La P, 7556; Geological Survey (Ireland), *Parliamentary papers,* 1845, vol. 45, pp. 147-56; Thomas Oldham (1816-78), *DNB*, professor of geology, TCD, 1845-50 and director of Irish Geological Survey 1846-50.

53. De la Beche to Phillips, 2 July 1840, PP; Phillips to De la Beche, 4 Feb 1842, 25 Jan 1845, DLB P; *PGS*, 1846, vol. 4, p. 529.

54. Phillips to De la Beche, 4 Jan, 6 March, 18 April 1845, DLB P; R. Davies to Phillips, 23 Jan; McCullagh to Phillips, 5 March 1845, both PP; Phillips to Ramsay, 3 Feb 1845, RP; TCD Board minutes, 22 March 1845.

55. Phillips to W.Hutton, 31 Aug 1845, Hu P; Phillips to De la Beche, 26 March 1845-10 Oct 1849, *passim,* DLB P; Jukes to Phillips, 2 Oct 1847; Smyth to Phillips, 14, 27 Dec 1848, all PP; Warington Wilkinson Smyth (1817-90), *DNB*, Survey mining geologist 1845-90.

56. Phillips to De la Beche, 7, 23 June 1846, 3 Oct 1847, 29 Oct 1848, 27 June 1849, DLB P; Phillips, Derbyshire July-Nov 1845, notebook 53; Derbyshire, Jan-June 1846, notebook 54; Derbyshire, 1847, notebook 55; Derbyshire, Leicestershire, 1847-8, notebook 56; Macclesfield, 1848, notebook 57; Nottingham, May 1849, notebook 58, all PP; Farey, 1811, pp. 240, 275-80; Hopkins, 1834; Green et al, 1869.

57. De la Beche to Ramsay, 10 July 1849, RP.

Chapter 8

Manifold Scientist

Even while Phillips was working for the Geological Survey and professing in Dublin, he and his sister retained St Mary's Lodge in York which they regarded as their home. From 1841 to 1849 Phillips spent several months a year away from home and far from London, which limited his local commitments and his involvement in metropolitan affairs. From summer 1849 to spring 1850 he was busy as a civil scientist reporting to government about the ventilation of coal mines. From spring 1850 Phillips spent much time in York, his chief institutional responsibility continuing to be assistant secretary to the British Association. In these years Phillips' emolument fluctuated, sometimes wildly. From 1841 to 1849 he usually received £600 pa (£300 Survey, £300 BAAS). When he lost his Survey income from October 1844 to March 1845, there was compensation from his £200 pa as professor in Dublin. In 1850 he enjoyed the highest income of his life to date: he received £750 as a government investigator and reporter and his usual £300 pa from the Association. From 1851 his only source of regular income was his salary from the BAAS.

Financially circumscribed yet again, several ways of making money were unpalatable to him. His experience as a professor had been unhappy so he did not apply for a chair and was not offered one. He had worked himself to the bone in the 1830s as a writer of geological works and encyclopaedist so he did not contemplate new or revised contributions in these genres. There was no place for him in De la Beche's geological empire, he had no wish to resume museum curating, and any post in London was ruled out because he found the capital uncongenial. He could not follow Lyell by writing about foreign travel and geology because it was expensive and not easily harmonised with his BAAS responsibilities. In part to make money he turned to writing popular works about Yorkshire, his adopted county. By summer 1853 several influential friends felt that he needed and deserved a civil list pension in order to help him make his wide-ranging talents tell fully. They applied to the prime minister on his behalf but the application was rejected. In this chapter I shall examine Phillips' multifarious activities from 1841 to 1853, other than those connected with the Survey.

8.1   Local commitments

When Phillips' sister accompanied him on his trips in the 1840s he let St Mary's Lodge to suitable tenants at a rent of 7 to 10 pounds a month. But he was always glad to return to his cherished medieval home which he had restored and externally cleaned. In the mid 1840s its environs were improved when the YPS acquired the ruined St Leonard's Hospital in the Museum grounds and rebought the lease of the Swimming Bath on the Manor Shore for £1475. The Society then spent £794 on the redesigning of its gardens, by Sir John Naesmith, the result of which for Phillips was that his front view was 'amazingly improved'. Also in 1845 Phillips gained control of two cottages next to his home when he acquired the tenancies of them from the YPS, as a result of negotiations with its treasurer, Robert Davies, a close friend.[1]

In 1850 Phillips' charming surroundings and his hitherto happy relations with the YPS about his home were disturbed. He was angry about the way in which the Society, having rejected the greatly increased charges proposed by the New Waterworks Company for connecting its properties to new mains, decided to pump water from the river using a steam engine and chimney erected adjacent to the Swimming Baths, just below his home. In August he complained formally to the YPS about the height of the chimney which spoilt his environment and that of the Museum. Later that month he petitioned the Society to suspend the works. In response the height of the chimney was reduced and the engine house built, but before the engine was delivered the Waterworks company and the YPS reached agreement. In this episode Phillips was protected by Kenrick but at odds with Thomas Meynell and Thomas Travis, the Society's secretaries. For the first time in twenty-five years he found himself disputing with the Society to which he owed so much.[2]

When Phillips resigned from the keepership of the YPS in late 1840 he hoped to give it 'some aid' in future. This phrase sums up well his contribution to the Society from 1841 to 1853. As an office-holder he served as a secretary to 1842 and from 1852 until his resignation in October 1853, as curator of geology 1841-5, as curator of mineralogy 1851-October 1853, as an ordinary member of Council 1842-5 and 1847-50, and as a vice-president 1845-7 and 1850-2. As councillor he presided at the Society's annual general meeting in 1848.[3]

In 1841 Phillips had definite views about the keepership and curatorships of the YPS. He rejected the possibility of a salaried keeper because the post would cost unaffordable money and required a range of knowledge nobody had. Instead he recommended that there be active honorary curators drawn from the membership, a scheme which he hoped would encourage commitment to the YPS and participation in it. The Society took his advice and appointed eleven curators who were supported by Baines as sub-curator with a salary of £100 pa. Later that year when Hewitson expressed a wish to be keeper, the Society stuck with Phillips' plan as wise and beneficial, but its optimism was misplaced. Though the increase in the number of curators was intended to give each a lighter load, research and papers were not stimulated. In 1844 Phillips' scheme was dropped because the bequest in 1843 of £9,000 from Stephen Beckwith, a long-standing member of the YPS, enabled it to appoint Charlesworth as keeper in July 1844 on a salary of £150 pa, with Baines as sub-curator retaining his of £100 pa. Phillips was not consulted about this reversion to having a keeper, though the Society hoped that, in wide knowledge, zeal and ability, Charlesworth would emulate Phillips.[4]

As curator of the geology collection, the Museum's largest, Phillips was not inactive. He rearranged all the fossils in order to exhibit some new Silurian specimens, accepted donations of fossils which illustrated the nature and distribution of ancient animal life in the palaeozoic era, organised a display of invertebrate fossils from Yorkshire arranged both stratigraphically and zoologically, and no doubt persuaded his sister to donate a large series of Silurian fossils collected by her in the Malvern area. Phillips thus helped to ensure that the Museum, the centre and mainspring of the YPS by the mid- 1840s, did not languish.[5]

As assistant general secretary of the British Association Phillips was supposed to be neutral about invitations to it but from 1842 he acted almost *ultra vires* in guiding the YPS's campaign to induce the Association to return to its birthplace. At the meetings of the BAAS held in 1840 and 1841 delegations from the YPS pressed unsuccessfully for 'a second inspection' by the 'national scientific assembly' in 1841 and 1842. By late 1841 some leading members of the YPS felt snubbed and were suffering from cold feet: they concluded that York had little scientific talent or taste and that the YPS, badly in debt, could not afford to receive the Association properly. These pessimistic assessments were countered by Phillips who convinced the YPS that it should enlist the co-operation of the Yorkshire philosophical societies in the invitation and in defraying the expenses of the meeting. In January 1842 the Society sent out a circular, drafted by Phillips, soliciting the support of eight such societies and the Yorkshire Geological Society for a Yorkshire meeting of the BAAS in York at an estimated expense of £500. Most of the Yorkshire lit-and-phils were indifferent, Leeds was hostile, and the YGS replied that it would provide money only if a union of the Yorkshire societies were to be formed. Ignoring these rebuffs, Phillips intimated to the BAAS Council in February 1842 that it would receive an invitation to visit York in 1843. The YPS, equally resolute, went ahead solo, gained the support of the Corporation, and secured promises totalling £432 for a BAAS subscription. A delegation made up mainly of Wellbeloved, Kenrick, and William Gray, junior, all close friends of Phillips, was instructed to employ an approach and arguments devised by him when pressing at the 1842 meeting for one next year in York. In the event the strong nationalist desire of the Irish for a second meeting in Ireland, this time at Cork, was irresistible.[6]

At the 1843 meeting of the BAAS it was unanimously agreed that it should visit York in 1844. For Phillips this gave a welcome opportunity to renovate the Yorkshire Museum, to fill its cases with specimens illustrating Yorkshire's natural history, to catalogue its collections, and to make it a leading feature of a well-prepared scientific reception. In line with Phillips' long-held view that one function of the BAAS was to stimulate local science, the YPS commissioned a cohort of its members to give papers on Yorkshire's natural history and antiquities, both of which were broadly construed. Phillips was asked to cover magnetism solo and geology, physical geography, and meteorology collaboratively. In the event he confined himself to three meteorological papers. On one matter Phillips failed: though he wished to restrict expenditure to plain necessities and not exceed £500 for the local reception fund, the YPS wanted the meeting to be brilliant and a credit to itself so £1,100 was raised. Overall Phillips' guidance ensured that the Museum was widely praised and that the Society, 'the nursing mother of the Association ... welcomed back her wandering children with hearty goodwill'.[7] Phillips himself welcomed many distinguished visiting savants, giving a few accommodation at St Mary's Lodge and entertaining others at breakfast, tea, and an evening conversazione at which the 'dense mass of learned persons' impressed Robert Chambers, the anonymous author of the sensational *Vestiges of Creation* published next month.[8]

As a performer at the ordinary meetings of the YPS Phillips was less active. He contributed on average less than a paper a year and was not significantly more active from 1850 when he spent more time in York. Only one covered geology (a newly discovered plesiosaurus, June 1852). Others were devoted to meteorology (shooting stars, October 1842; the aurora of 24 October 1847, November 1847), terrestrial magnetism (summary of his previous research on Yorkshire, January 1851) and astronomy (solar eclipse of 6 May 1845, May 1845). Two papers revealed his growing interest in geography (Ptolemy's map of Britain, Dec 1850) and archaeology (ancient metallurgy and mining in Britain, March 1848). His continuing concerns with meteorology and his evolving interests in astronomy were revealed to the BAAS and rarely to the YPS, even when they had a local resonance. The Society heard nothing about his 'rain-gauge plan' for a fire-fighting system at York Minster which had suffered big fires in 1829 and 1840. Using his knowledge of rainfall at the top of the Minster, Phillips proposed in 1842 to a sceptical Harcourt that it would be cheaper and more effective to use high- level reservoirs and cisterns filled by rain water than to pump river water through a special main. The YPS heard nothing about his discussions with Sir William Hamilton, an expert on refraction, about a new object lens to be made in York by Cooke for Phillips' achromatic refracting telescope which he had bought from Cooke in the late 1830s. Phillips did not reveal to the YPS, but to the Royal Society of London in June 1851, his successful repetition of Foucault's demonstration of the earth's rotation. This spectacular experiment, first reported in February 1851, was widely repeated in the next two years. Phillips was one of the first to do so. In collaboration with Cooke and William Gray, in May 1851 Phillips measured the changing direction of swing of a large pendulum set up in the north-west tower of York Minster, 'a very proper use for the glorious heaven-pointing church'.[9] He concealed from the YPS his pioneering drawings and photographs of the moon accomplished in 1852-3 using a new refracting telescope, of six-and-a-quarter-inch aperture and equipped with a clock drive, made in York by Cooke and erected in Phillips' garden.[10]

His involvement in the YPS's meteorology was sporadic and his contribution to its astronomy non-existent. In the early 1840s he was happy to encourage Ford, as the Society's meteorological man, to join the huge international scheme, organised by Quetelet in Brussels, of making meteorological observations on four stipulated days of the year. To that end Phillips persuaded the YPS in 1842 to permit him to procure a good barometer for it, which he failed to do. In 1851 he induced it to spend £25 on an anemometer and in 1853 he arranged for a YPS thermometer to be calibrated by John Welsh at Kew Observatory. He took no interest in Newman's astronomical work or the Society's Observatory which was hampered by poor instruments and, until 1851, a leaking roof. That year the YPS was forced to acknowledge that many of its honorary curators were sinecurists so it appointed a committee of Phillips, Harcourt, Ford, Newman, William Gray, and William Hey to be responsible for meteorology and astronomy. Phillips, it seems, made no attempt to animate it.[11]

As a lecturer to the YPS Phillips was even less prominent than before. In twelve years he gave only two short courses of lectures, one on geology and the other on the physical geography of Yorkshire (see Appendix 1).[12] In contrast he supported strongly the evening discussion meetings, until they expired, apparently about 1849, through lack of subjects and members. Phillips rescued the meetings from scientific twilight and acted as an authority in discussion. He spoke twice about recent results obtained by the Geological Survey, demonstrated an electrophorus, revealed from an analysis of simultaneous barometric observations made in 1844 that a great atmospheric pressure wave existed, evaluated different anemometers, argued that railway cuttings gave nothing more than nice geological illustrations, discussed Hopkins' recent work on the earth's interior, and gave the first version of his account of ancient metallurgy and mining in Britain. At best he attended about fifteen times a year. Earlier in the decade he encouraged the research of his friend Robert Davies, who accepted Phillips' advice that he should compose a book on the history of York in the fifteenth century, based on papers read to the evening meetings. Davies even sent proofs of his work to Phillips who scrutinised them minutely and beneficially. As a reconstructor of past life, Phillips encouraged Davies' publication about the manners, customs, language, and domestic habits of previous inhabitants of York.[13]

Phillips was the prime mover in a campaign which led to the publication in 1855 of a volume of the Society's *Proceedings*. Initially in 1846 he had in mind a more ambitious scheme for regular annual transactions; edited by Charlesworth, they would be appended to the annual report and focussed on the natural history and antiquities of Yorkshire. Phillips' aim was to revive the monthly meetings which were so sparsely attended that in 1847 they were moved from 1 pm to the evenings. The Society settled for proceedings, the opening numbers of which appeared in 1847. Phillips hoped they would stimulate future exertion among local and distant members who would provide valuable data for 'an accurate and complete natural and topographical history of this great county'. In 1851 the Society still hoped to produce an annual volume of transactions devoted to the natural history and antiquities of Yorkshire and to stimulate its honorary curators; but, at a time when it was increasingly difficult to find officers because the influx of new members was not replacing those who had died or resigned, it had to acknowledge that it could do no more than publish numbers of proceedings which eventually could be gathered to make a volume.[14]

In these years Phillips maintained his involvement in the mechanics' institute movement not as an organiser, propagandist, or regular lecturer, but as a speaker at important occasions especially in Yorkshire. In October 1845 he gave the opening speech at a soirée of the Doncaster Mechanics' Institute held in the Town Hall and attended by 500 people. Most of his themes were familiar ones: the dignity of labour with head or hand, science as social amelioration, and a providential view of nature. But one was new: that of the Earl of Rosse, a Yorkshireman and manufacturer of his own huge telescopes, as an exemplary working man and mechanic. Locally, Phillips was a supporter of the York Institute of Popular Science and Literature. Though he gave no lecture to it between 1841 and 1853, he proposed the odd toast at its annual meetings, chaired one of them in 1850 in lieu of the Lord Mayor, and crucially gave the address at the opening of its new premises in March 1846. His speech brought together his main views about the diffusion of knowledge. Ideally it should he communicated through all ranks of society, without offending any class of people and without shocking 'reasonable opinion'. Knowledge was valuable because it led to rational gratification and improvement of character in individuals and 'exaltation of social and national happiness'. Because God had endowed humans with minds, all individuals had a right to knowledge and a responsibility to improve themselves. One particular form of knowledge, that of nature, was accessible to everyone: there was, he averred, 'no one department of nature, however mysterious, which cannot be, in a considerable degree, brought within the comprehension of ALL'. As Phillips believed that the natural world was an expression of God's planning, superintendence, wisdom, and benevolence, it followed that the new premises of the York Institute constituted a consecrated temple of science. Phillips' sister supported these sentiments: next month she ran a stall at the Institute's bazaar which overall generated £195 profit in a week.[15]

In Leeds Phillips was regarded as a friend of the mechanics' institute movement but he found it impossible to attend any of the annual soirées of the Leeds Institute. Eventually in 1853 he gave four well-received lectures to it on the physical geography and geology of Yorkshire. The previous year, when staying with the Earl of Rosse at Birr Castle, Parsonstown, Ireland, he bothered to give a lecture at the local mechanics' institute on the geology of Ireland, with Rosse, then president of the Royal Society of London, in the audience.[16]

Though Phillips was by far the leading expert on the geology of Yorkshire, he gave only a modicum of support to the Yorkshire Geological Society, neither serving on its council nor subscribing as a member. Though he was its first honorary member, he gave to it only two papers, both in 1845, at successive meetings in Pontefract and Doncaster, as if testing the Society as a suitable vehicle for himself. Both papers revealed recent results he had obtained with his microscope. That of October concerned petrifaction of materials from Yorkshire; that of August reported on the remains in Yorkshire rocks of two of the simplest forms of life: animalcules, also known as infusoria, and foraminifera, single-celled creatures. Phillips was generally stimulated by the research of Christian Ehrenberg on organisms invisible to the unaided eye and in particular by Mantell, who in the mid-1840s was Britain's expert on live and fossilised animalcules and foraminifera which he studied microscopically. From 1844 to 1846 Phillips and Mantell exchanged information about microscopic techniques, compared their results, swapped materials, and gave each other specimens. Phillips was also advised by Andrew Pritchard, the well known microscopist and author of a standard work on infusoria. Phillips' chief procedure was to prepare thin slices of rocks which he examined with a microscope of magnification 350-400. Initially he had doubted Ehrenberg's views about the identity of past and present forms and about the richness of microscopic life in chalk. But further research soon led him to accept, cautiously and provisionally, that fossilised foraminifera in tertiary and newer strata were identical with living forms. Given Phillips' palaeontological beliefs, this result did not disturb him. He was, however, worried that micro-palaeontology, which he regarded as a useful additional resource, would fall into 'the rash hands of cosmogonists' who would 'build a system of extinct and living nature on microscopic infusoria'.[17] Phillips was probably unimpressed by the reception of these papers so after 1845 he attended only one meeting of YGS when it met in July 1846 at York. There he joined Murchison and Buckland in a discussion about whether coal was formed *in situ*. When the Society ran into difficulties from 1847 and almost expired in 1853, Phillips was totally indifferent to its fate. Its research on the Yorkshire coalfield had collapsed and its continuing focus on polytechnics did not inspire him.[18]

8.2   Metropolitan involvements

As a lecturer outside York and Yorkshire, Phillips confined himself to two favourite haunts, the Royal Manchester Institution (1842, 1849) and the Royal Institution, London (1844, 1853), giving to each just two courses of lectures on geology (see Appendix 2). To the latter he also delivered two Friday evening discourses. That of March 1844 argued from the denudation, flexing, laminations, and nature of fossils of the Mendip Hills that they were formed 'ages after the beginning of geological time, and as many before the commencement of historical time'. Lyell and his wife appreciated this discourse which was generally much praised. Phillips' second evening discourse, given in March 1853 and focussed on the geology of Ingleborough, was anti-Lyellian in that it stressed violent subterranean movements and enormous fracturing of strata. It is clear that Phillips liked the Royal Institution: he was inspired by its audiences and knew that he was popular there among the young, old, learned, and unlearned. It is also clear that in the early 1850s, when his regular income dropped, he did not resort to burdensome lecturing to make up the deficit. Thus in late 1850 he declined to lecture in Sheffield. Generally he charged 50 guineas for six lectures, 60 for eight, and 75 for ten; but for the Royal Institution he dropped his rate to 70 guineas for ten lectures (1844) and 50 guineas for nine lectures (1853). If he gave the same course at the same time to two different institutions in the same place, he reduced his total fee. Thus when he gave a course of six lectures to the Royal Manchester Institution and to the Manchester Athenaeum, he charged the Mancunians 70 guineas all told. In the 1830s Phillips had lectured in metropolis and province to increase his income and reputation. From 1841 to 1853 he was much more selective, giving on average just one course every three years. In 1846 he even declined an invitation to lecture at the Royal Institution.[19] Lecturing was no longer a staple of his career and of making money.

Phillips was not regularly involved in the affairs of the Geological and Royal Societies of London because from 1841 to 1853 he was not often within striking distance of London by road or rail. Though awarded its Wollaston medal in 1845 he was a peripheral figure in the Geological Society. He did not serve on its Council and gave no paper to it but, when he happened to be in London, attended the occasional function. Late in 1840 he contributed to a penetrating discussion by Whewell, De la Beche, Lyell, Agassiz, Daubeny, Greenough, Murchison, Buckland, and James Smith of Agassiz's theory about the extensive glaciation of Britain. As an expert on the distribution of boulders of Shap granite, Phillips was worried by the mechanical difficulties involved in the assumption that a glacier could have carried them up and over Stainmoor. Next day he dined at Greenough's with the Lyells, Murchisons, Buckland, Agassiz, and Charles Stokes. In March 1841 he joined Murchison, John Taylor, Sedgwick, Buckland, Greenough, Fitton, Lyell, Owen, and Sopwith at a Geological Society Club dinner.[20]

Though he was on good terms with individuals prominent in the Society, he was not a member of its governing coterie, which he called the Somerset House clique. That became clear in 1842 when he supported Charlesworth in his unsuccessful attempt to become the Society's curator. The affair revealed in an embarrassing way where the power lay in the Society. Generally it was ruled by its Council and not its fellowship which usually rubber-stamped the Council's deliberations. In turn the Council was often controlled by a self-selecting yet often dedicated coterie, an arrangement which had worked well for the Society and for geology. In July 1842, when the autocratic Murchison was president, a thinly attended Council rejected Charlesworth's application as inadmissible, not because he was an incompetent geologist but because he was judged privately to be a 'young belligerent' who was irritable, bad tempered, and indiscreet, and therefore unsuitable to be a paid servant of gentlemanly geologists. Charlesworth was furious about being declared ineligible without explanation for a post in the Society of which he was a fellow. He had indeed offended some of its leaders: in the 1830s he had defeated Lyell about the Suffolk crag formations and in 1842 was exposing a rash identification by the devious Owen. Having claimed that a skeleton found at Bacton, near Cromer, was that of the extinct anoplotherium, a hasty identification denied by Charlesworth who asserted it was a roebuck, Owen subsequently withdrew his initial identification and then claimed Charlesworth's as his own.[21]

In autumn 1842 Charlesworth accused the Council of the Geological Society of bias against him and challenged Lyell, Owen, and Buckland to a debate on tertiary strata. Many fellows felt that the Council had acted irregularly and foolishly in refusing to consider Charlesworth's application. They expressed their dissatisfaction by requisitioning a special general meeting and by supporting his candidature. Among them were Phillips, two members of the Council (Sedgwick and Hugh Strickland) and, even more embarrassingly, the Marquis of Northampton, president of the Royal Society of London. On 2 December 1842 at a special general meeting the Council admitted the irregularity of its decision of July and, having justified its motives by alluding to Charlesworth's manifestly intemperate conduct since then, secured the unanimous backing of the meeting. Charlesworth's initial supporters were so ashamed of his abusive behaviour, which involved calling the officers liars, that they were unprepared to divide the Society or to risk the resignation of leading members of its governing clique. Having finally excluded Charlesworth from the curatorship, the Council, guided by Murchison, Lyell, Fitton, and John Taylor, elected Edward Forbes to the post on a salary of £150 pa and secured the endorsement of a second special general meeting in mid-December. To Charlesworth's further chagrin, Forbes was not even FGS.[22]

Phillips' involvement in the affairs of the Royal Society of London was unsustained but not insignificant. He played no part in securing the reforms of February 1847 which made it an exclusively scientific society, with fifteen elections a year to its fellowship on the basis of distinguished research. He was, however, a founder member of its Philosophical Club established in April 1847 as a dining club which, unlike the purely social Royal Society Club, of which he was not a member, aimed to consolidate those reforms, stimulate its members, and promote the scientific interests of the Society. Founded mainly by zealous reformers, many of whom were Phillips' friends, its membership was limited to forty seven, as a reminder of the year in which reform had triumphed. The Philosophical Club was so effective as a ginger group that in 1848 thirteen of its members, including Phillips, captured control of the Society's Council on which he served for only a year. He was neither an activist nor a performer in the Philosophical Club, from which he resigned in 1859 presumably because he was busy enough with travelling to London to fulfil his duties as president of the Geological Society.[23]

Phillips' most important contribution to the Royal Society concerned the offer made to it in 1849 by the prime minister, Lord John Russell, of an annual grant of £1,000 to promote science. Having quickly accepted this unexpected bounty, the Society appointed a committee chaired by Murchison to decide how to handle what became known as the government grant. The Society thereby became a major patron of science but in its long history it had never distributed money annually for research. Murchison saw that the aims and mode of allocation of research grants given by the British Association could function as a model so he turned to Phillips for an explanatory and persuasive memorandum. This document, which Murchison thought excellent, was circulated to leading fellows of the Society, read out to the grants committee, and produced the required effect. In 1850 the Society began to administer the government grant, the purposes and mode of distribution of which were avowedly indebted to the BAAS precedent as adumbrated by Phillips.[24]

8.3 The BAAS factotum

In 1841 Phillips' salary as assistant general secretary to the British Association was raised from £200 to £300 a year in recognition of his effectiveness as its general factotum and as editor of its annual reports. His Association payment was valuable to him between 1841 and 1853: dependable and not negligible, it usually provided about half his total income. Though Phillips regretted that organising and attending meetings cut into his field research in summer, he never contemplated resigning, he attended every meeting from 1841 to 1853, and was less exhausted by them than in the 1830s. They were usually smaller, which made them easier to manage, but they were demanding: even the hale Sabine confessed to Phillips that after the 1848 meeting he was so tired that he rested on the steps of a house 'and had I not been relieved by a flood of tears I doubt whether I should not have fainted'. In these years Phillips enjoyed a comfortable working relation with Sabine, a general secretary for nineteen years except when president in 1852. Phillips knew that his services were valued. Various presidents praised him publicly as the indispensable and excellent organiser of the Association's machinery, as a key figure in its success, as the longest-serving and most valuable of its office-bearers, and for his loyal ready activity on its behalf. No wonder that in 1853 Sabine wanted him as general secretary.[25]

Phillips' onerous responsibilities remained mainly unchanged from the 1830s. He advised and informed presidents, though to his chagrin one of them, Brewster, persisted in riding his own hobby-horses in his presidential speech. Phillips continued to edit the annual reports which were published within nine to twelve months of the date of meeting. Such a tight schedule cost him much editorial work in autumn and winter. He also audited the sales of the annual reports to BAAS members and the public. The preliminary arrangements for meetings often involved much correspondence as well as visits to inspect local facilities. Just one evening lecture, given in 1844 by the fastidious Lyell who wrote five letters about it, involved Phillips in negotiating about its date, its duration, the transport of Lyell's maps and illustrations, and the means of displaying them.[26] In guiding William Grove about the preparations for the meeting held in Swansea in 1848, Phillips wrote nine letters to him, having previously visited it in spring 1847 and judged it suitable for a moderately sized meeting. He advised Grove not to agitate locally about the BAAS before spring 1848 because by then locals would be motivated and would not forget the broad arrangements laid down by himself. He was particularly concerned about transport to Swansea to which there was no railway. He discouraged expensive displays and encouraged the locals to produce papers on local topics. He insisted that the main meeting room be well ventilated and realised that Swansea's Methodist buildings were unsuitable for illustrated lectures and promenades. He hoped for a trouble-free thin meeting and was therefore alarmed by economic revival in south Wales. He arranged two evening lectures but vehemently rejected as extraneous Grove's suggestion of a concert by a band. Phillips was so concerned to guide the locals and the meeting that he spent two weeks in Swansea before it began.[27]

At meetings he even catered for individual foibles. In 1842 he located the unwell Friedrich Bessel where he could smoke freely and in 1844 found Hugh Falconer's watch left in York. After meetings he wrote many letters about reports, committees, grants, and lobbies. On one day in July 1845 he sent no less than six letters about Association matters to Herschel. Between the annual meetings he was almost ever present at the Association's Council meetings, the dates of which were sometimes fixed to ensure his attendance.[28] As a policy-maker he enjoyed one success and suffered one failure. The success was the change in membership rules adopted in 1841. Phillips wanted to secure a large captive market for the annual reports and to make life membership the financial anchor of the Association. In 1839 the Association had introduced for all members a book subscription of £5 as a compounded fee for all future reports. Thus compounded life membership (£5) and the compounded book subscription (£5) cost £10; the annual membership (£2) and compounded book subscription (£5) cost £7. In an attempt to induce more annual members to buy life membership and to render the Association more respectable, new membership regulations were introduced in 1841. Annual members received the corresponding report gratis but were prohibited from making a lifetime book subscription. Simultaneously that subscription was reduced to £2 so that from 1841 life membership and the book subscription cost only £7 which bought attendance at all meetings and all reports. This was a bargain compared to the annual membership fee of £2 which bought just one meeting and one report. It was such a snip that in 1846 the cost of life membership, including gratis reports, reverted to £10.[29]

In 1848 Phillips was also concerned that the annual meetings should not be restricted to places which invited the Association and guaranteed expenses. He felt that invitations from university towns and great cities tended to be commands and encouraged unnecessary and expensive displays which were diversions from the essential business of one scientific week. He feared that suitable places might not invite, that unsuitable invitations were difficult to refuse, and that invitations might dry up leaving the Association with nowhere to go. Phillips wished to abolish the local fund of around £500 raised in the place of meeting, and usually topped up by £250 from the Association, and to have it give him £500 for local expenses. He suggested that sections A and G (mechanical science) be combined in order to reduce the cost of rooms. He argued strongly that the system of raising local funds discouraged locals from joining the Association and reduced its income. He concluded that it should select places to visit, as well as accept invitations, and proposed Derby as the first place to be so honoured. His scheme was praised but not adopted because invitations from suitable places still abounded.[30]

Phillips was more involved than before with lobbies of government and in giving advice. In 1843 he took part in two successful applications to government. He helped Lyell and Edward Forbes to draft an appeal to government about paying for the publication of the results of Forbes' research on the marine biology of the Aegean Sea. Phillips wrote solo a memorandum on behalf of Rosse, Northampton, and John Taylor, about the desirability of adding contour lines to the Irish Ordnance maps. In 1853 he was a member of a committee which lobbied the prime minister, the Earl of Aberdeen, about establishing a big reflecting telescope in the southern hemisphere to observe nebulae, a proposal rejected by government in 1850. In 1847 he signed the memorial addressed to Lord Russell, the prime minister, by members of the Association and other scientific societies about the management and functions of the natural history department of the British Museum. In 1853 Phillips was consulted by Henry Darwin Rogers about his scheme for a geological society in the USA and its relation, if any, to the American Association for the Advancement of Science of which he had been a key founder member in 1848. Phillips told him that within the BAAS geology gained much from collateral sciences pursued in it and advised him strongly that any geological society formed in the USA should not leave the AAAS.

Rogers accepted Phillips' point: he worked for reform in the AAAS and dropped his scheme for a geological alternative.[31]

8.4 The BAAS polymath

In his contributions to the sciences pursued by the Association Phillips straddled the demarcations it authorised. He was as active as ever in section C devoted to geology. He continued to expatiate to section A (physics, astronomy, and mathematics) about terrestrial physics, especially meteorology. Generally he tried to use measuring instruments to produce data which he tried to summarise as empirical generalisations which he hoped would be a secure foundation for theory. He also evaluated rival anemometers and invented a couple of his own. In an attempt to compare the surface features of the moon and the earth, he ventured into lunar geology and in the early 1850s was a pioneer in using photography to supplement drawings made of the moon. He was also not inactive in section D which covered natural history. He was the solitary non-specialist zoologist involved in the working parties which produced in 1842 under the Association's auspices the important rules for zoological nomenclature which had been mainly devised by Hugh Strickland. In the late 1840s Phillips made his debut in ethnography, then a sub-section of section D. No other leading British geologist made such excursions into sections A and D of the Association. Phillips did so because he had long regarded physical and zoological sciences as collateral aids for geology and the Association provided for him the most propitious forum in which to exemplify this belief. His scientific scope and versatility were abundantly revealed at the meeting held in Hull in 1853. He gave two papers to section A (Yorkshire magnetism, photographs and drawings of moon), three papers to section C (Yorkshire unconformities, Yorkshire erratics, and new plesiosaurus in the Yorkshire Museum), and one paper to section D (a marine worm dredged at Scarborough). Additionally, he gave an invited evening lecture which covered not just the geology of Yorkshire but also its physical geography.[32]

In his contributions to section C Phillips covered three themes which had exercised him in the 1830s. On the vital question of how to use fossils to characterise strata, he continued to take the approach he had expounded in 1841 in connection with his *Palaeozoic fossils*. The least inadequate procedure was to rely on a large number of different forms of fossils. He continued to argue that particular species were valueless in characterising geological formations so in 1842 he attacked Agassiz for depending on them. Phillips maintained his scepticism about the wide validity of geological systems and nomenclature if, like Murchison's Silurian, they were based on a small area. On one occasion this scepticism shrank to nihilism. In 1853 Sedgwick bitterly defended his Cambrian system against what he regarded as the unjustified encroachments of Murchison's Silurian. Upset by such feuding between two of his oldest friends and wishing to avoid antagonising either of them, Phillips refused in the ensuing discussion to give a positive or even a provisional opinion about where to draw the line between the two systems. Indeed in his own paper on erratics given that year he used the compromise term of 'Cambro-Silurian.[33]

Phillips maintained his interest in the origin of coal, abandoning by 1842 his previous belief that it was drifted vegetable matter. Mainly on the basis of an examination with his microscope of the ashes of burnt coal, which showed that different plants were co-embedded, he concluded, albeit cautiously, that coal was formed from plants growing *in situ*. It was characteristic of Phillips that he used evidence acquired with an instrument to buttress that of the upright fossil trees near Bolton which members of the Association inspected in 1842. Instrumentation was also the key to Phillips' abandoning his diluvianism and his doubts about the theories of Lyell (icebergs) and Agassiz (ice-sheets) as explanations of the distribution of erratic blocks. The start of his conversion occurred in 1841 when Phillips saw the frozen river Ouse like a polar sea, full of solid and then broken and floating icebergs. In 1853 he reported his research made at Norber near Settle on the erratic blocks of Cambro-Silurian flags, perched on broad surfaces of mountain limestone which in the vicinity rested on those flags. Using the measuring methods he had employed in the Geological Survey, Phillips confirmed that these erratics occurred at a higher level than the parent rock. He concluded that much of Yorkshire lower than 1,500 feet above sea level had been covered by a glacial sea, that the erratics had been deposited not by water but by floating icebergs when they melted or overturned, and that higher ground might have been covered by glaciers. Though Phillips was still worried by the mechanical and climatic difficulties of Agassiz's theory, like most British geologists in the early 1850s he was prepared to admit both floating and terrestrial ice as agents of transport of gravel and erratics.[34]

There was one continuing issue, that of the relation between scripture and geology, in which Phillips was not involved. Given that in the 1830s Cockburn, the dean of York Minister, had depicted Buckland and the Association, with Phillips as their accessory, as mouthpieces for irreligious geology, it was not surprising that when the Association met in York in 1844 Cockburn yet again assailed modern geology as antiscriptural and portrayed the Association as attacking the Bible. In a paper to section C Cockburn presented a biblical cosmogony which proclaimed that the earth was about 6,000 years old and that nearly all the strata were formed by a deluge lasting a few weeks. His chief target was not Phillips, a local lay geologist, but the Bridgewater treatise of Buckland, a reverend professor who became a fellow dean in 1845. In Buckland's absence, Sedgwick as president of the section took ninety minutes to rebut Cockburn who refused to be silenced. He attacked Buckland and Sedgwick in a sermon in York Minster, issued a pamphlet entitled *The Bible defended against the British Association*, and harassed Sedgwick, Buckland, and Murchison with a series of letters, some of which were published in a fifth edition (1845). The controversy between Cockburn and Sedgwick aroused not just local but national interest and became an important moment in the debate about the relations between scripture and geology.[35]

Phillips was glad that he had not been sucked into this public controversy. He thought that the best way of dealing with Cockburn was to ignore him: as he told Whewell, Cockburn's 'random and rotten weapons do no mischief unless we, by too much attention to the scratches they make, aggravate a puncture into a sore'. In private Phillips maintained unchanged the views on the relation between Christianity and science that he had formed in the 1830s. As an irenic Anglican, who deplored jarring creeds, he gently revealed his Christianity to his friends in the 1840s. When Ramsay had been caught in a storm at sea and sought refuge in the Mumbles near Swansea, Phillips reprimanded him jocularly for not mumbling prayers as he would have done. Apropos Mantell's reproof of 'Bridgewater people' in his book on animalcules, Phillips told him it was not presumptuous to infer God's purposes from natural phenomena: 'they teach us, as all visible nature does, the *intention* of the Creator to fill all habitable space with beings capable of enjoying their own existence'.[36]

Several new themes were apparent in Phillips' contributions to section C. In response to the Association's request for a report on slaty cleavage, Phillips edged in 1842 towards a mechanical explanation on the basis of the way in which fossil trilobites in the Llandeilo flags were distorted longitudinally, transversely, and . obliquely, and thus indicated certain movements in them. For the railway sections committee, which spent £363 of the Association's money from 1841 to 1844 in procuring coloured sections of strata exposed by railway excavations, he prepared a pattern section from which a plate was engraved and extensively reproduced. In an attempt to accelerate his research on British belemnites, he was awarded a research grant of £50 in 1841. He hoped to produce a monograph on the general plan of the works by Voltz and Blainville, but the project languished because of difficulties of classification. From the mid-1840s he was stimulated by the ideas of Ramsay, his colleague in the Survey, about extensive marine erosion and denudation in south Wales, thought them to be widely applicable, and in 1847 used them against the view of Chambers that ancient sea terraces were produced by uniform elevation of land over much of the earth. Phillips rejected Chambers' view that these terraces were beaches. For him they were produced by the erosion of softer beds in horizontal strata. Phillips also ventured into experimental geology. He had long been interested in the mechanical state of waters and the seabed when organisms were deposited. In 1851, having gone out of his way to praise Charlesworth's valuable classification of the crag, he reported that on the basis of experiments done on the deposition of bivalves in disturbed water those fossil shells which settled with concave sides downwards had been frequently agitated in the water.[37]

In section A Phillips was as active as in the 1830s. As a magnetician Phillips confined himself to measuring dip in Yorkshire in 1850 and 1853, using a Charles Robinson dip circle, which was accurate to one minute of a degree, to confirm the result he had obtained in the 1830s that the lines of equal dip (isoclinal lines) were flexed according to the altitude of the ground. This conclusion, given to the Association in 1850 and 1853, was welcomed by James Forbes, by the German physicist Julius Plücker, and by Airy, the astronomer royal, who was notoriously critical of endless and mindless recording of magnetic data.[38]

As a meteorologist Phillips was regarded as an eminent expert not only in the BAAS but elsewhere. He was one of twenty-three individuals entitled to receive a copy of the meteorological and magnetic observations made at the Royal Observatory, Greenwich. Within the Association he was appointed in 1842 the reporting member of a research committee on the structure and colours of clouds and in 1849 the centre of intelligence of a committee on luminous meteors and another on auroral observations. These bodies turned out to be dead letters. Privately Phillips did nothing on clouds and meteors but in the 1840s he reported auroral outbursts to York newspapers and the *Athenaeum*. By 1850 he had developed the theory that an aurora was caused by electrical tension and polarities, he hoped to resume the research he had done in the 1830s on the magnetic features of the '*recusant*' aurora, and he discussed this project intensively with Faraday, who approved it. Sadly no appropriate opportunity occurred for several years.[39]

Phillips maintained his interest in temperature and rain measurements. The former did not rise above rampant empiricism: his 1841 paper used others' data from 1808-1811 to record the difference between the temperature inside and outside York

Minster; that of 1844 recorded graphically twenty- five years of data about the annual temperature of York. In contrast Phillips was a respected figure in pluviometry: in 1843 he and Wheatstone were asked by the Association to recommend methods for measuring rainfall and temperature at Kew Observatory, which it had acquired in 1842. Later in the decade Phillips acted as a referee of a rainfall paper for the Royal Society of London. But he gave only one paper to the Association on this topic when in 1844 he reported the results of two years' research with funnel gauges carried out by himself and Cooke. He refrained from commenting on the results and had not responded to his own call made in 1841 for three types of gauge to be compared. His work on waves of atmospheric pressure, revealed in 1844 and carried out in the limited district of northern England, followed the recommendations of Herschel and Quetelet; but it was a small coda to the sustained work of Herschel and William Birt on atmospheric pressure waves over much of Europe which the Association supported financially.[40]

Phillips' most sustained work in meteorology was on anemometry. In 1845 he was asked to report on instruments employed in anemometry, as an update of Harris' 1844 report which found Whewell's instrument less satisfactory than that of Osler because of friction in its mechanical parts. Phillips' 1846 report was mainly an evaluation of the instruments of Whewell and Osler, with both of whom he was in contact. Phillips' detailed suggestions about improvements to Whewell's device were so cogent that Whewell, by then master of Trinity College, invited Phillips to Cambridge to give meteorological advice. Generally Phillips was so concerned to take account of friction in existing anemometers that he invented an instrument which would eliminate the effects of mechanical movements, momentum, and friction. By 1846 he had devised an anemoscope which was a bulb of a thermometer covered with cotton wool which was immersed in water and then exposed to the wind which cooled it. He used his own lathes and railway journeys to calibrate the instrument which was not transformable into a self-registering anemometer. In 1849 he reported that he had devised an instrument called a therm-anemometer which was capable of measuring very light winds. It employed the principle of the cooling effect of air, not on a wet-bulb thermometer as before, but on the bulb of an ordinary thermometer, and was developed by him for his work on ventilation in mines. Phillips' devices were ingenious but did not seriously challenge the anemometer of Osler, significantly improved in 1846, and Thomas Robinson's cup anemometer completed that year.[41]

Phillips revealed for the first time to section A his interests in electricity and astronomy. In 1850 he reported on behalf of a committee appointed to examine the effects produced by lightning on a tree near Edinburgh. He showed that its trunk expanded and burst; characteristically he thought it worthwhile to record that a bit of it weighing two and a half hundredweight was carried 127 feet. Three years later his drawings and photographs of the moon aroused far more interest. Phillips' awareness of lunar geology was stimulated by Conybeare's 1833 dictum that observations of the moon might illuminate elevation, volcanoes, and craters. But the trigger was Phillips' visit in 1843 to Birr Castle, Ireland, at the invitation of the Earl of Rosse who had presided at the Association's meeting in Cork the previous month. Using Rosse's reflecting telescope of three feet aperture and normal magnification of 600, Phillips saw for the first time not only the colours of the planets and stars but also mountains on the moon. He also examined Rosse's model of one mountain, with its crateriform summit, cavernous sides, but no evidence of lava. In 1847 Phillips was impressed by Charles Brooke's success in applying photography to automatic registration by meteorological instruments. It was impossible for Phillips not to know about the equally successful research on the same topic carried out from 1845 at Kew Observatory by Francis Ronalds who reported regularly to the Association. At the 1850 BAAS James Nasmyth gave to Section A an illustrated paper on the lunar surface and its relation to that of the earth, and throughout he used terrestrial agents to explain lunar features. Nasmyth kept Phillips informed about this research, even inviting him to see the reflecting telescope, of twenty inches aperture and magnification of 500, specially made by Nasmyth for moon work. Next year Phillips was intrigued by the daguerreotypical photographs of the full moon taken by William Bond, of Harvard University Observatory, and exhibited to section A.[42]

At the 1852 meeting held in Belfast Sabine used his presidential address to argue that Rosse and two savants knowledgeable about physics and geology should examine the physical features of the moon, which he claimed was made of the same materials as the earth but possessed neither atmosphere nor sea. This suggestion, instigated by Phillips, led to the Association asking Rosse, Phillips, and Thomas Robinson to report on the Nasmythian topic of the physical character of the moon's surface compared with that of the earth. After the meeting Phillips, Robinson, and Sabine were invited by Rosse to Birr Castle where they looked at the moon with his powerful reflecting telescopes, including his newer one of six feet aperture and a normal magnification of 1,300. Using Rosse's older telescope, Phillips drew the mountain crater Gassendi but thought his drawing poor, perhaps because the telescope was not mounted equatorially so it did not track its moving object. It seems that he and Rosse then agreed that, on the basis of the map and treatise of Johann Mädler and Wilhem Beer, selected parts especially mountains of the lunar disc should be not only drawn to a magnification of a thousand but also photographed. Once back in York Phillips undertook in his garden a pilot project to test the feasibility of the scheme. In autumn 1852 he used his familiar Cooke refracting telescope, with 2.4 inches aperture and equatorially mounted, to produce what he regarded as good drawings of Gassendi. He also issued a leaflet on behalf of the Association about observing the moon. Early next year he acquired from Cooke a new refracting telescope of six and a quarter inches aperture which was mounted equatorially and driven by clockwork. With this instrument he made in May 1853 a drawing of Gassendi with the desired magnification of a 1,000 and followed it in July with photographs of the moon made by the wet collodion process which had been invented in 1851 (Fig. 8.1). It was more sensitive than the daguerreotype process and particularly appropriate for bright objects.[43]

At the 1853 meeting of the Association Phillips displayed his drawing and photographs. Working independently of Warren De la Rue, who had photographed the moon in late 1852, Phillips was the second in England to apply photography to astronomy. Given the success of his pilot scheme, he and Rosse recommended that the team of observers should focus on designated areas of the moon and combine their representations to produce a lunar chart about 200 inches in diameter, which would be far larger than that of Mädler and Beer (thirty-eight inches diameter). Phillips hoped that the chart, on a scale of one inch to 25 miles, would contribute to lunar geology by depicting lava streams and craters of eruption, upheaval, or explosion. The chart was not completed but in 1854 Phillips presented to section A not only micrometric eye drawings of the lunar surface by James Challis and Charles Piazzi Smyth, but also photographs taken by Nasmyth, John Lee, and himself. Phillips focussed on Gassendi, the fault lines and winding ridges of which fascinated

**Fig. 8.1   Phillips' second photograph of the moon taken 18 July 1853 at York.
Using a telescope made by Cooke, Phillips obtained an image of the
moon one and a quarter inches in diameter. His lunar photographs
had a diameter of 3 inches at best.**

him. He had done enough to show that in recording the lunar surface photography
was a useful supplement to drawing by eye, so without rancour he allowed De la
Rue to dominate lunar photography from the mid-1850s. Though the Association's
president in 1857 publicly praised Phillips as an eminent lunar astronomer, he was
also a lunar geologist. From observations of dry seas, lava streams, and oxidised
minerals, Phillips inferred that in the past the moon had been covered by water and
possessed an atmosphere containing oxygen, two assertions strenuously denied by
Nasmyth. In Phillips' time it lacked both water and oxygen so he concluded that like
the earth the moon was not unchanging but had a history of great vicissitudes.[44]

Phillips managed to find time to contribute to section D (natural history) because
he regarded zoology and botany as intimately connected with geology. In 1844 he
went out of his way to praise William Carpenter's report on the microscopic structure
of recent and fossil shells because it revealed affinities between recent and ancient
animals and the conditions of life enjoyed by the latter. In 1851 he contributed to
a discussion on molluscs which bore into rocks, arguing that they employ both
chemical and mechanical means. In 1853 he dilated on a species of sea-worm,
dredged at Scarborough, which he had kept alive for three weeks. In the late 1840s

he reported twice ethnological research which he claimed illuminated the history of natives and invaders in Britain. When working for the Survey in the Midlands, he observed people with black eyes, black hair, and dark complexion, whom he thought of Celtic origin. Rejecting the notion that black-eyed races come from blue-eyed ancestry, he concluded in 1848 that these Celts were derived from those who were not driven out by immigrated Germans or Scandinavians. Next year he reported that excavations of tumuli in the Yorkshire Wolds revealed skulls of Celtic and Teutonic races. Phillips was concerned to add the evidence of physical structure to the conclusions derived from philology so he argued that cranial differences could be used to study the natives and invaders of Britain. Curiously he did not mention that these excavations had been recently carried out by the Yorkshire Antiquarian Club of which he was a leading member.[45]

Phillips played a secondary role in the rules devised by his friend, Hugh Strickland, for a uniform and permanent nomenclature in zoology. These guides, published in 1843 by the authority of the BAAS Council, received the imprimatur of a twelve-strong committee appointed by it early in 1842. Having been asked by Strickland to evaluate the rules, both in general and in detail, Phillips was keen to become the solitary non-zoologist on that committee. Though he was more familiar with the problems of nomenclature in palaeontology, he was au fait with those in zoology from his lectures on living nature. Strickland advocated the law of priority, ie that zoologists should use the Latin binomial name given by the first person to describe a species. This conservative approach was arbitrary because it was based on historical contingency and it affronted reformers who wanted a totally new nomenclature based on fundamental principles. Strickland's rules were quickly accepted in Europe and the English-speaking world and enjoyed a long life precisely because they were unconnected with any zoological theory or system and recognised priority of discovery, two features which appealed to Phillips.[46]

8.5   Civil scientist

Phillips was asked on several occasions to act as a consultant to railway and water companies and landowners in northern England but with one exception was, it seems, too busy to accept. In 1845 the Leeds and Thirsk Railway Company wanted Phillips to rebut the claim of the Leeds Water Company that a tunnel north of Leeds would interfere with one of its springs. Two years later George Stephenson sought geological advice in order to buttress his own evidence to Parliament about the Derbyshire railway near Bakewell. In 1845 the Harrogate Waterworks Company was anxious that drawing on a spring west of the town would not interfere with the famous sulphur springs so it turned to Phillips for advice. In 1847 two landowners wanted him to inspect their land for minerals. One request, from a local, concerned veins at Roecliffe near Boroughbridge. The other, made by Buckland on behalf of no less than Sir James Graham, was to inspect his estate near Haltwhistle to determine the extent of coal in it. There was one offer which Phillips accepted: in the early 1850s he produced a map and sections of the Eskdale area in north-east Yorkshire for the York and North Midland Railway Company.[47]

For much of the period July 1849 to March 1850 Phillips undertook a role that was new to him. He became a civil scientist, using his expertise to advise a desperate government about how to solve or alleviate a pressing problem. At government's request he reported via a House of Commons paper on the ventilation of coal mines in Northumberland, Durham, Yorkshire, and Derbyshire. He was not the first scientist or geologist to be asked by government to investigate deaths in coal mines caused by explosion of fire damp (methane) and suffocation by choke damp (carbon dioxide and nitrogen left after an explosion). In 1844, after ninety-five males were killed at Haswell, county Durham, the home secretary asked Faraday and Lyell to investigate the disaster. Their report of October 1844, based on only a few days spent at Haswell and only one day underground, was written within two weeks of their visit. They focussed on the ventilation of goaves (underground chambers created by roof falls and the excavation of coal). Early in 1845 a report commissioned by the Newcastle-based United Committee of Coal Trade pointed out that there were no goaves in nearly all the local collieries where there had been serious accidents, so that the remedy proposed by Faraday and Lyell was irrelevant. It was made clear that experienced mine owners and managers thought that the two metropolitan savants knew little about the procedures and practical difficulties of mining. In March 1845 Faraday and Lyell were forced to admit that their proposed mode of ventilation, using iron pipes, was not feasible: having asserted in October 1844 that it was practically available, they confessed five months later that it was an expression of a principle and not practice. Their report, the local criticism of it, and their admission were all published in 1845 as a parliamentary paper, but only their report appeared in the *Philosophical magazine*. It was thus made clear to politicians, but not to scientists, that the use of scientific men as civil scientists was problematic.[48]

From the mid-1840s fatal accidents caused by explosions occurred in localities where such disasters had been previously unknown. Thus the problem, clearly associated with deeper and more extensive collieries, became a national one, having hitherto been confined to north-east England. In 1847 seventy-three people died after an explosion at a colliery near Barnsley in Yorkshire. Next year deaths caused by explosions occurred with sickening and apparently unstoppable regularity. In February, March, August and October there were twelve, twenty, fifteen, and thirty deaths at collieries in Warwickshire, south Wales, county Durham, and Cumberland respectively. In January 1849 a second disaster occurred near Barnsley where seventy-five were killed. The government's response, prompted by Lord Lincoln, a cabinet minister, was to turn, not to Faraday and Lyell, but to the staff of the Geological Survey and the Museum of Economic Geology as the most likely to produce useful solutions: it was possible to do this from 1845 when the Survey became a civilian enterprise run by the Office of Woods for which Lincoln was responsible. Lincoln was keen to show that his geologists were useful to the state. Three reports on explosions in collieries in five different coalfields were commissioned and published as parliamentary papers in 1846, 1847, and 1849, the investigators and authors being De la Beche, Lyon Playfair, chemist to the Survey, and Warington Smyth, its mining geologist. On one occasion they were aided by Seymour Tremenheere, who in 1843 had been appointed the sole inspector of mines to enforce the provisions of the Act which banned female labour and the employment of boys under ten years of age in mines, and prohibited anyone under fifteen years from tending machinery. Each report, based on visits to the appropriate collieries for several days and produced within a month of the disaster in question, drew attention to defective or non-existent ventilation as the chief cause of explosions. At the same time Tremenheere argued forcibly for an enlarged mines inspectorate.[49]

On 5 June 1849 thirty-three males were killed in an explosion at Hebburn in county Durham. Faced by the frequent and intolerable occurrence of serious accidents, government responded quickly in two ways. On 18 June a select committee of the House of Lords was appointed to enquire into the prevention of dangerous accidents in coal mines. This committee merely examined witnesses, two of whom (De la Beche and Ansted) were geologists; and it produced its report before the end of July. It stressed that ventilation was generally defective, favoured the appointment of inspectors with advisory powers, and advocated the establishment of mining schools. On 15 June 1849 Sir George Grey, the home secretary, who wanted a more thorough investigation than the Survey men could provide, appointed Phillips and John Blackwell to make an extensive general enquiry into the best means of preventing serious accidents in collieries and to focus particularly on ventilation. Unlike previous enquiries of the 1840s, details of their remit were public because published as a parliamentary paper. They were required to inspect mines and had the power to report to Grey any obstacles they met. They were asked to attend to twenty-six points, one of which called for actual experiments to measure the volume of air introduced by ventilation and its speed of movement through mines. Their investigations occupied about nine months and their reports were laid before Parliament in early May 1850. That of Blackwell, who had been unwell and was an owner of iron works, covered England south of Derbyshire and recommended better ventilation, more use of the Davy lamp, and better education. It was much shorter than Phillips' which contained coloured geological maps and columnar sections of the coalfields of his allocated area (north-east England, Yorkshire, and Derbyshire) and offered many plans showing the ventilation courses, modes of working, and ventilation furnaces of collieries. Publicly and privately the home secretary maintained that legislation should await the reports by Phillips and Blackwell.[50]

Phillips was appointed by Grey as a reporter to government, having been recommended by De la Beche. Subsequently De la Beche went out of his way to praise Phillips' qualifications for his important new job. In public he lauded Phillips' high character, extensive acquirements, and great talents to the House of Lords select committee. In private he praised Phillips' appropriateness to the Earl of Carlisle, formerly Lord Morpeth, the first commissioner of woods and a subscribing member and patron of the Yorkshire Philosophical Society. In fact, from 1824 Phillips had little interest in mining geology and technics, and little experience of collieries, so he was soon denigrated as a 'theoretical geologist' by mine owners and a tool of a procrastinating government by miners. He was, however, an adopted northerner and not a metropolitan intruder; he had several contacts in the mining industry in the north of England; he knew its general geology well; and his obsession with measuring instruments in general and anemometers in particular fitted him for the task in hand. The emolument was irresistible. He was offered and quickly accepted a salary of £600 pa plus £100 pa expenses. In the event in May 1850 he was paid £500 which in August was topped up by £250 on Grey's instructions. For work which spanned ten months he received £750 as salary and expenses combined. With £300 pa from the BAAS he had never been more affluent.[51]

Having studied the appropriate parliamentary papers of the 1840s which the home office had sent to him, Phillips prepared appropriate instruments, wrote to his local contacts for information, and in July and August 1849 visited collieries in Yorkshire and Derbyshire, with a break for the British Association meeting held in Birmingham. In autumn he spent almost two months in the north-east, where he stayed with Nicholas Wood, the well-known colliery viewer, ie superintendent and consultant. Phillips thought that his inspections should not threaten but strengthen the sense of responsibility of owners and managers of mines and he wished to induce them to accept inspection without seeing it as interference. He therefore met in Newcastle a group of local colliery viewers, with Wood in the chair, and explained that initially he wished to study the best practice in order to obtain a 'standard of excellence of a practical kind' for each district, whether the pits were deep (1,800 feet) or shallow and however ventilated. In response the viewers, some of whom vied with each other in the 1840s in improving the ventilation of the deep mines of the Newcastle area, submitted a list of fifteen collieries for Phillips to inspect. It included the best ventilated collieries and indeed began with Hetton which, with a ventilation rate of 190,000 cubic feet of air per minute, surpassed all its rivals. At Phillips' request the viewers had selected mines where he could inspect what they deemed to be the best ventilating practice. But working colliers, led by Martin Jude, secretary of the local Miners' Association, thought the selection unrepresentative and that Phillips' investigations would be 'partial and confined'. They objected to having no say in the selection of pits and to being prohibited from accompanying him on underground inspections. They wished to show Phillips the worst parts of the worst mines. Their reservations were so strong that Jude complained not just to Phillips but to the home secretary who agreed to their request that miners formally join Phillips' enquiry. Phillips was dismayed by Grey's concession: such miners would not facilitate or improve but would inconvenience his strategy of ascertaining for himself the best practice in ventilation. He therefore approached the home office and persuaded Grey to abandon his recently made concession. Phillips then met Jude and local miners in Newcastle in late October 1849 and allayed their fears by accepting their detailed criticism of the working of different pits as confidential evidence which he would not overlook. This meeting was useful to Phillips: it opened his eyes to the worst practices in the pits and showed him why Jude advocated the appointment of inspectors of mines with powers of interference and compulsion.[52]

Phillips not only saw for himself a great variety of collieries, safety lamps, and ventilation systems. He also had experiments done for him in mines and used his own thermanemometer, which he regarded as a perfect jewel, to measure small currents of air in them. His instrument was safer and less crude than the standard method of estimating air-speeds in collieries, which was to observe the speed of a puff of smoke from an ignited bit of gunpowder. His measurements were made as part of an unsuccessful attempt to produce what he regarded as an important contribution to 'science and practice', ie formulae concerning the loss of power in ventilation systems, from which practical rules for improvement could be derived. He also sent a forty-one point questionnaire to selected colliery managers and asked for six types of document, mainly plans, to be sent to him. He thought the returns were useful: they recorded much 'positive knowledge'.[53]

In his report Phillips condemned certain practices. He denounced pillar working, ie supporting the roofs of underground passages in coal seams by rectangular blocks of undisturbed coal. He criticised the use of bratticed shafts in mines as repulsive and ominous. In order to save money in sinking deeper pits just one shaft was sometimes made and was divided by wooden partitions into two to four compartments. This procedure, called bratticing, allowed pumping, ventilating, and the drawing of coal to be carried on in just one shaft but was objectionable: the partitions were inflammable and if disintegrated by an explosion, fell to the bottom of the shaft and cut off any communication with the surface. Apropos ventilation Phillips did not recommend new methods. He gave scant attention to mechanical fans which had been first used in 1849 in Wales, which was not his area. He did not suspect that the future lay with fans, which became widely used in the 1870s. In contrast he was au fait with Goldsworthy Gurney's invention of the high-pressure steam jet generated by an underground engine-boiler. The steam jet method was first employed from 1848 in a Northumberland colliery, which Phillips visited, and next year had some success in ventilating a pestilential sewer in London. From his data on the quality of the air as it returned to the upcast shaft, and as indicated by the safety lamp, Phillips was surprised but pleased to discover that strong and constant ventilation eliminated fire damp. He recommended that the common method of producing such ventilation, obtained from a furnace at the bottom of the upcast shaft, was the best. It was more powerful than the steam jet especially if the air channels underground were freely flowing to reduce frictional loss, if the furnace was crowned by a shaft used purely as a chimney and lined with fire-bricks, and if the in and out air currents were totally separated. Thus Phillips' general solution was based on the best ventilation practices he had seen in the north-east of England. For him explosions were mainly preventable by applying proved methods with skill and with 'unsleeping vigilance' in a disciplined environment. He was adamant that 'no encouragement should be given to colliers, or managers, in their often repeated assertion that there is too much air and that the men cannot bear it'.[54]

Phillips recognised that safety lamps were useful but it was clear to him that even the best, even when properly used, were inadequate in fiery mines. His conclusions were in harmony with the way in which in the 1840s better ventilation was increasingly seen as the chief way of preventing explosions and any form of safety lamp as defensive but not preventative. Phillips recommended as the best form of ventilation that occasioned by a furnace, a means adopted in north-east England by the great majority of viewers whom Phillips regarded as his friends. Though he was indebted to them for what they regarded as the best available practice, he was not just their tool. His experiments and collected data indicated that furnace ventilation was the best. His judgement was prescient. The use of steam jets fizzled out in the early 1850s and it was only in the 1860s that mechanical fans, mainly invented in the 1850s, began to displace furnaces.

Phillips recommended systematic inspection of mines under the authority of government because it would prevent accidents by discovering bad practice and advising about ways of improvement. As he told the home secretary privately, owners and mine managers would cordially co-operate only if inspections were 'friendly and suggestive', ie advisory. *In extremis,* however, he judged that inspectors should prohibit dangerous practices if remonstrating failed. Though Phillips favoured in private the establishment of a school of mines in London as the apex of provincial schools, he confined himself in his report to the latter which he deemed essential for managers of mines and desirable for colliers.[55]

The consequences of the reports by Phillips and Blackwell were soon apparent. Their recommendations corroborated those of the 1849 select committee and went some way to meeting those of various pressure groups, including miners' unions, and concerned members of Parliament. In August 1850 Parliament approved the appointment of coal mine inspectors, with power of entry and investigation; in November four were appointed, one of whom was Blackwell, to cover 2,000 collieries in which 200,000 workers were subject to an annual death rate of four per thousand. Phillips' report contributed therefore to the classic sequence by which intolerable situations were dealt with by government: firstly, government ordered an enquiry from an expert, or commission; then the resultant report, replete with evidence, was laid before Parliament which passed legislation which appointed appropriate government inspectors. Phillips' report also helped to prepare the way for the establishment in 1852 of the first mining institute. It was founded in Newcastle, under Wood's presidency, with the revealing title of 'The North of England Institute of Mining Engineers, and others interested in the prevention of accidents in mines and in the advancement of mining science generally'.

Phillips' career as a civil scientist ended as suddenly as it had begun. In autumn 1850 he was not consulted by government about the creation of mines inspectors. He neither lobbied to become one nor, unlike Blackwell, was made one. Presumably he ruled himself out in October when, in response to a request from the home office to survey again a dangerous colliery on which he had reported privately in June, he refused the terms of 5 guineas a day and actual travelling expenses because they were less than the fee he had stipulated.[56]

8.6   Popularising science

From 1849 until his move to Oxford Phillips supported two field clubs, the Yorkshire Naturalists' Club and the Yorkshire Antiquarian Club, both of which were founded in York in 1849. Phillips thought that the Yorkshire Philosophical Society was then not only in the doldrums intellectually but also exclusive, so he persuaded it to create associate membership at £1 pa and he supported the two clubs which tried to encourage more local participation in their respective pursuits. As the two clubs were devoted to the YPS's main foci, the natural history and antiquities of Yorkshire, Phillips took good care to persuade the YPS that they did not undercut but benefit it and, through his influence, they did not see it as a rival but fed its Museum with specimens and objects. In this respect the two Yorkshire clubs were different from others elsewhere which provided an alternative to philosophical societies. With no premises to maintain, field clubs charged a low subscription, recruited among the less well-off, spread meetings around their area, and of course promoted field work. By the later 1840s the field club movement was still mainly rural but the Tyneside Naturalists' Field Club, founded in 1846, had shown that a field club could span countryside and city.[57]

The idea of a Yorkshire Naturalists' Club had been floated unsuccessfully in 1845 in York by Beverley Morris, a local physician, who drew up some provisional rules. In January 1849 aspiration was transformed into action: Morris, Charlesworth, and Oswald Moore, a York doctor, met in Moore's house and founded the Yorkshire Naturalists' Club to promote the natural history of the county by bringing working Yorkshire naturalists into greater contact and by contributing specimens to 'public collections of local natural history'. The Club's membership fee was to be 5 shillings a year, it was to be run mainly by York residents, it aimed to meet monthly, and women were admissible as members though subsequently were not prominent in it. By March 1849 the Club had twenty-seven members, including Phillips and other prominent figures in the YPS, in response to a circular issued by Morris in February. In May 1849 Phillips chaired the first meeting of the Club, at which he was elected president and its rules amended, a move in which presumably he was not uninvolved.

Its first aim was changed to contributing specimens to remedy gaps in the collections of the Yorkshire Museum, which was to receive priority 'as being accessible to the whole county', Thus the Club, ostensibly devoted to Yorkshire's natural history, became a York institution devoted to collecting, donating, and subsequently even buying specimens on behalf of the relatively wealthy YPS. Consequently the organisation of field excursions was not the Club's chief activity.[58]

As president Phillips chaired the monthly evening meetings fairly regularly and generally promoted the Club. He went out of his way in October 1849 to convince the YPS that the Club, and the Yorkshire Antiquarian Club, though independent, would cooperate with the YPS. It seems that Phillips hinted that the Naturalists' Club had funds to buy desiderata for Yorkshire museums, with the YPS's own Yorkshire Museum taking precedence. In March 1850 Phillips hosted an evening meeting of the Club, preceded by hospitality given by his sister, at which he persuaded the Club to fund from a rented house collaborative dredging on the Yorkshire coast. This venture failed in 1851 because of the greater attraction of the Great Exhibition. Subsequently Phillips was neither a leading organiser in the Club, like Beverley Morris, its secretary until 1853, nor a leading performer, like Moore, Charlesworth, and F.O. Morris, all of whom were members of the YPS. But he identified with the Club to such an extent that even when he left York for Oxford he was retained in 1854 as president of its 120 members.[59]

Phillips' involvement with the Yorkshire Antiquarian Club was the product of his growing interest in archaeology, especially from 1846 when the Archaeological Institute of Great Britain and Ireland met in York, bringing with it a remarkable exhibition of antiquities and subsequently publishing a valuable report on the antiquities of Yorkshire and York. Phillips was drawn into the meeting by Albert Way, its secretary, who had met Phillips in York. Way wanted Phillips to back the proposed 1846 visit, to secure the co-operation of the YPS and the use of its lecture theatre, to arrange a place for the projected exhibition of antiquities, and to act as the local secretary. Phillips fulfilled most of Way's requests, thus helping to ensure that the second meeting of the Institute consolidated its position. The Institute had been formed and named under peculiar circumstances in 1845 when the British Archaeological Association, established in 1843, suffered secession by a small group which defiantly called itself the same name. To remove confusion, the larger existing Association, led by Way, renamed itself the Archaeological Institute.[60]

At the Institute's meeting in York Phillips sat on its general committee, on the sectional committee for early and medieval antiquities, and on the local committee. At its big dinner he was vice-chairman under Earl Fitzwilliam, the Institute's president, as chairman. Though Phillips gave no formal paper, in the discussion of Way's account of sepulchral lamps he claimed that chemistry could aid archaeology; and at Lord Northampton's request he argued extempore that the physical geography of Yorkshire had affected decisively the directions of ancient roads, the sites of ancient settlements, and the distribution of Roman power in Britain. This exposition functioned as an introduction to Charles Newton's paper on the British and Roman antiquities of Yorkshire, which was illustrated with a map showing roads, camps, entrenchments, and barrows.[61]

Next year Newton tried to encourage archaeology in the YPS. He donated a copy of his published map, to which YPS members had contributed by correspondence, to the Yorkshire Museum. In a letter to Phillips, read to the YPS, he urged the Society to undertake systematic excavation of barrows where they occurred in groups and in apparent relation to earthworks. In 1848 John Thurnam, a member of the YPS, responded to this call in a novel way by putting craniology at the service of archaeology. He was inspired by the research of Anders Retzius who had classified skulls found in prehistoric Scandinavian graves and concluded that the crania showed differences between peoples who seemed externally akin. Thurnam excavated a tumular mound in the grounds of the Retreat, the well-known York mental hospital of which he was superintendent. He reported his discovery of an Anglo-Saxon tumular cemetery to the YPS in June 1848 and nationally in 1849. His large 'diametrical section' of the mound, his woodcut of dead vegetation examined with a microscope of magnification 250, and his plate of crania measured with a craniograph, surely appealed to Phillips who had advised him about extinct mammals. In late 1848 Phillips also responded to Newton's plea when he proposed at an evening meeting of the YPS that a club be formed to study the sepulchral antiquities of Yorkshire.[62]

Next year the Yorkshire Antiquarian Club based in York was founded in June with Phillips and Thurnam as its leading members. Unlike the Naturalists' Club, it held few indoor meetings to which communications were made. Most of its effort was devoted to collaborative excavations of earthworks and especially barrows, from which crania were collected in the hope of providing physical evidence about the aboriginal and succeeding races of Britain. To pay for this field work the club levied a subscription of £2 pa so it was more exclusive than the Naturalists' Club and its membership of eighty smaller. By 1854 the Antiquarian Club had responded to Newton's urgings by excavating ten groups of barrows in east Yorkshire. Phillips jointed some of these excavating parties, recruited members, and above all ensured that the Club's relation with the YPS was harmonious. In October 1849 he promised the Society that the Club would try to augment the collection of antiquities in the Yorkshire Museum. Subsequently it did so by depositing there all the materials derived from excavations of ancient tumuli in east Yorkshire. Thus the Yorkshire Antiquarian Club gave an impetus to YPS's archaeology by shifting its focus from the antiquities of York to those of the county, via collaborative excavation of barrows and donation of the resultant finds to the Yorkshire Museum.[63]

It is clear that Phillips sympathised with the new historical archaeology which the Archaeological Institute promoted from the mid- 1840s. For him historical archaeology was like geology. Both could be advanced by studying change over time in a particular region, by using the widest possible range and variety of evidence, and by adopting inductive methods of enquiry as opposed to premature conjecture. Both shared the aim of reconstructing past life and ancient environments. Phillips became a subscribing member of the Archaeological Institute in the late 1840s and contributed to its journal two papers, in each of which he tried to provide a new sort of evidence to supplement literary sources. At its 1846 meeting Phillips had claimed that chemistry could aid archaeology. Two years later he dilated to the YPS on how a knowledge of the history of metallurgy and mining could illuminate the pre-Roman history of Britain. Fearing that this valuable contribution would be lost in the *YPS proceedings.* Way eventually persuaded Phillips to publish it in the Institute's *Archaeological journal* accompanied by a supplement by Way himself. Also at the 1846 meeting Phillips had argued that physical geography was useful in archaeology. In 1853 he discussed British and Roman roads in Yorkshire, in relation to the county's physical geography, and drew conclusions about the state of the British and their territory.[64]

Phillips' interest in popularisation of science was strikingly shown in his contribution in 1851 to the Great Exhibition of the Works of Industry of All Nations. His involvement began in October 1849 when he attended a public meeting in York which heard a deputation of Charles Dilke, Boscawen Ibbetson, and Edward Hailstone, explain Prince Albert's proposal. Phillips was enthusiastic about it, arguing that York had leisure if not objects to contribute. He was put on the local committee which raised £100 for the Exhibition and induced fourteen locals, of which he was one, to be exhibitors. Classed as an inventor, Phillips exhibited five instruments, an electrophorus, an improved rain gauge open at the top and on four sides, a maximum thermometer, a cardboard anemometer with a semi-circular vane (for collieries and hospitals), and a cheap air-barometer for collieries. For his rain gauge and anemometer he received honorary mention from the jury which judged philosophical instruments. One of his instruments even aroused the interest of Queen Victoria. It was entirely characteristic of Phillips that he used the opportunity of the Great Exhibition to show to a mass public his own ingenuity in devising and making instruments mostly used for measurements.[65]

From 1841 to 1853 Phillips focussed his general literary activities on two genres new to him, biography and tourist books. His solitary effort in the former was his dutiful but still useful memoir of William Smith, which appeared five years after his uncle's death. He was far more prolific and successful as a travel writer and mapper, with three sorts of publication in early 1853. His account of the rivers, mountains, and coast of Yorkshire enjoyed two editions within three years. The accompanying guide to railway excursions in Yorkshire fared even better with three editions in three years. His geological map of Yorkshire on a scale of five miles to the inch, which received a second edition in 1862, was a larger version of the geological map on a scale of twenty-five miles to the inch which graced the guide to Yorkshire. These maps, lithographed by Monkhouse in York, were viewed at the time as remarkable because they were the first geological maps in Britain to be produced by colour-printing, using chromolithography and not colouring by hand which was laborious.

William Smith was travelling to Birmingham for the 1839 meeting of the British Association when he was taken ill at Northampton while staying with George Baker and his sister, the well-known historians of their county. As soon as he heard this disturbing news, Phillips left Birmingham for Northampton and was with Smith for the last two days of his life. As Smith's executor, Phillips was responsible for Smith's wife who was insane and spent her last years in a lunatic asylum at York until her death in 1844. She was also destitute because the government rejected pleas that Smith's pension be transferred to her. Phillips inherited Smith's papers and, as his nephew and pupil, paid tribute to his uncle in an obituary in 1839. It stressed that in the 1790s Smith made the discoveries on which his reputation rested, that his fame was slowly earned, that in the 1820s he was excluded from the progress of science, but that in 1831 young English geologists drew him from retirement as the unforgotten leader of their science. Phillips was expected and encouraged by fellow geologists to write a full biography. He deferred the task until early 1843 when he began to work on Smith's voluminous and interminable manuscripts. At the same time he offered the biography to Murray, stressing that 500 copies would be enough, that it was being written as a sacred duty, and that he wished to avoid financial risk. Murray accepted and published it in March 1844 at 7/6d with a print run of 500. It received only one substantial review.[66]

The greatest value of the biography was that it used Smith's papers in Phillips' possession and documented many aspects of Smith's life, showing the nature of Smith's discoveries in the 1790s, culminating with the famous episode of 1799. Generally it corroborated Fitton's views about Smith's contributions to English geology. Knowing that he was in the unique position of having read all Smith's unpublished materials, Phillips made no attempt to secure letters or reminiscences from Smith's associates, even those who admired him. The book did not lack moral themes. Smith was portrayed as a disinterested researcher on whom recognition was finally bestowed as the father of English geology. He was depicted as possessing natural goodness of heart, serenity and nobility of mind, fortitude, and self-reliance, all of which compensated for his poverty. Indeed, Phillips claimed that Smith never yielded to depression. Consequently, Phillips deliberately softened the darkest parts of Smith's life because, to Smith, they were merely regretted though to others, especially Phillips himself, they were painful. Phillips was so concerned to represent Smith as internally serene that he did not reveal the details of some of 'the merely external circumstances' of his uncle's life. Thus Phillips said nothing about Smith's sojourn in jail in 1819 and very little about Mrs Smith, even though accounts of these torments would have emphasised Smith's amazing resilience. But to Phillips, who wrote as an orphan who had benefited from Smith's goodness and had witnessed Smith's poverty and domestic affliction, they were probably too melancholy and private for public consumption. Overall the biography tended towards hagiography but was saved from that fate because Phillips alluded to several unsaintly aspects of his uncle. Smith was a rambling lecturer, not methodical as a literary man, uninterested in geology's collateral sciences, had a narrow range qua geologist, found scientific co-operation difficult, and 'it required often the pressure of business and the lack of money to rouse him to needful exertion'.[67]

Phillips made his debut as a travel writer in 1845 when he contributed an essay on the geology of the district to the second edition of Adam and Charles Black's *Picturesque Guide to the English Lakes*. Phillips revised it for the thirteenth edition published in 1865, for which he received £10, and it continued to appear in various editions of the *Guide* published well after his death. His account, based on the views of Sedgwick tempered by his own observations, stressed the antiquity of the earth and the two violent disturbances which had divided the Lake District's geological history into three great periods before the era of humans.[68]

Phillips' guide book to Yorkshire was begun in 1849 when he began collecting materials for 'a popular view of the physical and antiquarian features of Yorkshire'. Its foci were thus exactly those of the YPS, at a time when the Society's membership was declining rapidly and its grounds becoming its chief attraction for local people. One aim of the book was therefore to incite Yorkshire's inhabitants, as well as visitors, to explore the natural history and antiquities of the county. Running to about 300 pages it was published as a library volume costing 15 shillings to the public and 10 to subscribers. Phillips wanted the second edition to be produced as a pocket volume for the voyageur at 10 shillings, but was defeated by technical printing considerations, so it sold under the same terms as the first edition.[69]

The work ranged widely, covering topography, physical geography, geology, climate, and magnetism; and, given Phillips' concern with archaeology, it spanned the pre-Roman, Roman, Anglo-Saxon, and Danish history and ethnography of the county. The book was mainly deuteronomic in that it drew on work already published by Phillips and others, including such classical authors as Herodotus,

Pliny, and Tacitus, but it contained revealing asides. Phillips portrayed Yorkshire, his adopted county, as a renewer of health and a strengthener of hopes. Ever the irenic Christian, he described York Minister as 'a bond of union amidst jarring creeds and warring opinions'. Though he deplored the spoliation of York, he welcomed the industrialisation of west Yorkshire as progress. In his account of the geological history of the county, he stressed that in the past the Ruler of Nature had employed the forces of nature to prepare for the comfortable existence of intelligent humans, who were chronologically recent. In his discussion of diluvium and erratics, he made a triple division into pre-glacial, glacial, and post-glacial periods, and was happier with icy water than with Lyell's icebergs or Agassiz's ice-sheets, as chief agent of what were widely recognised as glacial phenomena. Drawing on the research of Edward Forbes, he postulated that between Britain and Europe there had been long periods of land communication during which living creatures had migrated from centres of life; and he denigrated the notion of special creations of a given species in various localities.[70]

The book was adorned by thirty-five plates lithographed in York by Monkhouse mainly from Phillips' own sketches. Plate 30, showing the geology of Yorkshire, was coloured by chromolithography and reflected Phillips' long-standing interest in lithography. The small map, measuring four and a half by seven and a half inches, managed to show fifteen different strata, three traverse sections, and one columnar one. In his book Phillips publicised a separate and larger geological map, measuring twenty-two and a half by twenty-nine and a half inches. Costing 7/6d in sheet form and 10/- mounted, it was produced for Phillips by Monkhouse who again used chromolithography. Though broadly similar to the small map, the larger one contained extra detail but it did not offer any traverse sections. The columnar section showed Sedgwick's Cambrian rocks, revealed the different beds in the strata, depicted their relative thicknesses, and indicated unconformities. It became the standard geological map of Yorkshire for a generation. Through Phillips, Yorkshire was the subject and York the birthplace of the application of the new technique of chromolithography to geological maps.[71]

York enjoyed a focal position in the vast network of railway lines constructed from the 1830s to the 1860s. The main company operating from the city was the North Eastern, formed from four companies in 1854, an amalgamation noticed in the title of the third edition of Phillips' *Railway excursions*. York was one of the busiest provincial railway stations: in 1854 it was used daily by seventy-six trains. Not for the first time Phillips tapped local opportunity, on this occasion by describing rail trips from York, Leeds, and Hull in a companion volume to his *Rivers*. Though the text remained almost unchanged in the three editions, the second and third, the latter costing a shilling, were improved by the addition of illustrations. The work invited 'the children of toil and the denizens of towns' in west Yorkshire to escape from industrialised areas and to seek refreshment in scenic country, small towns, churches, and antiquities. This was some compensation for the burdensome aspects of industrialisation, which Phillips broadly accepted because overall it augmented the comforts of the people.[72]

8.7   A pension scandal

In 1852, in his second month as prime minister, the Earl of Derby granted civil list pensions to John Hind (£100 pa), Francis Ronalds (£75 pa), and Gideon Mantell (£100 pa), in response to lobbying by the Earl of Rosse, as president of the Royal Society, and by Lord Wrottesley and Sir Robert Inglis on behalf of the Parliamentary Committee of the British Association. Perhaps spurred by this success and thinking that he had pull with his fellow conservative, the Earl of Aberdeen, who had become prime minister in December 1852, Murchison launched a scheme in July 1853 for a civil list pension of £200 pa for Phillips. Murchison knew that in 1845 Peel, a previous conservative prime minister, had granted £200 pa for life to James Forbes, an old friend of Phillips. Murchison also knew that Phillips was less affluent than Forbes: he had no private income and in 1853 occupied only one regular post.[73]

Murchison and Sabine organised the lobby of Aberdeen and were supported by Rosse, Brewster, Lloyd, and De la Beche whom Phillips still regarded as a very faithful friend. In October 1853 Aberdeen rejected Phillips' case, informing Rosse that Phillips was not needy and that poverty had to attend intellectual merit. Wrottesley was dismayed by the way the prime minister had invoked out of the blue a confession of poverty as a necessary condition of any application for a pension and complained to him. In their correspondence of March 1854, which was published in the Association's *Report*, with Phillips named as the victim, Wrottesley claimed that the new and demeaning criterion had not been applied to successful applications in the past but Aberdeen was unmoved. Phillips put a brave face on the matter, telling Sabine that he would not repine because his love of science moderated selfish and vainglorious feelings. Yet two years later he still deplored 'the flinty hearts of Her Majesty's Ministers' and privately must have felt that his case was as deserving as that of Forbes and Mantell, his friends, and Owen (£200 pa, 1842).[74] The pension would have usefully topped up his salary of £300 pa from the British Association. But it was not vital for him because in late September 1853 he was given an intellectual and financial lifeline when he was appointed deputy reader in geology in the University of Oxford. Having spread geological enlightenment to the national intelligentsia he faced a similar task in one of the chief nurseries of national elites of various kinds.

Notes

1. Phillips to sister, 21 Sept, 21 Oct 1841, PP; Phillips to De la Beche, 18 March 1845 (q), DLB P; *YPS 1845 report*, pp. 8-9; YPS Council minutes, 16 Sept 1844, 3 Nov 1845; Sir John Murray Naesmith (1803-76).
2. Bell, 1969; *YPS 1850 report*, pp. 12-15; YPS Council minutes, 5 July-24 Sept 1850; Thomas Henry Travis (1810-80) barrister and York judge.
3. *YPS 1840 report*, p. 12; YPS Council minutes, 31 Jan 1848.
4. *YPS 1840 report*, pp. 12-15; *1843 report*, p. 9; *1844 report*, pp. 9-10; *1845 report*, p. 17; Goldie to Phillips, 17, 23 Dec 1841, PP; YPS Council minutes, 18 Jan 1841, 12 Feb, 15 April 1844; Stephen Beckwith (d 1843).
5. *YPS 1841 report*, pp. 9, 10, 14; *1842 report*, pp. 9-10; *1843 report*, p. 10; *1844 report*, pp. 14-17, 28; *1845 report*, p. 7.
6. BAAS General Committee minutes, 21 Sept 1840, 2 Aug 1841, 27 June 1842; BAAS Council minutes, 11 Feb, 28 Mar 1842, both BAAS archives; *YPS 1841 report*, p. 9 (q);

YPS Council minutes, 12 July 1841, 3, 18 Jan, 14, 28 Feb, 24 March, 6, 20 June 1842; Goldie to Phillips, 23 Dec 1841, PP; Phillips to Harcourt, 27 Dec 1841, HP.

7. BAAS General Committee minutes, 21 Aug 1843; Phillips to Harcourt, 23 Aug, 8 Oct 1843, HP; YPS Council minutes, 6 Nov 1843, 1 April 1844; *YPS 1843 report*, pp. 7-10; *1844 report*, pp. 8-9, 14, 17; Orange, 1973, pp. 63-9, (63 q).

8. Phillips to De la Beche, 26 Aug 1844, DLB P; Peacock 19 Sept, Brewster 21 Sept, Horner 23 Sept, Bayldon 25 Sept 1844, all PP; Phillips to Faraday, 14 Sept, Leibig to Faraday, 19 Dec 1844, in James, 1996, pp. 243, 298-300; Chambers, 1844a, pp. 322, 1844b; Robert Chambers (1802-71), *DNB*.

9. *YPS annual reports*, 1842-52; Phillips to Harcourt, 10 and 14 April 1842, HP; Phillips to Hamilton, 28 Sept 1843, 13 and 17 Jan 1844, Ham P; Hamilton to Phillips, 23 Dec 1843, 3 Jan 1844, PP; Phillips, 1851a; Phillips to Forbes, 1 May 1851, FP; Phillips to Murray, 5 May 1851 (q), Mu P; Jean Bernard Léon Foucault (1819-68), *DSB*.

10. See section 8.4

11. *YPS 1841 report*, pp. 12-14; *1842 report*, p. 13; *1843 report*, pp. 12-15; *1844 report*, p. 10; *1851 report*, p. 11; *1852 report*, p. 14; YPS Council minutes, 6 June 1842, 18 March, 30 Sept, 28 Oct 1851; Quetelet to Phillips, 12 Nov 1841, Welsh to Phillips, 14 Feb 1853, PP; John Welsh (1824-59), *DNB*, superintendent of Kew Observatory 1852-9; William Hey (1811-82), Anglican minister, headmaster of St Peter's School, York.

12. *YPS 1841 report*, p. 16; YPS Council minutes, 18 Jan 1842, 16 March 1853.

13. YPS evening meetings minutes, 3 Jan, 28 Feb 1842, 6 March 1843, 7 Oct 1844, 26 Oct 1846, 1 Feb, 15 Nov 1847, 24 Jan 1848; Goldie to Phillips, 17 Dec 1841, William Taylor to Phillips, 23 June 1841, Davies to Phillips, 21 Sept, 15 Oct 1841, 11 and 15 April 1842, all PP; Davies, 1843.

14. YPS Council minutes, 5 Jan 1846, 15 Feb 1847, 13 Feb 1849; *YPS 1846 report*, p. 17; *1847 report*, p. 8 (q); *1850 report*, pp. 20-1.

15. *Liverpool albion*, 2 Nov 1845, a reference I owe to Jim Secord; York Mechanics' Institute, Book listing lectures etc., York City archives, TC 49/1:15; Phillips, 1846a, pp. 5, 7, 14; Phillips to Ramsay, 6 April 1846, RP; *The Yorkshireman*, 4, 25 April 1846.

16. Lupton to Phillips, 20 Nov 1845, Baines to Phillips, 15 Sept 1846, PP; *Annual report of Leeds Mechanics' Institute presented 31 Jan 1854*; Phillips, notes for Parsonstown lecture, PP, box 102, fl. 7.

17. Phillips, 1848b, 284-5 (qs); 1848c; Dean, 1999, pp. 193-4, 216-18; Mantell, 1846; Phillips to Mantell, 9, 19, 31 Dec 1844, 11 Jan, 19 Aug 1845, 9 May 1846, Ma P; Pritchard to Phillips, 14 Dec 1844, 9 July 1845, PP; Christian Gottfried Ehrenberg (1795-1876), *DSB*; Andrew Pritchard (1804-82), *DNB*.

18. *Proceedings of Yorkshire Geological Society*, 1848, vol. 2, pp. 331-2; Morrell, 1983.

19. For Manchester lectures, Winstanley to Phillips, 7, 28 Sept, 20 Oct, 11 Nov 1841, 1 Feb 1842, Royal Manchester Institution Archives, M6/1/49/3; Ormerod to Phillips, 24 Jan, 27 April 1849, M6/1/49/5; syllabuses in M6/1/70, items 30 and 96; Phillips to De la Beche, 18 March 1842, 2 Dec 1849, DLB P. For Royal Institution lectures, syllabuses in folder 12, PP; Barlow to Phillips, 12 June 1844, 16 June, 1 Dec 1845, PP; Phillips to Royal Institution, 20 May 1852, RI archives, CL1/114; RI managers' minutes, 3 July 1843, 1 April 1844, 16 Nov 1846, 7 June 1852, 7 March 1853. For RI discourses, *Athenaeum*, 1844, p. 298; Lyell to Phillips, 23 March 1844, PP; Phillips, 1854a. For Sheffield, Phillips to Sorby, 23 Dec 1850, Sheffield LPS archives, 51-20, Sheffield City Archives.

20. Woodward, 1907, pp. 143-4; Wilson, 1972, p. 501; Geikie, 1875, vol. 1, p. 313; James Smith (1782-1867), *DNB*, Glaswegian merchant and expert on tertiary geology; Charles Stokes (1783-1853) London stock broker and fossil collector.

21. Woodward, 1907, pp. 121-2, 148; Phillips to De la Beche, 13 Nov 1842, DLB P; Charlesworth to Sedgwick, 27 Sept 1842, Phillips to Sedgwick, 15 Nov 1842, Se P; Charlesworth to Henslow, 11 Nov 1842, Hen P; Pyrah, 1988, pp. 60-2; Murchison to Whewell, nd [Nov 1842], WP a. 209^{132} (q).

22. GP, box 14, 3054, documents about Charlesworth affair; Darwin to Lyell, 5 and 7 Oct 1842, Darwin to Fox, 9 Dec 1842, Darwin to Henslow, 22 Jan 1843 in Burkhardt and Smith, 1986, pp. 336-8, 344-5, 348-9; Strickland to Phillips, 8 Dec 1842, PP; Wilson and Geikie, 1861, pp. 324-5.

23. Hall, 1984, pp. 63-91; MacLeod, 1983, pp. 74-6; Bonney, 1919, pp. 1-2, 53; Geikie, 1917, pp. 217, 357, shows that Phillips was a guest in 1835 and 1849 at the Royal Society Club.

24. MacLeod, 1971, pp. 325-9; Geikie, 1875, vol. 2, pp. 106-7; Murchison to Phillips, n d [spring 1850], EUL MS Gen 1999/1, green pencil number 60; Royal Society Council minutes, 7 March 1850; Lord John Russell (1792-1878), *DNB*, prime minister 1846-52.

25. Sabine to Phillips, 16 Feb 1849 (q), 4 Aug 1853, PP; for presidential acclaim by Peacock, Murchison, Brewster, and Hopkins, *BAAS 1844 report*, p. xxxiv, *1846 report*, p. xxxiii, *1850 report*, p. xxxii, *1853 report*, pp. xli, xliii.

26. Whewell to Phillips, 11 July 1841; Robinson to Phillips, 10, 30 Aug 1849; Brewster to Phillips, 20 Sept, 19 Oct 1850, Phillips to Brewster, 23 Sept 1850, all PP; *BAAS 1850 report*, pp. xxxi-xliv; BAAS Council minutes, 1 June 1849; Lyell to Phillips, 10 July, 5, 14 Aug, 16, 21 Sept 1844, PP.

27. BAAS Council minutes, 27 Feb, 14 April 1847; Phillips to Grove, 27 March, 2 June, 9 Dec 1847, 12 Feb, 3 March, 19 May, 3 June, 15, 20, 24 July, 2 Aug 1848, Gr P; William Robert Grove (1811-96), *DNB*, vice-president in 1848.

28. Sabine to Herschel, 18 June 1842, Her P; Falconer to Phillips, 22 Sept 1844, PP; Crowe, 1998, p. 310; Sabine to Phillips, 27 Sept 1841, PP; Friedrich Wilhelm Bessel (1784-1846), *DSB*, the famous German astronomer; Hugh Falconer (1808-65), *DNB*, expert on Indian palaeontology.

29. BAAS Council minutes, 13 March 1841; *BAAS 1841* report, p. v; Sabine to Phillips, 25 Feb 1841, PP; *1846 report*, p. vi.

30. Phillips to Forbes, 3 Feb, 24 Aug 1848, FP; BAAS Council minutes, 14 Jan 1848; *BAAS 1848 report*, pp. xvii, xxi-xxii.

31. BAAS Council minutes, 1 Dec 1843, 29 Feb 1844; *BAAS 1853 report*, p. xxv; *Parliamentary papers*, 1847, vol. 34, pp. 253-5; Rogers to Phillips, 7 Dec 1852, 8 Feb 1853, PP; Kohlstedt, 1976, pp. 173-80; George Hamilton Gordon, fourth Earl of Aberdeen (1784-1860). *DNB*, prime minister 1852-5.

32. *BAAS 1853 report*, pp. xl, 6-7, 14-18, 54, 70.

33. *Athenaeum*, 1842, pp. 615-16; 1853, p. 1135; *BAAS 1853 report*, pp. 54-61; Phillips, 1853b.

34. Phillips, 1842b, *Athenaeum*, 1842, p. 641; Secord, 2000, p. 210; Phillips to Murray, 25 Jan 1841, Mu P; Phillips to Sedgwick, 22 Dec 1851, Se P; Phillips, 1853b.

35. Morrell and Thackray, 1981, pp. 243-4; *Athenaeum*, 1844, pp. 903-4; Orange, 1973, pp. 66-9, Cockburn, 1844, 1845; Secord, 2000, pp. 232-3.

36. Phillips to Whewell, 6 Oct 1844, WP: Phillips to Ramsay, 13 June 1844, RP (National Library of Wales); Phillips to Mantell, 24 Feb 1846, Ma P; Mantell, 1846, pp. 89-90.

37. Phillips, 1843b; *BAAS 1841 report*, p. 331; *1842 report*, p. 213, *Athenaeum*, 1842, p. 615; Secord, 2000, p. 434; Phillips, 1851b.

38. Phillips, 1850a, 1853c; *Athenaeum*, 1850, p. 875, 1853, p. 1164; Thomas Charles Robinson (1792-1841), London instrument maker; Julius Plücker (1801-68), *DSB*.

39. *BAAS 1845 report*, p. 6; *1842 report*, p. 6; *1849 report*, p. xx; *Athenaeum*, 1847, pp. 1128, 1151-2; Sykes to Phillips, 9 Feb 1849, PP; Phillips to Faraday, 5 Dec (q), 14 Dec 1850, Faraday to Phillips, 7 Dec, 21 Dec 1850, in James, 1999, pp. 212-15, 219-21.

40. Phillips, 1841d, 1844b; BAAS Council minutes, 16 Aug 1843; Phillips to Weld, 17, 19 June 1848, Royal Society of London archives, RR.1.103-4; Phillips, 1844c and d; Morrell and Thackray, 1981, pp. 520-1; Jankovic, 1998; William Radcliff Birt (1804-81), a London teacher.

41. Phillips, 1846b, 1848d, 1849a; Phillips to Whewell, 13, 19 Aug 1843, WP; Whewell to Phillips, 4 Aug 1847, Osler to Phillips, 6 June, 21 Aug 1846, PP; *Athenaeum*, 1848, p. 887, 1849, p. 1015.

42. *Athenaeum*, 1850, p. 874; Conybeare, 1833, pp. 409-10; Phillips to Whewell, 5 Sept 1843, 6 Aug 1847, WP; Phillips to De la Beche, 5 Sept 1843, DLB P; *Athenaeum*, 1850, p. 876; Nasmyth to Phillips, 2 Dec 1850, 21 June 1851, PP; Charles Brooke (1804-79), *DNB*; from 1843 to 1852 Francis Ronalds (1788-1873), *DNB*, directed the Kew Observatory which was run by the BAAS from 1842; James Nasmyth (1808-90), *DNB*, engineer and astronomer; William Cranch Bond (1789-1859), *DSB*.

43. *BAAS 1852 report*, pp. xxxv, xliv-xlv; Phillips 1853d and e, 1863a, 1868a; Webb to Phillips, 18 July 1864, PP; Phillips to Sabine, 20 July 1853, Sa P; Johann Heinrich Mädler (1794-1874), *DSB*, and Wilhelm Beer (1797-1850), *DSB*, published in 1836 their map of the moon, 97.5 cm in diameter: next year their accompanying volume recorded the diameters of 148 craters and altitudes of 830 mountains. For the problems of representation in astronomy at the time, see Schaffer 1998a and b.

44. Phillips, 1853d and e, 1854c and d; *Athenaeum*, 1854, pp. 1177, 1204; *BAAS 1857 report*, p. l; De la Rue, 1859, pp. 132-3, 145; Phillips, 1859a, p. xlviii; James Challis (1803-82), *DNB*, professor of astronomy, Cambridge University; Charles Piazzi Smyth (1819-1900), *DNB*, astronomer royal for Scotland; John Lee (1783-1866) had a private observatory at Hartwell, Buckinghamshire; Warren De la Rue (1815-89), *DNB*, London printer, erected a private observatory at Canonbury in 1850.

45. *Athenaeum*, 1844, p. 930, 1851, p. 753; Phillips, 1848e, 1849b, 1853f; William Benjamin Carpenter (1813-85), *DNB*.

46. Morrell and Thackray, 1981, pp. 495-6; Strickland to Phillips, 28 Sept 1841, PP; Phillips to Stickland, 4 Oct 1841, Strickland Papers; *BAAS 1842 report*, pp. 105-21; McOuat, 1996, generously regards Phillips as an expert specialist zoologist.

47. Payne and Company to Phillips, 15 Jan 1845; Stephenson to Phillips, 14 June 1847; Powell to Phillips, 29 Dec 1845; Rawson to Phillips, 2 Feb 1847; Buckland to Phillips, 1 ? 1847, all PP; Phillips to Buckland, 3 July 1847, Buckland papers, Royal Society of London; Phillips to Hutton, 9 Nov 1853, Hu P; George Stephenson (1781-1848), *DNB*; Sir James Robert George Graham (1792-1861), *DNB*, home secretary 1841-6 and MP for Ripon in 1847.

48. Berman, 1978, pp. 174-86; Cantor, 1991, pp. 157-60; James, 1996, pp. xxxi-xxxiv, 250-1, 258-9, 342-6; James and Ray, 1999; Copy of the report of Messrs Lyell and Faraday to the Secretary of State for the Home Department on the subject of the explosion at the Haswell collieries in September last: also a copy of the report addressed to the United Committee of the Coal Trade by the special committee appointed to take into consideration the said report of Messrs Lyell and Faraday; and a copy of the reply of Messrs Lyell and Faraday thereto, *Parliamentary papers*, 1845, vol. 16, pp. 511-35; *Philosophical magazine*, 1845, vol. 26, pp. 17-35.

49. Galloway, 1882, pp. 234-40; *Parliamentary papers*, 1846, vol. 43, pp. 249-58; 1847, vol. 16, pp. 159-236; 1849, vol. 22. pp. 101-19; MacDonagh, 1967, pp. 64-6.

50. Report from the select committee of the House of Lords appointed to inquire into the best means of preventing the occurrence of dangerous accidents in coal mines, *Parliamentary papers*, 1849, vol. 7, pp. 1-712; Instructions issued by the Secretary of State for the Home Department to Professor Phillips and J.K. Blackwell to inquire into the state of collieries and ironstone mines in the principal coal districts, especially with reference to the system of ventilation, and to report thereon, *Parliamentary papers*, 1849, vol. 46, pp. 401-3; Blackwell, 1850; Phillips, 1850b; Blackwell to De la Beche, 29 June, 4 Nov 1849, DLB P; MacDonagh, 1967, pp. 74-5; Sir George Grey (1799-1882), *DNB*, home secretary 1846-52.

51. De la Beche to Phillips, 16 June 1849; Phillips to De la Beche, 17 June 1849; Waddington, Home Department, to Phillips, 15,19 June 1849; De la Beche to Carlisle, 30 June 1849; Paymaster General to Phillips, 3 May 1850, all in Correspondence and papers relating to

mines, BAAS archives, 61; Phillips to De la Beche, 29 Sept 1850, DLB P; *Parliamentary papers*, 1849, vol. 7, p. 27; MacDonagh, 1967, p.74.

52. Phillips to Fitzwilliam and Biram [his viewer], 29 June 1849; Jude to Grey, 8 Oct 1849 (q); Grey to Jude, 9 Oct 1849; Phillips to Jude, 10 Oct 1849; Phillips to Waddington, 15 Oct 1849 (q); Waddington to Phillips, 19 Oct 1849; Phillips, notes of meeting with Jude et al., 22 Oct 1849; Phillips to Lewis, Home Department, 25 Oct 1849; Phillips to Grey, 21 March 1850, all in BAAS archives, 61; Nicholas Wood (1795-1865).

53. Phillips to De la Beche, 2 Dec 1849, DLB P (q); Phillips, 1850b, p. 498; Phillips to Waddington, 14 Jan 1850, BAAS archives, 61 (q).

54. Phillips, 1850b, pp. 486, 494-8, 507-8, 511 (q), 520 (q); Goldsworthy Gurney (1793-1875), *DNB*, inventor; on ventilation, see Church, 1986, pp. 322-8, Hinsley, 1969-70, and Galloway, 1882, pp. 247-56.

55. Phillips, 1850b, pp. 500, 507, 520-2; Phillips to Grey, 21 March 1850, BAAS archives, 61; Phillips to De la Beche, 2 Feb 1846, 15, 30 Dec 1849, 14 Jan 1850, DLB P.

56. Church, 1986, pp. 423-6; Galloway, 1882, pp. 243-6; Phillips to De la Beche, 22 Oct 1850, DLB P.

57. YPS Council minutes, 2 Dec 1849; *Yorkshireman*, 9 Feb 1850; on field clubs, see Allen, 1978, pp. 158-75, and 1987, pp. 247-9.

58. Moore to North, 15 June 1855; Morris's circular, 23 Feb 1849; Minute book of the Yorkshire Naturalists' Club, 1849-55, 20 Jan, 5 May 1849, all in West Yorkshire archives, Leeds; Beverley Robinson Morris (1816-83).

59. YNC minutes, esp 7 Nov 1849, 6 March 1850, 15 Jan, 5 March 1851; YPS Council minutes, 2 Oct 1849; *YPS 1849 report*, p. 6-9; Francis Orpen Morris (1810-93) *DNB*, prolific author of natural history books, was Anglican vicar of Nafferton, near Driffield.

60. Archaeological Institute, 1848; Way to Phillips, 8 Nov, 1, 23 Dec 1845, PP; Albert Way (1805-74), *DNB*.

61. Archaeological Institute, 1848, p. xii; *Yorkshireman*, 25 July, 1 Aug 1846; Newton, 1847; Charles Thomas Newton (1816-94), *DNB*, then an assistant, department of antiquities, British Museum.

62. Newton, 1855; *YPS 1847 report*, pp. 22, 28; Way to Phillips, 12 Nov 1847, PP; Thurnam, 1849, 1855; YPS evening meetings minutes, 18 Dec 1848; Ramm, 1971; John Thurnam (1810-73), *DNB*; Anders Adolf Retzius (1796-1860), *DSB*.

63. Procter, 1855; *Archaeological journal*, 1850, vol. 7, pp. 34-5, 40, 1851, vol. 8, pp. 224-5; *YPS 1849 report*, pp. 8-10; YPS Council minutes, 2 Oct 1849; Tyrconnel to Phillips, 21 July 1849, PP

64. Van Riper, 1993, pp. 28-38; Phillips, 1853g, 1855a, 1859b; Way, 1859.

65. *Yorkshireman*, 3 Nov 1849, 4 June 1850; Great Exhibition, 1851a, vol. 1, p.454; Great Exhibition, 1851b, pp. 14,66; Anon, 1851, p. xx; Dilke to Phillips, 17 July 1851, PP; Charles Wentworth Dilke (1810-69), *DNB*, Levett Landon Boscawen Ibbetson (d 1869); Edward Hailstone (1818-90), *DNB*; Prince Albert (1819-61), *DNB*, chaired the Great Exhibition Commission.

66. Phillips, 1844a, pp. 123-4, on which see Torrens, 2003; Phillips to sister, 27, 28 (4 letters) Aug, 2 Sept 1839, PP; Buckland and Phillips to Treasury, 4 Sept 1839, Buckland to Phillips, 12 Dec 1839, Buckland papers, Devon Record Office; Phillips, 1839a; Sanders to Phillips, 1 Feb 1840; PP; Phillips to Sedgwick, 21 Jan 1844, Se P; Phillips to Ramsay, 10 Feb 1843, RP; Phillips to Murray, 20 Jan, 11 May 1843, Mu P; for the review see Trimmer, 1845; George Baker (1781-1851), *DNB*; Anne Elizabeth Baker (1786-1861), *DNB*.

67. Phillips, 1844a, pp. vii, 28-31, 67, 109-10, 127, 130, 132-3 (qs), 144; Fitton, 1832, 1833.

68. Black, 1845, 1888, pp. 231-71; Black to Phillips, 9 March 1844, 4 Sept 1864, 29 March 1865, PP.

69. Phillips to Murray, 4 April 1849 (q), 8 Oct 1852, 28 Feb, 7 March 1853, Mu P; *YPS 1852 report*, p. 11; YPS Council minutes, 5 April 1852.

70. Phillips, 1853a, pp. vi, 73 (q), 94, 168-92.

71. Butcher, 1983; Phillips, 1853a, pp. 288-9, plate 30; Phillips, 1853i.

72. Phillips, 1853h, pp. 3 (q), 83; Feinstein, 1981, p. 128.
73. *BAAS 1852 report*, p. lxi; Geikie, 1875, vol. 2, p. 148; Phillips to Sabine, 20 July 1853, Sa P; Forbes to Murchison, 19 Sept 1845, MP; John Hind (1796-1866), *DNB* mathematician; John Wrottesley, second Lord (1798-1867), *DNB*, president Royal Society of London 1854-8; Sir Robert Harry Inglis (1786-1855), *DNB*, conservative MP for Oxford University 1829-54.
74. Phillips to Sabine, 20, 21 July, 12 Aug, 2 Nov 1853, Sa P; Sabine to Phillips, 4 Aug 1853, PP; Phillips to Murchison, 2 Oct 1853, MP; BAAS Council minutes, 20 Sept 1854; *BAAS 1854 report*, pp. xliii-xlv; Phillips to Forbes, 1 Nov 1856, FP (q); MacLeod, 1970.

PART III
THE OXFORD PROFESSOR
1853-1874

Chapter 9

The Oxford Chair

Phillips became an Oxford don and was saved for science in a bizarre way. On 14 September 1853 his friend Hugh Strickland, who had been geologising with him during the BAAS meeting at Hull, became a martyr of science when he was struck and killed by a train while examining a railway cutting near Retford. From 1850 Strickland had been Buckland's deputy as reader in geology at Oxford. Phillips' old friend, Daubeny, saw this tragedy as an opportunity for serving the University and Phillips by bringing him to Oxford as Buckland's deputy. Fifteen days after Strickland's death, Phillips was appointed his successor.

At Oxford Phillips was no sinecurist. As deputy reader, reader, and professor of geology he taught regularly for twenty years. As a keeper he was in charge of the Ashmolean Museum from 1854 to 1870 and the new University Museum from 1857 until his death. Though he was not uninterested in the former, he gave far greater priority to the latter. In his opening years he lived at Magdalen Bridge courtesy of Daubeny but from 1858 he occupied rent-free the newly built keeper's house adjacent to the Museum. The largest part of his regular income came from the University. It was usefully topped up by the rent of £40-60 pa received from tenants of his York home and significantly increased until 1862 by his salary of £300 pa from the British Association. From 1854 to 1858 Phillips received an average of £72 pa as keeper of the Ashmolean Museum and £300 pa it seems, as 'professor'. Next academic year he enjoyed a rent-free official University residence, a salary of £80 pa as keeper of two museums, and £300 pa for professing. From 1859 until his death these rewards were unchanged, whether he was a dual or single keeper, except that his professorial salary was increased to £400. In the 1850s he charged some students the class fee of £1 but from the early 1860s admission to his lectures was gratis. Thus over a period of twenty years his annual regular University income rose from a little less than £400 to £480.[1]

In the years 1858 to 1861 Phillips was exceptionally busy: in addition to his regular responsibilities at Oxford and for the Association, he was president of the Geological Society of London 1852-60 and Rede lecturer at the University of Cambridge in spring 1860. About this time his sister's health declined and top-class medical advice from Sir Benjamin Brodie and Joseph Hodgson was unavailing: in spring 1862 she died from a tumour. For over thirty years she sustained and nurtured the career of her bachelor brother who was like her an orphan. Just as such geological wives as Mary Buckland, Charlotte Murchison, Mary Lyell, and Emma Darwin supported their husbands' work, Anne Phillips helped her brother geologically and provided a secure and happy home for him. He recognised that she was his chief companion in the voyage of life. In her Oxford years she accompanied him on a geological trip to the Auvergne in 1855 but her chief roles were to run their home and promote their social life. For instance, she soon established cordial relations with the family of Francis Jeune, master of Pembroke College and vice-chancellor 1858-62. In 1856 on the seventeenth birthday of a Jeune daughter, Anne Phillips sent her a magnificent bouquet and Phillips added a geological work of his own. When the British

243

Fig. 9.1 Anne Phillips, a photograph of 1860, copies of which were sent to Faraday and Sedgwick, both of whom thought highly of her. No other representation of her is known.

Association met in Oxford in 1860, she organised a large tea party in her home for the Jeune family, the Harcourts, and visiting philosophers. Next year, when William Pengelly was in Oxford in connection with his collection of Devonian fossils, she accompanied Phillips to a vice-chancellorian dinner party attended by Pengelly, the Jeune family, Henry Smith, Obadiah Westwood (professor of entomology), and William Thomson (provost of Queen's College who became archbishop of York in 1862) and his wife. Anne Phillips was on good terms with Phillips' Oxonian and geological friends, especially 'dear old Sedgwick' as she called him. She made him a stool to ease his notorious gout and gave him an 1860 photograph of herself which he treasured (Fig. 9.1). He was well aware of her importance: in his view she was the kindest, dearest, and best friend whom God had given to Phillips.[2]

After his sister's death Phillips tried to inure himself to loneliness. He was fortunate in liking Oxford where his activities and colleagues fended off desolation. Loyal friends elsewhere went out of their way to help him, especially in the mid-1860s, by staying with him and by inviting him to be their guest at Christmas and New Year. The most considerate and persistent of them was Henry Wood, vicar of Holwell, near Sherborne, Dorset, who had been an Oxford pupil of Phillips and had secured him as godfather to one of his sons. Other inviters and stayers were William Symonds, rector of Pendock, Worcestershire, Robert Davies, John Edward Lee, and Harcourt who from 1861 lived at Nuneham Park near Oxford.[3] From the mid-1860s age began to take its physical toll on Phillips though he avoided persistent illness and major surgery. He was usually unwell for no more than two or three days, except when afflicted by neuralgia and lumbago. By 1873, however, the two fiends of neuralgia and rheumatism made him unsteady on his legs. These physical infirmities were accompanied by recurrent sadness at the deaths of such old friends as Whewell (1866), Daubeny (1867), James Forbes (1868), Harcourt (1871), Murchison (1871), and Sedgwick (1873); for three of them (Daubeny, Harcourt, Sedgwick) he wrote formal obituaries. Though Phillips became more conscious of life's transitoriness and melancholy, he was sustained in his final years by his Christian belief in God's redeeming power and in an afterlife.[4]

The year of his sister's death was a climacteric in his life. He reduced his participation in the Geological Society of London and ended his administrative involvement in the British Association; but he did not degenerate into a drooping dodo reduced to perpetual torpor by the miasma of the Thames valley. He carried on working productively until his unexpected death. In the remaining five chapters of this book I shall examine the final two decades of Phillips' long career by looking at his teaching, research, and keepering at Oxford, connections with other institutions, and his views on evolution and related matters. Though he and his contemporaries sometimes viewed some of these activities as inter-connected, the nature and importance of his various types of contributions in his Oxford period become clearer if one takes a thematic approach to them while not ignoring the effects of the passing of time.

9.1 The Oxford appointment

Phillips was appointed in 1853 as deputy reader in geology in the University of Oxford in place of Buckland who did not teach after 1849 because of a mental breakdown from which he never recovered. In his prime Buckland was renowned as a flamboyant lecturer who cleverly presented geology as a branch of history, of natural theology, and as a subordinate part of a liberal education based on the classics. Until about 1830 he drew big crowds not only to his lectures on geology (average of sixty) but to those on mineralogy (average thirty-four). By the 1840s both classes had dwindled, the former to about ten on average and the latter around seven. Buckland was so depressed about the prospects of science at Oxford that in 1845 he was glad to assume the deanship of Westminster Abbey. When it became sadly clear that he was incapable of lecturing, the vice-chancellor, Frederick Plumptre, acted on Daubeny's advice and appointed as deputies two Oxford graduates who had been Buckland's pupils. In 1850 Nevil Story-Maskelyne and Strickland became deputy readers in mineralogy and geology respectively. Each fulfilled Daubeny's two criteria, an Oxford connection and competence. So did Lyell but he declined to be considered for the geological post. [5]

In his inaugural lecture of 1850 Strickland revealed his high aspirations for geology at Oxford. He presented geology as an inferential science whose laws of nature led to a belief in Creative Wisdom and Benevolent Providence. Moreover the facts of extinction and the sudden appearance of new species, as revealed in the fossil record, indicated for Strickland special interpositions of creative power which to humans appeared supernatural and miraculous. Thus geology would not only provide refreshing exercise but also supplement revealed theology for all those Oxford students destined to become clergymen. Strickland also stressed the newness of humans in the earth's long geological history and the speciousness of the evolutionary views expressed in *Vestiges of creation*. He judged that Oxford had advantages for geology: Buckland's huge collection could be exploited; the Radcliffe Science Library was well stocked; and the many strata in the area, underfoot treasures all marked lithologically and by fossils, provided propitious opportunities for field work.[6]

Strickland soon became disenchanted. There was no money to deal with collections, such as that of Buckland, much of which was still unpacked. It occupied four rooms in the Clarendon Building where Strickland lectured to empty benches: in 1850 he attracted only seven attenders who paid 1 guinea each. He discovered that the Radcliffe Science Library, like his salary of £100 pa, was inadequate. He felt that geology was oppressed by the sinister dominance of classics and by the colleges which squeezed professorial lectures into the short period of 1-3 pm. He told the Royal Commission on Oxford, which reported in 1852, that 'the science of geology presents but little attraction to the members of the University at present'.[7] His despair was suddenly ended by his accidental death in September 1853.

Strickland's death put the vice-chancellor, Richard Cotton, an evangelical clergyman who was provost of Worcester College, in a quandary. There was no Oxford graduate known to be capable of deputising for Buckland as reader in geology so Cotton took the highly unusual step of looking at a non-Oxonian for the post. It seems that Daubeny and Robert Walker, an evangelical clergyman who was reader in experimental philosophy, proposed to Cotton that Phillips be considered and that Daubeny lobbied in Oxford on Phillips' behalf. When Phillips learned that Cotton was well disposed towards him but neither suggested nor required testimonials, he asked Whewell, as master of Trinity College, Cambridge, and Sedgwick, as professor of geology at Cambridge, to write on his behalf to Cotton. Whewell thought that Phillips was 'eminently fit' for the post and offered total support but spotted the snag that Phillips was a 'non academical person'. It is not known whether Whewell wrote to Cotton but Sedgwick sent a letter which proved to be decisive. Encouraged and satisfied by Sedgwick's valuable testimony, Cotton appointed Phillips on 29 September 1853 as deputy reader in geology.[8]

Neither an Oxford man nor a graduate of any university, Phillips had to become an academical person according to Oxford's usages. In late October he visited Oxford not just to make arrangements for teaching and to inspect the geological collections, but also to become a graduate of the University. Through the good offices of Daubeny, a fellow of Magdalen College, he matriculated at Magdalen on 25 October at the unusual age of fifty-two. Two days later as a result of Cotton's initiative, Phillips became an MA of the University by a decree of Convocation, its supreme body. Unlike some of his future professorial colleagues, such as Rolleston, Henry Smith, and Benjamin Brodie, who were Oxonians and good to brilliant classicists when they were undergraduates, Phillips was a new sort of professor: he was an outsider appointed entirely on his 'European reputation'.[9]

Having received a friendly welcome in Oxford, Phillips was gratified by his appointment. It offered not only immediate opportunities but an attractive prospect. Given the sad decline of Buckland's health, he could expect to succeed Buckland in the near future as reader in geology if he did well as deputy reader. If he continued to find Oxford agreeable, he could anticipate ending his career there. Phillips' appointment was widely welcomed by geologists including Lyell and Edward Forbes, both of whom differed from Phillips on specific questions. Though Lyell did not apparently contact Cotton, he promised to lobby privately on Phillips' behalf in appropriate quarters. As an Oxford graduate Lyell was ashamed of his University for neglecting natural science and had denounced it to the Royal Commission as a theological seminary dominated by ecclesiastical parties. He wanted Oxford to be 'a great national seat of learning and science'. He therefore supported Phillips' application warmly because, if successful, it would break the monopoly of Oxonians filling chairs and would produce the 'beginning of the resuscitation of Oxford'. As president of the Geological Society, Forbes publicly congratulated Phillips, 'our illustrious associate', on his appointment and Oxford for promoting the true interests of science. Two of Phillips' friends, Sanders and Henry Rogers, privately echoed Forbes' sentiments: the former saw Oxford's conversion to modern geology as a blessing; the latter rejoiced that Oxford was committed to 'a severe and scientific cultivation of geology'.[10]

At Oxford Phillips' appointment was welcomed by the local bishop, Samuel Wilberforce. As a vice-president of the British Association when it visited Oxford in 1847 he had presumably been impressed by Phillips. Among local scientists Phillips' nomination was agreeable to Powell, an old associate, and to Henry Acland and Walker, who were local secretaries, at the 1847 meeting. Above all Daubeny was jubilant, both publicly and privately, that the University had ventured beyond the circle of Oxford graduates in appointing his candidate, who was a close friend. Clearly Daubeny wanted to lure Phillips to Oxford because of their personal similarities: they were bachelors who were not strident but gentle in their behaviour; both remained publicly calm amid jarring creeds and contending parties. Additionally, Daubeny was a quiet but resolute promoter of science at Oxford: his greatest success had been the examination statute of 1850 which gave science at Oxford some recognition for the first time. He saw Phillips as an ally. Phillips was a fluent lecturer who spoke from the fullness of his knowledge. He was a leading researcher and an accomplished expositor of geology via his books. As an experienced museum keeper, a savant with an exceptionally wide range of scientific interests, and proved administrative ability, he was exceptionally well qualified to become keeper of a new University Museum devoted to science, for which Daubeny and Acland had been pressing since 1847. Daubeny surely hoped that Phillips would animate the Ashmolean Society, the local scientific society founded in the Ashmolean Museum in 1828.

It is not fanciful to suggest that Daubeny also realised that in the Oxford context Phillips had three other advantages. As an irenic Anglican he would avoid any conflict between science and religion, avoid sectarian disputes, and thus avoid associating geology with any particular ecclesiastical party. As a classicist who was capable of quoting Homer and Lucretius in the originals, he would demonstrate that scientific and classical learning were quite compatible and he would not alienate classicists by making aggressive claims about the superiority of science as a mode of liberal education. As a historian and antiquarian, as well as expert geologist, he was a generally cultivated man who would not pit the professoriate and University, which recognised experts, against fellows and their colleges, which were homes for the ideal of the educated man. It is significant that when Faraday's niece met her uncle and Phillips for tea in February 1853 she enjoyed his conversation on antiquities and literature as well as science. She concluded that Phillips had 'taken pleasure in the general cultivation of his mind and tastes'.[11]

9.2   Oxford in 1853

The University of Oxford in 1853 was socially exclusive. From about 1820 to about 1860, there were about 400 new undergraduates per year. Thus the total undergraduate population remained steady while the population of Britain grew. Oxford's reputation deterred some parents able to pay for their offspring. For them it was a place of extravagance, dissipation, and aristocratic social values; of poor teaching by tutors in some colleges; and of excessive specialisation in classics which was allegedly a vehicle of liberal education. Parents had to look elsewhere for professional training in medicine and law. Oxford's expensiveness deterred others such as clergy who could not afford to pay out about £120 per year per son. And wealthy dissenters could not even enrol, ie matriculate, their offspring until 1856.[12]

Oxford was a collegiate university with almost as many colleges as professors and readers. There were obvious contrasts between the professoriate and college fellows, and between the University and the colleges. About twenty-five professors, not necessarily bachelors, lectured to small classes and received such modest salaries from the University that some held more than one chair and others were non-resident. Compared with the University the eighteen teaching colleges were wealthy, powerful, and statutorily independent. Most of their fellows were non-resident, leaving in the colleges a total of about fifty resident fellows who acted as college tutors. Always bachelors and often graduates of their own colleges, these tutors usually taught for ten to fifteen years and then departed to marry and become ecclesiastics and schoolmasters. None of these features was fundamentally changed by the Oxford University Bill of 1854, so that the power of the colleges, of college fellows, and of the associated aim of liberal education remained relatively undisturbed.[13]

Even so, for the first time in Oxford's history the claims of science were recognised in its examinations when the new regulations of 1850 permitted undergraduates to take their final degree examinations in natural science. This innovation was, however, hedged with restrictions. Candidates for a degree were required to obtain a pass or honours in literae humaniores, colloquially known as Greats, and then to pass in one of three other schools, ie mathematics, which was not new, and law/modern history and natural science which were. Thus any undergraduate keen on science had to spend about three years on compulsory classics, with a little mathematics and theology, before embarking on his preferred studies. If happy with a pass degree, he was examined in at least two of mechanical philosophy, chemistry, and physiology. For an honours degree, he was required to show a more extended knowledge of these three core sciences and an exact acquaintance with one or more subsidiary sciences, one of which was geology. Thus science was given token and half-hearted recognition. It was an addendum or complement to the study of classics whose dominance was not disturbed. Science had gained peripheral recognition as a vehicle of liberal education and not for its utilitarian or occupational value. Its peripheral place in Oxford's examinations was associated with the absence of tutoring in science in most colleges. Geology, like botany and mineralogy, was optional and supplementary because only those who sought honours in science were permitted to decide whether it was one of the subsidiary subjects in which they wished to be examined. The examination regulations of 1850 did, however, give some encouragement to geology in that all candidates for a degree were required to attend two professorial lecture courses. Even so, geology as an examination subject was peripheral in the new final honours school of natural science, which itself was peripheral in the University's schemes of study.[14] Phillips presumably realised that at Oxford geology could not flourish as part of professional training in engineering, as had happened at King's College, London, and Trinity College, Dublin. It would have to prosper mainly as liberal education for future clergymen, schoolmasters, and gentlemen, as it had at Cambridge under Sedgwick and, for a time, at Oxford under Buckland.

Phillips' prospective colleagues in the science professoriate were not united about the position and prospects of science at Oxford. The most pessimistic was William Donkin, professor of astronomy, who had no University observatory in which to research and teach so he created a small one at the top of his home where he taught about three students a year. Donkin's bizarre position was the result of a disagreement in 1839 between the University and the Radcliffe Trustees who had established an observatory in Oxford in 1772. From then until 1839 the first three Radcliffe Observers were also professors of astronomy in the University. That year the University and Trustees disagreed: on the death of Stephen Rigaud, the former elected the reverend George Johnson as professor and the latter appointed Manuel Johnson to direct their observatory. This breach left the University without an observatory and enabled the Radcliffe Observer to be a full-time researcher. George Johnson occupied the astronomy chair for only three years before assuming that of moral philosophy. He was succeeded by Donkin whose situation made it impossible for him to combine serious research and teaching in astronomy.[15]

Three other colleagues were very critical of the place of science at Oxford but their proposed remedies were impracticable. Bartholomew Price, elected professor of natural philosophy in 1853, wanted a greatly strengthened professoriate, with several professors per subject, supported by University lecturers; and he believed that professors, not college tutors, should prepare undergraduates for examinations. Powell, professor of geometry, had attracted in the 1830s and 1840s to his gratis lectures at best seven students and at worst (on three occasions) none. He concluded that mathematics and physical sciences were regarded as peculiar and extraneous compared with classics and moral subjects which were seen as essential parts of a liberal education. In sum, mathematics was neglected because nothing was gainable by it. As Henry Vaughan, professor of modern history, said: 'mathematics in Oxford are a bad investment for intellectual, physical, and pecuniary capital'. Powell's solution was to make compulsory for graduation mathematics, astronomy, natural philosophy, and one of chemistry, geology, and physiology. Story-Maskelyne, deputy reader in mineralogy from 1850, lectured in the Clarendon Building and had a laboratory in the basement of the Ashmolean Museum where he also lived. His salary was £100 pa until 1852 when it was raised, it seems, to £300 pa, perhaps to compensate for the derisory fees from the handful of undergraduates who paid a pound to attend his lectures. He wanted more and better-paid professors, the abolition of professorial cumul, improved facilities, and research fellowships to remedy the disheartening position of science. He deplored the way in which little cognisance was taken of science 'as a gymnasium for the mind', or even for 'its usefulness as mere marketable knowledge in the world'.[16]

Three of Phillips' future colleagues were almost as critical but were not entirely without hope because as incremental reformers they had achieved two important successes. Daubeny, professor of chemistry, had supplemented his income by assuming the chairs of botany (1834) and agriculture (1840). He had seen the size of his chemistry class decline from about thirty in the 1820s to around twelve in the 1840s. He had paid out of his own pocket for books, apparatus, and materials in chemistry and for the transformation of the Botanic Garden where he lectured on botany and agriculture. In 1848 he took the drastic step of moving his chemical activities from the basement of the Ashmolean Museum to his own college, Magdalen, for whom he built at his own expense a lecture room and chemical laboratory. Generally Daubeny realised that at Oxford science was not a passport to honour and emolument but he hoped that it would be possible to provide better accommodation and higher salaries for professors and to persuade colleges to appoint tutors in science. These remedies were also advocated by Walker, whose lectures in experimental philosophy given in the Clarendon Building were exempt from the general decline in attendance at lectures on science in the 1840s. In the early 1850s he gave in each year three courses, each of which attracted about thirty students. As Lee's reader in anatomy Acland was in a peculiar position. At Christ Church, where the anatomy collection was kept, he gave the University's solitary course of lectures on anatomy and physiology to an audience of between twelve and twenty, but he was not regarded by the University as equivalent to a professor. Acland's response to Oxford's discountenancing of science was to develop the ideal that the University should offer a liberal education and basic medical sciences to a small cohort of medical students, who would afterwards pursue clinical training elsewhere and attain distinction in their careers.[17]

Before Phillips assumed his post, Daubeny, Walker, and Acland had combined to secure two important innovations at Oxford. From 1848 Daubeny was the prime mover of the campaign which led to the new examination statutes of 1850 which recognised science, not for its utilitarian or occupational value, but as part of a liberal education and hence as an addition to the University's traditional concerns. In his agitation, firm but temperate, Daubeny was vigorously supported by Acland and Walker. From 1847 Acland took the lead in pressing for a University Museum. Well supported by Daubeny, Walker, and the bishop of Oxford, Acland advocated bringing together under one roof not only the scattered resources available in science but also teaching and research via lecture rooms, laboratories, and a library. This vision of a science centre, pursued as the best means of rendering effective the new examination statutes, was supported by the Royal Commission which reported in 1852. After much debate, the University took the decisive step in December 1853 of paying £4,000 for a site for the new Museum on the edge of the University Park.[18] Thus when Phillips arrived in Oxford early in 1854 to give his first lecture course the University was committed to a grandiose scheme for a new Museum, even though few undergraduates showed any interest in science.

9.3 Début in Oxford

When Phillips and his sister arrived in Oxford in January 1854, Daubeny had arranged that they could live at Magdalen Bridge in a cottage on the river Cherwell, with a picturesque view of Magdalen Tower. On 3 February Phillips gave his inaugural lecture in the Clarendon Building in a room crowded with Buckland's specimens. Having presented geology as a long-settled and honoured subject at Oxford through the labours of Buckland and Strickland, Phillips revealed considerable knowledge of classical literature while surveying the history of life on earth divided into palaeozoic, mesozoic, and cainozoic periods. He stressed that there had been many successive systems and renewals of life, only the most recent of which resembled present-day forms, that man was recent, and species immutable. He emphasised that the great antiquity of the earth was untranslatable into years and that great physical changes had occurred on it. Drawing on his own research, he maintained that at the end of the palaeozoic period there was an epoch of such extraordinary violence, shown by folding and breaking up of the bed of the sea, that there was a corresponding change in the system of life. Phillips did not feel it necessary or desirable to discuss the relation between Genesis and geology or to dilate extensively on natural theology. He merely alluded to the general laws of nature as being appointed by the Maker. His inaugural lecture showed generally that he would treat geology with 'strict and severe dignity, suited to its present character'.[19]

His first lecture course, advertised by a printed sheet, covered the philosophy of geology. It embraced familiar anti-Lyellian topics, such as a comparison of past agents of displacement of land and sea with 'known *disturbing causes* now operating in nature' and the past and present conditions of the interior of the earth. It concluded with a discussion of ancient life in three different environments (the sea, fresh water, and land) and of the relation of geological to historical time. After each lecture Phillips buttressed it by discussing for half an hour with keen students the answers to pertinent questions taken from a printed list (Fig. 9.2). They ranged from the nature of inductive science to such difficult matters as why the earth has mountains and how British mountains could be covered with glaciers. To help his students and demonstrate his pedagogic ambition he issued a small bibliography. It included two publications by Phillips himself: the latest edition of his geological map of the British Isles and the fourth edition of his *Guide* published early in 1854 at 5 shillings. The first three editions were dedicated to the Yorkshire Philosophical Society, but the fourth was dedicated to the vice-chancellor and members of the University of Oxford. It differed from its predecessor of 1836 only in its treatment of geological systems, chiefly the Cambrian, Devonian, and Permian, and in the greater emphasis given to the heat of the earth and metamorphism of rocks. After over a decade in which he had not considered updating any of his textbooks, Phillips returned to revising one of them in order to take commercial advantage of his new audience and new high station at Oxford. [20]

Phillips' first course of lectures indicated that there was hope for geology at Oxford. Whereas Buckland lectured in the 1840s to empty benches, Phillips' room was crowded to overflowing. One auditor was amazed by the way in which Phillips made ripple and worm marks on rocks into historical events. Such an approach attracted an audience of about fifty, including heads and fellows of colleges. Phillips regarded his début as successful and agreeable, though few students paid the attendance fee of £1. He rejoiced that the lectures of his friends Daubeny and Walker

ON THE PHILOSOPHY OF GEOLOGY.

' Ecfinge aliquid et excude quod sit perpetuo tuum.'

THE DEPUTY READER IN GEOLOGY, thinking it probable that many Gentlemen in his Class, and specially those who propose to be examined in Physical Science, may derive benefit from having recalled to their minds some of the topics treated of in his lectures, begs to propose for their consideration a few questions arising out of these lectures. By reflecting on the subjects thus selected, each student may better understand the progress he has made, and perceive the steps in which further assistance would be useful to him. To facilitate the efforts of such students the Deputy Reader will have the pleasure of attending in the Class Room, for the space of half an hour, after the close of each Lecture.

1. Geology is classed among Inductive Sciences—Why?

2. Our knowledge of the composition and structure of the exterior parts of the Earth.—On what is it founded?

3. By what method can we interpret the Ancient Monuments of Nature, so as to refer to appropriate causes, and place in historical succession, the Facts observed in the Structure of the Earth?

4. *Geological Theory* is guided and advanced by the progress of other branches of science—as Astronomy, Mechanics, Mineralogy, Chemistry, Zoology, Botany.—Give examples.

5. What is known as to the *Mean Density of the Earth*—its spheroidal figure—its interior structure—solidity—fluidity? Distinguish between the present and any former states of the Earth.

6. What is known of the *greatest depth of the Sea?* How is it known?

7. The *Crust of the Earth*. To what depth is it known or inferred?

8. Distinguish between *stratified* and *unstratified* rocks—in regard to structure, mode of occurrence in the Earth, organic contents, and process of formation—Draw a *section* representing the modes of occurrence of these rocks.

9. The Series of Strata constitute a scale of geological Time. Explain this, and compare it with successive events in History, anterior to Chronology.

10. The Strata have been *displaced*. Give proofs of continued *depression* of the sea bed, and other proofs of the *elevation* of land.

11. Make remarks on the *form* and *magnitude* of these displacements, their effect on *Physical Geography*, their occurrence in *geological times*.

12. Displacements of Land and Sea. Have such occurred within *historical periods?* In Volcanic Regions? At a distance from Volcanos?

13. Answer geologically the question of De Luc—*Why has the Earth any Mountains?*

14. Examine the dictum, that—*Valleys were formed by the streams which run in them.*

15. What shares, in the moulding of the beautiful and varied surface of the Earth, are to be ascribed to subterranean movements—action of the sea—operations of the atmosphere—rains—rivers—glaciers, &c.?

16. Estimate the annual or secular effect of *Diurnal Causes* in wasting the earth's surface, and filling up the bed of the sea.

17. The temperature of the different zones of the globe. On what *external conditions* does it depend?

18. How might it be affected by great displacements of land and sea? *The British climates* in particular—How could our mountains be covered with *glaciers?*

19. The temperature of the *Interior of the Earth*. How is it known or inferred? Give data from volcanos—thermal springs—artesian wells—mines—collieries, &c.

20. Do the inferences arising from the data suggested under the preceding heads, lead to any 'general view' or *hypothesis* of great physical changes in the condition of the interior and exterior parts of our Planet? Can this be tested by another series of inquiries into *Ancient Life on the Globe?*

Publications on Geology suited for Students.

ANSTED—Manual of Geology.
BUCKLAND—Bridgewater Treatise, 2 vols.
DAUBENY—On Volcanos.
DE LA BECHE—Geological Observer.
LYELL—Elements of Geology; Principles of Geology, 4 vols.
MORRIS—Catalogue of British Fossils.
PHILLIPS—Guide to Geology (1854); Treatise on Geology, 2 vols. (1853).

Phillips's Geological Map of the British Isles; published by the Society for the Promotion of Christian Knowledge.

Fig. 9.2   A polite invitation and list of twenty questions, first issued to his students by Phillips in 1854 and showing his pedagogic commitment. The Latin tag, from Pliny, *Letters*, means 'Fashion and hammer out

were also well attended and felt that this 'burst of success ... in physical science' would greatly advance the plans for the new Museum. He found the University agreeable, lively, and free in thought, quite different from 'the blind old bigotted Oxford ... painted in ... newspapers'. Clearly Phillips disagreed with Murchison's view of 1853 that science was still very much depressed at Oxford.[21]

In early summer 1854 Phillips organised geological field work at two famous local sites, Stonesfield and Shotover Hill. At the latter, Phillips' class investigated one of Buckland's favourite strata, the iron sands at its summit. Drawing on the distinction between freshwater and marine deposits made in Phillips' lectures, the class concluded that the sands were estuarine, a result published by him in 1858. Phillips was pleased with these early field trips, reporting jubilantly to Lyell that his class of forty had undertaken 'six hours' walk and *work*'. Next year in May he added to day-long excursions: he took his class to Malvern where it studied for four days the phenomena described in his 1848 *Memoir*.[22]

In mid-1854 Phillips took a decisive step which indicated his commitment to Oxford. In order to take more fully his place there, he let St Mary's Lodge for a year. At King's College, London, and Trinity College, Dublin, he had been as a professor nothing more than a visiting lecturer. At Oxford, in contrast, he felt as deputy reader that he had an agreeable future. Buckland was terminally ill so Phillips could hope to succeed him as reader in geology. Though Phillips felt that the University could do more for science, its position was nevertheless favourable so he did not contribute to polemics in Oxford about reform. He liked Oxford's pleasant haunts and felt at home with old and new friends there, especially with those who were neither pessimists nor utopians but steady incremental reformers.[23]

Phillips' growing alignment with Oxford was confirmed in other ways. Firstly, he contributed an essay on the geology of the Oxford area to the first of a four-volume collection of Oxford essays, written by members of the University and apparently the brainchild of Thomas Sandars. Phillips proclaimed that Oxford aspired to the 'highest triumphs of piety and knowledge'. Accordingly he argued that the geology of Oxford provided evidence of one constant plan, one universal power, one beneficent mind, one universal language, and one great Author: and that man, geologically recent, was on an earthly pilgrimage. As in his inaugural lecture there was no mention of the Bible. Apropos knowledge, Phillips focussed on Stonesfield which was remarkable because, as well as fossils of molluscs and fish, those of plants, insects, reptiles, and land mammals occurred there. He wanted a complete census of all these fossils which he thought had been deposited there in a tranquil lagoon.[24]

Secondly, Phillips brought out in 1855 another expository work, his *Manual of geology* which ran to over 600 pages and sold for 12/6. It was a moderate revision of his *Encyclopaedia metropolitana* article except for the chapters on palaeozoic geology which were rewritten. It was reshaped to give a main focus which was unique at the time. Most of the book was devoted to stratified rocks, their succession, and their associated fossils, many of which were illustrated by Phillips' own drawings. According to Seeley, who revised it in the 1880s, the *Manual* was the best of its type then written, with an excellent structure and an awareness of the mutual dependence of the principles of geology, so it was worthy of Oxford. With its emphasis on stratigraphy and the convulsive movements which had disturbed strata, from the lower palaeozoic to the tertiary, Phillips' book was quite different from the second edition of Morris' *Catalogue of British fossils* (1854) which merely listed 4,000 species and 1,280 genera. Indeed, convulsions and dislocations had two uses for Phillips. He could continue to argue that disruptions of strata were part of the Creator's general plan of terrestrial adaptations, with special provision made in favour of man. By invoking these disruptions, widely spread in geological time, he strengthened his arguments contra Lyell that in the earth's history there had been periods of repose interrupted by great and widespread violence. Generally the *Manual* maintained the anti-Lyellian positions set out in Phillips' books of the 1830s. It soon faced but survived tough competition not only from a new edition of Lyell's *Elements* but also from Jukes' *Manual* of 1857.[25]

It is significant that in 1854 and 1855 Phillips did not consider applying for two posts for which he was well qualified. The first was the chair of natural history at Edinburgh, rendered vacant by the sudden death of Edward Forbes in November 1854. Phillips' old friend James Forbes tried to lure him to the Athens of the North, stressing the good financial rewards, the existence of an established museum, agreeable society, and the superiority of the Royal Society of Edinburgh to the Ashmolean Society. Phillips' reply was revealing: Oxford provided everything that could be reasonably wished.[26] The second post was the directorship of the Geological Survey, occasioned by the death of De la Beche on 13 April 1855. Sedgwick quickly told Phillips he was the man for the job but he did nothing. Unbeknown to Phillips, Lyon Playfair, joint secretary with Henry Cole of the government's Department of Science and Art, wanted the School of Mines to be separated from the Geological Survey and to be developed as a national centre for science education. To direct the proposed broadened School Playfair thought of Phillips, 'an enthusiastic educationalist, acquainted with the mining proprietors … an excellent lecturer, and a most amiable man'. This scheme was quickly dashed: at the end of April, Murchison was appointed as director of the Survey, with responsibility for an unchanged School of Mines and Museum of Practical Geology.[27]

Prima facie Phillips' closest colleague at Oxford was Story-Maskelyne who, as deputy reader in mineralogy, lectured in the Clarendon Building and taught mineralogical and chemical analysis in his small laboratory in the Ashmolean Museum. In 1856, after Buckland's death, he became reader in mineralogy, a promotion supported by Phillips. Their close working relation enabled them to demarcate their subjects for teaching purposes amicably and quickly but it did not last long. The mineralogist was dissatisfied with his prospects at Oxford. In 1855 he had been defeated by Brodie in the election to the chair of chemistry, rendered vacant by Daubeny's resignation the previous year. Story-Maskelyne feared that his laboratory teaching would be taken over by Brodie. Moreover the mineralogist's emolument was so small that projected marriage was impossible. In 1857 he solved his personal problems by assuming another post, that of keeper of minerals at the British Museum, to which Phillips had drawn his attention. Story-Maskelyne then withdrew from his laboratory in the Ashmolean Museum, moved to London, married in 1858, and became a pluralist giving greater allegiance to London than to Oxford. As a non-resident reader and professor, he deprived himself of Phillips' daily friendship and Spenserian courtesy, which he valued, and he reduced his lectures and audience at Oxford to the minimum. He was not on the spot even before he gave his first course of lectures in the new Museum in May 1861, so he asked Phillips to have the mineralogy collection ready for display. Story-Maskelyne's non-residence, which induced the University to pay him a salary of only £100 pa, precluded any possibility of collaborating with Phillips in teaching and maybe research even though from 1861 they shared teaching accommodation in the new Museum where Maskelyne had a small laboratory. Two features of the younger man's career suggest what might have developed at Oxford. Story-Maskelyne taught practical mineralogy to a few promising Oxonians not at Oxford but in the British Museum. Though primarily a mineralogist, he became a distinguished geologist. Elected a fellow of the Geological Society in 1854, he was awarded its Wollaston medal in 1893 for his research on crystallography and petrology.[28]

9.4 Geological lecturing

Until 1861 Phillips's lectures were delivered in the Clarendon Building. Subsequently he lectured in the room allocated to geology and mineralogy in the University Museum. On three walls it was fitted with pulleys for suspending drawings and diagrams. It seems that two walls were covered by murals, one of the Mer de Glace and the other a lava stream, painted by a local pre-Raphaelite, the reverend Richard St J. Tyrwhitt. Though Phillips' facilities for teaching were improved, the place of geology in the University's examinations remained doubly peripheral. It was not strengthened in 1859 when the University abolished the requirement for all intent on graduating to attend two courses of professorial lectures. It was marginally improved from 1864 when those graduating with honours in science were no longer required to have passed finals in classics but still had to take an examination in classics in or after their seventh term after matriculating. One immediate consequence of this new regulation was the disappearance of students studying science for a pass degree. But most science professors were prepared to face the embarrassment of fewer undergraduates because those who attended were committed and undiluted by passmen. These new arrangements, which ensured that all honours graduates in science were still well grounded in classics, did not disturb the centrality of physics, chemistry, and physiology in the school of natural science. In the early 1870s Phillips began to feel that the examinations did not encourage geology which was ignored by all except a few resolute undergraduates. He thought that, as geology was a large and comprehensive subject which drew on other sciences, it should no longer be merely an optional special subject tagged on to the three core sciences. He wanted geology to join botany, geology, and mineralogy to form an additional group of subjects in which undergraduates could take honours after they had studied in detail physics, chemistry, and physiology. Phillips remained happy with Oxford's emphasis on producing scientists, like himself, who knew the classics and possessed literary culture. But eventually he thought that Daubeny's emphasis on three core sciences, enshrined in the examination regulations of 1850, gave insufficient encouragement to other important sciences, such as geology, and demoted them to a subordinate position.[29]

Despite this discouragement, Phillips usually gave two courses, each of twelve lectures, per academic year. He covered a wide range of topics in polite geology, only one of his courses being devoted to its economic aspects (1865). As a university teacher of geology he was perhaps unique in England in embracing both physical and fossil geology, in both of which he was expert. Several of his courses covered the principles and theories of geology, his last effort of this kind embracing the interior of the earth, the successive conditions of land and sea, physical geography, the succession of life on earth, and geological time (1873). Several were expositions of his books: in 1871 and 1872 he gave successive courses on the geology of Oxford and its vicinity; and he dilated several times on the succession of life on the earth and the conditions of land and sea in the past. Not unexpectedly he lectured several times on fossils (general, plant and animal, vertebrate, extinct, and characteristic) and a few times on stratigraphy. Sometimes he took advantage of local opportunity by expatiating on the geology of Oxford and the fossils in the Museum. Towards the end of his life he gave more attention to the physical geology, having launched this aspect of his subject with a course on igneous rocks in 1857 and one on physical geography in 1858. In 1870 he lectured on the composition and structure of the external parts of the earth; and in 1872 on the heat of the interior of the earth, its effect on ancient climates, and its manifestations in earthquakes and volcanoes.[30]

It is clear from a bibliography of 1872, designed 'to guide the student to a right general view of the several subjects enumerated' that Phillips' teaching was wide in scope and challenging. Much of it was devoted to leading themes in physical geology, i.e. rocks in general, unstratified rocks, metamorphic rocks, divisional structures (cleavage and foliation), mineral veins, earthquakes, volcanoes, physical geography, and the cooling and temperature of the earth. This section of the bibliography was full, with the part on physical geography alluding to fourteen authors. It was up to date: it included not only the 1872 editions of Lyell's *Principles* and Jukes' *Manual* but Joseph Everett's latest report on subterranean temperatures. It was neither parochial nor Europhobic because it alluded to over twenty foreign authors, sometimes in English translations but more often in the original German and French. In the references for unstratified rocks, mineral veins, and volcanoes, there were more foreign authors cited than natives. In those for rocks in general and metamorphic rocks, only foreigners were cited. From these it is clear that Phillips had a high regard for the standard works of Gustav Bischof and Bernhard von Cotta on chemical and mineralogical geology and for those of Gabriel Daubrée and Achille Delesse on experimental chemical geology.[31]

The section of the bibliography devoted to palaeontology was equally full, up to date, and non-Anglocentric. Though Phillips drew attention to Palaeontographical Society memoirs by Morris and John Lycett, Thomas Davidson, Thomas Wright, and Edward Binney, he cited as authorities on Stonesfield flora and fauna works in German by Oppel and Friedrich Quenstedt, two experts on jurassic strata. The part on Cambro-Silurian fauna was neatly balanced: it not only included standard works by Sedgwick and Murchison but also the recent compendium by John Bigsby and Joachim Barrande's ongoing survey of the Silurian system in Bohemia, published in French. As if all these sources were not enough, Phillips advised those students interested in more references and in researches still in progress to consult him.[32]

Phillips' lectures were buttressed in several ways. Given the absence of tutoring in geology in the colleges, Phillips initially met his students for half an hour after each lecture to discuss questions arising from it. By 1863 he was setting aside a day a week for what he called 'explanations'. In the 1870s he sometimes met students informally on two days at noon, having lectured on two other days at the same time. In the terms when he did not lecture, he still made himself available twice a week at noon to meet members of the University and to assist geology students. Phillips also ran practical classes which for him were essential because though them students learned the grammar of the science. Practical work took two forms. Firstly, Phillips taught from specimens in the geological collection. He ranged from discoursing on the fragmented and reconstructed remains of saurians to reconstructing the history of a bit of fossil wood in flint from the upper chalk, beginning with a forest of carboniferous trees. Secondly, he organised field work at such propitious local sites as railway cuttings north of Oxford, Shotover Hill, Culham, and Stonesfield, which could be reached on a day trip, and at more distant ones such as Malvern where students stayed for three or so days. All this teaching was done solo without assistance of any kind. By the early 1870s Phillips, who generally favoured the appointment of adjunct professors to assist the head professors at Oxford, yearned for an assistant who would conduct class teaching in lithology and palaeontology under himself as a 'comprehensive influence' to back up his lectures.[33]

Phillips used as a course book where appropriate the fourth and fifth editions of his own *Guide*. The former had sold well: by summer 1862 only thirty copies of the print run of 1,000 remained unsold so Longman called on Phillips to produce yet another revision. The fifth edition (1864), of which 1,000 copies were printed, was 50% longer than its predecessor with old material re-ordered and new material added, sometimes as a new chapter. For example, metamorphic strata were discussed in an account of the strata in the earth's crust while lapse of time gained a chapter to itself. Still easily portable, it cost just 4/- and by the early 1870s was almost sold out. Jukes, a literary rival of Phillips, was favourably impressed: 'it may be the grandfather of all guides, but if so it is its own great grandson with all the vigour and grace of youth instead of the greyness and seediness of age'.[34] In contrast Phillips found it impossible to revise his *Manual* which he recommended at Oxford as a useful general work. By the early 1870s he thought it needed to be split into two substantial volumes but felt that revision on that scale was beyond him. Charles Griffin, the publisher, was happy for an editor chosen by Phillips to revise the *Manual*. But the project lapsed because Phillips' choices, Rupert Jones and Thomas Wiltshire, were too busy or daunted to undertake such a large task either solo or with Phillips.[35]

The attendance at Phillips' Oxford lectures never equalled that of his first year, even though in the 1860s he was seen, with Brodie and Rolleston, as the best of an impressive cohort of professors of science. Until 1859 all those graduating had to attend two lecture courses and pay the appropriate fee which for geology was one pound. In these years it seems that Phillips attracted between ten and twenty per course. From 1859, when compulsory attendance at two professorial courses was abrogated, until 1862, his class size decreased and on two occasions probably dropped to single figures. From 1862 Phillips charged no fee, in an attempt to increase the size of his class, and began to attract women to it. By the mid-1860s his audience numbered fifteen to twenty, of whom over half were women, the rest being undergraduates and male strangers. Through women his class became on occasion the largest of any in science. In autumn 1866 he drew in eighteen members of the University, eight male strangers, and no fewer than thirty- eight women. The attendances at other lectures were chemistry forty-five, geometry thirty-two, physiology twenty-four, natural philosophy fifteen, astronomy seven, and experimental philosophy five. Three years later Phillips attracted what was probably his largest class, of fifty-six men and thirty women, which was considerably bigger than Sedgwick's thirty-five at Cambridge and again the best attended lecture course in science at Oxford. It seems that Phillips' lectures were the only ones in science open to women, who relished the opportunity. From 1862 his lecture courses, given in the University, were available gratis to any member of the public and the University. His long experience of public lecturing about geology to mixed audiences enabled him to bridge the gap not just between gown and town but also between the sexes, at a time when the University was widely criticised for its social and intellectual exclusiveness. Though not a charismatic performer like Sedgwick at Cambridge, Phillips appealed to women who thought his lectures wonderful, even though they were designed for undergraduates and adepts in geology: without condescension he communicated in a lucid and genial way his varied and deep knowledge of his subject.[36]

9.5 Geological pupils

Though we lack systematic information about all the men who attended Phillips' lectures, we have attendance registers for 1857 to 1862 and several letters to him from pupils. From such limited data and excluding those who were awarded the Burdett-Coutts scholarship in geology, the career patterns of members of his audience are clear. Not unexpectedly the great majority became clergymen, almost always Anglican vicars in Britain. For them geology was part of a liberal education and in no way hostile to their calling or faith. Quite a number of Phillips' audience became lawyers and barristers, the best known being Robert Wright, a well-known judge, and Thomas Brassey a prominent member of Parliament. A handful of his pupils pursued careers in well-known public schools. Three became arts teachers at Eton, Westminster and Winchester. A fourth, Mark Ward, taught science at Clifton College, Bristol, 1869-72, before becoming an inspector of schools. Two became headmasters: Arthur Butler was a reforming head at Haileybury while George Bell served for over thirty years at Christ's Hospital and Marlborough. Five pupils became well-known writers and intellectuals. John Nichol, son of the astronomer who was a friend of Phillips, became the first professor of English in the University of Glasgow and a man of letters. As well as pursuing a legal career, James Morison became an advocate of positivism and a biographer. Auberon Herbert, after a short career as an member of Parliament, became a professional eccentric: he was an agnostic, vegetarian, and promulgator of the doctrines of Herbert Spencer. John Green attended Phillips's lectures on the advice of William Dawkins, the star pupil of Phillips, and according to Dawkins was profoundly influenced by them. Certainly Green was well read and informed in geology which he regarded as a glorious science. His historical works about the English people were famous and were republished for decades. Perhaps drawing on Phillips' emphases on populations and on processes of change, Green did not focus on individuals and the 'drum and trumpet' approach to their deeds. Instead, he examined the ways and movements by which the English nations had been made intellectually, constitutionally, and socially. For a time Green even thought of having an opening chapter on the geology and geography of England and their bearing on its industry and people. A fifth pupil, John Earwaker, was precocious as an undergraduate. He wrote favourably about Oxford's facilities for science and with Phillips directed a two-day visit of the Geologists' Association in 1871. Earwaker became a well-known antiquary in north-west England and north Wales.[37]

A goodly smattering of Oxford's academic staff attended Phillips' lectures. Among the college fellows, who were usually ordained, were John Burgon, a leading high churchman and unrepentant reactionary in University politics, and Thomas Fowler, logician and man of letters who supported facilities for science at Oxford in the 1870s. Only one college fellow, the reverend Henry Wood, made his mark geologically: as a pamphleteer and reviewer, who had been a lecture and field pupil of Phillips, he opposed Lyell's views on the antiquity of man and derided Lyell's 'unreasoning apotheosis of the uniformity of nature'. One of Phillips' professorial colleagues, Joseph Bosworth, a distinguished Anglo-Saxon scholar, sat at his feet as did five heads of colleges, all of whom were clergymen. Two of them, Edward Hawkins and Robert Scott, were also successive professors of the exegesis of holy scripture but heard little from Phillips about the relations between geology and scripture. The other three, James Thompson, Edward Cradock, and James Sewell, probably attended out of general interest and to judge what Phillips offered to the undergraduates in their colleges. There is no evidence that any of these clerical academics at Oxford found Phillips' lectures to be theologically subversive. Indeed, in 1867 a leading Cambridge cleric, Henry Cookson, turned to Phillips for private tuition. A keen mathematician and geologist, Cookson was master of Peterhouse, had just served as president of the Cambridge Philosophical Society, and as vice-chancellor of the University promoted the cause of science in it. For him inductive geology led to love of nature and honouring nature's Lord.[38]

Compared with the large cohort of pupils who became clergymen, lawyers, teachers, and writers, only a handful became physicians and academic scientists. Gilbert Child, who prospered as a physician in Oxford, was an authority on sanitary and ecclesiastical matters. After a few years as Lee's reader in anatomy at Oxford, William Church pursued a career as a physician at St Bartholomew's Hospital, London, and rose to be president of the Royal College of Physicians and the Royal College of Medicine. Augustus Shepherd became a London physician who was an expert on lung disease. For such prospective physicians, knowledge of some geology was compatible with their membership of a learned profession. The future scientists who were Phillips' pupils attained only modest distinction. Having taught natural science at Jesus College from 1857 and served for several years as Walker's deputy in the chair of experimental philosophy at Oxford, George Griffith failed to be elected to it in 1865 and turned to school teaching at Harrow. William Donkin, son of the professor of astronomy, taught natural science at Keble College in the 1870s and then moved to London where he was professor of practical chemistry at St George's Hospital. For Griffith and Donkin, geology was part of a wide scientific education which led to lectureships in science, broadly construed, in Oxford colleges.[39]

Like Sedgwick at Cambridge, Phillips taught geology as part of a broad and liberal education to a wide range of undergraduates, few of whom became full-time scientists. Though Phillips' lectures were more technical and less discursive than Sedgwick's, he was again like Sedgwick in that he led no research school and had very few pupils who became full-time and distinguished geologists. Ignoring Dawkins, a Burdett-Coutts scholar, Richard Tiddeman was the only pupil of Phillips who pursued a productive full-time geological career. Having graduated in 1862, Tiddemann continued to be indebted to Phillips who recommended him in 1863 for a post with the Geological Survey where he remained for all his working life. Phillips also supported him in his successful bid to be an FGS. Tiddeman not only learned about the geology of northern England from Phillips' publications; but also, when he began field work with the Survey in north-west England, wanted Phillips with him and asked for hints about surveying there, even though he had as colleagues the experienced Aveline and the promising Thomas McKenny Hughes. It was Phillips who in 1865 advised Tiddeman to examine the limestone around Clitheroe. There Tiddeman saw for himself a phenomenon which Phillips had repeatedly told Tiddeman was baffling, ie the sharp contrast between the carboniferous strata north of Malham and Settle and those around Skipton and Clitheroe. Using Phillips' notion of the Craven faults and his ideas about ancient marine conditions, Tiddeman eventually explained this contrast. He postulated in the late 1880s that, on the downthrow side of the faults in the Clitheroe/Skipton area, there were among the carboniferous rocks sporadic conical mounds of limestone which were quite different from the massive limestone north of Malham and 'Settle. These mounds, he argued, had grown like coral reefs in a slowly subsiding shallow sea south of the faults. It was in part for this Phillipsian research, on what Tiddeman called knoll-reefs, that he was awarded the Murchison medal of the Geological Society in 1911.[40]

In 1860 geology at Oxford was boosted when the University accepted an endowment of £5,000 for a geological scholarship from Angela Burdett-Coutts, the famous philanthropist and richest heiress in England. She had wide interests in science, especially in natural history, some of her collections eventually finding their way to the Royal Botanic Gardens, Kew, and the British Museum (Natural History). She used her wealth not only to give soirées after Friday evening lectures at the Royal Institution but also to promote geology. At Torquay she attended the lectures given by William Pengelly whose geological research she regarded highly. From 1858 she supported financially his exploration of the bone cave at Brixham near Torquay, which showed that man was contemporary with extinct mammals. Subsequently she paid for his important research on fossil plants in the unique Bovey Tracey beds near Torquay. In summer 1859 she revealed to her friend, the bishop of Oxford, her intention of giving to the new Museum at Oxford Pengelly's collection of Devonian fossils from Devon and Cornwall which he had collected at her request and sold to her. She thought the Museum would be a safe location and regarded Phillips, its keeper, as distinguished. In autumn 1859 she extended her scheme by offering to endow a geological scholarship in connection with donating the Pengelly collection. Her main motive was theological. She felt that Anglican priests educated at Oxford knew less science than dissenting ministers. In particular she regarded geology as an essential part of the education of Anglican clergy because it provided evidence of God's providence. Having taken the advice of Acland and Phillips, whom she met socially, the Burdett-Coutts scholarship was established in 1860 as the University's second-oldest postgraduate award in science. The first was the Radcliffe travelling fellowship in medical or natural science founded in 1715 and by the 1860s worth £200 pa, tenable for three years provided eighteen months were spent abroad. The Burdett-Coutts scholarship, worth about £60-75 pa, was tenable for two years by a scholar, elected each year, who had to be an Oxford graduate. The candidates faced a searching examination by three professors, one of whom was Phillips. Miss Burdett-Coutts publicised and maintained her interest in the scholarship, welcoming changes in its rules. She liked her scholars, especially Dawkins, and was so impressed by Phillips that in 1864 she invited him to stay with her. In return he reported regularly about the successful candidate; and, having made her in 1860 one of only three recipients of bound presentation copies of his *Life on Earth*, dedicated his *Vesuvius* to her in 1869.[41]

Of the ten Burdett-Coutts scholars in Phillips' lifetime, one (Edward Langdon) disappeared into obscurity but three (George West, Edmund Jermyn, and Charles Taylor) presumably made her happy because they became clergymen au fait with recent geology. They published nothing on geology but attained various degrees of distinction as antiquarians. Two scholars who had graduated in natural science from Magdalen, Phillips' college, and had been Radcliffe travelling fellows after being Burdett-Coutts scholars, rose to eminence as medical men. Joseph Payne became a well-known epidemiologist and pathologist at St Thomas' Hospital, London, while William Corfield, a geological protégé of Daubeny, was appointed in 1869 the first professor of public health and hygiene at University College, London. One scholar pursued for a time a career with some connection with geology. From 1876 to 1884 Edward Cleminshaw was a science master at Sherborne School, Dorset, where he galvanised the School Field Society and the School Museum to which he gave important geological specimens. In 1880 he acquired literary reputation when his translation of Adolphe Wurtz's famous treatise on atomic theory was published. After leaving Sherborne Cleminshaw prospered as an analytical chemist at Oldbury near Birmingham. Two scholars became academic scientists. Thomas Wyndham who analysed specimens chemically for Phillips, died young in office as tutor in natural science at Merton College and as a demonstrator in chemistry in the University. The long-lived and pugnacious Edwin Ray Lankester, who held a Radcliffe fellowship after the Burdett-Coutts scholarship, occupied chairs in zoology at University College, London, and Oxford before directing the British Museum (Natural History) and gaining a knighthood. In the 1860s Lankester was a precocious and productive geologist who published regularly from 1862 when he was fifteen, beginning with a paper on fossil fish of the old red sandstone. While a Burdett-Coutts scholar he worked on his monograph on that subject for the Palaeontographical Society. Though Lankester respected Phillips, there is no evidence that his research and career were promoted by Phillips.[42]

The opposite was true of William Dawkins, the first recipient of the scholarship. He was so enthused by Phillips' lectures that in 1860 he became the first undergraduate to take geology as part of the final degree examination in natural science. As an undergraduate Dawkins was given free access by Phillips to Buckland's geological collection, which had been bequeathed to the University. It contained countless specimens from bone caves and in 1859 stimulated Dawkins to begin excavating a hyena den at Wookey Hole near Wells, Somerset, which he found contained remains of humans and extinct mammals. This discovery launched Dawkins' sustained interest in geological archaeology. Phillips always thought highly of Dawkins whom he regarded as by far his best pupil, who in some ways had outstripped his teacher. When Phillips was looking after his dying sister in May 1862, he entrusted to Dawkins the supervision of his field class on Shotover Hill which they had often visited together. Dawkins was happy to represent himself as Phillips' geological son and accordingly invited his geological father to his wedding. On many occasions he was indebted to Phillips for patronage. Phillips let Dawkins use specimens at the University Museum, supported his candidature as FGS, joined him in excavations of caves near Wells in April 1862, was prevented by his sister's illness from accepting Dawkins' invitation to be present at his first paper at the Geological Society, helped him to wriggle into the Geological Survey, gave him introductions to its leading English members, and as an FRS secured for Dawkins permission to consult and borrow rare books in the library of the Royal Society. Phillips persuaded the Palaeontographical Society to invite Dawkins to write a monograph on British pleistocene mammals, which was Dawkins' first book. In 1866 Phillips tried unsuccessfully to induce the Royal Society to accept a paper by Dawkins on fossil rhinoceros; next year he promoted successfully Dawkins' application to be FRS. In the early 1870s, Phillips supported Dawkins' important exploration of Victoria Cave, Settle, which became renowned not only for its artefacts but also for its mammal bones, many of which had been gnawed by hyenas and a few scratched or cut by tools. Phillips' last act of patronage was to give Dawkins a glowing reference when he applied unsuccessfully in 1873 for Sedgwick's chair at Cambridge.[43]

Though Dawkins was suspicious of Phillips' beliefs in designed adaptations and the novelty of man, he drew heavily on his technical expertise. He consulted Phillips about whether to adopt a stratigraphical or zoological approach in his monograph on pleistocene mammals. He asked for Phillips' advice on bone hunting of such mammals in Yorkshire, a task he fitted successfully into his honeymoon in 1866. His ideas of that time about the fauna and geology of caves were derived, he said, from Phillips' lectures. When defeated by some saurian teeth in 1870, he asked Phillips to identify them. In Dawkins' *Cave hunting*, dedicated to Baroness Burdett-Coutts, he summarised his research on pleistocene mammals and palaeolithic man. It was indebted to Phillips for private information about Kirkdale Cave and for published descriptions and explanations. Indeed Dawkins went out of his way to acknowledge that his own research corroborated Phillips' view that pleistocene mammals lived in Britain before the period of extensive glaciation. Dawkins was thus indebted to Phillips for 'frequent help and prudent counsel'. In short Phillips enabled Dawkins to develop a successful career as a researcher in geological archaeology which culminated in a knighthood. Though Dawkins is remembered as a cave hunter and industrial geologist, he was well prepared to be the first professor of geology at the University of Manchester because he derived from Phillips a wide view of his subject which embraced physical geology.[44]

9.6 The geological collection

Phillips was in charge of the collection of geological specimens housed until 1861 in the Clarendon Building and subsequently in the less cramped conditions afforded by the new Museum. In the former locale the collection was used mainly by Phillips to illustrate his lectures. In the latter it continued as a teaching aid but was also used by non-Oxonians, such as leading researchers, geologically minded visitors, and geology students from other teaching institutions. The collection thus mirrored Phillips' lectures in the Museum in that it became a vehicle for public geological enlightenment.

When Phillips went to Oxford he inherited the geological collections assembled by his predecessors. Strickland's was small but Buckland's was a huge, rich, labelled but unarranged affair. Phillips immediately appreciated the importance of Buckland's collection for teaching and research and resolved to do justice to it by arranging it properly; but had not made significant progress when it was moved to the new Museum. From 1854 to 1860 the general collection grew steadily. Phillips provided his own teaching specimens, maps, sections, and models. Members of the University, Phillips' friends, and geological well-wishers made useful donations. Most donors gave fossils but John Enys donated jointed rocks and John MacCulloch's widow donated her late husband's rock collection via John Gray of the British Museum.[45]

At the Museum Phillips continued to look after the expanding geological collection solo. Inevitably he fell into arrears of sorting, arranging, documenting, storing, and displaying specimens. Generally Phillips' priority was to organise a creditable and instructive exhibition of specimens, after which he tried not to neglect the labelling and cataloguing of stored and reference material. Initially the geological display was composed mainly of fossils, arranged by zoological class and geological age, so that the student could study a particular class of organisms in different geological periods and the organisms of a particular geological period. Where possible specimens were arranged so as to show their conditions of origin. An obvious theme of the display was progression in part of the fossil record, 'an upward progress' typified by invertebrates, fishes, reptiles, and mammals. Ten years later the display still contained fossils arranged stratigraphically. Adjacent to the fossil saurians and mammals were complete skeletons of analogous living genera, so that comparisons could be made. Two additions had been made to the display: local Oxford material, including the unique assembly of remains of ceteosaurus; and illustrations of geological phenomena such as dykes and mineral constitution. Phillips' successor, Prestwich, was agreeably surprised by what Phillips had done with the geological collection, judging its general arrangement to be excellent but its labelling incomplete. It seems that as curator of the collection Phillips worked solo until about 1867 when Henry Caudell became his assistant. Caudell specialised in disinterring fossils embedded in matrixes and in restoring fossil bones from fragments.[46]

Donations, mainly of British fossils, continued to pour in during the 1860s. By the end of the decade, some donors began to try to attach conditions to their gifts, wishing them to be kept apart, intact, and accessible, or displayed or catalogued as they wanted. Some benefactors sorely tried Phillips' patience and tact. For example, he received in 1873 the offer of a collection of fossil saurians from Thomas Hawkins, the author of famous apocalyptic books about the earth's history. It needed money spent on it and required the mediation of Liddell, the vice-chancellor, to do justice to it and to satisfy Hawkins' yearning for fulsome acknowledgement of his gift.[47]

Phillips relied mainly on donations to augment the geological collection but on occasion he resorted to purchase of significant finds. The best known buy he made was of a local bone bed of a saurian known as cetiosaurus, a 'whale lizard' so called because its spongy bones were like those of a whale. The discovery began in 1868 when a huge fragmented thigh bone, sixty-four inches long, was found by quarrymen in the great oolite at Enslow Bridge, nine miles north of Oxford. It was pieced together and in 1869 identified and publicised by Phillips as a bone of cetiosaurus. Next year a bed of bones of this saurian was found in the same place. Aided by James Parker, his friend, and by Rolleston, his professorial colleague, Phillips completed a general plan of the ossuary and then paid almost fifty pounds for the purchase and transport of all the bones to the Museum. With the help of Henry Caudell, his technical assistant, and drawing on Rolleston's knowledge and specimens of crocodile skeletons, Phillips steadily pieced the fragments together to produce a unique and sensational assembly of the remains of what he called cetiosaurus oxoniensis.[48]

This species was important in the debates about evolution which were still raging. In his 1871 book Phillips showed that in spite of its name cetiosaurus was not a whale-like aquatic reptile as Owen, its namer, had claimed. It was terrestrial: from its large claws and its bone arrangements Phillips inferred that it walked vertically in particular directions and did not have the sprawling gait of a crocodile. From a mutilated fragment of a tooth Phillips concluded it was not carnivorous but vegetarian. He estimated that it was fifty feet long and ten feet high, which made it the largest terrestrial creature then known. No wonder that Huxley wanted to see 'Phillipsio-cetiosaurizans' and 'Frankensteinosaurus'. He always welcomed powerful intellectual assaults on Owen's palaeontology. Moreover Phillips' interpretation of ceteosaurus oxoniensis as terrestrial gave Huxley another genus to add to his dinosaurs, which in his Darwinian view were generally intermediate in character between ostrich-like birds and reptiles.[49]

Phillips was happy to allow visitors to see the geological collection. Sometimes professors of geology elsewhere brought their students to the Museum. In 1869 John Morris and his pupils from University College, London, heard Phillips expatiate on items in the collection and then did field work at Shotover Hill. In 1872 Rupert Jones' students at the Sandhurst Military College were charmed by Phillips and his 'scientific paradise'. The collection became a mecca for the Geologists' Association, founded in 1858 to encourage mutual help among amateur enthusiasts via meetings and field excursions. These field trips, led by an expert, soon became popular and distinctive. The Association visited the Museum and Oxford in 1861, 1864, 1869 (in conjunction with University College London), 1871, and 1874. By the 1870s each excursion lasted two days and was organised by Phillips and a local helper, Earwaker in 1871 and James Parker in 1874. The standard format was a lecture by Phillips, on specimens in the geological collection or a site to be visited, followed by field trips to Shotover Hill, Enslow Bridge, Charlbury, and Stonesfield, some of which he led. In 1874 Phillips also offered lunch, tea, and an evening conversazione in his home. The Association, which elected him an honorary member in 1872, appreciated the cordiality of the veteran geologist and exulted in the 'rare opportunity of learning from the great master of mesozoic geology himself'.[50]

From the late 1860s leading palaeontologists visited Oxford mainly to examine saurian specimens in the geological collection. Of those who paid one visit, William Carruthers examined alleged fossil plants from Stonesfield, while Sir Philip Egerton and Etheridge looked at saurian bones. Those who came twice were Lord Enniskillen, Owen, John Hulke, and Huxley. Owen was interested in Phillips' 'latest grand acquisition', the remains of cetiosaurus. Phillips also aided Owen's research by sending him drawings and lending some specimens, but he would not let Owen borrow unique, very rare, or fragile specimens. In return for the feast of body and mind which Hulke enjoyed, he raised with Phillips queries about identifications of plesiosaurus and cetiosaurus.[51] The most significant palaeontological visitor was Thomas Huxley, then acting as Darwin's bull-dog in the fight about evolution. Huxley was particularly interested in Buckland's specimens of megalosaurus, a genus discovered by Buckland himself.

In 1867 Phillips and Huxley met accidentally and discussed saurians. As Phillips had doubts about the place of some bones in a skeleton of megalosaurus he had drawn, he invited Huxley to the Oxford Museum to look at the saurian specimens and especially the megalosaurus. Accompanied by Phillips, Huxley examined the megalosaurian remains and noticed in the shoulder girdle what he thought was a misplaced bird-shaped hip bone. They then re-built the pelvic girdle, which reminded Huxley of that of an ostrich. He concluded that megalosaurus was a bird-like dinosaur and soon developed his theories about the affinity between dinosaurian reptiles and birds and the evolutionary descent of the latter from the former. In his famous paper of 1870 on this topic, Huxley went out of his way to cite a long letter of 1868 from Phillips to himself, in which Phillips acknowledged that megalosaurus, though twenty-five feet long and weighing two tons, was nevertheless sib with primeval birds. Huxley, who had privately condemned Phillips in the early 1860s as a cautious fence-sitter, craftily enlisted Phillips under his Darwinian colours in 1870. Phillips himself asserted publicly in 1871 that megalosaurus Bucklandi was essentially reptilian but, as Huxley had shown on the basis of its pelvis, it possessed 'avian points of structure' and 'a curious analogy, if not some degree of affinity, with the ostrich', a pedestrian bird. Phillips himself pointed out that the great reptile's shoulder-girdle was bird-like and resembled that of apteryx, a living New Zealand bird with rudimentary wings and no tail, also known as the Kiwi (Fig. 9.3). In 1868 Huxley had argued that even separate classes of animals could have had common ancestors. In 1871 Phillips publicly contemplated this possibility. Privately he was surprisingly jocular about megalosaurus. He wondered what dear old Buckland would have thought of his carnivorous giant reptile being a cousin of warm-blooded air-swimmers, a result of 'morganatic union', but by early 1870 he was charmed by Huxley's fine and curious results and reasoning.[52] In short, the Oxford collection and Phillips enabled Huxley to adduce British fossil evidence to reinforce his claims

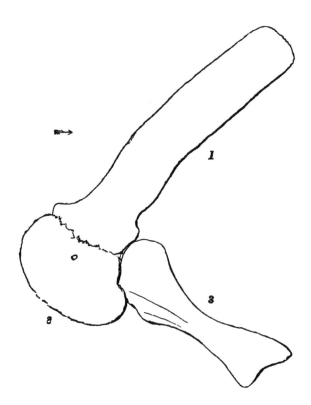

Diagram LXIII.   Megalosaurus.   Scale one-tenth of nature.

The left aspect of the shoulder girdle is here restored in outline from specimens in the Oxford Museum, which are complete except in regard to the lower end of the humerus. It will be remarked how bird-like in the general arrangement and the forms of the bones is the humero-scapular structure, and specially how closely it resembles Apteryx.

1. Scapula.   2. Coracoid.   3. Humerus.

Fig. 9.3   Phillips' drawing of his reconstruction of the shoulder girdle of megalosaurus, published in Phillips, 1871a. The caption shows that Phillips presented evidence for mixed zoological relationships.

that some ancient reptiles were bird-like, that forms intermediate between birds and reptiles had existed, and that such different classes of animals as birds and reptiles were related by descent from common ancestors.

9.7   Schools and extra-mural lecturing

As a teacher of a science which at Oxford was made peripheral by the examination system, Phillips was not uninterested in the place of science in schools. In the mid-1860s he knew that the public schools, which were the main feeders of Oxford, taught little science, and he thought that this omission was particularly disadvantageous for botany, zoology, mineralogy, and geology at Oxford. Subsequently, Ward and Henry Wood, his former pupils, told him about the difficulties of teaching science at Clifton College, Bristol, and the success of geology lectures at Sherberne School, Dorset. It is inconceivable that Cleminshaw, another pupil and geologically precocious boy at Rugby School, did not talk to Phillips about the pioneering efforts of James Wilson as science master there. Phillips was therefore not uninformed about science teaching in schools but avoided campaigning on the matter. He found acrimonious or strident public controversy involving scientists demeaning; and he wanted to avoid pitting the claims of science against those of classics. His remedy, privately revealed to Lord Wrottesley perhaps in connection with the 1865 report of the Select Committee of the House of Lords on the Public Schools Bill, was unfussy piecemeal reform. For Phillips the key agents were visionary headmasters who would provide facilities for pupils and appoint full-time well-trained science teachers who had social standing and attainments in classics and mathematics. Such innovations, he thought, would discourage unseemly rivalry between 'the glorious fundamental studies' of classics and mathematics and 'the newer treasures of science'.[53]

Phillips therefore did not take advantage of various opportunities of contributing to public debate about science teaching in schools. He did not give evidence on behalf of Oxford to any of the royal commissions which reported on this topic in the 1860s. In Phillips' presidential address of 1865 to the British Association he ignored the subject, he was not involved in the Association's report of 1867 on promoting science in schools, and nothing came of Lord Wrottesley's wish that he should serve on the Royal Commission on Endowed Schools which reported in 1867. Characteristically Phillips' views about the desirable features of science teaching were revealed as an appendix to a work on the subject by the reverend William Tuckwell, an Oxford friend who in 1864 became head of Taunton College School. Drawing on his own teaching based on specimens, Phillips advocated observation of natural objects and phenomena and, where appropriate, experimentation on them. He believed that talk growing out of these experiences was more effective than didactic or deductive exposition. He urged the formation of museums containing, not trifles or exotics, but well-arranged objects preserved from mischief and decay, and of laboratories containing simple apparatus. Generally Phillips thought it fatal to introduce science in a pressurised way. He supported his beliefs by donating fossils, thermometers, and cash to help the cause of science at Tuckwell's school; but he declined to give his name to Tuckwell's project for a new school, where science would be central, presumably because he opposed any side-lining of classics or literary culture. Phillips' other contribution to science teaching in schools, apart from teaching undergraduates who became science masters, was as examiner at Cheltenham Ladies College.

Using Thomas Wright, a Cheltenham doctor and geological friend of Phillips, as an intermediary, the formidable headmistress, Dorothea Beale, a mathematician who herself taught science, persuaded Phillips in the mid-1860s to set and mark papers on physical geography and physics, and subsequently on geology. She had widened the curriculum for girls and tried to raise standards in her school by using as examiners distinguished academics from Oxford, the nearest university.[54]

Phillips gave few individual lectures and courses away from Oxford. His heavy responsibilities there and increasing age left him with little time or energy to diffuse science elsewhere. He lost contact, it seems, with the mechanics' institute movement. He gave single lectures in places such as Worcester (1854), Leeds (1866), and York (1866), which he liked, only when he was invited by geological friends such as the reverend William Symonds, Thomas Teale, a Leeds surgeon, and by his old York chums, Davies and Kenrick. These occasional lectures covered familiar topics such as the Malvern Hills and the physical geography of Yorkshire.[55]

In twenty years Phillips gave only three lecture courses outside Oxford, all at institutions of which he approved and where he was appreciated (see Appendix 2). His six weekly evening lectures at the London Institution in 1853-4 were delivered between his appointment at Oxford and his inaugural lecture. Though he wanted to give no more than eight lectures at the Royal Institution in 1857, he was persuaded to give ten, weekly, for 75 guineas. Additionally, in March he delivered a Friday evening discourse about the Malvern Hills, in which he stressed not only the repeated upwards and downwards movements of the crust of the earth in one area, but also lateral pressure and thrusts. Phillips' last lecture course outside Oxford was given to the Royal Manchester Institution which, after two failed attempts in 1856 and 1858, secured him in 1860 to give four afternoon lectures in a fortnight for 50 guineas. Phillips welcomed the opportunity to expatiate on his own *Life on the earth*. His long career as an itinerant lecturer, spanning thirty-five years, ended appropriately in Manchester, one of his favourite haunts.[56]

Notes

1. RCO (1850), p. 108; RCOC (1872), pp. 46, 406, 419-20; Ovenell, 1986, pp. 219-20.
2. Phillips to Sedgwick, 20 Aug 1855, 22 Nov 1860, 21 Feb 1862, Se P; Sedgwick to Phillips, 7 March, 23 Oct 1864, PP: Gifford, 1932, pp. 64, 105: Benjamin Collins Brodie (1783-1862), *DNB*, president of Royal Society of London 1858-61; Joseph Hodgson (1788-1869), *DNB*; Mary Buckland (1797-1857); Charlotte Murchison (1788-1869); Mary Elizabeth Lyell (1808-73); Emma Darwin (1808-96); Francis Jeune (1806-68), *DNB;* William Pengelly (1812-94), *DNB;* John Obadiah Westwood (1805-93), *DNB*, Hope professor of entomology 1861-93; William Thomson (1819-90), *DNB*, provost of Queen's College 1855-61, archbishop of York 1863-90.
3. Phillips to Sedgwick, 25 Sept 1862, 20 April 1872, Se P; Wood to Phillips, 21 Dec 1865, 13 Nov 1866, 22 Jan, 12 Feb, 20 Nov 1867; Symonds to Phillips, 3 Nov 1865; Lee to Phillips, 23 Oct 1866; Davies to Phillips, 22 Dec 1866, 13 Nov 1867, all PP; Phillips to Forbes, 21 Oct 1866, FP; Henry Hatton Wood (1825-82); William Samuel Symonds (1818-87), *DNB*.
4. Phillips to H.O. Coxe, 26 Jan 1868, Bodleian, Eng. lett.d 418, 93-7; Galton, 1874, p. 100; Phillips to M. Harcourt, 21 July 1873, HP; Phillips, 1868b, 1872a, 1873a.
5. Rupke, 1983, 1997; Jardine, 1858, pp. ccxlvii-ccli; Lyell to Phillips, 21 Sept 1853, PP; Frederick Charles Plumptre (1796-1870), master of University College 1836-70, vice-chancellor 1848-52; Mervyn Herbert Nevil Story-Maskelyne (1823-1911), *DNB*, on

whom see Morton, 1987, became reader in mineralogy and then professor, resigning in 1895.

6. Strickland, 1852, pp. 13-16, 23-4, 30.
7. RCO (1850), pp. 99-106 (106 q).
8. Wilberforce to Daubeny, 10 Oct 1853, DP, Ms 400; Whewell to Phillips, 22 Sept 1853, PP; Phillips to Whewell, 17 Sept 1853, WP; Phillips to Sedgwick, 22, 27 Sept 1853, Se P; Cotton to Sedgwick, 29 Sept 1853, Se P; Richard Lynch Cotton (1794-1880), *DNB*, provost of Worcester College 1839-80, vice-chancellor 1852-6; Robert Walker (1801-65), *DNB*, reader in experimental philosophy 1839-60, professor 1860-5.
9. Phillips to Sedgwick, 31 Oct 1853, Se P; Macray, 1894-1911; Hebdomadal Board minutes, 17 Oct 1853; Tuckwell, 1900, p. 56 (q); Benjamin Collins Brodie (1817-80), *DNB*, professor of chemistry 1855-72.
10. Phillips to Murchison, 30 Sept, 2 Oct 1853, MP; Phillips to Sabine, 29 Oct 1853, Sa P; RCO (1852), p. 123 (q); Lyell to Phillips, 21 Sept 1853, PP (q); Forbes, 1854, p. xlv; Sanders to Phillips, 29 May 1854, H.D. Rogers to Phillips, 9 Sept 1855 (q), PP.
11. Wilberforce to Daubeny, 10 Oct 1853, DP; Daubeny to Harcourt, 11 Nov 1853, HP; Daubeny, 1853, pp. 56-7; Morrell and Thackray, 1981, pp. 392-5; diary of Margery Reid, 5 Feb 1853, Royal Institution Ms F13B, a reference I owe to Frank James; Samuel Wilberforce (1805-73), *DNB*, bishop of Oxford 1845-69; Henry Wentworth Acland (1815-1900), *DNB*, Lee's reader in anatomy 1845-57, regius professor of medicine 1857-94, librarian of Radcliffe Science Library 1851-1900.
12. Stone, 1975.
13. Engel, 1983.
14. Curthoys, 1997; Taylor, 1952.
15. RCO (1850), pp. 107-10; Thackeray, 1972, p. 9; Guest, 1991; William Fishburn Donkin (1814-69), *DNB*, professor of astronomy 1842-69; Stephen Peter Rigaud (1774-1839), *DNB*, professor of astronomy, 1827-39; George Henry Sacheverell Johnson (1808-81), *DNB*, professor of astronomy 1839-42, of moral philosophy 1842-5, dean of Wells Cathedral 1854-81; Manuel John Johnson (1805-59), *DNB*, Radcliffe Observer 1839-59.
16. RCO (1850), pp. 59-67, 191-5, 257-60, 90 (q), 185-91, 286 (q); Bartholomew Price (1818-98), *DNB*, mathematics lecturer Pembroke College 1845-53, professor of natural philosophy 1853-98; Henry Halford Vaughan (1811-85), *DNB*, professor of modern history 1848-58.
17. Hutchins, 1995a; Simcock, 1984, pp. 7-10; RCO (1850), pp. 267-8, 284-5, 292, 235-7, 282-4; Atlay, 1903, pp. 130-60.
18. Fox, 1997, pp. 641-50.
19. Phillips to Sedgwick, 27 Jan, 8 Feb 1854, Se P: Phillips to De la Beche, 1, 8 (q) Feb 1854, DLB P: Phillips, lecture notes, Hilary term 1854, including inaugural lecture, PP, box 112, folder 1.
20. Phillips, 'On the philosophy of geology', lecture list and twenty questions, PP, box 78, folder 8; Phillips, 1850c; Phillips, 1854b.
21. *Athenaeum*, 1854, pp. 213-14; Phillips to De la Beche, 8 Feb 1854, DLB P: Phillips to Grove, 9 Feb 1854, Gr P (q); Phillips to Sedgwick, 19 Feb 1854, Se P: Murchison to Sedgwick, 15 June 1853, Se P.
22. Phillips, lecture notes, Trinity term 1854, Trinity term 1855, PP, box 112, ff. 2, 4; Phillips, 1858b; Phillips to Lyell, 10 Aug 1854, 9 June 1855, LP.
23. Phillips to Murray, 1 Sept 1854, Mu P: Phillips to Sedgwick, 19 April 1855, Se P.
24. Phillips, 1855b, pp. 192 (q), 197, 212, 201-9; Sandars to Phillips, 4 Aug 1855, nd [1855?], PP; Thomas Collett Sandars (1825-94), *DNB*, legal editor.
25. Phillips, 1855c, pp. 38-9, 466-7, 610-11, 628-9; Phillips, 1885, vol. 1, pp. v-vi; Morris, 1854; Lyell, 1855; Jukes, 1857.
26. Forbes to Phillips, 20 Nov 1854, PP; Phillips to Forbes, 25 Nov 1854, FP.
27. Sedgwick to Phillips, 16 April 1855, PP; Layton, 1973, 155 (q); Geikie, 1875, vol. 2, pp. 182-90; Henry Cole (1808-82) *DNB*.

28. Story-Maskelyne to Phillips, 10 Feb, 5 April, 25 Sept 1855, 25 Sept 1856, 9 Sept 1857, 22 April 1861, PP.

29. Acland and Ruskin 1893, pp. 35-6; OUM,HA, box 2, folder 2; Curthoys, 1997; Phillips to vice-chancellor, 27 Nov 1873, Gu P, 65; Richard St John Tyrwhitt (1827-95), *DNB*, vicar of St Mary Magdalen, Oxford, 1858-72.

30. Phillips, lecture notes, 1854-74, PP, box 112, 113; *Oxford University gazette*, 1870, vol. 1-, *passim*.

31. University of Oxford, 1872, pp. 16-20; Lyell, 1872; Jukes, 1872; Everett, 1872; Bischof, 1854-9; Cotta, 1866 a and b; Daubrée, 1859; Delesse, 1858; Joseph David Everett (1831-1904), *DNB*, professor of natural philosophy, Queen's University, Belfast, 1867-97 and key figure in BAAS's underground temperature research; Carl Gustav Christoph Bischof (1792-1870), *DSB*; Carl Bernhard von Cotta (1808-79), *DSB*; Gabriel August Daubrée (1814-96), *DSB*; Achille Ernest Oscar Joseph Delesse (1817-81).

32. Morris and Lycett, 1850-4; Davidson, 1851-86; Wright, 1864-82; Binney, 1868-75; Oppel, 1863; Quenstedt, 1858; Murchison, 1867; Sedgwick and McCoy, 1855; Bigsby, 1868; Barrande, 1852-1911; John Lycett (1804-1882); Thomas Davidson (1817-85), *DNB*; Thomas Wright (1809-84), *DNB*; Edward William Binney (1812-81), *DNB;* Friedrich Quenstedt (1809-89), *DSB*; Frederick McCoy (1823-99), *DNB*; John Jeremiah Bigsby (1792-1881) *DNB;* Joachim Barrande (1799-1883), *DSB*.

33. Phillips, flier 'on the philosophy of geology', Hilary term 1854, PP, box 78, folder 8; Phillips to Sabine, 27 Oct 1863, Sa P; *Scientific opinion*, 1870, vol. 3, p. 381; *Oxford University Gazette*, 1873, vol. 4, p. 7; Phillips to vice-chancellor, 27 Nov 1873, Gu P, 65 (q); Huxley, 1870, p. 14; Phillips, 1865a; Phillips, 1871a, p. 8; Phillips, notes for practical classes, 1862, 1865, 1868, PP, box 113, ff. 3, 10, 17.

34. Phillips, 1864a; Longman to Phillips, 9 July, 1862, 31 Dec 1863; Jukes to Phillips 9 Jan 1864, all PP.

35. Griffin to Phillips, 26 Feb 1872; Jones to Phillips, 17 March 1872; Wiltshire to Phillips, 15 May 1872, all PP; Charles Griffin (1819-62); Thomas Rupert Jones (1819-1911), *DNB*, professory of geology, Royal Military College, Sandhurst, 1862-80, an accomplished reviser of general works; reverend Thomas Whiltshire (1826-1902), heavily occupied as secretary of the Palaeontographical Society 1863-99.

36. Phillips to vice-chancellor, 26 April 1866, PP; Phillips, register of attendance, 1857-62, PP, folder 18; Phillips, attendance at lectures and practical classes of professors within the Museum precincts 1867, OUA,UM/F/4/3; Devonshire Commission, first and second reports, *Parliamentary Papers*, 1872, vol. 25, p. 202, and appendix 9, pp. 33-8, which gave attendances at science classes 1869-70 at Oxford and Cambridge; Evans, 1875; Henry Fry to Phillips, 18 May 1872; Thelma Richard to Phillips, 1 April, both PP.

37. Robert Samuel Wright (1839-1904), *DNB*; Thomas Brassey (1836-1918), *DNB*; Mark James Barrington Ward (1844-1924) on whom Phillips to Murray, 15 April 1869, Mu P; Arthur Gray Butler (1831-1909), *DNB*, headmaster, Haileybury, 1862-7; George Charles Bell (d 1913), headmaster, Christ's Hospital 1868-76, Marlborough 1876-1903; John Nichol (1833-94), *DNB*; James Augustus Cotter Morison (1832-88), *DNB*; Auberon Edward William Molyneux Herbert (1838-1906), *DNB*; Stephen, 1902, pp. 22, 41-4, 86, 106, 211; John Richard Green (1837-83), *DNB*; William Boyd Dawkins (1837-1929), *DNB;* Earwaker to Phillips, 27 Nov 1869, PP; Earwaker, 1870-1; John Parsons Earwaker (1847-95), *DNB*.

38. John William Burgon (1813-88), *DNB,* fellow of Oriel College 1846-76; Thomas Fowler (1832-1904), *DNB*, fellow of Lincoln College 1855-81; Wood, 1865, 1867, p. 392; Wood, fellow of Queen's College 1846-57; reverend Joseph Bosworth (1789-1876), *DNB*, professor Anglo-Saxon 1858-76; Edward Hawkins (1789-1882), *DNB*, provost of Oriel College 1828-82, professor of exegesis of holy scripture 1847-61; Robert Scott (1811-87), *DNB*, master of Balliol College 1854-70, professor of exegesis of holy scripture 1861-70, and lexicographer; James Thompson (1801-60), rector of Lincoln College 1851-60; Edward Hartopp Cradock (1810-86), principal of Brasenose College 1853-86; James

Edward Sewell (1810-1903), *DNB*, warden of New College 1860-1903; Henry Wilkinson Cookson (1810-76), *DNB*, master of Peterhouse 1847-76; Cookson to Phillips, 22, 30 Oct 1867, 17 Oct 1871, PP; Phillips to Murray, 25 Sept 1867, Mu P.

39. Gilbert William Child (1832-96); William Selby Church (1837-1928), *DNB*, Lee's reader in anatomy 1860-9, president Royal College of Physicians 1899-1905, president Royal College of Medicine 1908-10, knighted 1902; Augustus Burke Shepherd (1838-85); George Griffith (1833-1902); William Frederick Donkin (d 1888).

40. Tiddeman to Phillips, 7 Dec 1863, 18 July, 19 Oct, 1864, 20 Oct 1865, 21 Jan 1869, 13 April 1870, 30 May, 14 Dec 1872, PP; Porter, 1982b; Tiddeman, 1890; Richard Hill Tiddeman (1842-1917), Geological Survey 1864-1902, obituary in *QJGS*, 1918, vol. 74, pp. liv-lvi, presentation of Murchison medal, *QJGS*, 1911, vol. 67, p. xliii.

41. Pengelly, 1897, p. 93; Pengelly and Heer, 1862; Phillips to Pengelly, 30 Sept [1859], Torquay Museum, HP.P.27; Burdett-Coutts to bishop of Oxford, 6 July 1859, and to ?, 2 Nov 1859, in letters from Miss Burdett-Coutts respecting her geological scholarships 1859-60, OUA, W. P. B 2(13); Murchison to Phillips, 21 Oct 1859, Phillips to Burdett-Coutts, 15 March 1862, in Burdett-Coutts scholarships, PP, box 114, f.4; Burdett-Coutts to Phillips, 25 April 1860, 20 Feb, 16 March 1863, 21 April 1864, 29 April 1865, 3 March 1866, 30 Feb 1867, 9 March 1868, 20 Feb, 16 Mar 1869, PP; on Angela Georgina Burdett-Coutts (1814-1906), *DNB*, see Healey, 1978.

42. Edward Langdon (b 1841), scholar 1864, has defeated even Hugh Torrens; George Herbert West, 1868, published on Gothic architecture; Edmund Jermyn, 1870, an ecclesiastical antiquarian, held posts in India; Charles Samuel Taylor, 1871, wrote on Gloucestershire and Somerset antiquities and was sometime editor and president of the Bristol and Gloucestershire Archaeological Society; Joseph Francis Payne (1840-1910), *DNB*, 1863; William Henry Corfield (1843-1903), *DNB*, 1866, held his UCL post until his death; on Edward Cleminshaw (1849-1922), 1873, see Torrens, 1878, pp. 37-8; Thomas Heathcote Gerald Wyndham (1842-76), 1867, on whom see Phillips, 1871a, pp. 511-12; Lankester, 1868-70; Lankester to Phillips, 24 Oct 1871, PP; on Edwin Ray Lankester (1847-1928), *DNB*, 1869, professor of zoology UCL 1874-91, professor of zoology, Oxford, 1891-8, director of British Museum (Natural History) 1898-1907, knighted 1907, see Lester and Bowler, 1995.

43. Gordon, 1894, pp. 53, 56; Dawkins to Phillips, 27 Dec 1861, 24 Oct 1864, nd [Oct 1864], 21 Jan, 15, 16 March 1865, 25 March, 18 June, 15 Nov, 8 Dec 1866, 7 Nov 1871, PP, Phillips to White, 25 Oct 1864, Royal Society of London, MC.7. 128; Phillips to Royal Society of London, 27 March 1866, Royal Society of London, MC.7. 314; Phillips to Royal Society of London, 22 Nov 1866, Royal Society of London, MM, xxi, 19; Phillips to Dawkins, 25 Oct 1864, 15 March 1865, Wellcome Historical Medical Library, autograph letters; Phillips to Dawkins, 9 March 1865, Torquay Museum, H. P.P.27; Dawkins and Sandford, 1866-72; Phillips to Dawkins, 21 Jan [1862], 24 Jan 1862, 26 May [1862], Buxton Museum, pasted into Dawkins' copy of Phillips, 1860b; Phillips to Dawkins, 5 Feb 1873, in Dawkins, 1874a.

44. Dawkins, annotations on Phillips, 1860b; Dawkins to Phillips, nd [Oct 1864], 16 March 1865, 16 July 1866, 6 Feb 1870, PP; Dawkins, 1874b, pp. x (q), 28, 36-40, 53, 284, 405; Dawkins, 1870a; on consultancies of Dawkins, lecturer in geology, Owens College [University of Manchester] 1872-4, professor 1874-1908, knighted 1919, see Tweedale, 1991.

45. Phillips to Sedgwick, 31 Oct 1853, 29 Nov 1853, Se P; Phillips to Murchison, 19 May 1854, MP; Phillips to De la Beche, 1, 8 Feb 1854, DLB P; Phillips to Murchison, 21 June 1856, EUL 523/4; Gordon, 1894, pp. 51-3; Hebdomadal Council reports, 7 May 1858; Enys to Phillips, 26 Feb, 19 March 1859; Gray to Phillips, 21 Dec 1857, 29 March 1858; Phillips to vice-chancellor, 31 March 1858 all PP; John Samuel Enys (1797-1872).

46. Phillips, 1863b, 91-2 (q); *Proceedings of the Geologists' Association*, 1873, vol. 2, p. 243; Phillips to vice-chancellor, 27 Nov 1873, Gu P, 65; Prestwich to Smith, 3 April 1875,

OUA, UM/M/1.3; Dawkins to Phillips, 3 Dec 1867, PP; Phillips, 1871a, pp, viii, 248, 384.

47. Hawkins to Phillips, 11 Sept, 20 Dec 1873; Hawkins to Liddell, 28 Oct 1873, 18 Feb 1874; Phillips to Hawkins 17 Nov 1873; Phillips to Liddell, 2 Feb 1874, all PP; Thomas Hawkins (1810-89) *DNB*.

48. *Athenaeum*, 1869, i, p. 310; Phillips, 1870, 1871a, pp. 245-94; Phillips to vice-chancellor, 4 May 1872, OUA, UM/M/5/1; James Parker (1833-1912), Oxford publisher, bookseller, geologist, and antiquary.

49. Phillips, 1871a, pp. 284, 293-4; Huxley to Phillips, 28 March, 2, 28 April, 1870, PP.

50. Morris to Phillips, 13 May, 4 June 1869; Jones to Phillips, 7, 17 (q) March 1872; Wiltshire to Phillips, 14, 22 April, 15 May 1871, all PP; Jones, 1883; *Proceedings of the Geologists' Association*, 1865, vol. 1, pp. 155-7; 1873, vol. 2, pp. 243-4; 1876, vol. 4, pp. 91-7 (92 q); *Geological magazine*, 1874, vol. 11, pp. 330-1; Holmes and Sherborn, 1891.

51. Carruthers to Phillips, 26 Nov 1871; Enniskillen to Phillips, 2, 4 May 1872; Sanders to Phillips, 28 April 1871; Owen to Phillips, 6 July (q), 30 Aug 1870, 24 Sept 1872; Hulke to Phillips, 5 Nov 1868, 24, 28 July 1871, 26 March, 2 April 1872, all PP; William Carruthers (1830-92), keeper of botany, British Museum (Natural History); John Whitaker Hulke (1830-95), *DNB*, London surgeon and president Geological Society 1882-4, died as president of Royal College of Surgeons 1893-5.

52. Desmond, 1994, pp. 357-9; Huxley, 1870, pp. 13-16, reproduced Phillips to Huxley, 1 Jan 1868; Phillips, 1871a, pp. 196-7 (qs), 217, 208; Huxley, 1868; Rudwick, 1976a, pp. 249-52; Phillips to Huxley, 14 Oct (q), 19 Nov 1869, 21 Feb 1870, Huxley papers.

53. Phillips to Wrottesley, 11 April 1865; Ward to Phillips, 26 March 1872; Wood, 7, 31 March 1870, all PP; Torrens, 1978, p. 37; James Maurice Wilson (1836-1931), *DNB*, science and mathematics master, Rugby School, 1859-79.

54. 'Report of a committee appointed by the Council of the British Association for the Advancement of Science to consider the best means for promoting scientific education in schools', *Parliamentary Papers*, 1867-8, vol. 54, pp. 3-14; Wrottesley to Phillips, 30 May 1865, PP; Tuckwell, 1865; Tuckwell to Phillips, 4 May, 15, 29 June, 13 Dec 1865, 17 May 1866; Wright to Phillips, Thursday evening [1866]; Beale to Phillips, 14, 17 May 1867, all PP; William Tuckwell (1829-1919), fellow New College 1848-58, master New College School 1857-64, head Taunton College School 1864-77; on Tuckwell's efforts at Taunton see Devonshire Commission, 6th report, *Parliamentary Papers*, 1875, vol. 28. pp. 66-8, 73-7, 95-7, 175-7, 199; Dorothea Beale (1831-1906), *DNB*, head Cheltenham Ladies College 1858-1906.

55. Symonds to Phillips, 12 Sept 1854; Teale to Phillips, 19 Oct 1865; Davies to Phillips 11 Jan, 26 March 1866; Kenrick to Phillips, 8 Feb, 3 April 1866, all PP; Phillips, notes on Malvern Hills for Worcester lecture, 10 Oct 1854, PP box 107, f. 4; Clark, 1924, p. 181; *YPS 1866 report*, p. 13; Thomas Pridgin Teale (1800-67) on whom see Davis, 1889, pp. 54-8.

56. Phillips, 'Syllabus, 1824-52' for London Institution and Royal Manchester Institution, PP, folder 12; RI managers' minutes, 14 July 1856, 16 March 1857; Phillips to RI, 8, 10 July 1856, RI archives, CL1/115,116; Phillips, 1858c; Aspden to Phillips, 14, 23 July 1856, 13 Aug 1858, 5 June 1860, Salomons to Phillips, 16 May 1860, RMI archives, M6/1/49/6.

Chapter 10

Professorial Research

In his Oxford period Phillips continued to believe that geology needed the aid of the collateral sciences of not just zoology, botany, and chemistry but also natural philosophy and astronomy. His first lecture course at Oxford began with an account of general data furnished by chemical, mechanical, and astronomical science regarding the mass of the earth. In his first presidential address to the Geological Society in 1859 he asserted that some of geology's highest generalisations were based on astronomy. As president of the British Association in 1865 he revealed his view of the natural world as a unified totality: 'the greater our progress in the study of the economy of nature, the more she unveils herself as one vast whole; one comprehensive plan; one universal rule, in a yet unexhausted series of individual peculiarities. Such is the aspect of this moving, working, living system of force and law'.[1] Such sentiments underlied his belief that the study of the earth's past overlapped with terrestrial and cosmical physics, a view not acted on or held by any other of that small band called heroic geologists. In mid-Victorian Britain Phillips was a unique geologist in that he worked on terrestrial magnetism, meteorology, and especially astronomy, in which he focussed on the physical features of the moon, Mars, and the sun. He hoped that the observation of these features would illuminate the history of the earth and of the solar system. His astronomical contemporaries thought he was an experienced productive observer: in the 1860s one feature of the moon and two of Mars were named after him. He brought to astronomy a combination of skills which in the 1850s and 1860s was unusual among geologists. Like several of them he was an excellent draughtsman, he had long experience of detailed surveying and mapping, and his visual imagination enabled him to represent three-dimensional objects and phenomena in both two and three dimensions. But he was unique among geologists in two respects. He was adept in using measuring instruments in geology and physical science, some of which he had devised and made himself. One of these, his self-registering maximum thermometer, was commercially successful in the 1860s. Phillips himself enjoyed a long friendship with the well-known telescope-maker, Thomas Cooke, who in 1862 provided for him a suitable instrument for astronomical research, at a time when most geologists used only the established instruments of field work and a few petrologists were beginning to exploit the microscope.

In this chapter I shall consider Phillips' chief and characteristic efforts in polite geology via his research on slaty cleavage, belemnites, glaciation, volcanoes and earthquakes, and regional geology. Then I shall examine his contribution to economic geology as a consultant and end with an analysis of his research in archaeology, astronomy, and terrestrial physics, all of which he regarded as cognate with geology.

Though much of this chapter considers research topics mainly chosen by him, he did undertake two commissioned publications which required an up-to-date and synoptic grasp. Phillips regarded the production of a geological map as a severe test of a geologist's knowledge and judgement. In 1862 he was therefore happy to accept 10 guineas from the Society for Promoting Christian Knowledge for the final edition of his Map of the British Isles and adjacent coast of France, which was engraved by Lowry. It was the first of the series to acknowledge that in part it was based on Smith's 1815 map of England and Wales.[2] Phillips' authority as a geologist was also revealed in the article on geology which he wrote for the fourth edition of Herschel's *Manual of scientific enquiry* published in 1871 under the editorial supervision of Robert Main, an Oxford friend of Phillips. It was no less than Darwin who suggested to Herschel that his old piece of 1849 on geology could be revised by Phillips, seven years his senior. In Phillips' revision he retained Darwin's original headings but added six new ones which revealed characteristic concerns, namely, a bibliography, instruments, rules for collecting, methods for observing, fossil footsteps, and erratic boulders.[3]

10.1 Cleavage and belemnites

Slaty cleavage, that is the tendency of some rocks to split in a particular direction into thin sheets or slates, independently of the plane of stratification, intrigued geologists because it indicated a change of internal structure in stratified rocks after their deposition. In the mid-1830s Sedgwick put the subject on a new footing by distinguishing formally between jointing, stratification, and cleavage of rocks. Joints, he claimed, were fissures at definite distances from each other, with the intervening rock having no tendency to cleave parallel to these fissures. In contrast, cleavage indicated the potential indefinite subdivision of rock along planes. From field work in Wales Sedgwick showed that the cleavage planes were different from those of stratification and that over large areas they preserved a constant dip and direction even in strata which were greatly contorted. He concluded that the strike of cleavage often coincided with that of stratification. Sedgwick's explanation of cleavage was that it was a form of crystallisation, a conjecture which was adopted in the 1840s by De la Beche and Hopkins.[4]

An alternative approach to cleavage was developed in the late 1840s by Daniel Sharpe whose theory of mechanical pressure was indebted to Phillips in two ways. In a paper of 1843 Phillips was the first to postulate a connection between distortions of symmetrical fossils and the cleavage of rocks containing them: he ascribed such distortions to the creeping movement of the particles of rocks along the planes of cleavage. Sharpe expanded this view to claim that the cause of contortion was the cause of cleavage, ie mechanical pressure perpendicular to the planes of cleavage. Phillips also claimed in 1843 that the strike of cleavage was parallel to the main direction of axes of elevation of any strata which were bent; it was this notion which induced Sharpe to relinquish his former suspicions of the mechanical pressure theory. In the early 1850s Sorby used his new technique, of examining with a powerful microscope thin slices of rock, to confirm Phillips' observations and to support Sharpe's theory of great lateral pressure. Shortly afterwards Sharpe's theory was strengthened by John Tyndall, who showed by experiments on wax that cleavage planes could be produced by pressure perpendicular to them and by Samuel Haughton, who came to the same conclusion from his calculations about distortions of fossils.[5]

In 1855 the British Association asked Phillips to report on the contentious subject of cleavage and the related phenomenon of foliation. He ignored the latter but was well qualified to deal with the former, having first studied the differences between cleavage and stratification in 1821 in the Lake District. In the late 1820s he had measured around Kirkby Lonsdale the change of the dip of a plane of cleavage as it passed through different beds, but in his 1836 monograph confessed that cleavage remained a very obscure subject. He tried to remove that obscurity by field work in Wales (1836 and 1843), Devon (1839) and Cork (1843) which resulted in his 1843 paper. Before writing his 1856 report he visited Skiddaw in the Lake District to check Sharpe's notion of axes of cleavage; having written his report he sent it to Sedgwick for comments, which he found beneficial. Phillips' report defended Sedgwick's priority in establishing the distinction between planes of cleavage and stratification and in postulating a great general cause for the former. But Phillips went on to argue, on the basis of his 1843 paper, that compression at right angles to the plane of cleavage caused cleavage, a view not universally accepted at the time but one he continued to hold in the 1860s. He thought that the 'imitative experiments' of Sorby and Tyndall showed that mechanical pressure was an efficient cause of cleavage, whereas Sedgwick's notion of crystalline polarity merely re-worded the phenomenon into an alleged explanation. Sedgwick remained unconvinced: he doubted whether pressure alone was responsible for changes of strike and dip of cleaved rocks.[6]

The Palaeontographical Society was founded at a meeting of the Geological Society in 1847 when, in a discussion of Prestwich's paper on London clay, James Bowerbank suggested the creation of a society devoted to publication about tertiary geology. The idea was immediately expanded to encompass the publication of plates of fossils from all British deposits. The plan appealed particularly to members of the London Clay Club, formed about 1836 to promote tertiary palaeontology. In 1848 the Palaeontographical Society began to publish two types of monographs devoted to taxonomic descriptions and illustrations of British fossils. Some of them were devoted to organisms in certain strata, such as Searles Wood on molluscs in the crag, Rupert Jones on entomostraca (lower crustaceans) in the cretaceous, Owen on reptiles in the mesozoic, and Thomas Wright on echinodermata of the oolite. Others were concerned with a particular organism in all the strata in which they occurred as fossils. Witness Henri Milne-Edwards and Jules Haime on corals, Davidson on brachiopods, and Salter on trilobites. Some of these palaeontologists devoted much or all of their work to the fossilised creatures of just one phylum, Davidson, Wright, and Wood being examples of this sort of palaeontological specialisation which gathered force from the 1850s. The monographs published by the Palaeontographical Society were often issued in instalments and some took years before they were finished: Davidson's monograph was published over a period of thirty-five years.[7]

As author in the mid-1860s of a work on British belemnites for the Palaeontographical Society, Phillips was not exclusively a specialist or general palaeontologist. But his wide view of geology embraced knowledge of many fossils found in palaeozoic and mesozoic strata. He had been interested in belemnites for over thirty years and was familiar with the jurassic and cretaceous strata in which they occurred. But figuring and describing belemnitidae for the demanding Palaeontographical Society involved highly detailed research on a notoriously troublesome family of creatures. In his 1864 memoir on their structure, Huxley admitted that it was still very difficult to identify species of belemnites.[8] They were preserved as fossils almost always in the shape of a solid bullet, known as the guard; and they were differentiated according to its form, including the number and position of grooves on it. The paucity of unambiguously distinguishing features produced taxonomical difficulties, and illustrating their small differences required highly skilled drawing.

Phillips was familiar with belemnites as a teenager but his sustained interest began in the late 1820s. In his book on the Yorkshire coast he noticed nine species and in 1829 visited the Strasbourg Museum where Voltz, a specialist on belemnites, taught him much about them. Voltz's sustained encouragement induced Phillips to list eighteen and illustrate six species in the second edition of the Yorkshire coast book and to plan a monograph on British species. By 1835 he had identified thirty-four and had produced a draft text but no plates. Confident that he could cope with problems of identification and nomenclature, in 1836 Phillips advertised his *Figures and descriptions of British belemnites* as in preparation. By late 1837 he had identified forty species, had plates engraved, and hoped to have the work ready in April 1838 and to publish it by subscription.[9]

In 1841 the British Association tried to encourage him to complete it by giving him a grant of £50. Next year he had proof plates ready and had informed Buckland that he was to be the book's dedicatee, but he was still wrestling with the problem of classifying belemnites. Spurred by the encouragement of Buckland and Stutchbury, he was sufficiently near completing the long-promised book in 1843 that he approached Murray about publishing it in four parts, each of which would appear quarterly and contain about ten plates. Murray did not jib at this proposed mode of publication, which was novel to him, but he received nothing from Phillips. Phillips was busy with other matters and knew that the standard works on belemnites by Blainville and Voltz, on which he intended to model his book, were being challenged in the early 1840s by Orbigny's research on jurrasic palaeontology.[10]

Once Bowerbank had settled into being first secretary and editor to the Palaeontographical Society, he invited Phillips in 1848 to produce a monograph on belemnites and offered his full assistance, but to no avail. In March 1864 Wiltshire, who had become the Society's secretary and editor in 1863, persuaded Phillips to agree to complete the work before 1865, an impossible deadline. The mode of publication of the monograph helped Phillips to finish it: the descriptions of almost seventy species and the accompanying thirty-six plates were issued in five parts between 1865 and 1870. Even more crucial was Wiltshire's editorial contribution. In a steady stream of almost fifty letters to Phillips from 1864 to 1868 he asked very detailed questions about the naming of species, checked and corrected references, looked after the plates which were lithographed by Lackerbauer in Paris, advised about the important cross sections of the guards of belemnites, queried the sizes of belemnites shown in drawings (life-size or not), and dealt endlessly with the proofs. In short, Wiltshire ended Phillips' procrastination and ensured that he finished and published on British belemnites.[11]

The monograph was collaborative in other ways. Even on the Yorkshire coast specimens, Phillips needed help: from autumn 1864 he himself surveyed in detail a thirty-mile stretch but for specimens from Robin Hood's Bay he paid Peter Cullen of Scarborough 13 shillings for two days of collecting from sites which Phillips was physically incapable of reaching. For information about specimens outside Yorkshire Phillips sent a circular to appropriate individuals in spring 1864 asking for details of collections of belemnites. He stressed that he wished to give full descriptions of each species by including examples '*of all ages*': he was aware that other species of fossil molluscs showed considerable differences between juvenile and mature forms, for example in the shapes of the whorls of ammonites. Phillips had a useful response to his appeal from twenty people, with Sanders, Charles Moore, Walter Stoddart, Etheridge, George Morton, and Samuel Woodward, whom Phillips regarded as the

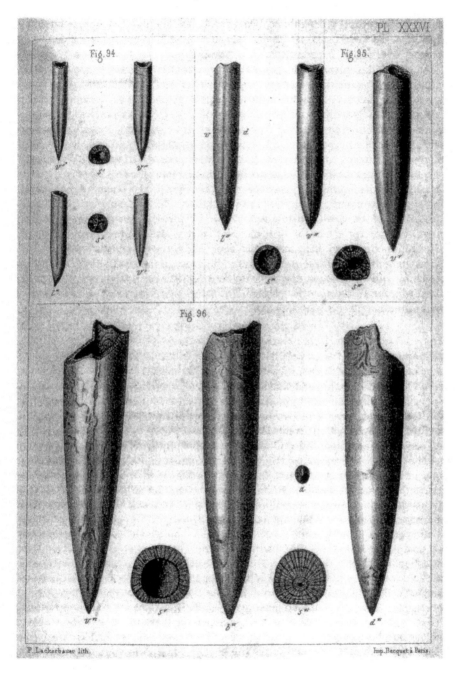

Fig. 10.1   Phillips' drawings of a particular species of belemnite showed
subtle differences between young specimens (fig. 94), growing ones
(fig. 95), and a grown individual (fig. 96), all of which look alike to
the untutored eye. V, l, and d signified ventral, lateral, and dorsal
aspects; a showed the apex and s cross sections. For b in fig. 96 read l.

leading British expert on belemnites, being especially helpful. In his research on the Oxford clay, in which most specimens were juveniles, Phillips was indebted to his University pupils who found mature ones for him.[12]

Though Phillips' monograph was ostensibly about belemnites in jurassic strata, he concentrated on the lias because it was the richest stratum for them. Identifying a species was not easy: in order to do so securely he had to see many specimens in all periods of growth. A single specimen of a species was useless because it was unique and unrepresentative. Young specimens of species, though sometimes abundant, were also suspect as an exclusive basis of classification because maturing and grown specimens, with different features, were ignored. In line with his attitude to identifying species, Phillips' plates often depicted them in three stages of growth, thus enhancing the figures as sources for reference (Fig. 10.1). Though the belemnite monograph was mainly descriptive and illustrative, Phillips exploited the notion of fossil zones, as developed via ammonites by Oppel, Quenstedt, and Thomas Wright. Phillips correlated the distribution of belemnites in the beds of the lias of the Yorkshire coast with their ammonite zones. He thus showed that belemnites as well as ammonites could be used to mark out zones in the lias, a result useful in correlating widely separated deposits via their fossils. The monograph pleased the Palaontographical Society who elected Phillips as a vice-president in 1867 and it remained for several decades a standard work.[13]

10.2 Glaciation

Phillips' position at Oxford enabled him to visit the Continent from which his various responsibilities had debarred him after 1830. In 1855 he went to the Auvergne, in 1863 to St Acheul near Amiens, in 1864 to Switzerland, and in 1868 to Italy to examine Vesuvius in action. On the three longest journeys he was accompanied by his long-standing friend, John Edward Lee, who in 1864 brought along his wife and two daughters. For seven weeks they enjoyed the antiquities and scenery of Switzerland, doing only a little rough walking and not undertaking experimental work on the physical properties of glaciers, as Phillips' close friend James Forbes had done. But Phillips saw enough of glaciers and glacial action to conclude that in the past there had been vast glaciers in mountainous or elevated areas of Europe including Britain. He accepted that glaciers could carry big boulders, polish and score the underlying rock, and pile up moraines. But, like many of his contemporaries, he rejected three contentious views all advocated in Britain by Ramsay by the early 1860s. The first was that the glaciers of upland Britain had gouged out valleys and rock-basins, now occupied by lakes. Phillips had inferred from his Swiss trip that even the big and thick glaciers there had only a limited gouging capacity. It seemed to him that any British glacier, smaller and less thick than those in Switzerland, would not have been able to generate sufficient pressure to excavate deep basins and valleys, such as Wastwater in the Lake District. At the end of his life he adduced a further argument based on crushing coefficients. He believed that a column of ice 1,000 feet thick would shear or disintegrate under its own weight and would be incapable of excavating rock.[14]

Phillips also rejected Ramsay's second postulate that there had been more than one recent ice age, and even one as long ago as the Permian period, a claim initially based on his examination of the conglomerates and pebble-beds of the Malvern area. For Phillips his own explanation given in his *Malvern memoir* was still valid: they had been produced on a shore by violent sea movements on a fault line associated with the elevation of the Malvern Hills. In general Phillips could not conceive of several ice ages punctuated by temperate interglacial periods. It was difficult enough for him to accept that there had been even one glacial period in the earth's recent past because this major climatic change was incompatible with his long-held conviction that overall the earth was cooling. It was impossible for him, and many others, to square this conviction with Ramsay's view that there had been several ice ages. Phillips never wavered in his view that there had been only one ice age. Indeed in the 1860s he defended the terminology of pre-glacial, glacial, and post-glacial that he had first enunciated in 1853.[15]

Thirdly, Phillips rejected Ramsay's notion that vast sheets of land-ice had covered much of Britain. It conflicted with his belief in the gradual cooling of the earth; and his study of erratic blocks, gravel, and boulder clay, all of which occur in east Yorkshire, led him from the 1840s until 1873 to explain lowland diluvium as mainly the result of diluvial currents and to invoke change of land level and floating icebergs, broken from highland glaciers, to explain the movement of large erratics. In 1868, however, the sustained research of Searles Wood junior on diluvium began to give Phillips pause. Wood claimed that in south-east Yorkshire there were three sorts of boulder clay. The lowest contained erratics from Scandinavia, the next layer which was purple contained those from Shap, and the uppermost deposit (known as Hessle) contained those from the Southern Uplands of Scotland. It was Wood's research in conjunction with Phillips' interest in the big erratics of Shap granite found in east Yorkshire which led him, shortly before his death, to modify his previous view. He realised that he had to explain not only the transport of the big Shap blocks but also 'the mixture of stones of different sorts brought in different directions', a fact about Yorkshire diluvium which led to inconclusive controversy for decades after his death.[16]

Having deemed violent surface currents of water or extensive land- ice to be inadequate explanations, Phillips concluded that the ice age had been associated with considerable change of level of land and sea which produced an extensive and deep north European sea. In an arctic climate there was extensive highland glaciation, with ice sheets pushing into the sea, gathering erratic blocks, and breaking off to form icebergs which were moved by sea currents. Then the sea slowly subsided, icebergs eventually dropped large erratics, and the sea currents gradually deposited gravel and small erratics on submerged land and emerging shores. While the arctic climate endured there was an extension of land-ice which sometimes moved and churned the seashore deposits to produce the disordered accumulations characteristic of boulder clay. This explanation avoided polarisation between ice and water. It was eclectic in that it drew on Lyell's icebergs, on the old diluvial theory in the new form of deep glacial water, and on Agassiz's notions of an ice age and extensive lowland land-ice, the last of which Phillips had previously rejected.[17]

10.3 Vesuvius and earthquakes

The excursions of 1855 and 1868 allowed Phillips to pursue field work at continental sites famous for their igneous phenomena and culminated in his book on Vesuvius (1869). Phillips and Lee planned to go to the Euganean Hills in north Italy in 1869 to see volcanic cones, but found themselves unable to undertake their fourth joint continental trip. Even so it is clear that Lee greatly facilitated Phillips' vulcanological research by acting as a geological research assistant, providing historical and classical allusions to Vesuvius, and paying some of his expenses.[18]

Until his visit to Vesuvius in 1868 Phillips pursued field research in igneous geology only in passing or on odd occasions when he visited important locations. In summer 1855 he went with his sister and Lee to the Auvergne, where he was specially interested in lava flows and dykes; and in December he studied the granite and serpentine of south west Cornwall. As a teacher and text writer he was au fait with igneous geology. Through his close friend, Daubeny, and papers given at the Geological Society and the British Association, Phillips was well informed about controversies and new approaches in vulcanology. One of the longest-running disputes concerned the cause of the shape of volcanoes. The supporters of the theory that they were craters of elevation believed that a volcano was formed when molten material in the earth's interior elevated previously horizontal strata around a centre of maximum elevation to give a dome structure, through the top of which lava burst and then flowed down the sides of the volcano. Advocates of this view, who included Humboldt, von Buch, and Daubeny, were impressed by the steep slopes of volcanoes, the high inclination of which they claimed was independent of any eruption. The alternative theory, advocated by George Scrope and Lyell, postulated that a volcano was a cone of eruption formed by successive ejections of lava and ash which eventually built up a big cone with highly inclined slopes. Scrope and Lyell believed what their opponents doubted, ie that red hot lava could solidify at a high inclination. Sometimes this approach was also known as the crater of denudation theory because Lyell in particular argued that the circular escarpments of craters of volcanoes were remnants of cones of eruption, the central parts of which had been destroyed by what he called engulfment.[19]

In the mid-1850s new approaches were taken to the nature of vulcanicity. Daubeny continued to be the most able advocate of the chemical theory of volcanic action, a view which prompted continuing research on the gases such as ammonia discharged through the flue of what he regarded as a subterranean laboratory. In 1862 he described to Phillips an eruption of Vesuvius that he had witnessed, stressing the preceding earthquake, the release of carbon dioxide, and local elevation of land by about four feet. Phillips was also familiar with Sorby's research on the structure and composition of igneous materials, which he examined microscopically using his standard technique. In 1855 Sorby reported to Phillips his discovery that fluid cavities existed in granite, which suggested an aqueo-igneous origin. Two years later Sorby read to the Geological Society his famous paper on the origin of minerals and rocks, including blocks ejected from Vesuvius. He concluded that they had been formed at dull red heat at great depth under great pressure, from liquid water, gases, and melted rock. In the ensuing discussion Horner, the aged chairman, averred that Sorby's paper tested his credulity more than anything else he had heard there but Phillips stoutly defended the method, accuracy, and results of his friend Sorby, the main originator of petrology. As an exploiter of instruments Phillips did not share the view of Daniel Sharpe and Ramsay that there was no point in looking at mountains with microscopes. Sorby's emphasis on the importance of water was also supported in the late 1850s by the work of Daubrée and Delesse on the way in which superheated water under great pressure could dissolve and alter some minerals at a temperature below their fusion point. Their research, which impressed Phillips greatly, enlarged the notion of metamorphism from Lyell's original meaning of alteration by heat to alteration irrespective of agency.[20]

Phillips was also familiar with the experimental approach to vulcanology adopted by Robert Mallet, an expert on earthquakes who thought that too much attention had been given to chemical analysis of volcanic products. In 1862 he was given a grant of £100 by the British Association to make physical measurements at Vesuvius, especially the temperatures at the mouths and in the craters of Vesuvius, the physical properties of the steam evolved, and the extreme and mean velocities of discharged material. He stressed that Vesuvius had many advantages as an experimental station devoted to vulcanology as a branch of terrestrial physics. In autumn he sent to Phillips his printed desiderata concerning physical data about Vesuvius. Phillips also knew Mallet's book of 1862 on the Neapolitan earthquake of 1857, in which Mallet reported his visual estimate, by comparison with furnaces, that five 500 feet down in the mouth of Vesuvius the temperature was 2000 °F and that from it dry superheated steam was evolved.[21]

Late in 1867 Phillips realised that there was a good chance of the eruptions of Vesuvius continuing into 1868 so he decided to go with Lee to Italy in the Easter vacation to see them for himself. Friends were generous with advice and introductions. Kenrick in York introduced Phillips by letter to Henry Wreford, the brother of Kenrick's brother-in-law. Wreford lived in Naples, wrote for the *Athenaeum*, and was the long-established Neapolitan and Roman correspondent of *The Times*. Wreford's help was crucial for the success of Phillips' trip. He not only engaged for Phillips the best local guide, Giovanni Cozzolini, who had long experience of escorting savants up Vesuvius; he also gave Phillips introductions to Guglielmo Guiscardi and Luigi Palmieri, the leading local experts on the geology and earthquakes of the area, and to the observatory 2,000 feet up Vesuvius where Palmieri recorded the physical phenomena which preceded, accompanied, and followed eruptions. Phillips received travel advice from several sources. The publisher Murray sent his latest Italian handbook gratis; Main provided a detailed itinerary from Paris to Turin; and McKenny Hughes warned about the danger of brigands. Helpful introductions came from surprising quarters. Pusey provided one to Palmieri and no less than John Henry Newman, who did not personally know Phillips, to the English College, Rome. Newman stressed that Phillips was a leading without being misleading geologist because he did not show irreverence to a higher province of thought. The most persistent geological advice came from Lyell who was keen to recruit the ever cautious Phillips to his cone-of-eruption cause as expounded in the tenth edition of his *Principles*. Lyell urged Phillips to visit the crucial north side of Vesuvius, wrongly neglected by the advocates of the crater-of-elevation theory; advised him not to be diverted by the current fireworks and lava; and recommended as useful contacts in Naples Guiscardi and Arcangelo Scacchi, a leading collector of Vesuvian minerals.[22]

It took Phillips and Lee eight days to travel from Oxford to Naples via Paris, Nice, Genoa, and Florence. They stayed sixteen days at Naples and were aided by Wreford, who reported Phillips' doings to the *Athenaeum*, and by Palmieri, Guiscardi, and Scacchi. Phillips' main purposes, the investigation of lava currents of different ages and the features of the north side of Vesuvius, owed more to Lyell than Mallet partly because fulfilling the latter's desiderata was physically beyond him. Phillips and Lee made three important excursions. The first was the ascent from Pompeii to the semi-circular ridge of Monte Somma, which for Lyell was a remnant of the ancient cone of Vesuvius. They saw old and new lava streams, hot cinders, red hot lava near the surface, and lava tunnels (Fig. 10.2). The second was the ascent from Resina to the

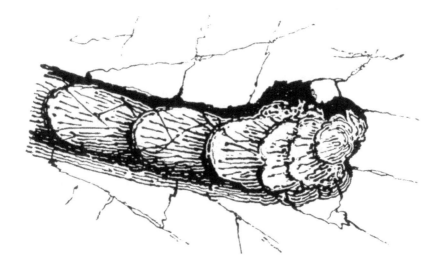

Diagram XVIII.   Canal in which Lava is in the last stage of
motion above Pompeii.   (*Original.*)

**Fig. 10.2   Phillips' drawing of 1868 of a Vesuvian lava flow. Having considered
for about forty years the volcanic phenomena of the past, Phillips at
last saw a volcano in action when he was sixty-seven.**

Atrio del Cavallo, immediately north of Vesuvius and south of Somma, from which
they saw dykes in the Somma cliffs, made of lava and consolidated ashes, and the
steep and loose slope of ashes and cinders which made up the cone of Vesuvius.
They examined a large lava stream which came down from the edge of the crater
of Vesuvius but failed to reach its source because they were hindered by showers of
stones from Vesuvius and then driven back by a thunderstorm which brought hail and
snow. Their third excursion was to Monte Nuovo in the so-called Phlegraean Fields.
All its features indicated to Phillips that it was formed by successive eruptions and
not 'upheaved in a solid form by the local effect of a general subterranean pressure'.
These excursions induced him to agree, albeit cautiously, with Lyell's theory. At
night time Phillips used a Cooke telescope, with a magnification of twenty five to
fifty, to observe the Vesuvian fireworks. On two evenings he saw fiery lava flowing
down the north side of the cone of Vesuvius. He also witnessed incandescent bombs
thrown up vertically from Vesuvius. Like some previous observers, he measured their
times of fall and calculated that they fell at maximum almost 2,000 feet, a result that
convinced him that explosive forces existed near the upper surface of the lava.[23]
    On his return to Oxford via Rome, Florence, Venice, Munich, Stuttgart, and
Heidelberg, Phillips gave a course of lectures about his trip to Vesuvius and then
produced a rapidly written version which was published in 1869 by the Clarendon
Press, probably through the influence of Price, a key figure in the delegacy which
oversaw it. An octavo volume selling at 10/6 and making about £200 profit for
Phillips, it was well illustrated, mainly by woodcuts by Phillips himself and De
Wilde, which were supplemented by a few engravings by Lowry which were

lithographed by Newbald and Stead of York. The illustrations often depicted dykes, lava currents and volcanic ashes, the frontispiece being a coloured map of the Vesuvian area showing four ages of lava flows. The book was written for the alert or even adept general reader and was more demanding than the recent works of James Lobley on Vesuvius and Richard Peacock on earthquakes and volcanoes. Phillips also wrote for his geological confrères, particularly when he discussed volcanic energy in its physical, chemical, and mineralogical aspects, with a characteristically cautious inductive approach. Firstly, he gave a historical account of Vesuvius in eruption and repose, which drew extensively on ancient Latin authors and was fuller than anything given by Lyell and Daubeny. Secondly, he stated the main facts and phenomena, including the key point that Vesuvius had increased in bulk as recently as late 1867 when a large new cone appeared, grew rapidly, and produced lava which filled the crateral area and overflowed from it. Lastly, he drew conclusions, founded on these observations and in harmony with the laws of nature.[24]

Phillips' interpretation was based on the theory of the cooling and contracting earth which he had cherished for thirty years. He also believed contra Hopkins that the interior of the earth was hot and molten sufficiently near the earth's crust, assumed to be no more than forty miles thick, to affect it geologically. Volcanoes, he thought, occurred when water from seas or lakes met fused material, via fissures formed by internal pressure, and then erupted into the air or sea. Thus Phillips' explanation of vulcanicity was consonant with the cone-of-eruption theory but not with the crater-of-elevation theory. He also rejected the chemical theory of Daubeny because it followed from his own theory that chemical reactions were the effects and not causes of eruptions. He ended the book by claiming that even 'the variable might of subterranean fire' was part of a system of cosmical change embracing the solar system, stars, and nebulae, the last of which were possibly forming new suns and planets. Phillips' volume was welcomed by three savants whom he respected as experts on volcanoes. In private and public Mallet thought it by far the most important book in English about Vesuvius: though it was not a treatise about vulcanology or seismology, it said enough to countervail popular trash. Scrope, who welcomed its debts to his writings, including the 1862 edition of his work on volcanoes, judged it scientific and authoritative, though he thought Phillips had treated too tenderly the theory of crater-of-elevation with its 'miserable notion of volcanic mountains having all risen in a night like mushrooms'. Lyell simply agreed with almost all Phillips' conclusions and was happy to have produced a convert.[25]

At Murray's request Phillips made one of his few contributions to a general periodical when he wrote for the *Quarterly review* an essay about earthquakes which, like *Vesuvius*, was published early in 1869. Phillips' main line of argument was that the experimental/observational-cum-mathematical approach to earthquakes of Mallet and Haughton was superior to that adopted by Lyell. Mallet, a prominent civil engineer who in the 1850s regularly reported his researches to the British Association, studied earthquakes in four ways. In the late 1840s he published his first attempt to assimilate the facts of earthquakes to the known mathematical laws of wave motion in solids and fluids. From 1850 he used clever instrumentation, such as self-registering seismometers and electric clocks, to determine by experiment the speed of propagation in various solid materials of waves produced by explosions. In 1858 he summarised data about 7,000 earthquakes and produced a seismic map of the world in his *Earthquake catalogue*. That year he visited Naples to study the effects of the Neapolitan earthquake of 1857. In 1862 in his book about it, he set out the principles of observational seismology, a word he coined. Privately and publicly Phillips regarded Mallet's book as a great work because it revealed a new way of studying the interior of the earth, using instruments and mathematics. Mallet dealt with the physical aspects of earthquakes which he considered to be transits through the earth's crust of waves of elastic compression. Phillips accepted Mallet's view that these earth waves were incapable of generating permanent elevation of land, and even mountains, which Lyell postulated had been produced by earthquakes. The essential mathematical formula used by Mallet were devised by Haughton, by 1867 a close friend of Phillips who regarded him as 'the *greatest geological* dynamicist'. The research of these two Irishmen led Phillips to conclude that earthquakes preceded, accompanied, and followed volcanic eruptions. His explanation of earthquakes was that internal pressure in the crust of the cooling earth produced fissures through which water from the surface penetrated to a great depth, giving the primary shock; the heating of the water generated steam which delivered the secondary shock and in a volcanic district could produce an eruption.[26]

10.4 Topographical geology

In Phillips' last two major geological works he reverted to the regional geology of England, a genre in which he had established his national reputation with his book on the Yorkshire coast. In 1871 the Clarendon Press of the University published his *Geology of Oxford and the valley of the Thames* and in 1875 Murray brought out the posthumous third edition of the Yorkshire coast book, edited by Etheridge. Though these tomes examined topographically distant areas, they overlapped geologically. The Yorkshire coast volume covered strata from the new red sandstone to the chalk, ie the full range of mesozoic strata. Though the Oxford book embraced what Phillips regarded as the whole period of geological time, from the pre-Cambrian period to the pleistocene, three-fifths of its pages were devoted to mesozoic strata. The two books can therefore be regarded as companion volumes dealing with mesozoic geology. They were, however, not parallel volumes because they had different structures. The Oxford book examined the area and its fossils, both organised chronologically, from the oldest to the newest deposits; whereas the Yorkshire coast book retained the topographical approach of the first edition by discussing the inland areas and then taking the reader along the coast from Spurn Point to Redcar. Phillips was well aware of recent specialist research on both areas, including that of the Geological Survey on the Oxford region. He knew from his own experience that the maps and memoirs of the Survey, from 1855 directed by his friends Murchison and Ramsay, were superior in detail to anything an individual could achieve. But he thought that the solo geologist was not entirely superseded if he assumed a synoptic role and gave a connected and complete view of the structure of an area and its fossils. Well aware of the limitations of the individual researcher, Phillips solicited and received help from appropriate specialists for his Oxford book and even more for that on the Yorkshire coast.

Phillips' *Geology of Oxford* took a decade and a half to prepare. As early as 1854 he used his University class to search for and find freshwater shells in the iron sand at the top of Shotover Hill. Next year in his essay on the neighbourhood of Oxford and its geology he summarised his views and adumbrated desiderata, especially about a complete census of Stonesfield fossils. While president of the Geological Society, he published two papers on traverse sections of strata at key sites near Oxford. At the 1860 meeting of the British Association held in Oxford, he dilated on his new geological map of the Oxford area; and next year described post-glacial gravels of the Thames valley. It was, however, the reinterpretation by Huxley in 1867 of the megalosaurus in the University Museum and the first find of ceteosaurus next year which prompted Phillips in 1869 to propose to the Clarendon Press a book about *Oxford fossils*. It aimed to make more widely known the geological collection in the University Museum, by giving drawings and descriptions of remarkable local fossils curated in it. By late 1869 the title was changed to the *Geology of Oxford and its neighbourhood*. In spring 1870, when Phillips began to acquire more bones of ceteosaurus, he decided to give great emphasis to descriptions and drawings of saurian remains.[27]

The book's title revealed its synoptic ambitions. Whereas most geologists were content to study limited periods and areas, Phillips' octavo volume, costing 1 guinea, covered all the geological formations from the oldest fossiliferous rocks up to alluvium. His notion of what constituted the Thames valley was exceptionally wide: it extended from the Malvern Hills to London and embraced the feeders of the Thames as far south as the river Kennet. The account of each geological period, which was considered mineralogically and palaeontologically, was accompanied by lists of fossils, seventeen plates, mainly of fossils, drawn by Phillips and engraved by Lowry, and no fewer than 207 woodcuts, again mainly of fossils, nearly all drawn by Phillips and executed by De Wilde.

The book portrayed the University of Oxford favourably. Like other works by Phillips' colleagues it confirmed that the Clarendon Press was taking science seriously. It was dedicated to Magdalen College, of which Phillips was an honorary fellow from 1868, for its firm support of science in Oxford. It was indebted to the knowledge and anatomical collections of Rolleston, Phillips' professorial friend. Apropos some of the books consulted and listed in the bibliography, Phillips gladly acknowledged that the Radcliffe Library, housed in the University Museum, was magnificent. And the many illustrations of specimens preserved in that Museum showed the continuing vitality of Oxford's long tradition of supporting public collections devoted to natural history.[28]

The central feature of the book was its extensive treatment of mesozoic geology, especially the Bath, Oxford, and Portland oolitic periods. It became a widely praised reference work on them just as Murchison's *Siluria* functioned as an authoritative account of palaeozoic rocks. Phillips' book was ostensibly a geological 'county history' so it gave a full account of the remarkable ensemble of fossils found at Stonesfield. For many readers, including expert palaeontologists such as Huxley, Egerton, and Etheridge, the highlights of a book they regarded as great, grand, and delightful were the descriptions and illustrations of oolitic vertebrates, especially specimens of megalosaurus and ceteosaurus in the University Museum (see section 9.6). The piecing together of the skeleton of the latter from fragments was a tour de force of Cuvierian reconstruction. Though Phillips was a septuagenarian who moved among his fellow geologists 'with a kind of antique glory' and was revered by them because he embodied the traditions of the past, the work on ceteosaurus was for him a new departure. It was the leading feature of a book which devoted only a few pages to economic geology, just as in his lectures at Oxford he gave only one course on it. Though he was not uninformed about Oxford's water supply, he gave scant attention to it and did not cite or exploit Prestwich's pioneering work of 1851 on the water-bearing strata around London.[29]

The revised Yorkshire coast book had a long gestation. From the mid- 1850s until 1873 Phillips prepared for it by making several trips to that coast and its hinterland, recording on blank Sopwith forms lots of columnar sections, some of which he coloured. He paid particular attention to key sites such as Speeton cliffs which he visited in 1855, 1866, 1870, 1872, and 1873 in order to try to subdivide the Speeton clay into three zones, characterised by different fossils. Early in 1872, with his *Geology of Oxford* out of the way, he decided that he would update the 1835 edition of his Yorkshire coast volume, which had sold out, not by writing a new book but by adding a supplement to a reprint of it. He soon realised that it would be better to produce a re-written, enlarged, and final edition of the work which had made his national reputation in 1829. This plan for his 'last effort in *topographical geology*' enabled him to take account of new facts revealed by the erosion of the coast, progress in palaeontology, new approaches to superficial deposits, the recent economic importance of some strata, especially the ironstones in the Middlesbrough area, and the anticipated importance of submerged coalfields. Above all he wished to bring before the geological public his final views on 'the great mesozoic section here so plainly cut', the study of which had been for him a life-long labour of love.[30]

Phillips knew that he had no spectacular discovery, analogous to ceteosaurus oxoniensis, that would adorn his book on the Yorkshire coast. He also realised that he had to relate the results of his own research to recent work on the correlation of Yorkshire strata with those elsewhere, some of which had been published by men half his age. One important correlation concerned the Speeton clay which was particularly difficult to work on because of landslips, faults, contortions, and mining operations. In 1829 Phillips had tentatively ascribed the upper Speeton clay to the cretaceous gault of south England and the lower to the oolitic Kimmeridge clay. In the late 1860s James Judd extended the work of John Leckenby on the Speeton fossils and reached an anti-Phillipsian conclusion. Though paying tribute to the success of Phillips' bold generalisations in his remarkable 1829 book, Judd divided the Speeton clay into seven divisions, namely upper, middle, and lower neocomian, Portlandian, and upper, middle, and lower Kimmeridge clay. Judd retained the essence of one of Phillips' correlations but rejected that concerning the gault, then regarded as upper cretaceous. Judd referred the upper divisions of the Speeton clay to the lower cretaceous strata found at Neuchatel, which he called neocomian, a conclusion which Phillips gladly accepted.[31]

Another correlation, involving oolitic strata, was contentious. In 1829 Phillips had referred the oolitic strata at Cave and Gristhorpe, south of Scarborough, to the great oolite. In 1860 this correlation was vigorously challenged by Wright who referred them to the inferior oolite on the basis of zones of life labelled by a particular ammonite. Wright was supported in 1864 by Leckenby, who used newly discovered fossil plants to characterise strata; but Wright was attacked in 1862 by Edward Hull, a great admirer of Phillips' work on the lateral variation of the composition of carboniferous limestone. Hull supported Phillips' correlation with the great oolite and argued that identity of fossil remains did not necessarily indicate contemporaneity of strata. In his book Phillips leaned towards Wright's view. He did so as a result of his own field work from the mid-1850s which steadily revealed the pervasiveness of lateral variation in oolitic strata. For instance, in much of England the inferior oolite was mainly limestone of marine origin but in Yorkshire it was mainly coal-bearing sandstone, shale, and ironstone, with interposed shelly marine beds. Phillips explained this local peculiarity by invoking alternating marine and estuarine deposits. On the basis of their stratigraphical position and fossils he postulated that the three estuarine deposits were equivalent to forest marble, Fuller's earth, and the inferior oolite in south-west England. He thus accepted Wright's conclusion based on fossil zoning and resolved the paradoxes in it by invoking yet again his own notion of lateral variation in a stratum.[32]

Phillips was therefore well aware that it would be difficult to make his new book on the Yorkshire coast up to date and authoritative, so he was happy to be helped from many distinguished quarters, especially from 1872. From Scarborough Lycett stressed that the palaeontology needed revision, of the kind exemplified in his 1872 monograph on British fossil trigoniae (mussel-like molluscs). Leckenby accompanied Phillips on a coastal survey by boat and with Dawkins and Gwyn Jeffreys, an expert conchologist, revised the list of cainozoic fossils. Peter Cullen of Scarborough continued as a useful local collector of fossils, especially those of plants. The ever-vigorous Lee, who had worked on the Yorkshire coast with Phillips in 1867 and 1870, undertook further shared field work in 1872. Williamson, a leading palaeobotanist, and Simpson, of the Whitby Museum, provided drawings of fossil plants. William Gray, an old York friend, supplied a corrected list of heights of cliffs and hills. From 1867 Judd wanted Phillips to be au fait with his research on Speeton clay and stressed to Phillips the importance of the alternation of marine and estuarine conditions in the jurassic period. In return Phillips encouraged Judd's important work on Lincolnshire geology, which clarified its relation with the oolitic strata of south-west England and the neocomian strata of the Yorkshire coast. Jeffreys provided a list of post-tertiary fossils at Bridlington while Rupert Jones and William Parker, who were authorities on foraminifera, did likewise for oolitic foraminifera of England. Two members of the Geological Survey were generous helpers. William Whitaker supplied a bibliography of about 600 items covering the geology of Yorkshire. Above all, Etheridge, who wanted Phillips to live to revise his mountain limestone monograph, revised the nomenclature of Phillips' list of figured fossils. After Phillips' death, he edited the nearly complete text and ensured that it was quickly published by Murray.[33]

The posthumous work, dedicated to the memory of Smith, was defective in two important respects; the plates were mainly those of 1829 and the listed fossils were not always figured. It soon faced formidable competition from specialists on jurassic strata. In 1876 Ralph Tate and John Blake brought out their work on the Yorkshire lias which they divided into ammonite zones as described and used by Oppel. Though they gave credit to Phillips for dividing the lias decades ago into upper, middle, and lower divisions, which gave the main outlines of their classification, they found little new on it in his 1875 book. Crucially they differed from him on the line of demarcation between the lower and middle lias. His was in part lithologically based whereas they followed Oppel and Wright by ignoring lithology and putting the dividing line lower down because of a faunal break. In the late 1870s the Geological Survey began to publish maps of parts of east Yorkshire at six inches to the mile and in the next decade mapped it rapidly at one inch to the mile, both scales being larger than that of five miles to the inch employed by Etheridge in 1875. Also in the 1880s Charles Fox-Strangways produced several memoirs about parts of east Yorkshire. In 1892 he brought out as a Survey stratigraphical memoir his two big volumes on the jurassic rocks of Yorkshire. Like Geikie, the director general of the Survey, Fox-

Strangways was deferential to Phillips' work on the Yorkshire coast and retained the main features of his divisions of strata.[34]

10.5 Consultancy

As a geological consultant Phillips acted as an expert witness before government commissions and more frequently as a paid expert employed by individuals, institutions, and firms. In the first role he gave evidence at two days' notice in 1865 to the Commission on preventing the pollution of rivers, especially the Thames, strongly advocating the use of natural and artificial reservoirs to store spring water. In February 1869 he appeared before committee D of the Commission on coal in the United Kingdom. This committee was charged to enquire into the probability of finding coal under the new red sandstone strata. In his evidence Phillips considered several areas, including that between the exposed Leicestershire and Warwickshire coalfields, arguing that it contained accessible but concealed coal in parallel hollows and troughs, formed by folding of the carboniferous strata and running south-east to north- west, the direction of the large axes of folding.[35]

As a paid expert Phillips dealt with water supply, mineral surveying, tunnelling under a river, iron ore, and coal prospecting. As soon as he was settled in Oxford, he advised the Buckinghamshire County Asylum that it was feasible to sink a well for water from a local sandhill. In 1857 he advised the Whitby Waterworks Company which in 1862 asked him to support its claim that contamination of one of its wells by a new and nearby cemetery was impossible. One of his former Oxford pupils, Charles Tomlinson, vicar of Denchworth, near Wantage, wanted Phillips' advice in 1869 about installing pumped water in the village. As an expert on the Yorkshire coast Phillips was asked in 1867 by the owner of the defunct Ravenscar alum works to report on all the minerals on his estate; and in 1872 by a London engineer about the depth and thickness of the chalk under the river Humber in connection with a projected tunnel beneath it three miles west of Hull.[36]

Phillips' knowledge of liassic iron ores was exploited a few times. In 1855 he was asked to report on iron ore at an estate near Woodstock; and in 1857 he advised his colleague Donkin about buying property, including an ironstone mine, in Eskdale in north-east Yorkshire. Two years later Phillips pocketed fifty pounds for reporting to the Duke of Sutherland that his estate at Stittenham, near Castle Howard, contained ironstone four feet thick and for instructing about the shallow bores he thought appropriate. In the mid-1860s the Leeds, north Yorkshire and Durham Railway Company asked him about the character and extent of ironstone in the Middlesbrough area. Such consultancies reminded Phillips that in 1829 he had drawn attention to the ironstone in north-east Yorkshire but commercial exploitation followed only in the 1850s and apparently independently of his announcement. As Phillips pointed out as president of the Geological Society, in that case practice was silent when geology foresaw.[37]

Phillips' views about the relation between geology and coal-prospecting remained unchanged from the 1830s. Geology discouraged those trials for coal which were encouraged by mining lore but denounced as pointless by geologists; and it encouraged but did not guarantee efforts to find coal when the expectations of practical men were discouraging. It could indeed initially buttress optimism but eventually provide dismaying conclusions for entrepreneurs who had turned to it and forsaken lore as a guide. This was the fate of Thomas Frewen who established contact with Phillips in 1859 and then from 1864 to 1869 with Phillips as his consultant employed John Alleyne Bosworth, a mining engineer, to bore at Sapcote, Leicestershire, through the new red sandstone strata in the hope of finding coal underneath them. In early 1870 Phillips advised that boring be ended, leaving the disappointed Frewen to die in October.[38]

The saga began in 1859 when Frewen, stimulated by Phillips' 1837 report on coal in the Lancaster area, approached Phillips as a 'scientific geologist' to advise him about boring for coal at Sapcote, five miles east of the exposed Warwickshire coalfield. Frewen was not sanguine because he knew that the coal strata at Nuneaton rose in an easterly direction but, having visited the area, Phillips reported that they might fall at Sapcote. Frewen accepted this possibility, guessed that coal might be found at a depth of 1,500 feet, but decided that for the time being boring to that depth was financially beyond him.[39]

In 1864 he changed his mind and employed Bosworth to bore at Sapcote from February of that year until December 1865 when the work was temporarily suspended, having reached a depth of about 520 feet at a cost of almost £2000 without finding workable coal. Phillips acted as a consultant, being paid firstly 10 guineas and then £15 when he visited Sapcote and £10 when Bosworth came to Oxford. The boring was troublesome: wet materials seeped into the bore and the chain connected to the rod broke underground, doing great mechanical damage. Boring was not fast: in 1865 it took two hours to break up just one inch, using a rod four inches in diameter and weighing one and a quarter tons which was dropped three inches forty times a minute. No wonder that Bosworth needed Phillips' moral support to repel the jeering and badgering he endured from the locals. Frewen, too, used Phillips' reputation as a first-rate geologist to quell local ridicule, though Phillips did not explicitly reinforce Frewen's hope, from analogies with the features of Warwickshire collieries, that at Sapcote coal occurred at about 1,000 feet depth and not at 3,000 feet as Jukes had averred. By late 1865 Phillips' chemical analyses of the shales found under the new red sandstone indicated that they were carboniferous and that therefore the coal measures had been reached. He thought these shales were above the coal measures and hoped they were not impossibly thick. His optimism was increased when a thin seam of coal was found.[40]

In spring 1866 Frewen resumed boring with Phillips still his scientific pilot. But in summer 1866 Bosworth expressed to Phillips his fear that the underlying coal measures and associated shale were vertical and therefore the former would not be reached by boring through the latter. He tried to reject this awful possibility on the grounds that at 500 feet depth the shale was horizontal. Phillips immediately rejected Bosworth's suspicion of a subterranean trough of coal shale and measures with vertical sides: so far there was no evidence of disturbance, flexing, or faulting so patience and further boring were required. In February 1867 Frewen began to lose his nerve: he suspected that either there was no coal or there was indeed a subterranean trough which Bosworth had reached at 840 feet. Bosworth and Phillips rejected both possibilities but by summer 1867 Bosworth was so discouraged that Frewen urged Phillips, his huntsman, to encourage Bosworth, his leading hound, in the coal chase and to still the jibes and taunts of the Leicestershire disbelievers. For his part, in June 1867 Phillips assured Frewen that coal would be found at 1,050 feet, his evidence being his comparison between the cores from the bore and the Nuneaton shales which were adjacent to the coal there. Two months later the bore reached for the first time inclined shale, so in October Phillips reviewed the situation for Frewen. Using analogical arguments from what was known about the Leicestershire and Warwickshire coalfields, Phillips advised that the existing bore be continued below 1,000 feet, hoping that workable coal would be reached at 1,200 feet depth. Though he was aware that the shale recently reached was inclined at 45° and that only new bores could show its thickness, he did not recommend that they be made to find the 'lay' of the strata because they would have bankrupted Frewen.[41]

Phillips' verdict was supported by Etheridge, whom Frewen had consulted, so boring continued, sometimes at just an inch per day. Bosworth had an unhappy 1868: he found no coal, at about 1,200 feet depth the shale was vertical, and in autumn he broke his ribs when thrown out of his gig. Frewen was equally dismayed by the vertical shale and a second opinion was no help. During Phillips' absence in Italy in spring 1868, Frewen had consulted Etheridge who argued that the fossil content of the cores at about 850 feet was like that of the shales at one of the Nuneaton collieries. This encouraging discovery was soon qualified when in June 1868 Etheridge confessed that the fossils did not necessarily indicate the presence of productive coal measures. As Frewen ruefully noted, 'if I get no coal I shall at least get some fun out of two such great guns'.[42]

The bore was continued and reached a depth in late 1869 of almost 1,400 feet, without discovering productive coal. Having reviewed with Bosworth and Etheridge all the cores from the bore, Phillips advised Frewen in January 1870 to abandon exploration at Sapcote. He concluded that with increased depth the previously horizontal shales became more and more inclined until at about 1,200 feet they were vertical, then their inclination reduced to 60°, followed by an increase until at about 1,400 feet they were again vertical. As these shales were so steeply inclined or vertical, Phillips concluded that to bore another 200 feet, ie 100 feet deeper than the original target, was pointless because the bore would not be cutting across the shales. He also pointed out that even if continued boring revealed coal at 1,600 feet, no colliery could be established in such highly inclined strata. Thus Bosworth's prognostication of upturned strata, which Phillips had rejected earlier, was finally admitted by him; and he was eventually forced by the evidence to reject his own prediction that coal would be found at about 1,000 feet depth. In 1870 Phillips learned that in concealed coalfields the bending of strata to produce steep-sided deep troughs vitiated analogical arguments from neighbouring exposed coalfields, a conclusion he had not stressed to the coal commission the previous year. This salutary experience did not deter him in 1874 from negotiating about advising about boring through new red sandstone strata to productive coal measures at Elstob, near Sedgefield, county Durham.[43]

10.6 Archaeology and astronomy

Phillips' interest in archaeology was muted compared with that shown in his last years at York. He issued no publications on it and was not conspicuous in promoting it institutionally. As keeper of the Ashmolean Museum at Oxford he never tried to make it a centre for teaching and research in archaeology under himself or anyone else. As keeper he was appointed in 1855 the Oxfordshire secretary of the Society of Antiquaries, of which he was not a fellow. Like many other county secretaries of the Society, he treated the post as a sinecure. Rather surprisingly he ignored the request of John Edward Lee, an old and close friend, to write an introduction to Lee's translation of the important account of the lake dwellings of Switzerland by Ferdinand Keller, whom Lee regarded as a friend of Phillips.[44]

Most of Phillips' archaeological interests were shown privately. He was a confrère of Henry Wellesley, an Oxford antiquary. He advised occasionally on expeditions to excavate local barrows. He gave valued insights and opinions on craniology to his professorial colleague, Rolleston, who saw archaeology as mainly the study of skulls and skeletons found in British barrows and Anglo-Saxon cemeteries. In the early 1870s Phillips was still interested in the relation between the circumference, volume, and brain weight of crania: he measured them with tape and calipers and estimated their volume by weighing sand put in them. Though he thought crania were useful indicators of race and culture, he viewed pottery as better. Only to Elizabeth Johnes, whom he called the true Muse of Wales because she was a Welsh-speaking expert on Welsh language and myth, did he reveal that as well as crania and shards he was knowledgeable about the history of language, especially Welsh, and Welsh placenames, an interest of his Oxford friend, Charles Williams, who introduced Phillips to her. Shortly before his death he reported to her his finds in Oxford of a manuscript pertinent to the Welsh variant of the legend of Merlin.[45]

Phillips gave a modicum of support to the Archaeological Institute and its leading spirit, Albert Way. In 1859 he was induced by Phillips to publish his research on Roman metallurgy, in parallel with Phillips' slight reworking of an 1847 paper on ancient metallurgy and mining. He continued to consult Phillips, whom he regarded as a master. In 1866 he was disappointed that Phillips was unable to be president of the Institute's new section of primeval antiquities, supported by John Evans as secretary. An adequate substitute for Phillips was found in the form of Sir John Lubbock who had just brought out his popular *Prehistoric times*. In 1869 Way was delighted to report to Phillips that a recent excavation at Sidmouth had validated Phillips' private suggestion that lathes were used to fashion early British pottery. In the early 1870s Phillips persuaded his friend, James Clutterbuck, an expert on the gravel and drainage of the Thames Valley, to make contact with Way.[46]

In his opening years at Oxford Phillips had no facilities for making astronomical observations; but he continued to mull over the pros and cons of drawings and photographs of celestial objects. His interest in this question was reinforced in 1856 when he saw and admired the drawing of the lunar crater Copernicus which Angelo Secchi had sent to the Royal Society. When Phillips compared it with his own drawing of another crater, Gassendi, he deemed himself Secchi's equal as an astronomical draughtsman. Even so, the large scale of ten miles to the inch employed by Secchi enabled Phillips to postulate in 1856 that Copernicus was not a truncated cone of a single volcanic eruption but had risen in stages when lava currents poured out. By the 1860s Phillips still believed that drawings were essential for understanding the creation and erosion of lunar mountains and the volcanic surface of the moon. He acknowledged that photographs revealed lunar forms but he thought that drawings made at different times of the day, using a telescope with a micrometer attached to the eyepiece, gave measured details of features and their shadows which could not be seen even in De la Rue's best photographs.[47]

It was this conviction which led Phillips in 1860 to order from Cooke a new achromatic refracting telescope of six inches aperture and magnification of 200 to 400. By early 1862 it was mounted on a portable solid stand and with its clock-drive was housed under a wooden dome and in a wooden house which were also transportable. This small movable observatory involved considerable financial outlay. The telescope cost 320 guineas, additional apparatus 30 guineas, and the housing over £100. Phillips then tried early in 1862 to persuade the Royal Society of London to buy the telescope for 320 guineas and the housing for £50 and to let it out to observers. He offered his services gratis for two years. This ruse was informally rejected by Sabine, De la Rue, William Sharpey, and George Stokes, acting on behalf of the government grant committee of the Society. They urged Phillips to have reliable material in print on selenography before applying formally. He followed their advice, reading a paper to the Society on 20 March 1862 about how to attain systematic representation of the physical aspects of the moon. Impressed by Phillips' scheme for 'giving a new and well considered impulse to selenography' the Society soon awarded him £100, which helped to defray the cost of his private observatory. By July 1862 it was installed about ten feet from the west wall of his keeper's residence in Oxford. In September he began to use it for astronomical research, most of which was published by the Royal Society, whose president from 1861 to 1871 was his old friend Sabine. In the 1860s a bizarre contrast developed in the University of Oxford. Research in observational astronomy was pursued by a geologist-astronomer, who was not a fellow of the Astronomical Society, using his own private observatory funded mainly by himself. Yet nothing equivalent was done by Donkin, the astronomy professor, who lacked the appropriate equipment even in the new University observatory, built in 1860 for £168 and situated east of the University Museum.[48]

Phillips' research at Oxford on selenography developed that of his last years at York in that he continued to focus on ring mountains, especially Gassendi and Theophilus, and produced more sensitive eye-drawings (Fig. 10.3). Driven by the hope that lunar observations would advance the understanding of volcanic action, his chief aim was to describe, illustrate, and explain the features of lunar craters. From their shadows, recorded five times a day, he inferred the existence of very steep crater walls unknown on the earth. He then explained their steepness as the result of low gravity on the moon and the absence of water and air which would have eroded them. Though Phillips speculated about elevation of the crust of the moon to produce mountains and about cones of volcanic eruption, and tried to trace analogies between apparent volcanic formations on the moon and earth, it was his competence as an observer which impressed his contemporaries: in 1864 a ring mountain was named after him.[49]

Phillips also continued to promote selenography in the British Association. In 1860 Birt's account of the broken annuli of some lunar craters prompted the Association to establish but not to give money to a committee composed of Rosse, Robinson, Birt, and Phillips, its secretary, to observe the moon's surface and compare it with the earth's. Next year Phillips revealed to the Association his thoughts about the ways in which the moon's surface was the result of previous physical phenomena such as faults, deltas, and dried sea beds. In 1863 he reported on the physical bearings of larger lunar mountains and invoked the idea of the displacement of the solidified part of the moon's crust. Next year Birt drew Phillips' attention to the possibility of very recent explosions on the moon, which were not shown on Beer and Mädler's map, and then induced the Association to create and finance a new lunar mapping committee which would implement Birt's plan of preparing standard forms for registering craters and for constructing a lunar map on a scale four times that of Beer and Mädler's. Birt was the driving force of this committee, whose members were specialist astronomers except Phillips. By 1869 four areas of the moon had

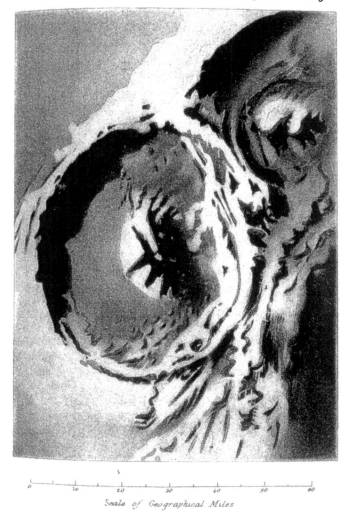

THEOPHILUS & part of CYRILLUS.

26 March 1863. 8.p.m. Power 200÷400.

Moon's Decl. N. 22. 7.   Sun's Decl. N. 1.44.   Age 6.23 days.

Scale of Geographical Miles

NOR?

Fig. 10.3   Phillips' drawing of 1863 of the lunar ring mountains, also known as craters, Theophilus and part of Cyrillus, for Phillips, 1868a. He recorded that the former had a maximum diameter of 55 miles, its west crest was 15,000 feet high, and the ten or so central ridged mountains were 5,000 feet above the central plain. He noted that Theophilus intruded with a continuous outline into the older Cyrillus, like a blister pushing another aside and then bursting, which gave double folding on the line of the junction.

been surveyed, mapped, and published. Phillips helped to organise the thirty-five observers who contributed to the project, which cost the Association £355 until 1870 when it ended; and in 1865 he successfully suggested that the map be increased in diameter from seventy-five to 100 inches.[50]

When it became clear in late 1868 that the Association's lunar committee was unlikely to be made permanent, Birt consulted Phillips who proposed the formation of a Selenographical Society. Birt responded enthusiastically, discussing the project at length with Phillips whom he saw as a key figure in its formation. Birt and Phillips agreed about the importance of selenography, which they defined as: the study of the physical processes, past and present, continually or occasionally, on the moon; the application of the laws of their action to explain observed features on its surface; and the drawing of conclusions about its past history and present state. Phillips' suggestion of a Selenographical Society was realised only posthumously: it was established in 1878 by Birt who was its first president.[51]

Phillips' observations of Mars were prompted by its highly favourable position in autumn 1862 when it was in opposition, ie exactly opposite to the sun when seen from the earth. For two and a half months he used his six-inch Cooke telescope to sketch the planet in order to settle the disputed question about whether Mars possessed permanent features, shown by the consistency of observed surface markings, or whether it was covered by a shell of drifting clouds. When he arranged fourteen of his sketches taken in order of the meridian lines of longitude on the face of Mars, he concluded that there were permanent bright and dusky tracts on Mars which were covered by variable clouds which modified the permanent features of land, snow, and water. He also used this sequence of sketches to construct the first-ever globe of Mars on which its main features were shown. In February 1863 he favoured the Royal Society with a display of this globe and another constructed from the drawings of Norman Lockyer, whose own observations had led him to conclude that Mars' main features were permanent. When Phillips considered how this globe looked at different periods in its revolution, he was able to explain how the drawings of Mars by eminent observers, who had all ignored the planet's lines of longitude, differed so much.[52]

In November and December 1864 Phillips took advantage of another favourable opposition of Mars to produce a series of observations from which he confirmed that the outlines of land and sea did not change and that the planet's surface was like that of the earth. From the extent of snow on Mars he surmised that its climate was like that of the earth's continents. In 1865 Phillips published two maps of Mars, which were the third of their kind. One was based on his own observations of 1862 and 1864. The other reflected his belief that the true boundaries of land and sea on Mars were traceable if appropriate observations were made by different people at different times and were then compared. This second chart of Mars, drawn by Phillips and engraved by Lowry, drew on the 1862 observations of Lockyer, Phillips, and Main, those of 1864 by Phillips and William Dawes, and the engravings of De la Rue and Secchi (Fig. 10.4). No wonder that Dawes, who in 1865 established that the red tinge of Mars was not like a sunset and not atmospheric in origin, regarded Phillips as the commander in chief of a co-operative attempt to map Mars. In 1869 Richard Proctor produced a chart of Mars on which the south polar sea and the central island were named after Phillips. In this way his pioneering research on Mars was eponymously and enduringly commemorated, other features of the planet being named after such well-known living astronomers as Dawes, Lockyer, Rosse, Huggins, De la Rue, and

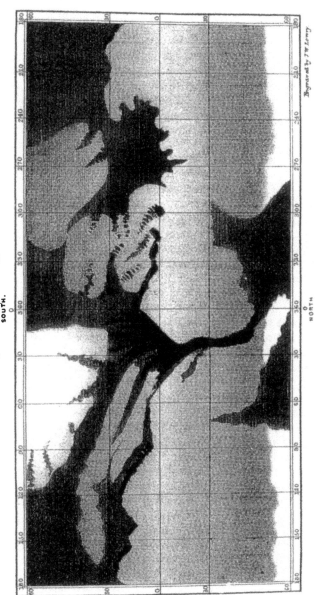

Chart of the Planet Mars.
By Professor Phillips.

SOUTH.

Engraved by J.W. Lowry.

This Chart is founded on the observations made in 1862, separately, by Mr. J.N. Lockyer, and Prof. Phillips; repeated in 1864 by Prof. Phillips. Drawings made in 1862 by Mr. Main, and in 1864 by Mr. Dawes, have also been consulted. For the Northern circumpolar regions, and extent of southern snows, the engravings of Mr. De la Rue & Padre Secchi have been employed. The earlier designs of Herschel & Mädler have also been referred to.

Quarterly Journal of Science, No. 7.

Fig. 10.4 **Phillips' chart of Mars for Phillips, 1865c. Based on his skill in mapping, including map projection, this chart illustrated and confirmed the consistency of surface markings of the planet.**

Secchi. It impressed no less than François Terby, a Martian expert in Louvain, and it even reached the publisher Longman who in 1867 wanted Phillips to vet an intended book on the planet by Proctor, provided Phillips himself was not writing one.[53]

For many years Phillips had been interested in the physical features of the sun but could not observe them regularly and draw them accurately until he acquired in 1862 his equatorial telescope made by Cooke. In the mid- 1860s he focussed on sun spots, perhaps because they were often associated with violent eruptive actions which reminded him of terrestrial volcanoes. In any case he was intrigued by the way in which in 1859 a flare on the sun had caused a huge magnetic storm on the earth. Though he was well aware of the advantages of the Kew Observatory's photoheliograph, which was used from 1862 to study the motion and size of sunspots, he remained adamant that eye drawings were better than photographs for the form, arrangement, and intestine motions of the spots; and for the dotted, areolar, granular, and crested features of the whole face of the sun. Well aware of the personal equation in his observations, in his case caused by head movements, he limited them by gripping the moving telescope so that his drawings were in his view trustworthy representations. His two chief results corroborated those of the experts. He agreed with the recent notion of Dawes, Herschel, and Secchi that the centre of a spot was an opening in the sun's bright photosphere through which the dark, hotter, and gaseous sun could be seen. He confirmed a phenomenon which had been represented photographically by De la Rue and Balfour Stewart at Kew Observatory, namely, that the umbra (centre) of the spot was below the general surface of the penumbra. Phillips' research on the distribution of spots added nothing to that of Richard Carrington who, after eight years' systematic observation, had produced in 1863 empirical laws about the positions and the movement over time of zones of spots. Though Phillips had a spectroscope from the mid-1860s, he played no part in the solar physics which was so energetically promoted in Britain in that decade by Lockyer, Stewart, De la Rue, and Huggins.[54]

Phillips did not neglect to observe such interesting phenomena as favourable eclipses of the sun and meteors. In 1858 he joined other Oxonians and William Lassell at Steeple Aston, near Oxford, to observe an annular eclipse, the central line of shadow of which passed near the village. In 1867 he used his old Cooke telescope, of 2.4 inches aperture and placed east of his keeper's house, to observe a partial eclipse, having had his chronometer checked by Main. In both cases Phillips' aim was not to observe physical features of the sun but its heating effect, so he recorded temperature changes during the eclipses; his results, expressed on graphs, were sadly inconclusive. In 1870 he devised a new method for observing an eclipse, which was an unacknowledged adaptation of Carrington's technique for studying sun spots. Phillips threw the image of the sun on to a screen on which he marked with a pencil various features and noted the time. He derived the apparent diameters of the sun and moon from measurements of the arcs on the screen. His results agreed closely with those obtained by Main at the Radcliffe Observatory for the diameter of the moon relative to that of the sun. Phillips was also interested in luminous meteors which were often known as shooting stars. In 1856 he observed those of 7 January and set out desiderata for observing them. Ten years later he noted the colour, brightness, apparent length, and duration of individual members of a remarkable meteor shower he observed in the early hours of 14 November. His article reporting his results was sent that morning to the *Athenaeum* and quickly translated for publication in *Les mondes*. He did not bother to mention that Hubert Newton's prediction of 1864 about a remarkable meteor shower on 13-14 November 1866 had been punctually fulfilled and that observers were numerous because forewarned. Though Phillips realised that the meteors seemed to come from the same part of the sky, ie from the constellation Leo, and moved from east to west with enormous apparent velocities like rockets, he left the interpretation of such features to specialist astronomers.[55]

Phillips was not just a widely respected astronomical observer: he also promoted observatories at Oxford. In 1854 he persuaded Daubeny to order from Cooke an equatorially mounted telescope of 5.5 inches aperture which Daubeny gave next year to Magdalen, his college. By 1857 Magdalen had spent £195 on building an observatory to house the instrument which was used for teaching its undergraduates. There is no record of Phillips using it. In 1860, after the death of Manuel Johnson, the Radcliffe observer and friend of Phillips, the Radcliffe Trustees consulted Airy, Robinson, Acland, and Phillips before finally deciding the future of their observatory. Phillips' response not only helped to shape its future but also revealed his concern about the 'privations' of astronomy in the University. In March 1860 there were no observing facilities with which undergraduates could see a double star, nebulae, Saturn's inner ring, or even a lunar crater or a solar spot. That situation had not changed in February 1870 when, after Donkin's death in 1869, the reverend Charles Pritchard was elected professor of astronomy.[56] Yet by 1875 the University exulted in a new observatory which suddenly put Oxford on to the international map for astronomical research. That transformation, as Hutchins has shown, was in considerable part promoted by Phillips.

Pritchard respected Phillips as a fellow religious scientist, as a useful astronomer, and as an effective administrator within the University and the British Association. By summer 1870 Phillips had enabled Pritchard to secure honorary doctorates for four astronomical friends, including John Gassiot, Huggins, and crucially De la Rue, all of whom had lobbied in Pritchard's favour for the Oxford chair. Initially Pritchard made piecemeal improvements to the small University Observatory built in 1860 but soon became frustrated with his facilities. In spring 1872 Pritchard learned about Phillips' wish to bequest his own private observatory to the University. In the summer Phillips approached the Royal Society to obtain its permission to do so. It agreed in June that his astronomical research with his 'beloved instrument' ought to be continued 'con amore' as an effective aid to astronomy in the University.[57]

Stimulated by this imprimatur, Phillips enlarged his ambition: he persuaded Pritchard to put to the University in December 1872 a scheme for a school of astronomical physics based on a new observatory with new equipment. Pritchard obtained an estimate (£1,200) from Howard Grubb for a large refracting telescope of 12.25 inches aperture. In January 1873 Phillips worked out that it would be feasible to house this telescope in the north-east corner of the University Museum site provided that the observatory floor was high enough to allow the telescope's line of sight to clear the roof of the Museum to the south-west. He and Pritchard then unofficially approached Charles Barry, junior, for a design for an observatory with appropriate dimensions. He responded in February 1873 with an ornate octagonal tower and dome, suitably high (fifty feet) and in Gothic style.[58]

Early next month the University, which certainly knew about the proposed site, decided to invest substantially in astronomy. It allocated £2,500 for buying and housing the Grubb instrument, offered another site not so near the Museum, and commissioned from Barry a design for a vaguely Italianate square building to carry the revolving dome. Barry had soon to extend his design of June 1873 because in that very month De la Rue offered to the University his famous thirteen-inch reflecting telescope and associated equipment. Having taken the unanimous advice of an *ad hoc* committee, on which Phillips sat, the University accepted De la Rue's gift in late 1873 and provided an extra £1,500 for an extended building to be designed by Barry. When the University Observatory was completed in 1875, at a cost of £5,100, it had two square towers, one containing the Grubb refracting telescope and the other De la Rue's reflecting telescope, joined by a corridor which housed the instruments from the 1860 building.[59]

The new Observatory was founded in two stages. The catalyst for the first was Phillips, the initiator of the second was De la Rue. It was Phillips' example and resolute intervention which prompted Pritchard to be ambitious in 1872, it was Phillips who in 1873 provided a feasible site for a proposed new observatory, and it was Phillips who in 1870 secured an honorary degree for De la Rue who in return felt impelled in 1874 to donate his celebrated instrument to the University. Thus Phillips helped to found the new University Observatory for Astronomical Physics, as Pritchard called it, which complemented the University's new Clarendon Laboratory, the first purpose-built physics laboratory in Britain, which was opened in 1870 and cost about £12,000.[60]

10.7 Magnetism and meteorology

Phillips maintained his interest in terrestrial physics through his research in magnetism and meteorology. The former was resumed in 1856 when the British Association asked Sabine, Lloyd, Phillips, Ross, and Fox to repeat their British magnetic survey for 1837, in order to discover what changes had occurred over two decades. When Sabine reported in 1861 the results of the second survey, it was clear that Lloyd and he were persistently active, with the other three making only minor contributions. Phillips' was short-lived. Initially Ross wanted him to survey much of north England but Sabine told him that this huge remit was not only impossible but also undesirable because Phillips intended to use his old instruments of 1837. All the other observers were equipped with modern instruments, including a new circle, made by Henry Barrow, which was essential to Lloyd's new method of measuring force. Phillips realised that instrumentally he was the odd man out so he confined himself to measuring dip and intensity in York in autumn 1857 with his old instruments. When his results were compared with Ross's, those for dip at York differed by ten minutes. Sabine and Lloyd thought this unacceptable, urged Phillips to go to Kew Observatory to compare his instruments with the standard ones there, and informed him that satisfactory standardisation of instruments was more demanding than he thought. Having been chastened in autumn 1857, Phillips was told next spring by Lloyd that his dip results were strange. In summer this strangeness was explained. Ross's new circle was sent to its maker, Barrow, for cleaning, after which it was found that readings with it changed consistently by ten minutes. By this time Phillips accepted the validity of the points made by Lloyd and Sabine about the necessity of intercomparison of instruments and methods employed, so that his magnetic flurry of 1857-8 was briefer and much less important that that of the late 1830s. By not using what Sabine and Lloyd regarded as the most modern instruments of approved construction, Phillips showed that he was out of touch. He was relegated to such a minor role in the second magnetic survey of Britain that Sabine's report did not include any of Phillips' results.[61]

Phillips was well aware that as a science meteorology was in its infancy. Many phenomena were ephemeral or always changing so it was difficult to infer their causes and the laws of their operation. The results of much observation by many people were not impressive: there were no empirical laws, graphs were merely empirical and not summarisable in mathematical form, and true causes remained elusive. Yet Phillips shared with his friends Lloyd and Price the Baconian hope that accurately observed facts under standard conditions would in future lead to empirical laws, as they had in terrestrial magnetism. He remained convinced that even local observations of passing phenomena were valuable supplements to the regular observations carried out at observatories. By reading a barometer, a maximum and minimum thermometer, and a hygrometer, nine times a day, and employing a pluviometer, rural observers could provide data which for Phillips were essential for public health, agriculture, and weather forecasting. Yet his previous experience of the difficulties of fruitful collaboration in meteorology led him to have little contact with the British Meteorological Society, founded in 1850, and with James Glaisher and George Symons, its leading figures. The latter's standard work of 1867 on *Rain* illustrated different gauges but not any of Phillips'. Though Phillips was a member of the British rainfall committee established in 1865 by the British Association, he failed to implement his promise made in 1867 to examine funnel rain gauges and he took no part in the huge voluntary co-operative enterprise organised by Symons and financially supported by the Association between 1864 and 1876 with £890 to pay for instruments. From 1862 until his death Symons published annually *British rainfall* based on data supplied by local observers, numbering 500 in 1862 and 2,000 in 1876. Phillips probably suspected Symons' mindless tabulation of data generated by observers who did not always record under standardised condition or were inaccurate, so that the size of Symons' enterprise did not indicate its scientific value.[62]

By this time Phillips was indifferent to a national audience for his meteorological papers, most of which were given to the local Ashmolean Society. He dilated on the temperature of York Minster (1854), of the north sea (1855), and of his own house (1872); he reviewed the various anemometers available in 1861 including his own, and the related anemoscope and thermanemometer he had devised; and he explained the application of formulae for computing rainfall (1861). In 1867 he reported on the rainfall of the Lake District, using others' measurements, and produced a map which showed the relation between rainfall and height of the land. Next year it was delivered to the BAAS and became the sole meteorological paper by him also given outside Oxford.[63]

At Oxford Phillips never lost his interest in meteorological instruments, especially those which could be used in geological research. He had little success with another projected thermanemometer.[64] But at a time when the reliability of the aneroid barometer was suspect he tested the instrument to show it could deliver accurate results. In the 1860s and 1870s an improved version of his self-registering maximum thermometer, first made public to the British Association in 1832, was made and sold by leading British instrument-makers including Casella, Negretti and Zambra, Browning, and Pastorelli. He was the only leading British geologist who was commemorated eponymously in his lifetime by having a meteorological instrument named after him.

When the aneroid barometer became widely popular in Britain in the 1850s Phillips welcomed it as useful up to a height of 1,000 feet but he knew that it had drawbacks. Its readings changed over time so that it had to be reset regularly against a standard barometer; and it suffered from considerable and unequal variation when the temperature changed. Even so Glaisher was satisfied with it on his famous balloon ascents in 1862. In 1863 Phillips began to use a miniature aneroid, uncompensated for the effect of temperature, only 1.9 inches in diameter, and made for him by Cooke. By comparing it with a standard barometer, he produced empirical formulae which enabled him to correct his instrument and measure heights which the Ordnance Survey evaluated in autumn 1863 as very near the truth. Buoyed up by this imprimatur, Phillips carried his instrument to Switzerland in 1864, read it to three decimal places on Vesuvius in 1868, and from 1869 used it in his geological research on the Oxford area and the Yorkshire coast. In 1872 he and James Parker tested a compensated aneroid made by John Browning as well as Phillips' own uncompensated instrument. The consistent results obtained confirmed Phillips' view that the aneroid could measure heights accurately and led him to dissent from the expert opinions expressed in 1869 by Stewart and in 1871 by the meteorological committee of the Royal Society that it was unreliable. In 1872 Phillips claimed before the British Association that his uncompensated but corrected instrument was accurate provided it was tapped but not jolted, used horizontally, enclosed in a poorly conducting box, and not exposed to sunshine and wind. Shortly afterwards the Ordnance Survey informed him that in two key cases its levelling results were identical with those he had obtained with his Cooke aneroid. This corroboration no doubt pleased Phillips not least because it rebutted the pessimistic view expressed by Cooke in 1865 that the construction of any aneroid made it an imperfect and unindependent instrument, subject to variations which could not be tabulated or reduced to a formula.[65]

After revealing his invention of a maximum thermometer in 1832 Phillips made several of them mainly for measuring underground temperatures. At the Kew Observatory Welsh constructed an improved version of Phillips' device and used it from 1851. Even so it was still a delicate instrument, with problems concerning the air bubble, so in 1852 Negretti and Zambra produced their own rival instrument which they claimed was impossible to derange except by breaking. Both thermometers were used at Kew where it was believed in 1854 that in careful hands Phillips' was the better, more sensitive but not as robust as that of Negretti and Zambra. Next year both instruments were displayed at the Paris Exhibition and in his standard work on *Practical meteorology* John Drew pointed out that, though simple and reliable, Phillips' instrument had not received the attention it deserved.[66]

Fortified by such publicity and recommendations, and informed by Welsh about instrument-makers, Phillips displayed his instrument to the British Association in 1856 and arranged with Louis Casella for it to be produced and sold for research in meteorology and physics (Fig. 10.5). Phillips was pleased with Casella's work: the instrument was excellent in construction and had done full justice to its principle. By 1860 Casella was producing no fewer than seven different maximum thermometers, all constructed 'on Professor Phillips' principle' and possessing, according to Casella, unrivalled precision, hardihood, and portability. One type was specially arranged by Casella for David Livingstone's exploration of the river Zambesi. Another, the cheapest at 8/6 and suitable for use in gardens, was produced by Casella 'expressly for the object of *popularising* the admirable and important principle of maximum thermometers invented by Professor Phillips'. Casella's Phillipsian thermometers sold so well that in 1865 Casella thanked Phillips for promoting his sales. In the late 1860s Phillips' instrument received further publicity when William Thomson and his former pupil, Joseph Everett, began to use a long-running grant from the British Association to supervise measurements undertaken to determine the rate at which the temperature of the earth increased with depth, a topic dear to Phillips. For measurements in water Thomson designed a modified Phillips' maximum thermometer which Casella made. To protect its bulb it was hermetically sealed in a strong glass tube. It was employed from 1868 to 1871 when it was replaced by the maximum thermometer made by Negretti and Zambra which withstood jolts better.[67]

The chief objection to this form of thermometer is that the bubble of air is likely to become displaced by passing back into the bulb if the temperature falls very

FIG. 2.

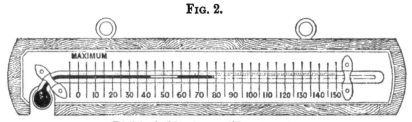

Phillips's Maximum Thermometer

low. When this happens the instrument loses its registering properties, and becomes an ordinary thermometer.

Fig. 10.5 Phillips' maximum thermometer, invented in the early 1830s and manufactured commercially in the 1860s, was still highly regarded in the 1880s by the director of the Meteorological Office though he did note an important defect.

Other instrument-makers besides Casella sold Phillips' maximum thermometer. Though Negretti and Zambra claimed that their own instrument was superior, in the 1860s they still made and sold Phillips' device. In the early 1870s John Browning and Francis Pastorelli did likewise. Browning was not alone in thinking it the best of its kind. In 1872 in a discussion at the British Meteorological Society it was agreed that Phillips' instrument, which usually sold at between 7/6 and just less than a pound, was preferable for use at sea and that of Negretti and Zambra better on land.[68]

Not surprisingly Phillips acted as a meteorological adviser. As president of the Royal Society, Sabine regretted that Phillips was often unable to attend Council and committee meetings concerned with meteorology. When baffled Sabine consulted

Phillips on such matters as how to derive the most important periodic variations from meteorological data given by self-reading instruments and by seven observatories whose data would be tabulated at the Kew Observatory and sent to the Meteorological Office of the Board of Trade. In 1869 Sabine even wanted Phillips to look at the meteorological instruments during a visitation to the Greenwich Observatory, run by the formidable astronomer royal, Airy.[69]

Phillips was probably happier advising Robert Scott, who was appointed director of the Meteorological Office in February 1867 on the recommendation of Sabine. Prompted by Sabine, Scott consulted Phillips privately in late 1867 about the desirability of sending to colliery owners telegraphic notices of the approach of low pressure so that fiery seams would be worked with greater care because of the greater likelihood of effusion of dangerous gases. Encouraged by Phillips, Scott tried to interest government in correlating colliery explosions with atmospheric conditions but his scheme was soon squashed by the Board of Trade and the Home Office. In 1872 Scott sought and accepted Phillips' advice on a paper on this subject in which he noted that half of 528 explosions from 1868 involved serious barometrical disturbance. He agreed with Phillips that more air was the best preventative, on the model of the South Hetton colliery as explained by Phillips. Later that year Scott asked Phillips whether he was still opposed to the introduction of measuring instruments into coal mines because their managers were not well enough educated to use them properly. A few months later Phillips was happy to test for Browning several types of aneroid barometer and reported on their value. In 1874 Phillips was doubtless flattered to be asked by Lockyer to give evidence about meteorology to the Devonshire commission on scientific instruction and the advancement of science. Phillips' death prevented him from doing so, and from applauding Scott's success in the mid-1870s in persuading the Board of Trade to permit the Meteorological Office to warn a few collieries by telegraph about rapid decreases of atmospheric pressure.[70]

Notes:

1. Phillips, syllabus, On the philosophy of geology, spring 1854, PP, box 78, f. 8; Phillips, 1859a, p. xxvi; Phillips, 1865b, p. lviii.
2. Phillips 1862a; Douglas and Edmonds, 1950, pp. 372-4; J. Evans to Phillips, 28 Jan, 21 Feb 1860, PP.
3. Phillips, 1871b; Darwin, 1849; Darwin to Herschel, 6 Oct 1870, American Philosophical Society; Darwin's article appeared in the first (1849), second (1851), and third (1859) editions of Herschel's *Manual*; the third and fourth editions were superintended by the reverend Robert Main (1808-78), *DNB*, director of Radcliffe Observatory, Oxford, 1860-78.
4. Sedgwick, 1835a; strike, a term coined by Sedgwick, meant the direction of horizontality of the tilted plane of a stratum.
5. Phillips, 1843b; Sharpe, 1847, 1849; Sorby, 1853; Tyndall, 1856a and b; Haughton, 1856; Higham, 1963, pp. 41-5; Daniel Sharpe (1806-56), *DNB*, president Geological Society 1856; John Tyndall (1820-93), *DNB*, professor of natural philosophy, Royal Institution, 1853-87; reverend Samuel Haughton (1821-97), *DNB*, professor of geology, Trinity College, Dublin, 1851-81.
6. *BAAS 1855 report*, p. lxv; Phillips, 1856a; Phillips, 1829c; Phillips, 1836a, pp. 5-8; Phillips to Sedgwick, 8 and 30 Sept, 15 Dec 1856, 16 March 1857, Se P; Sedgwick to

Phillips, 1 Sept 1856, PP; foliation occurred when rocks were crystalline in texture but stratified in structure.

7. Woodward, 1907, pp. 162-4; Wood, 1848-82; Jones, 1850; Owen, 1851-89; Wright, 1857-80; Milne-Edwards and Haime, 1850-5; Davidson, 1851-86; Salter, 1864-83; James Scott Bowerbank (1797-1877), *DNB*; Searles Valentine Wood, elder (1798-1880), *DNB*; Henri Milne-Edwards (1800-85), *DSB*; Jules Haime (1824-56).

8. Huxley, 1864.

9. Phillips, 1865-70, pp. 7, 109; Phillips 1829b, p. 174; Phillips, 1835a, pp. vi, 166; *Athenaeum*, 1835, p. 646; *BAAS 1835 report*, p. xxvi; Buckland to Phillips, 22 May 1835, Belemnite letters 1835-69, PP, box 50; Phillips to De la Beche, 1 April 1835, 4 Nov 1837, DLB P; Phillips to Mantell, 11 Nov 1836 Ma P; Griesback to Phillips, 6 Feb 1838, PP; YPS, Scientific communications, 3 Mar 1835, 2 Jan 1838; advertisement at back of Phillips, 1836b; Voltz to Phillips, 15 June 1833, 25 Oct 1835, 7 Feb, 21 Nov 1836, PP.

10. *BAAS 1842 report*, p. 213; *Athenaeum*, 1842, p. 615; Buckland to Phillips, 21 May, 22 Sept 1842, Stutchbury to Phillips, 25 Feb 1843, all PP; Stutchbury to De la Beche, 11 Sept 1841, DLB P; Phillips to Murray, 20 and 26 Jan 1843, Mu P; Phillips to De la Beche, 19 Dec 1844, DLB P; Blainville, 1827; Voltz, 1830; Orbigny, 1842-9; Alcide Dessaline d'Orbigny (1802-57), *DSB*.

11. Bowerbank to Phillips, 24 March, 15 July 1848; Wiltshire to Phillips especially 16 and 18 March 1864, 1865-8, all PP, box 50.

12. Phillips, 1865-70, pp. 12, 28, 57, 91, 110; Phillips' circular, 1864 [late March/early April] and responses, Cullen to Phillips, 20 April 1865, all in PP, box 50; Charles Moore (1815-81), *DNB*, of Bath; Walter William Stoddart (1834-80) of Bristol; George Highfield Morton (1826-1900), *DNB*, of Liverpool.

13. Phillips, 1865-70, pp. 12, 25-31, 93; Phillips, 1871a, p. 133; Wiltshire to Phillips, 31 Oct 1867, PP, box 50.

14. Phillips, Switzerland: geological journal 1864, PP, notebooks 82 and 83; Phillips to Murchison, 10 Aug 1864, MP; Forbes, 1843, 1859; Ramsay, 1860, 1862; Phillips 1865c; Phillips, 1873b, p. 73; Mallet to Phillips, 29 Dec 1872, PP.

15. Ramsay, 1855; Phillips, 1848a, p. 112; Phillips to Lyell, 9 June 1855, LP; Phillips, 1853a, pp. 183-92; Phillips, 1863c; Rudwick, 1969, illuminates the unenthusiastic reception of the theory of glaciation.

16. Phillips, 1864b; Phillips, 1871a, pp. 456-63, Wood and Rome, 1868: Phillips, 1868c; Phillips, 1875, p. 166 (q); Searles Valentine Wood, junior (1830-84), *DNB*, specialist on gravel of east England.

17. Phillips, 1875, pp. 162-7.

18. Lee, 1881, p. 62; Lee to Phillips, 6 Feb 1869, PP.

19. Phillips, Auvergne, 1855, PP, notebook 70; Phillips, Cornwall, December 1855, PP, notebook 72; Friedrich Wilhelm Heinrich Alexander von Humboldt (1769-1859), *DSB*; Leopold von Buch (1774-1852), *DSB*; George Julius Poulett Scrope (1797-1876), *DNB*.

20. Daubeny, 1848; Daubeny to Phillips, 20 Jan 1862, PP; Sorby to Phillips, 26 April 1855, followed by letters July 1855- Sept 1858, PP; Sorby, 1858, pp. 480-4; Higham, 1963, pp 54-7; Geikie, 1895, p. 343; Sorby, 1897, pp. 8-9; Daubrée, 1859; Delesse, 1858; Phillips to Harcourt, 26 March 1861, HP.

21. Mallet, 1862a, 1862b, vol. 2, pp. 313-14; Mallet, flier, 29 Sept 1862, Vesuvius documents, PP, box 100a.

22. Kenrick, 22 Dec 1867; Wreford, Saturday (twice) and Monday; Murray, 30 Dec 1867; Main, 30 Dec 1867; Pusey, 23rd, Monday; Lee, 1 Feb 1868; Lyell, 3 Feb, 2 Mar, 1868, all to Phillips, PP; Newman to Neve, 14 March 1868, PP; Lyell, 1868, vol. 1, pp. 620-54, esp 634-8; Guglielmo Guiscardi (1821-85), professor of geology, Naples University; Luigi Palmieri (1807-96), professor of physics; Arcangelo Scacchi (1810-93), professor of mineralogy; John Henry Newman (1801-90), *DNB*.

23. Phillips, Italia 1, 6 March-17 March 1868, notebook 88, Italia 2, 18 March-3 April 1868, notebook 89, both PP; Wreford reports, *Athenaeum*, 1868, i, pp. 461-2, 528-9; Phillips, 1869a, pp. viii, 115-31, 203-46 (219 q), 263-4.
24. Phillips, Italia 3, 4-21 April 1868, notebook 90, Germany, 16 April 1868 notebook 91, both PP; Phillips, financial estimates, Vesuvius documents, box 100a, PP; Lobley, 1868; Peacock, 1866; Phillips, 1869a, pp. ix, 112-13; James Logan Lobley (1834-1913); Richard Atkinson Peacock (b.1811).
25. Phillips, 1869a, pp. 172-202, 319-20, 335-7 (336 q); Mallet to Phillips, 25 Jan 1869, Scrope to Phillips, 5 Feb 1869 (q), Lyell to Phillips, 5 Feb 1869, all PP; Palmieri, 1873, pp. 2, 15; Scrope, 1862, 1869.
26. Phillips to Murray, 15 Nov (q), 23 Nov, 18 Dec 1868, Mu P; Phillips, 1869b, pp. 95-6, 106-10, 115; Mallet, 1848, 1861, 1862b; Mallet and Mallet, 1858.
27. Phillips, 1871a, p. 412; Phillips, 1855b; *BAAS 1860 report*, p. 90; Phillips, 1861a; Phillips, proposal, 24 Feb 1869, Price to Phillips, 23 April, 24 Nov 1869, PP; Phillips to Sedgwick, 14 March 1870, Se P.
28. Phillips, 1871a, pp. v, vii-viii, 1, 228, 251, 480.
29. Phillips, 1871a, pp. 137-294 on Bath oolitic period, pp. 167-237 on Stonesfield fossils, pp. 245-94 on ceteosaurus; pp. 493-504 on economical applications; Huxley, 10 Nov, Egerton, 28 Dec, Etheridge 23 Nov 1871, to Phillips, PP; Geikie, 1872 (q); Prestwich, 1851.
30. Phillips, notes and sections Yorkshire, box 58, 59; Phillips, Yorkshire coast and Wolds, 1872, notebook 71; Yorkshire coast, October 1857, notebook 74; Yorkshire, Sept-Oct 1864, July-Aug 1866, notebook 85; Yorkshire coast, Aug-Sept 1867, notebook 86, all PP; Phillips, 1875, pp. viii (q), 98-104; Phillips to Murray, 20 March 1872, 30 Sept 1873 (q), Mu P; Lee to Phillips, 25 March 1872, PP.
31. Phillips, 1829b, pp. 46, 76, 124-5; Leckenby, 1859; Judd, 1868; Phillips, 1875, pp. 49, 98-9; John Leckenby (1814-77), treasurer Scarborough borough; John Wesley Judd (1840-1916), professor of geology, Royal College of Science and School of Mines, London, 1876-1905, president Geological Society, 1886-8.
32. Phillips, 1829b, pp. 39-40; Wright, 1860; Leckenby, 1864; Hull, 1862; Phillips, 1858a; Phillips, 1860a, pp. xli, xlviii; Phillips, 1875, pp. 29-30, 109-23; Edward Hull (1829-1917), director of Irish Geological Survey, 1869-91.
33. Lycett to Phillips, 5 April 1972, PP; Lycett, 1872; Phillips, 1875, pp. vii-ix, 101, 113, 137-8, 151, 193-4, 204, 235, 254, 273-321; Judd to Phillips, 30 July 1872, PP; Judd, 1870; Lee to Phillips, 18 Dec 1867; Etheridge to Phillips, 7 April 1873, PP; John Lycett (1804-82); John Gwyn Jeffreys (1809-85), *DNB*; William Kitchen Parker (1823-90), *DNB*; William Whitaker (1836-1925), president Geological Society 1898-1900.
34. Tate and Blake, 1876, pp. 6-10; Fox-Strangways, 1892, vol. 1, pp. iv, 15-17; Ralph Tate (1840-1901), teacher at Redcar until 1875; reverend John Frederick Blake (1839-1906), teacher and assistant chaplain, St Peter's School, York, 1865-74; Charles Edward Fox-Strangways (1844-1910).
35. Lushington to Phillips, 7 Nov 1865, PP; 'First report of the Commissioners appointed to enquire into the best means of preventing the pollution of rivers (river Thames)', *Parliamentary papers*, 1866, vol. 33, pp. 1-419 (158-62); 'Report of the Commissioners appointed to enquire into the several matters relating to coal in the United Kingdom. Volume II. General minutes and proceedings of committees A, B, C, D & E', *Parliamentary papers*, 1871, vol. 18 (716-27).
36. Phillips, 'Report on the probability of obtaining a supply of water from the sandhill near the Buckinghamshire County Asylum, and the means by which it should be effected', 20 May 1854, PP, box 102, f. 9; Amos to Phillips, 18 June 1862; Tomlinson to Phillips, 15 Nov 1869; Hammond to Phillips, 25 Sept 1867; Farlie? to Phillips, 14 Nov 1872, all PP.
37. Maskelyne to Phillips, 5 April 1855; Donkin to Phillips, 23 July, 1857; Loch to Phillips, 5 Aug 1859, with note by Phillips, 8 Aug 1859; Phillips, report on Stittenham, 25 Aug, Dec 1859; Nelson to Phillips, 1 March 1865, all PP; Phillips, 1875, pp. 174-5; Phillips, 1859a, p. xli; George Granville Leveson-Gower, second Duke of Sutherland (1786-1861).

38. Phillips, 1871a, p. 494; Thomas Frewen (1811-70), conservative member of Parliament for South Leicestershire 1835-6; John Alleyne Bosworth (b. 1812) belonged to a Leicester family of land surveyors and inventors (information kindly supplied by Hugh Torrens).

39. Frewen to Phillips, nd [early 1859] (q), 29 Jan, 23 March, 31 May 1859; Phillips, incomplete report, 29 March 1859, all PP (some in Sapcote bore letters, box 47).

40. Frewen to Phillips, 14 June 1864, 16, 18, 27 March, 5 Oct, 2, 9 Nov, 7, 25 Dec 1865; Phillips, reports, 28 Oct, 20 Nov 1865, all PP.

41. Frewen to Phillips, 2 July, 28 Sept, 28 Oct 1866, 19 Feb, 4 March, 27 March, 4, 18 June, 10, 29 Sept 1867; Bosworth to Phillips, 2 Jan 1867; Phillips, reports, 10 July 1866, 8 Oct 1867, all PP.

42. Frewen to Phillips, 25 Oct, 16, 30 Dec 1867, 8, 21, 29 Feb, 7 May, 13 June (q), 14 July, 5 Sept, 10, 12 Dec 1868, all PP.

43. Phillips, report, 25 Jan 1870; Johnson to Phillips, 24, 26, 28 March, 1 April 1874, all PP.

44. Evans, 1956, p. 303; C.K. Watson to Phillips, 5, 23 May 1865; Lee to Phillips, 28 Dec 1864, 9 July 1865, PP; Keller, 1866; Ferdinand Keller (1800-81).

45. Way to Phillips, 20 July 1865, 6 Feb 1866; J. Wilson to Phillips, 31 May, 4 June 1871; Rolleston to Phillips, 20 Dec 1872; Kenrick to Phillips, 13 Jan 1873, all PP; Rolleston, 1875, p. 123; Rolleston, 1884, p. 636; Greenwell and Rolleston, 1877, pp. 679, 711; Phillips to Elizabeth Johnes, 30 May, 25 Sept, 24 Dec 1867, 23, 27 Sept 1868, 30 May 1873, 28 March 1874, Dolaucothi MSS, National Library of Wales; Phillips to Ramsay, 31 May 1867, RP; Henry Wellesley (1791-1866), *DNB*, principal of New Inn Hall, Oxford, 1847-66; in 1882 Elizabeth (known as Betha) Johnes (1834-1927) married Sir James Hills (1833-1919) who in 1883 assumed the name of Hills-Johnes.

46. Way to Phillips, 15 Oct, 9 Nov 1858, 20 July 1865, 6 Feb, 21 Mar 1866, 1 June 1868, 13 Feb 1869, 19 March 1872, PP; Way, 1859; Phillips, 1859b; *Archaeological Journal*, 1866, vol. 23, pp. 190-211, 309, 332; Lubbock, 1865; John Evans (1823-1908), *DNB*, neolithic and bronze age archaeologist; Sir John Lubbock (1834-1913), *DNB*; James Charles Clutterbuck (*c* 1800-85), vicar of Long Wittenham, Berkshire, 1830-85.

47. Phillips to Sabine, 29 Sept, 1 Nov 1856, 29 June 1857, Sa P; Phillips to Forbes, 11 April 1856, FP; Phillips, 1857a, 1863a, p. 35; Pietro Angelo Secchi (1818-78), *DSB*, director of observatory of Collegio Romano.

48. Hutchins, 1994, pp. 210-12; Phillips, 1863a, pp. 35-6; Phillips, 1868a, p. 334; Phillips to Sharpey, 2 June 1872, Royal Society archives, M.C.9, 381; Sabine to Phillips, 1 (q), 3 March 1862, PP; William Sharpey (1802-80), *DNB*, and George Gabriel Stokes (1819-1903), *DNB*, were secretaries of the Royal Society.

49. Phillips, 1868a; Hutchins, 1994, pp. 212-13; Neison, 1876, pp. 41, 343-4, 457-8; Chacornac to Phillips, 28 Sept 1864, PP.

50. *BAAS 1860 report*, pp. xlvii, 34-5; Phillips, 1861b, 1863d; Birt to Phillips, 6 June 1864, PP; *BAAS 1864 report*, p. xlviii; *BAAS 1866 report*, p. lx; Hutchins, 1994, pp. 212-13.

51. Birt to Phillips, 14 Dec, 31 Dec 1868, and c, PP, box 93, f. 6.

52. Hutchins, 1994, pp. 213-14; Phillips, 1863e; Joseph Norman Lockyer (1836-1920), *DNB*, used a Cooke telescope of 6.25 inches aperture and was encouraged by Phillips to publish on Mars: Meadows, 1972, pp. 44-5.

53. Phillips, 1865d and e; Dawes to Phillips, 13 Feb, 24 March, 5 April 1865, PP; Proctor, 1869; Terby to Phillips, 8 Jan 1872; Longman to Phillips, 9, 28 July 1867, all PP; William Rutter Dawes (1799-1868), *DNB*; Richard Anthony Proctor (1837-88), *DNB*; William Huggins (1824-1910), *DNB*; François Joseph Charles Terby (1846-1911).

54. Phillips, 1865f, g, and h, 1867a; Carrington, 1863; Cooke to Phillips, 22 Nov 1866, PP; Clerke, 1887, pp. 157-209, 239-59; Balfour Stewart (1828-87), *DNB*, director, Kew Observatory, 1859-71; Richard Christopher Carrington (1826-75), *DNB*.

55. Phillips, 1867b; Phillips, notes of eclipses, PP, box 93, f. 4; Phillips, 1871c; Phillips, 1856b, 1866a, b; Clerke, 1887, pp. 374-91; William Lassell (1799-1880), *DNB*, perfector of reflecting telescope and discoverer of satellites of planets; Hubert Anson Newton

(1830-96), *DSB*, professor of mathematics, Yale University, showed in 1866 that the Leonid meteors followed a cometary path.

56. Hutchins, 1990; Phillips to Acland, 7 March 1860, Acland papers, Bodleian Library, d. 94. ff. 36-7; Guest, 1991, pp. 275-82; Manuel John Johnson (1805-59), *DNB*, Radcliffe observer, 1839-59; Charles Pritchard (1808-93), *DNB*, professor of astronomy 1870-93.

57. Hutchins, 1994, pp. 220-2; Phillips to Sharpey, 2, 6 (q) June 1872, Royal Society archives, M.C.9, 381, 385; Sharpey to Phillips, 4, 17 June 1872; Stokes to Phillips, 21 June 1872, all PP; John Peter Gassiot (1797-1877), *DNB*, scientific benefactor.

58. Hutchins, 1994, pp. 223-8; Fox, 1997, pp. 683-4; *Oxford University Gazette*, 1873, vol. 4. pp. 42-3; Howard Grubb (1844-1931), *DSB*; Charles Barry, junior (1823-1900).

59. Hutchins, 1994, pp. 228-37.

60. Hutchins, 1994, pp. 238-42; Fox, 1997, pp. 676-82.

61. *BAAS 1856 report*, p. xli; Ross to Phillips, 26 June, 1 July, 4, 13 Oct 1857; Sabine to Phillips, 26 June, 19, 26 Oct, 2 Nov 1857; Lloyd to Phillips, 19 May 1858; Welsh to Phillips, 25 June 1857, 8 June 1858, all PP; Phillips to Sabine, 3 July, 17 Oct 1857, Sa P; Sabine, 1861.

62. *BAAS 1857 report*, pp. lv-lvi; *BAAS 1860 report*, p. 3; Phillips to Acland, 1 Sept 1870, Acland Papers, d. 64, ff. 275-6; *BAAS 1867 report*, p. 448; Howarth, 1931, pp. 177-80; Symons, 1864, 1867; James Glaisher (1809-1903), *DNB*; George James Symons (1838-1900), *DNB*.

63. Phillips, 1854e, 1855d, 1872b; drafts 1861, PP, box 96, f. 15; Phillips, 1868d.

64. Turberville to Phillips, 8 Nov 1865, PP.

65. Phillips, 1855c, pp. 639-41; Middleton, 1964, pp. 398-441; Sanders to Phillips, 8 Oct 1863, Ordnance Survey Office to Phillips, 2 Oct 1863, 12 Oct 1872, PP; Phillips to Galton, 29 May 1864, Galton papers, UCL; Phillips, Italia II, PP, notebook 89, entry for 24 March 1868; Phillips, Aneroid 1869-72, PP, notebook 93; Phillips, notes on aneroid, Aug 1872, PP, box 79, f. 2; Phillips, 1872b; Stewart, 1868; Cooke to Phillips, 2 June 1865, PP; John Browning (1835-1925).

66. Phillips, 1832b, 1856c; Negretti and Zambra, 1864, pp. 73-4; *BAAS 1854 report*, p. xxxiv; *BAAS 1855 report*, p. xxxiii; Drew, 1855, p. 59; Drew to Phillips, 10 Nov 1855, PP; John Drew (1809-57), *DNB*; Enrico Angelo Ludovico Negretti (1818-79); Joseph Warren Zambra (1822-97).

67. Welsh to Phillips, 14 Dec 1854, PP; Phillips, 1856c; Casella, 1860, pp. 17-18, brought to my attention by Neil Brown, Science Museum; Casella to Phillips, 12 June 1865, PP; Everett, 1869, pp. 176-7, 1870, p. 37, 1871, p. 25; Louis Paschal Casella (1809-97); David Livingstone (1813-73), *DNB*; William Thomson, Lord Kelvin (1824-1907), *DNB*.

68. Negretti and Zambra, 1864, price list, p. 4; Middleton, 1966, p. 155; Browning to Phillips, 26 Oct 1872, enclosing rough proof of Browning's Set of meteorological instruments, Nov 1872, PP.

69. Sabine to Phillips, 21 March 1866, 21 Dec 1867, 23 Jan 1869, all PP; Sabine to Phillips, 21 May 1869, EUL Gen 784/1.

70. Scott to Phillips, 19, 24 Dec 1867, 30 Jan 1868, 22 Jan, 10, 17, 19 Feb, 15 March, 5 July 1872; Browning to Phillips, 22, 26 Oct, 18 Nov 1872; Lockyer to Phillips, 30 March 1874, all PP; Scott and Galloway, 1872; information from Jim Burton based on Burton, 1988.

Chapter 11

Keepering

At Oxford Phillips was keeper of the Ashmolean Museum from 1854 to 1870 and keeper of the University Museum from 1857 until his death. He treated these posts quite differently. At the Ashmolean he was the thirteenth keeper of the oldest public museum in Britain, opened in 1683. As such he was not a zealous innovator and he kept only a light hand on the administrative tiller. Once others had created a new function for the Ashmolean in the 1860s, he was happy to resign. The keepership of the new University Museum was far more demanding. As a paid official of the University he was not entirely a free agent and had to respond to its requests and requirements. Once again he was in part a general factotum, a role with which he was familiar from his responsibilities with the British Association and the Yorkshire Philosophical Society. Yet as the first keeper of the new Museum and the first occupant of the purpose built keeper's house, Phillips had no precedents to follow. Cautiously and gradually he defined the functions of his post so that by the mid-1860s he was regarded as the lord keeper of the Museum. He presided over the University's biggest building which not only displayed collections but also possessed laboratories, lecture rooms, and a library. He was in charge of a science centre, the facilities of which were superior to those then existing at Cambridge. Though he was not, like Acland and Daubeny, a polemicist in Oxford on behalf of science and the Museum, he implemented their ideals which he shared.

In his dealings with colleagues he became a centre of union among the Museum professors and prevented the personal and departmental friction and jealousy which could so easily have marred its early years. There were several reasons for his acting as a unifying figure. All his Museum colleagues were younger than he, with the majority having been born in the 1810s and 1820s. Phillips' age, allied to long experience of the Museum and scientific worlds, helped him to gain from his colleagues the respect owed to a father figure. His range of interests was so wide that he was at the least cognisant with all the sciences pursued in or near the Museum. As a publishing researcher he did not become a dodo in the 1860s and was still active when he died. He was broadly content with his positions at Oxford so he was the opposite of a grabbing, jealous, vain, or malevolent careerist. On the contrary, as one of his colleagues wrote, he had 'the divine gift of entering into and understanding the wants, the requirements, the position, of the persons with whom he had to do, and this without any conscious effort, but by the mere force of habitual and intelligent kindliness'. With his calm and genial manner, he was a disinterested facilitator of improvements in the Museum and was the trusted spokesman within the University for it. It has been alleged by Engel that science at Oxford ran into difficulties in the mid-1870s because of the agricultural depression. But one might also argue that the death of Phillips in 1874 removed from the University the only person capable of uniting the science professors and of eliciting their general will. In this chapter I briefly examine Phillips' modest contribution to the Ashmolean as its keeper. Given the paucity of documentation I do not add much to the standard accounts of Phillips' reign at the Ashmolean given by Ovenell and MacGregor. In happy contrast I clarify for the first time the nature of Phillips' keepership of the University Museum. In so doing I am also responding to Pickstone's call for more study of museological science in Victorian Britain.[1]

11.1 The Ashmolean Museum

The old Ashmolean building in Oxford, founded by Elias Ashmole and opened in 1683, is the oldest public museum in Britain. Previously known as the Ashmolean Museum, it is now the University's Museum of the History of Science. For its first 175 years this modest three-storied building was an important site for science in the University via its collections and the research and teaching which took place there. Under Philip Duncan, its keeper 1828-54, its zoological and especially ornithological collections became well known and a basis for the lectures he gave in it; and its chemical laboratory remained in use for teaching and research by Daubeny and then Story-Maskelyne. During Duncan's regime the Ashmolean's collections of antiquities, coins, medals, books, manuscripts, paintings, and ethnological specimens, were deemed to be of secondary importance. By late 1853, however, the University had committed itself to a new University Museum which would be a science centre, embracing collections, teaching, and research. Aware that the Ashmolean could not long continue as primarily a museum for natural history and feeling his advanced age (eighty-two), Duncan resigned in spring 1854.[2]

Phillips, who had just arrived in Oxford as deputy reader in geology, was soon appointed his successor. Having had charge at York of a museum devoted to Yorkshire science and York antiquities, he was well qualified to be keeper of another mixed museum. He was eminent in science and not uninterested in antiquities, archaeology, ethnology, and drawing. As Ashmolean keeper he was, however, soon diverted and dismayed. The diversion was the planning of the new University Museum, in which he was formally involved from February 1854. In March 1857 he was elected its first keeper and late next year moved into the keeper's house adjacent to it. Not surprisingly he increasingly saw the new Museum as his primary responsibility, leaving the Ashmolean as a secondary consideration. His dismay concerned his salary which was paid from the Rawlinson Trust. Having been paid £55 in 1854, he received nothing for three years: those who appointed him had failed to notice that Phillips was not eligible to be paid from this Trust on two counts, one understandable but the other bizarre; he was not a regular Oxford MA and he was an FRS. After Phillips' election as keeper of the new Museum, his dual keepership was regularised by the University. His Ashmolean post was freed from Rawlinson's restrictions, in 1858 he was paid £309 as arrears of salary due to him for four years, and from 1859 his combined salary as keeper of both museums was £80 pa (paid mainly from the Rawlinson Trust) with rent-free accommodation in the keeper's house.[3] The financial shemozzle from 1855 did not endear Phillips to the Ashmolean.

From the start of his keepership Phillips knew that the eventual transfer of the natural history specimens to the new Museum would deprive the Ashmolean of its chief function. In 1857 he saw another role disappear when chemical teaching and research ended. Negotiations between the University and the Ashmolean visitors, begun in 1858, led in 1860 to the transfer to the Bodleian Library of the Ashmolean's books, manuscripts, coins, and medals. It was probable by 1858 that the Ashmolean would soon be left with only its antiquities, ethnological specimens, and pictures. In order to forestall schemes for using part of the building for examinations and as an extension of the Bodleian, James Parker wrote an open letter to Phillips in March 1858 in which he proposed that the Ashmolean should become primarily an archaeological and historical museum. Parker was the librarian and, with his father, John Henry, a leading spirit of the Oxford Architectural Society which in 1860 added history to its formal remit. He was a keen collector of fossils and certainly by the 1860s was a friend of Phillips and collaborator in local geological research with him. Amidst all these changes and plans Phillips, it seems, let others make the most of the running. He was still interested in archaeology as an adjunct of geology; and, perhaps drawing on his experience with the Yorkshire Antiquarian Club, he instigated in the late 1850s the practice of depositing in the Ashmolean collections of Anglo-Saxon material derived from local excavations. The first major fruit of this policy was the donation from 1857 of the finds made by John Akerman at Brighthampton and Standlake, two villages near Oxford. Though Phillips agreed with Parker's proposal, Phillips shunned the Oxford Architectural and Historical Society which backed it; he did not lobby in the University in its favour, and it seems contributed nothing to decisions about the structural alterations, approved by the University and completed in 1864, which gave to the Ashmolean a definite archaeological role.[4]

From 1864 the basement housed mainly the Arundel collection of classical sculpture, the upper floor was used by the University for examinations, and the archaeological and ethnological museum occupied no more than the ground floor. Inevitably many of its specimens could still not be displayed. Given the problem of overcrowding, which was exacerbated by many donations, Phillips gave low priority to attracting collections to the Ashmolean. He was content to continue receiving local Anglo-Saxon remains, which culminated in the important donations in 1865 of specimens from graves at Fairford, Gloucestershire, discovered and described by William Wylie. On only one occasion did Phillips urge the University to try to secure an important collection: in the mid-1860s he wanted and gained part of the ethnological collection of the late Henry Christy.[5]

Phillips was too busy elsewhere to do much in the Ashmolean so he continued to supervise George Rowell, his underkeeper, with no more than a light touch. Rowell arranged collections in geographical order, compiled a register of accessions, and in 1870 produced a published list of archaeological and ethnological donations since 1836, as a supplement to Duncan's catalogue of that year. When these basic tasks were almost complete, Phillips was given a welcome life-line to rescue him from his Ashmolean post. In autumn 1868 John Henry Parker, by then retired and aged sixty-two, proposed to the University that he become keeper of the Ashmolean. As bait he offered to endow the post himself with £250 pa from money left at his disposal by Robert Stephenson, the famous civil engineer. After the University had accepted Parker's scheme, Phillips was happy to relinquish his Ashmolean post to Parker in March 1870. The new keeper received £325 pa (£250 from himself and £75 from the Rawlinson Trust), leaving the University to pay Phillips' salary as keeper of the new Museum.[6]

Phillips' reign witnessed some successes and it had the distinct merit of avoiding three features which marred Parker's. Either resident in Rome or a chronic invalid in Oxford, the latter was an absentee keeper who died in post. From 1878 he fought a bitter and public controversy about catalogues with the prickly and opinionated Rowell, which led to Rowell's resignation in 1879. Thirdly, in 1881 Parker's regime was ferociously denounced in public by Greville Chester who from 1865 had been one of the Ashmolean's biggest benefactors. In contrast Phillips was not an absentee keeper. He was sufficiently on hand to guide Rowell without alienating him. Finally, through Rowell Phillips did just enough about arranging, labelling, and listing collections to prevent criticism from such interested parties as donors.[7]

11.2 Keeper in waiting

On 3 February 1854 Phillips gave his inaugural lecture. Next day he made his first appearance at a meeting of the delegacy (a University committee) concerned with the projected University Museum. Within two months the delegacy had produced a general design for it, with the aid of Rohde Hawkins, an architect who specialised in educational buildings. The Museum was to have a large central court, roofed with glass, for displaying collections, and the court was to be surrounded on three sides by lecture rooms and laboratories. Hawkins convinced the delegates that such a building could be erected for about £30,000. Their proposals were approved by Convocation on 30 March.[8]

On 8 April a new delegacy, which included Acland, Daubeny, Johnson, Bartholomew Price, and Phillips, was appointed to obtain detailed architectural plans and financial estimates and to recommend to Convocation the best design. The delegates' statement of requirements revealed their commitment to the connection of the sciences: the Museum, two storeys high and costing no more than £30,000, was to contain the various natural history collections displayed in a glass-covered central court, seven lecture rooms, laboratories, dissecting rooms, private work rooms, professorial sitting rooms, and a library; and they proposed a detached house for a keeper. Most of these facilities were to be arranged around three sides of the quadrangular court, with the fourth side being designed to allow for future extension. These interior matters were seen as more important than exterior decoration and there was no restriction on architectural style. Phillips supported strongly this highly ambitious plan for a University science centre which was far more than a space for displaying collections and out of proportion to the current state and future prospects of science in the University. He also made the useful suggestion that an architectural referee be appointed to help the delegates judge the designs which were to be submitted in open competition.[9]

After public display of the thirty-two designs, which bore identifying mottoes, the delegates short-listed six which were sent for evaluation to Philip Hardwick, the well-known London architect. He reported that all the designs cost more than £30,000, the least expensive being Fiat Justitia in Palladian style (by Edward Barry) which sadly violated another condition because it had three storeys. Though worried by Hardwick's report, the delegates nevertheless set up a working party of Henry Wellesley, Phillips, and the reverend George Butler, which on 2 December 1854 narrowed the field to two designs, Fiat Justitia and Nisi Dominus in Rhenish Gothic style (by Thomas Deane and Benjamin Woodward of Dublin). The working party thought Fiat offered a superior court for displaying collections; but Nisi was expandable and had an aspiring outline. Acland, who approved these views, reported that Nisi had a contractor who could do the job for £30,000.[10]

When the full delegacy met on 4 December, Acland and Johnson argued strongly in favour of Nisi's Gothic style, which offered unique capacity for irregular adaptation and asymmetric extension. After carefully evaluating the designs, Phillips told the delegacy that he was happy to accept either design subject to alterations. He was familiar with Gothic and classical styles from his own life: in York he had lived in a Gothic home and worked in a classical museum. For him there was no battle of architectural styles and no style necessarily appropriate for a scientific building. Above all he wanted the University to adopt one of the designs for a new Museum, to which there was still considerable opposition; and he did not wish this cause to be lost by indulging in diversionary architectural battles. Acland had a different view. On 11 December he issued privately a pseudonymous pamphlet arguing that Nisi's Gothic design was harmonious with the collegiate associations of Oxford. Next day Convocation voted in favour of Nisi's design. Within four days Wellesley, Acland, and Phillips were asked to report on necessary changes in Deane and Woodward's design. Their conclusions, in Phillips' hand and agreed to by Woodward, concerned professorial accommodation, a more detached chemistry laboratory, and better internal communication between departments. This was Phillips' first experience of the way in which the Museum as eventually built was the result of revisions of the competition drawings.[11]

In 1855 Phillips began to act as its keeper elect. Before Convocation voted on 8 May 1855 to have the Museum built by Lucas Brothers of Lambeth for £29,041, he correlated the tenders from fourteen building firms. He had some influence on the arrangements for the laying of the foundation stone on 20 June 1855 by Lord Derby, chancellor of the University. Phillips wished to promote the Museum in the colleges by having all their heads on the platform. Several of them did attend but it seems that he failed to implement his scheme for granting open access to the ceremony to all undergraduates. At the ceremony, at which Duncan but not Phillips spoke, the anthem, collects, and prayers revealed beliefs which Phillips shared. The new Museum would glorify God's name and show via study of His great and glorious works His wisdom, power, and goodness; and instil humility, not pride, in teachers and learners. No doubt Phillips also welcomed the way in which, at an accompanying ceremony, honorary doctorates were conferred on leading scientists, including colleagues and friends of his such as Lyell, Sabine, and Lloyd.[12]

11.3 Ruskinesque decorator

From 1855 Phillips was heavily involved in schemes for the Museum's interior decorative stonework which was paid for by private donations solicited by a printed appeal dated 1 June 1855 which, it seems, was written and distributed by him. The decoration envisaged was not merely ornamental: on the contrary it was needed 'to augment the scientific and artistic expression of the edifice'. The appeal focussed mainly on three sorts of decoration. Firstly, the shafts of 128 columns of the cloister round the court, each costing five pounds, were to be of British rocks selected to illustrate geological epochs. Secondly, the 192 carved capitals of the 128 columns and sixty-four piers, costing five pounds each, would illustrate the flora of various epochs, climates, and regions. Finally, there were to be seventeen statues, each costing seventy pounds, of six ancient philosophers in the porch or entrance and eleven moderns in the court.[13]

Phillips was officially in charge of the shafts scheme as soon as it became public and nurtured it sedulously. Initially in May 1855 he had thought of thirty-two igneous shafts of columns on the ground floor but was undecided about the ninety-six on the first floor. In the appeal next month it was stated that those on the first floor would be of sedimentary rocks. But Phillips quickly discovered that it was difficult to obtain large enough igneous specimens so instead he settled for all the more important kinds of marble, by which he meant polishable stone. In selecting and arranging the marbles, Phillips found that the intended allocation of igneous and sedimentary rock specimens to the ground and first floors respectively, was inhibiting. His final scheme had igneous, metamorphic, and sedimentary rocks on both floors. For example on the north of the ground floor there were eight shafts, all from Ireland and Devon, which were composed of palaeozoic rocks, some sedimentary and some metamorphic. Sometimes marbles of a given geological age were not grouped together. Overall the materials of the shafts were not haphazardly chosen pretty stones: in Phillips' view they were '*really* and *obviously useful*, as a part of the exhibition of natural objects' and as lessons in geology.[14]

This combination of 'grace with utility' was in harmony with the insistence of John Ruskin that true decoration in a gothic revival building had to inform the observer about nature. There is also no doubt that Ruskin, a pupil and helper of Buckland and a fellow of the Geological Society, knew a lot about geology; and that he thought the Museum a highly significant building. He subscribed no less than £300 for its decoration and, with his close friend Acland, published in 1859 the first book about it even before it was structurally complete. It has been well argued by Blau that, in their concerns with the Museum's decoration, Deane and Woodward exemplified Ruskinian Gothic. O'Dwyer has made the even stronger claim that Ruskin also contributed to Woodward's design for the Museum, principally in its elevations. Ruskin might well have suggested or approved the shafts scheme but, sadly, there is not a scrap of evidence that he did so. The only recorded contact between Phillips and Ruskin in the 1850s occurred in 1855 or 1856 when Ruskin corresponded with Phillips, whom he regarded as a kind friend, about the precipitous shales opposite Bolton Abbey, Yorkshire, and in the fourth volume of his *Modern painters* (1856) cited Phillips' *Manual* (1855) on them. Phillips' scheme was first discussed in May 1855 in the home of Acland and may well have owed something to Ruskin, as mediated by Acland, or to Acland himself. It is, however, certain that the Museum of Economic Geology, London, opened in 1851, was familiar to Phillips. He knew that the shafts of the columns in its interior were composed of polished British rocks and were therefore both functional and didactic. Phillips' notion of having shafts doubling as geological specimens was therefore not then unique in England, as Garnham has claimed.[15]

Phillips' project was immediately backed by leading local promoters of the Museum. Daubeny, Walker, Acland, the reverend Frederick Plumptre, and Phillips himself gave two shafts each. The reverend David Williams, the incoming vice-chancellor in 1856, indicated his enthusiasm by paying for no fewer than five. Among Phillips' geological associates, Walter Trevelyan (five) and Murchison (four) were early contributors to the project. The latter was keen that Phillips, as Rex Carbonaricus, should have 'a marble palace of all colours'. By June 1860 125 shafts costing £831 were in place. The shafts scheme was the only decorative project which was complete when the Museum was structurally finished in 1860. Partly for this reason various users of the Museum from Tuckwell onwards have regarded this resplendent petrographic display as an attractive and successful part of its decoration.[16]

It is significant that Phillips and not Daubeny, the professor of botany, was entrusted with the scheme for botanical capitals. It was probably devised by Acland, who was very keen in May 1855 that the ornament of the capitals should illustrate zoology and if possible geology. Next month the appeal for subscriptions dropped geology and within natural history emphasised flora. Though Ruskin and Woodward approved this plan, it seems that nothing was done until August 1858 when Phillips was asked by the Museum delegacy to furnish a list of plants for the capitals of columns and piers. He responded with a scheme for a unique series of carved plants, in which each column was to be crowned by a different plant, and with some sketches to show what he proposed. The plants were to be carved naturalistically, to show faith in nature, and were to be arranged in orders according to the natural system of classification. Thus the carved plants were to be not just pretty flowers randomly arranged: they were to be in a definite sequence which would help to teach a particular mode of classification, so that like the shafts they would in part be didactic.[17]

The capitals were carved by the famous O'Shea brothers, James and John, brought by Woodward from Ireland where they had worked as sculptors under him at the Museum of Trinity College, Dublin. At Oxford they were employed by him directly, especially from October 1858 when Lucas's contract ended. They often went to the University's Botanic Garden, chose a plant according to Phillips' arrangement, took it back to the Museum, and there copied its living form to show foliage in what Phillips called its 'natural habit'. On occasion they included animals associated with certain plants, such as sparrows with gramineae, ie wheat, barley and oats; but it was presumably Phillips who primed them to carve the marsupial didelphys peeping from zamiaceae (tropical dwarf palms). Thus Phillips' plan for the capitals implemented one key principle enunciated by Ruskin apropos Gothic revival architecture, ie that decoration should be founded on natural forms and be true to nature. Phillips' scheme was consonant with another principle dear to Ruskin, namely that ornamentation should be entrusted to the inventive power, even genius, of unlettered craftsmen. Sadly, lack of subscriptions led to the abandonment in the early 1860s of the capitals project, leaving thirty carved on the ground floor and sixteen on the first, a situation which persisted well beyond Phillips' death.[18]

Phillips made a minor contribution to the scheme of statues of great founders and improvers of natural knowledge, to which Queen Victoria gave £350. He had, it seems, nothing to do with abandoning the plan for six statues in the entrance and substituting a scheme for all seventeen (increased in 1856 to eighteen) to be placed on the edges of the court. He was, however, concerned about the size and mounting of the statues. He was successful in his plea that they should be life-sized but his 1855 plan for having them mounted on two levels was not adopted. By the early 1860s a dozen statues, not always of those originally envisaged, were in place in the court, mainly on supports projecting from the piers of the ground floor arcades. Again, they were not merely decorative: they were intended to induce contemplation and arouse inspiration.[19]

In 1856 Phillips continued to show to his colleagues on the sub-delegacy for the Museum his skill as an administrator and his expertise on museums. He supported Ruskin's view that interior ornamentation should be completed before attempting to decorate the exterior. He discovered that Woodward was cavalier in seeking approval of a deviation from the original design without submitting a financial estimate. As the solitary expert on the problematic glass roofing of the court, Phillips argued that it was feasible because that of the Museum of Economic Geology did not leak and the

Oxford roof was designed to be stronger than the roofs of the Crystal Palace and of railway stations. Finally, he began to co-ordinate the desiderata for lecture rooms and laboratories and drew up rough estimates for spending £1,000 pa in the Museum. He soon had his reward. On 24 March 1857 he was unanimously elected keeper of the Museum and could anticipate living rent-free in a spacious detached house designed by Woodward as part of the ensemble of buildings.[20]

11.4 Keeper 1857-1860

Discussion about the desiderata for the keeper and his accommodation had begun in February 1854 when the Museum delegates, of whom Phillips was one, agreed that the Museum needed a keeper who might be a professor who would occupy 'apartments suitable for a gentleman'. Two years later various committees of the Hebdomadal Council, which was the chief administrative body of the University, concluded that there should be a keeper, who could be the professor of zoology, geology, or mineralogy, paid between £80 and £100 pa and living in a rent-free Museum House. These notions were supported by the sub-delegacy and delegacy of the Museum, with Plumptre and Williams, past and present vice-chancellors, and Walker arguing strongly that the keeper should be a professor and not, as advocated by Burgon, an 'inferior person, like a college butler'. The delegacy and sub-delegacy also agreed that the keeper's house should be built on site to the south-east of the Museum, as Woodward envisaged in 1855.[21]

Woodward's design for the keeper's house, approved by the University in February 1857, departed considerably from his original notion of a three-storeyed tower house to counterbalance the tower of the Museum, then planned to be a truncated pyramid. His revised and enlarged design incorporated a wedge-shaped spire, echoing that of the Museum as built, a polychromatic façade, and Venetian Gothic ogee arches on the ground floor. Woodward concealed the rising cost of his revisions and, once building began in March 1857, the expense exceeded the University's limit of £700 which he had accepted, so that Phillips himself had to pay for some internal fittings and decoration of his house. Completed by late 1858 it was more finished-looking and more comprehensively carved than the Museum (Fig. 11.1). Its accommodation was generously suitable for a professor: on the ground floor it had a dining room, morning room, kitchen, and substantial outbuildings grouped around a yard; on the first floor, a long drawing room with two balconies on the south side, and three bedrooms; and on the second, three attic bedrooms. On the south side of the house, there was a large open porch with marble shafts. Phillips liked the architecture of the house, into which he moved late in 1858, but found it draughty. As his front porch was only fifty yards from the south-east corner of the Museum, he enjoyed the convenience of living on the job.[22]

From March 1857 until spring 1860, when the Museum was structurally complete, Phillips' responsibilities as keeper tested his emollient judgement and administrative effectiveness. He spent much time and effort on the fabric, fittings, and services of the building. In 1858 two aspects of the design were troublesome. Woodward wanted iron columns to support the heavy iron and glass roof of the court. The contractor for the ironwork, Francis Skidmore, chose slender wrought iron tubes for the columns, partly to secure the maximum amount of natural light for the court and partly to apply the new technique of tubular construction to Gothic architecture. In February

Fig. 11.1   The exterior of the Oxford University Museum from the south-west, an engraving of 1860 by J.H. Le Keux. It emphasises the size of the plant which Phillips, as keeper, looked after. The Museum is on the left, the chemistry laboratory in the middle, and the keeper's house of three stories on the right.

his tubes buckled and part of the roof collapsed. Skidmore consulted the structural engineer Fairbairn who visited Oxford and advised that plate iron be used for girders, that four shafts of cast iron tied together be used for columns, and that wrought iron be used only decoratively. His recommendations, costing around £2,000, were adopted. It is inconceivable that Phillips, the only Oxonian who knew Fairbairn well, did not have a hand in choosing him as an expert on strength of materials and tubular construction.[23]

The second problem of 1858 concerned the possible use of the tower for scientific research. Woodward's original design for the tower had dormer windows and a platform at the top, which Phillips welcomed as suitable for meteorological observations and experiments. In autumn 1857 it became clear that without authorisation Woodward had changed the design to a tall spire without dormers and without a roof platform. Given this fait accompli, Phillips and Acland proposed in April 1858 the construction of an iron gallery costing £255 at least halfway up the spire roof and enterable from it. This proposal, supported by Johnson and Donkin and agreed to by Woodward, was rejected by Convocation in June 1858 on aesthetic and financial grounds. To his chagrin Phillips failed to establish a place for regular meteorological work in the Museum: Woodward denied him the lead platform of the original design and the University denied him the iron gallery of Woodward's doubly revised design.[24]

The contract with Lucas covered only the shell of the Museum. It excluded ventilation, lighting, heating, water supply, drainage, the paving of the court, doors and floors of main rooms, painting, varnishing, glazing, display cases, fittings, apparatus, and furniture. Most of these necessary features were paid for piecemeal by the University via Convocation which between May 1856 and June 1860 was asked for fourteen separate grants of money by the Museum delegacy, of which Phillips was secretary from 1858. Inevitably he was involved in various contingencies which arose. In 1859, when the contractor for paving the court defaulted, Phillips provided for the Museum delegacy a summary of the circumstances leading to its request to Convocation for an extra grant to supplement the £800 already allocated. In 1858 he suggested a layout of the tall cases in the central court, a scheme which was adopted almost unchanged. In 1859 he not only monitored the cost of fittings for the rooms of various professors but explained to no less than Acland that any requests for future fittings should go to the Museum sub-delegacy which ought to know also the claims of other professors. As the Museum neared completion, Phillips wrote the occasional reports on what still needed to be done. He interested himself in the gas and water supply and drainage. Aware of the importance of the Museum's surroundings, he was concerned with the preparing and planting of the grounds.[25]

Phillips became de facto a treasurer and accountant to the Museum delegacy in late 1850s. He authorised payment of bills and dealt with the insurance of the building and contents of the Museum. He also joined Plumptre and William Thomson, future archbishop of York, in a sub-delegacy for Museum accounts set up in autumn 1858 to try to control its runaway expenditure which by spring of that year had reached £45,000. From then on Phillips contributed to the sub-delegacy's financial statements, which grew out of the first estimate of annual expenditure he had prepared in December 1857. But neither he nor anyone else devised a way of controlling the costs of piecemeal changes in the building plan: by 1860 £57,000 had been spent on the Museum, and by 1867 £87,000.[26]

By summer 1858 Phillips was conscious of his developing role as not only the collector but also the co-ordinator of information and desiderata pertaining to the Museum. His aim was to ensure that the various professors, the Museum delegacy, and he were 'well harmonised without confusion'. He thought that professors would retain total control of their time and the character of their lectures but hoped that 'systematic combination' might occur. He judged that as keeper he should protect the professors of the Museum from 'useless interference' while he looked after 'concerted arrangement, proportion, and harmony'. In order to fund the Museum adequately, Phillips hoped that for large applications professors would first put their requests to the Museum delegacy, who would examine and evaluate them with the keeper before passing approved requests to the University authorities backed by the 'whole force of the Museum'. For small grants, he thought that £300 pa put at the disposal of the Museum delegacy would suffice. By 1860 some of these high ideals had been put into practice. Late in 1857 he collected and summarised for the first time professorial statements about regular and special expenses of their departments so that the Hebdomadal Council would know the estimated expenditure for the next year. In May 1860 he correlated the requests from the professors for furniture for their rooms.[27]

11.5 The Museum's meanings

The structurally complete Museum embodied ideals dear to Acland, Phillips, and Daubeny. A single building provided for teaching, learning, and research in the sciences, which were assumed to be philosophically related. It catered for specialisation and a measure of vocational training in different subjects while proclaiming the higher visions of the connections of individual sciences and science as liberal education. These beliefs were palpable in the Museum's physical layout, which Forgan has called the integrated museum plan. All the sciences enjoyed immediate access to the large central arcaded court which was devoted to collections. Around it on both the ground and first floor were lecture theatres, laboratories, and other facilities for the various sciences, with the whole of the west front on the first floor devoted to the Radcliffe Library. Communication between the science departments in the core of the building was easy via two storeys of arcaded corridors and, for those attached to it, short ground-floor corridors were provided. These arrangements facilitated bilateral cooperation in research, such as that between Phillips and Rolleston on Oxford geology and archaeology; and on one occasion trilateral co-operation, when Henry Smith and Donkin advised Brodie about the mathematics of his so-called chemical calculus which he had devised to replace chemical atomism. Only one important science, botany, was not included in the Museum, much to the disappointment of Acland and Phillips: under Daubeny and his successors the base for botany remained the Botanic Garden.[28]

The architecture of the building of which Phillips was keeper was important nationally and locally (Fig. 11.2). Since the rebuilding of the Houses of Parliament begun in 1840 there had been in England no major public building and no museum erected in Gothic style. Occupying a site of four acres, the Museum was the largest building erected by the University, an honour previously held by the Bodleian Library quadrangle completed in the 1620s. The lecture rooms and laboratories which surrounded the Museum's court were indeed the scientific and Victorian equivalents of the rooms for arts subjects grouped around that quadrangle. Moreover in Oxford Gothic style was visible in many colleges and churches; in the Museum it was associated with the University and science. It is, however, wrong to see the Museum as an entirely secular building. Though its façade was like that of a Flemish town hall and the glass and iron roof of its court resembled that of the Crystal Palace, parts of it looked and felt like a monastery. The chemistry laboratory was explicitly modelled on the Abbot's Kitchen at Glastonbury Abbey. The arcaded corridors on two floors around the court were like the cloisters of a monastery, with the professorial rooms entered from the corridors resembling monks' cells.[29]

These monastic resonances were not unacceptable to Acland, Daubeny, and Phillips because for each of them the Museum had a natural theological function. For Acland it symbolised the value of science construed as the study of the material works of the Great Artificer, an activity which for him necessarily reinforced Christianity. For Daubeny the court of the Museum was 'the sanctuary of the temple of science, intended to include all those wonderful contrivances by which the Author of the universe manifests himself to His creatures'; and the surrounding laboratories and lecture rooms devoted to science were 'chambers of the ministering priests, engaged in worshipping at her altar, and in expounding her mysteries'. Phillips was less effusive but just as firm in his belief that the botanical sculpture of the capitals of columns in the Museum gave a glimpse of how 'the GREAT ARTIFICER moulds the

Fig. 11.2   The interior of Oxford University Museum photographed *circa*
1890 but little different from its appearance in Phillips' time.
The court, crammed with specimens, was divided by cast-iron
columns with wrought-iron decoration, covered by a glass roof, and
surrounded on two stories by open arcades, the columns of which
were composed of different rocks chosen by Phillips.

lilies of the field and the leaves of the forest'. No doubt all three of them appreciated the significance of the carving on the keystone of the arch of the Museum's entrance. It depicted an angel bearing in one hand the Bible and in the other a biological cell. It told those entering the Museum that the study of the works of nature was a holy act, the result of which was not hostile to revealed Christianity. Of Phillips' professorial colleagues in science in 1860 only Brodie, a notorious atheist, and Story-Maskelyne, a free-thinker, rejected the angel's message. The other eight, made up of five non-clerics (Rolleston, Henry Smith, Acland, Donkin, and Daubeny) and three clergymen (Price, Walker, and Main), broadly accepted it. These Christian philosophers ensured that the Museum was anything but a nest of atheism, infidelity, and materialism.[30]

The first organisation to use the Museum was not the University but the British Association which paid its third visit to Oxford in 1860. The Museum provided a commodious and appropriate location for most of the Association's sections and for two conversazione and two soirées, one embellished by experiments and the other by microscopes. To visitors the Museum showed spectacularly the University's commitment to observational and experimental sciences. It was also shown by the way in which the University's chancellor and the vice-chancellor (Jeune, an early supporter of the Museum project) were vice-presidents of the Association. The established church in the locality also showed its devotion to the sciences pursued by the Association: Liddell, dean of Christ Church and a mentor of Acland whom he had taught, and Wilberforce, the bishop of Oxford, were vice-presidents. Wilberforce, it should be noted, was a fellow of the Royal and Geological Societies, an early backer of the Museum, and a life member of the Association of which he had been vice-president in 1847 when it met in Oxford. Most of the University's scientists were prominent as office-holders in the Association, showing to visitors that Oxford had enough distinguished men to make the Museum effective and to promote science successfully in the University. Daubeny, Acland, and Donkin were vice-presidents. Phillips was assistant general secretary and at the meeting Walker was elected general secretary. The triumvirate of Rolleston, Henry Smith, and Griffith acted as local secretaries. In the Association's sections, Price presided over section A with Main as a vice-president, Brodie over B, and Rolleston over the sub- section of physiology in D. In sections B and D two promising young Oxonians, Augustus Harcourt and Church, acted as secretaries. Walker delivered one of the two evening lectures. Thus all the Museum professors except Story-Maskelyne, who was merely a member of the committee of section B, were conspicuous at the 1860 BAAS.[31]

It was in the zoology and botany section that there occurred in the Museum on 30 June 1860 the allegedly momentous conflict between Wilberforce and Huxley apropos Darwin's theory of evolution by natural selection. Though the exact details are unclear it seems that Wilberforce, a lover of dinner-table banter, facetiously asked Huxley whether he was descended from an ape on his grandmother's or grandfather's side. Huxley declared for the grandpaternal ape and, with pompous solemnity, attacked Wilberforce for prostituting his talents and office by introducing ridicule into a grave scientific discussion. The exchange has too often been seen as a victory for Darwinism over clerical obscuration and, less crudely, of aggressive scientific professionalism over declining clerical power. On the first view Wilberforce was an interfering ignoramus who deserved to be mocked as Soapy Sam. As Brooke has rightly stressed, this interpretation of the Huxley-Wilberforce affair is inadequate because it ignores other themes which underpinned their differences.[32]

For Phillips and Daubeny the conflict had several resonances. They regretted the absence in Huxley's contributions to the meeting of a Christian or deistic tone or of any allusion to contrivance in nature, a theme emphasised by Wrottesley in his presidential address. They feared that the commotion, by giving ammunition to the opponents of science at Oxford, might retard its progress there and affect adversely the future of the Museum. They were unhappy about a debate which departed from the Association's ideology of science as calm neutral ground and about the regrettable deportment of the protagonists, particularly Huxley and especially in front of a large audience which was mixed intellectually, socially, and sexually. Moreover, Daubeny himself had been subjected to Huxley's rudeness on 28 June when in section D he had given a paper on the sexuality of plants, which gave a fair and in part supportive hearing to Darwin's theory. Huxley dismissed Daubeny's contribution but Darwin, who was not present, thought it '*very* liberal and candid'. Daubeny's reaction to Huxley's deportment and views was shown at an evening party he gave in the Botanic Garden on 30 June: he climbed into and sat silently in a monkey puzzle tree. Phillips' reaction was less comical. As editor of the *BAAS 1860 report* he applied his standard rule that discussions of papers were not to be recorded: the Association's own formal record of the meeting did not mention the unseemly altercation between Huxley and Wilberforce.[33]

In the academic year 1860-1 Phillips witnessed the installation of the Radcliffe Library in the Museum and oversaw the transfer of various collections to it, thus making it fully operational. Though Acland's importance in the fight for the Museum has sometimes been exaggerated, it is indisputable that he had long pleaded that a scientific library should be an essential part of the new Museum; and that, as the Radcliffe Librarian from 1851, he launched in 1856 a scheme to transfer to the Museum the scientific books kept in the Radcliffe Camera, which could then be used as a reading room by the Bodleian Library. In 1860 the Radcliffe Trustees and the University agreed to Acland's proposals. The books were moved in August 1861 to the upper storey of the west front. It then became possible for books and collections to be used together: readers were allowed not only to consult books but to take them down to the court where the specimens they described and illustrated were displayed. In this way what Rudwick has called proxy specimens could be compared with actual specimens. Phillips welcomed the arrival of the Radcliffe Library in the Museum as an important enhancement of the facilities for studying science at Oxford, though he was worried about the transfer before it happened.[34]

11.6 Collections

The unification of collections in the new Museum went smoothly. Phillips' meticulous yet emollient efficiency ensured that there were no problems with the transfers of geology, mineralogy, and experimental philosophy from the Clarendon Building, of anatomy from Christ Church, which took six years, and of zoology from the Ashmolean Museum. It seems that Phillips decided where this collection should be located and by February 1861 had secured the services of Rowell at a salary of £90 pa to be assistant keeper in the new Museum, with particular responsibility for zoology. Greater difficulty was presented by the huge collection of entomological specimens and the accompanying library, both of international status, which had been donated, along with natural history specimens and engravings, to the University in 1849 by the reverend Frederick Hope. It soon became apparent that there was an accommodation problem. As a temporary and unworthy measure, in 1850 the engravings were accepted by the Bodleian and all the zoological material and the library were put in the Taylor Institution, opened in 1848 to promote the study of modern European languages. Hope was so concerned about the state of his collection that in 1857 he persuaded a rather obstructive University to accept his offer of £250 pa for a special curator, provided his protégé, Obadiah Westwood, be the first occupant of the post. Next year the collection was further split when most of the natural history specimens, some of which had shown signs of deterioration, were moved to the Ashmolean Museum. They were reunited with the entomological collection and library in 1861 in the new Museum but were not under Phillips' immediate care because in January that year Westwood was nominated by Hope to be the first Hope professor of zoology, after the University had accepted in 1860 Hope's endowment of the chair with a capital sum of £10,000.

Westwood was responsible for the entomological collection and library, which together were known as Hope's entomological museum, but he was very unhappy about its location: though it contained the star collection in the Museum, it was allocated just one room about thirty feet long on the first floor of the south side of the Museum. This accommodation was considerably less than the whole of the first floor of the west front, which Acland had envisaged in the mid-1850s; and it was increasingly inadequate, as donations including those from Hope's widow poured in. Even so the Hope museum and the displays of collections in the court and in some corridors indicated that in the Museum those subjects in which displayed or stored specimens were central would not be subordinate to those pursued mainly in laboratories.[35]

The improved facilities provided by the Museum for the display of specimens encouraged so many donations that the Museum became more and more devoted to natural history. Most gifts were easily processed but some were not, especially when donors wished to preserve the integrity of their gift and laid down conditions for acceptance. The most common difficulties Phillips faced were finding space for donations, cataloguing them, and ensuring that they were compatible with the Museum's purposes as defined by himself. The fate of three zoological collections is instructive. One proposed donation ended in embarrassment. In 1854 the widow and father-in-law, Sir William Jardine, of Phillips' friend the late Hugh Strickland, offered to the University his large ornithological collection subject to certain stringent conditions, such as suitable accommodation. After Jardine had rejected the Ashmolean Museum as unsuitable, protracted but inconclusive negotiations ensued. Even the availability of the new Museum did not lead to agreement. In 1864 Phillips offered wall space in a lecture room, a proposal vehemently rejected by Jardine. Two years later Phillips reported to the vice-chancellor that he was not in favour of the desperate expedient of housing the collection in a room recently allocated to mathematics, that there was no space anywhere in the Museum for it, and that the only possible location was the top floor of the Ashmolean Museum. Mrs Strickland withdrew her offer and, no doubt to Phillips' chagrin, found it easy to donate her husband's collection to Cambridge in 1867. The affair of the Strickland collection induced Phillips in late 1866 to prepare a circular about gifts offered under conditions which conflicted with the Museum's practice.[36]

In contrast another protracted saga ended successfully. The gift of the collection of bird skins by Sir Harford Jones-Brydges was announced in 1858 but its arrival at the Museum was postponed for years. In March 1874 Phillips advised Jones-Brydges to defer depositing the collection because of the shortage of space for zoological specimens and the limited staff available for cataloguing. After Phillips' death Jones-Brydges ignored this advice and as a loyal Oxonian soon presented the collection.[37]

Phillips' determination to accept donations on his terms and not on those of donors was tested apropos the important zoological collection amassed by William Burchell during his explorations of Africa and Brazil, each of which lasted five years. After his suicide in 1863, his sister offered it in August 1864, provided it was properly displayed as a single collection and occupied separate special rooms. She wanted a lasting undivided memorial to her brother, whose collection was the history of his life. Phillips' response was that in the Museum there was no room to store the whole collection, never mind display it. His preferred solution was to have no conditions attached to bequests and to group existing zoological specimens, which were typical forms, according to their natural affinities and then to these typical series add donated specimens when suitable, the donor's name being recorded. In April 1865 Burchell's sister capitulated: her brother's collection, including the types of Burchell's zebra and white rhinoceros, arrived at the Museum that year.[38]

In trying to cope with the sheer size of the collections and additions to them, Phillips accepted that the desirable enlargement of the Museum, especially for zoological space, would not occur while he was keeper. He had limited money to pay for cabinets for new accessions and for staff to catalogue them, not least because the Museum turned out to be expensive to maintain: repairing the roof, heating the court, and lighting it after dark were all costly. In these constraining circumstances it seems that, as keeper, Phillips thought his chief responsibility was to oversee a creditable exhibition in the court and arcades of selected and labelled specimens which would be useful for teaching the appropriate sciences to undergraduates and for their private study. For Phillips this specialist teaching function, totally appropriate for a university museum, took precedence over its other pedagogic function, which was to edify non-specialist members of the University and on occasion the general public. He took this view partly because his resources were limited and partly because he thought that the way in which specimens in museums were exhibited for the public was incompatible with serious study.[39]

11.7   Unifier of the Museum 1861-1874

From 1861 until his death Phillips continued as general factotum of the Museum. In 1866 he acted as security officer when he permitted the Oxford Philharmonic Society to rehearse in the big lecture room. Next year, as information officer, he responded to the enquiry from Joseph Greenwood, principal of Owens College, Manchester, about physical science at Oxford. Phillips continued to care for the fabric of the Museum and to act as its accountant. He was particularly concerned about problems with the glass roof. It admitted so much natural light that on sunny days the court became quite hot and anatomical and zoological specimens began to suffer. The glass tiles, of which it was composed, were not air proof so that cold air leaked into the court and on cloudy days lowered the temperature considerably. Phillips' solutions, such as varnishing the glass and puttying the joints of the glass tiles, were ingenious but did not eliminate these undesirable effects. He also dealt with changes to ventilation and lighting in various departments of the Museum and was responsible for dealing with the contractor who built the Clarendon Laboratory. As the accountant for the Museum he prepared estimates and financial statements, explaining them in 1866 to Henry Wall, professor of logic and a virulent opponent of the Museum. By the early 1870s Phillips was dealing with the University Chest on behalf of the Museum.[40]

These useful services to the Museum were undertaken not just as humdrum responsibilities of the keeper but also because Phillips had a clear vision of his ambitions for it. That was re-affirmed in 1873 when he and Acland led an unsuccessful campaign in favour of moving the Botanic Garden and the activities of the professor of botany to a site adjacent to the Museum. Phillips thought it essential to bring together in or near the Museum teaching and research in all the main branches of natural science because they and mathematics were 'legally sisters' and constituted a 'family circle' of which botany was a member. It followed that it was desirable to have in one centre all teaching in science and mathematics so that senior and junior members of the University would gain 'that larger and more comprehensive view of natural phenomena which is sometimes likely to be forgotten amidst the unceasing toil of special teaching'. Phillips wanted to unite in the Museum botany and palaeobotany, vegetable and animal physiology; and to have a botanic garden, which he viewed as a living museum, near other collections and the Radcliffe Library. Then it would be possible for all students to study botany more effectively, while still 'conversing with God in the garden of creation'.[41]

Phillips' vision for the Museum was allied to his familiarity with all the sciences cultivated in or near it. In the 1860s he published research on geology, astronomy, and natural and experimental philosophy; he had more than a passing acquaintance with mineralogy, mathematics, chemistry, and zoology; and he was not entirely ignorant of anatomy and physiology. No scientist at Oxford came remotely near him in breadth of scientific accomplishments. He was ideally qualified by his exceptional range of competence to be keeper of a museum in which the main sciences were brought together and the unity of the universe studied. Phillips also enjoyed easy relations with the professors who worked in or near the Museum. Via overlapping research interests, temperamental affinity, or shared concerns he was on good terms with Rolleston, Henry Smith, Story-Maskelyne, Walker, Price, Westwood, Donkin, Pritchard, Acland (who taught in the Radcliffe Infirmary), and Main (who worked in the Radcliffe Observatory). Though Brodie and Robert Clifton, appointed professor of experimental philosophy in 1865, were institutionally demanding, Phillips was not at odds with them. His Museum colleagues respected his wide scientific interests, unfailing courtesy, irenic temperament, and calm judgement. Unlike keepers elsewhere, such as Owen who superintended the British Museum (Natural History), he was not an aggressive careerist. On the basis of all these characteristics, Phillips became in the 1860s the dependable spokesman within the University for the Museum and its professors, helping them to agree among themselves and to avoid personal friction and departmental jealousy. He was trusted by individual professors to present their views even in their absence and there was no instance of a report by Phillips to the Museum delegacy being rejected. His abundant notes and annotations on pertinent documents show that he prepared meticulously for its meetings. As Acland averred, Phillips cheered the Museum on by acting as its wise president who gave it stability of view and purpose.[42]

His role as lord keeper, a vice-chancellorian sobriquet, was shown when he dealt with matters which in other hands could have led to acrimonious controversy. He produced the printed time-table of lectures in the Museum, not an easy task when only the period of noon to 2 pm six days a week was available. He co-ordinated requests from professors for grants and allocated space. When in 1866 Henry Smith wanted to be granted an unused room in the Museum by the delegates, he insisted on Phillips's assent not least because Phillips saw mathematics as the strongest link between the sciences pursued in the Museum. Above all Phillips was trusted to put to the University authorities, especially the vice-chancellor, a convincing and agreed overview of professorial requirements.[43]

In the mid-1860s Phillips launched the proposals which led to the building of the Clarendon Laboratory, opened in 1870, and to Oxford increasing its lead over Cambridge in providing for science. The campaign had its origin in the election of Clifton in November 1865 to the chair of experimental philosophy in preference to Griffith, who had acted as deputy to the ill Walker. Phillips had brought Griffith to national prominence as his successor as assistant secretary to the British Association and wrote a testimonial for him. Next year Phillips reported on the pressing problems of accommodation in the Museum and processed the requirements that Clifton and Brodie had set out for experimental philosophy and chemistry. While the latter desired improvements to his existing laboratory, the former was so dissatisfied with his laboratory facilities that he wanted a new suite of laboratories for practical teaching, with particular emphasis on precise measurement. Writing on behalf of the Museum professors to the vice-chancellor, in November 1866 and February 1867, Phillips urged that the Museum be extended eastwards at a cost of £6,000 to satisfy Clifton's ambitions and attached a detailed statement from him. Having supported Clifton's chief opponent in 1865, Phillips was no curmudgeon: within a year he backed Clifton's ambitious aims for experimental philosophy because he realised the inadequacy of the current facilities for the subject and had a passion for measuring instruments.[44]

Fortunately for Clifton, in late 1867 the Clarendon Trustees agreed to instantiate his dreams by offering to pay for a new laboratory. The University accepted this unexpected gift so that by 1870 it possessed a two-storeyed laboratory for experimental physics which cost the Trustees £10,700. Built in Gothic style to Deane's design on the north-west of the Museum, the two buildings were connected by a corridor, which made clear that this model laboratory, then unique in Europe for its various facilities, was nevertheless still part of the Museum and did not vitiate the Phillipsian vision. The Clarendon Laboratory crowned the Museum's provision for science at Oxford which by the early 1870s confounded the critics of the University. Even Huxley had to admit that 'there was nowhere in the world a more efficient or better school ... for teaching the great branches of physical science than was at the present time to be found in the University of Oxford'.[45]

Notes

1. Rolleston and Smith, 1874, p. 9 (q); Engel, 1983, pp. 217-30; Pickstone. 1994, 2000.
2. On the regime of Philip Bury Duncan (1772-1863), *DNB*, see Ovenell, 1986, pp. 198-210; Simcock, 1984, pp. 11-13; MacGregor and Headon, 2000; Duncan, 1836.
3. Ovenell, 1986, pp. 211-29 (219-20 q); Richard Rawlinson (1690-1755), *DNB*, disliked FRSs.
4. Ovenell, 1986, pp. 216-19; MacGregor, 1997, pp. 602-604; Parker, J. 1858; Parker, J.H., 1870, pp. 6, 7, 32-4; for J. Parker's geological research, collections and assistance to Phillips, see Gunther, 1925, pp. 245-6, Phillips, 1871a, pp. 187, 194-6, 249, 251, 319,

and Woodward, 1907, pp. 165-6; John Henry Parker (1806-84), *DNB*, Oxford bookseller and antiquary, keeper of Ashmolean 1870-84; John Yonge Akerman (1806-73), *DNB*, secretary of Society of Antiquaries 1853-60.

5. Ovenell, 1986, pp. 224-5; Wylie, 1852; Phillips to vice- chancellor, 19 June 1865, Lubbock to Phillips, 9 July 1865, PP; William Michael Wylie (1811-87); Henry Christy (1810-65), *DNB*.

6. Ovenell, 1986, pp. 220-29; Rowell, 1870, prepared under Phillips' direction; OUA, Ashmolean keepership correspondence 1868-70, W.P.B./2/14; Phillips to vice-chancellor, 13 Feb 1869, UM/M/1.3; George Augustus Rowell (1804-92), underkeeper of Ashmolean Museum 1854-62, 1864-79; Robert Stephenson (1803-59), *DNB*.

7. On Parker's reign, see Ovenell, 1986, pp. 230-49; Chester, 1881; Greville John Chester (1830-92).

8. UM/M/1.1, 4,7, 13 Feb, 25, 28 March 1854; Major Rohde Hawkins (1820-84), *DNB*.

9. UM/M/1.1, 27 April, 2 May 1854; OUA, N.W.2.1, Statement of the requirements of the Oxford University Museum ... by the delegates who were appointed ... 8 April 1854; Fox, 1997, p. 651; Colvin, 1983, p. 125.

10. UM/M/1.1, 17, 28 Oct, 11, 14, 29 Nov, 2 Dec 1854; Philip Hardwick (1792-1870), *DNB*; Edward Middleton Barry (1830-80), *DNB*; George Butler (1819-90), *DNB*, then a curate in Oxford, principal of Butler's Hall, Oxford, 1856-8; Thomas Deane (1792-1871), *DNB*; Benjamin Woodward (1815-61), *DNB*.

11. UM/M/1/1, 4, 11, 15, 18 Dec 1854; OUM/HA/box 1, f. 3, iii, Phillips' notebook 'Oxford Museum, 1854 –', which contains his evaluations of the designs; O'Dwyer, 1997, pp. 171-2.

12. OUM/HA/box 1, f. 4, tender documents, April 1855; box 3, f. 1, report from Walker and Phillips about foundation stone ceremony, 8 June 1855; UM/M/1.1, 11 June 1855; Fox, 1997, p. 641; OUA, N.W.2.1, Form of prayer ... 20 June, 1855; Atlay, 1903, p. 213; *The Times*, 21 June 1855; Edward George Geoffrey Smith Stanley, fourteenth Earl of Derby (1799-1869), *DNB*, chancellor of the University, 1852-69.

13. O'Dwyer, 1997, p. 221; OUA, N.W.2.1, flier, 1 June 1855 (q).

14. OUM/HA/box 2, f. 7 , Phillips' documents relating to shafts; UM/M/1.1, 18 May, 29 June, 24 Oct 1855; Phillips to Acland, 21 Jan 1859, in Acland and Ruskin, 1893, pp. 91-100 (92 q) which reprinted the main text of the first edition of 1859.

15. Acland and Ruskin, 1893, pp. 51, 100 (q); UM/M/1.1, 18 May 1855; Blau, 1982, pp. 48-81 on OUM; O'Dwyer, 1997, pp. 176, 224; Ruskin, 1856, pp. 306-7; Phillips, 1855c, p. 177; Yanni, 1999, pp. 51-61, on Museum of Economic Geology; Garnham, 1992, p. 5; John Ruskin (1819-1900), *DNB*.

16. OUM/HA/box 2, f. 5, Phillips' lists of donors and gifts, 1855, 1856; Murchison to Phillips, 15 May [1856?], PP; O'Dwyer, 1997, p. 224; for appreciation of the shafts see Tuckwell, 1900, pp. 48-9, Vernon and Vernon, 1909, pp. 82-3, and Vincent, 1994, p. 3. Frederick Charles Plumptre (1796-1870), vice-chancellor 1848-52; David Williams (1786-1860), vice-chancellor 1856-8; Walter Calverley Trevelyan (1797-1879), *DNB*, Oxford-educated naturalist.

17. UM/M/1.1, 18 May 1855, 14 Aug 1858; Acland and Ruskin, 1893, pp. 25, 83, 99; Blau, 1982 pp. 66-7; OUA, UM/F/7/7, capitals documents, Phillips' memorandum and sketches, 5 Nov 1858; OUM/HA, box 2, f. 6, Phillips' notes and sketches for capitals; Atlay, 1903, p. 214.

18. Vernon and Vernon, 1909, p. 79; Acland and Ruskin, 1893, pp. 29-31, 51-2, 95-9 (99 q).

19. UM/M.1.1, 29 June 1855; Acland and Ruskin, 1893, p. 25; OUM/HA/box 2, f. 5, matters relating to statues and shafts, Phillips' list of statues, sketches of possible positions, and Phillips to Acland, 11 Jan 1857.

20. UM/M/1.1, 3, 19 March, 2 May, 7 Nov 1856; OUM/HA, box 3, f. 1, administrative matters, 1855-7, Phillips' estimates for annual expenditure, 13 Nov 1856; UM/M/1.2, 24 March 1857.

21. UM/M/1.1, 4 (q), 13 Feb 1854, 20 Oct (q), 21 Nov, 6 Dec 1856; HCR, 5 March, 4 Nov 1856.
22. UM/M/1.1, 28 Feb 1857, 12 Nov 1858; O'Dwyer, 1997, pp. 193-5, 422-32 is the only architectural historian who discusses the keeper's house at length.
23. UM/M/1.1, 23 April 1858; OUM/HA, box 5, f. 3, documents concerning roof and tower, Fairbairn's report, 10 April 1858, Plumptre circular, 2 June 1858; O'Dwyer, 1997, pp. 262-3; Francis Alfred Skidmore (1816-96).
24. UM/M/1.1, 24 April 1858; OUM/HA, box 2, f. 8, Phillips' memo about tower, nd [April 1858]; OUM/HA/box 5, f. 3, documents about roof and tower, Donkin to Plumptre, 25 May 1858; Fox, 1997, p. 655; O'Dwyer, 1997, pp. 198-203.
25. Vernon and Vernon, 1909, pp. 67-71; OUM/HA/box 1, f. 13, documents about paving of court, Phillips' document, 23 Nov 1859; OUM/HA, box 2, f. 2, cases and fittings, Phillips' sketches, diagram, and notes on cases and layout, 16 Oct 1858; OUM/HA, box 2, f. 3, professors' fittings, especially Phillips to Plumptre, 16 Dec 1859; OUM/HA/box 1, f. 11, letters about completion of Museum, Phillips' draft report, nd [after Nov 1858]; OUM/HA/ out of sequence box 1, f. 8, Phillips' sketch and notes on water supply and drainage, 23 Oct 1857; OUM/HA, box 1, f. 17, grounds documents, Baxter to Phillips, 4 Sept 1858, 18 Nov 1861.
26. UM/M/1.2, 18 Dec 1857, 2 Nov 1858; Fox, 1997, p. 655; OUM/HA/box 3, f. 6, four bills authorised for payment by Phillips, 1859-60; OUM/HA/box 4, f. 3, statements of accounts, Phillips' summary for Oct 1858 to Oct 1859; OUM/HA, box 5, f. 3, Phillips' notes and calculations on statement of accounts to 1 Oct 1858 by Plumptre, Phillips, and Thomson, 26 Nov 1858.
27. OUM/HA, box 3, f. 2, administrative matters, 1858-60 (Phillips), memo, 30 June 1858 (qs); OUM/HA, box 3, f. 1, Phillips' report on professorial expenses, 12 Dec 1857, memo to Museum delegacy 18 Dec 1857, Phillips to Cradock, 19 Dec 1857; UM/M/1.2, 18 Dec 1857, 2 Nov 1858, 1 Jan, 9 Nov 1859, 9 May 1860.
28. Fox, 1997, pp. 660-1; Acland and Ruskin, 1893, p. xxvii, 62; Forgan, 1989, pp. 413-17; for bilateral research, see sections 10.4 and 10.6; for trilateral research, Brock, 1967, pp. 70, 99, 106-8, 111-13.
29. Colvin, 1983, pp. 128-34, is good on Woodward's architecture but wrong (p. 132) in asserting that Oxford's scientists were trying to free themselves from the dead hand of classical learning; Yanni, 1999, pp. 78-80, 85-7.
30. Atlay, 1903. pp. 213, 215, 219, 305-8; Daubeny, 1856, p. lxix (q); Phillips in Acland and Ruskin, 1893, p. 99.
31. *BAAS 1860 report*; Augustus George Vernon Harcourt (1834-1919), *DNB*, Lee's reader in chemistry at Christ Church, 1859-1902, nephew of W.V. Harcourt and pioneer in chemical kinetics; when twenty-three Church succeeded Rolleston in 1860 as Lee's reader in anatomy.
32. For recent accounts see Desmond and Moore, 1992, pp. 492-9; Desmond, 1994, pp. 277-81; Fox, 1997, pp. 657-9; Secord, 2000, pp. 513-14; Brooke, 2001a; Browne, 2002, pp. 114-28.
33. Morrell and Thackray, 1981, pp. 395-6; *BAAS 1860 report*, pp. lxxiv-lxxv (Wrottesley); Daubeny, 1860; *Athenaeum*, 1860, ii, pp. 18-19, 26; Darwin to Lyell, 11 Aug 1860 (q) in Burkhardt and Smith, 1993, pp. 319-21; Hutchins, 1995b, pp. 25, 58.
34. Aclay, 1903, pp. 308-13; Guest, 1991, pp. 185-7; UM/M/1.2, 8 June 1861.
35. UM/M/1.2, 6 Feb, 8 June 1861; Fox, 1997, pp. 657, 664-5; Smith, 1986, pp. 1-46; OUA, MR/6/1/4, Acland and Phillips, 23 May 1856, report on Hope collection, in Phillips' hand; Frederick William Hope (1797-1862), *DNB*, Oxford graduate, entomologist, and collector.
36. Fox, 1997, p. 673; HCR, 26 April 1860; Ovenell, 1986, pp. 214-15; UM/M/1.2, Phillips to Jardine, 12 April 1864, Jardine to Phillips, 14 April 1864, Phillips to vice-chancellor, 1 Nov 1866, and minute of 14 Dec 1866; Sir William Jardine (1800-74), *DNB*, expert ornithologist.

37. Davies and Hull, 1976, pp. 78-9; OUA, UM/C/3/2, Museum correspondence, Phillips to Jones-Brydges, 21 March 1874; Sir Harford James Jones-Brydges (1808-91).

38. UM/M/1.2, Jackson, executor for Burchell, to Museum, 20 Aug 1864, Phillips to Jackson, 23 Aug, 29 Oct, 15 Dec 1864, Jackson to Phillips, 23 Nov 1864, minute of 13 May 1865; Davies and Hull, 1976, pp. 79-80; on the collections of William John Burchell (1782-1863), *DNB*, see Pickering, 1997, 1998.

39. Phillips to Jones-Brydges, 21 March 1874, OUA, UM/C/3/2; Phillips, evidence to Devonshire commission, pp. 194-6.

40. UM/M/1.2, 8 June 1861, 17 March 1863, 3 Feb 1866, letter in response to Walls, 12 Dec 1866; UM/M/1.3, 10 May 1867, 27 June 1868, Phillips to University Chest, 23 March 1871, 14 Nov 1873; OUM/HA/box 5, f. 6, Delegacy matters 1862-6, Phillips, detailed accounts under four heads to 31 May, 1866, dated 1 June 1866; OUA N.W.2.1, 1 May 1863; OUM/HA/box 2, f. 11, Phillips' account book, 1860-74; Joseph Gouge Greenwood (1821-94), *DNB*, principal of Owens College 1857-89; Henry Wall (1810-73), professor of logic 1849-73.

41. Phillips, on the suggested removal of the Botanic Garden to a site adjacent to the University Museum, 16 May 1873 (qs), in M S Gunther 64, Museum of History of Science, Oxford; Fox, 1997, pp. 661-2.

42. Rolleston and Smith, 1874; Vernon and Vernon, 1909, p. 60; Acland, report to Radcliffe trustees, 11 July 1876, p. 8, in M S Gunther 65; Acland to Phillips, 9 Aug 1858, PP, 21 Feb 1868, OUM/HA/box 3, f 5; Robert Bellamy Clifton (1836-1921), professor of experimental philosophy 1865-1915.

43. Leighton to Phillips, 2 March 1867, PP; Phillips' synopsis of lectures, Hilary term, 1868, 28 Jan 1868, PP; UM/M/1.2, 30 Oct 1866, Phillips to vice-chancellor, 1 Nov 1866; OUM/HA/box 3, f. 4, Smith to Phillips, 21 June, 29 Oct, Phillips' note on Smith's room, 30 Oct 1866; Francis Knyvett Leighton (1807-81), vice-chancellor 1866-70.

44. Fox, 1977, pp. 676-9; UM/M/1.2, Phillips to vice-chancellor, 1 Nov 1866; UM/M/1.3, Phillips to vice-chancellor, 1 Feb 1867; OUM/HA/box 3, f. 4, Clifton to Phillips, 1 Nov 1866, Brodie to Phillips, 10 Nov 1866, Phillips' notes, 1, 10 Nov 1866; OUM/HA/box 1, f. 12, Museum departmental matters, Walker to Phillips, 20 Feb 1862, Griffith to Phillips, 21 Feb 1862.

45. Fox, 1997, pp. 679-82 (682 q); Gooday, 1989, chapter 6, illuminates the young Clifton and the early Clarendon Laboratory.

Chapter 12

Voluntary Commitments

At Oxford Phillips was very busy as teacher, researcher, and keeper. He was by no means a free agent so he had limited time and energy to devote to the many voluntary scientific bodies of which he was a member. His varying involvements in them depended not just on his ambitions for his career but also on what he regarded as obligatory or desirable for the promotion of science. From the early 1860s he became increasingly concerned about the feasibility of certain activities for an ageing man resident in Oxford, especially if they involved frequent travel.

It seems that he withdrew from mechanics' institutes, which he had previously supported when opportunity arose. Though a member of many local scientific and geological societies, he played no role in the latter and was involved with only a few of the former. He was the leading figure in the Ashmolean Society at Oxford. He showed occasional interest in the societies at Bristol, Leeds, and in Worcestershire when corralled by friends. Always grateful to the Yorkshire Philosophical Society and still regarding himself as an adopted Yorkshireman, he retained his allegiance to it. Nationally he continued to be deeply involved with the British Association as a voluntary member and until 1862 as a paid official. In London he had dealings with only those metropolitan societies, the Geological and the Royal, of which he was a fellow. In the 1850s he served the former mainly through office-holding; in the 1860s he used the latter as the most propitious outlet for his research on astronomy. To these various and varied institutional involvements we now turn.

12.1 Local scientific societies

Phillips was connected on occasion with local scientific societies in Bristol, Worcestershire, and Leeds because friends there recruited him. It was Sanders who persuaded him to attend a meeting in Bristol in 1868 at which a scheme costing £12,000 for the improvement of the Bristol Institution was publicised.[1] It was Symonds and his associates in the Worcester area who persuaded him to lecture to the Malvern Field Club in 1854 and Teale who induced him to perform in Leeds in 1866.[2] But Phillips' main involvement with a local society was with the Ashmolean Society founded in Oxford in 1828 to promote discussion of natural history, experimental philosophy, and other branches of so-called modern research. Appealing mainly to senior members of the University, it met in the Ashmolean Museum until the early 1860s and afterwards in the new University Museum. With a small membership of between forty and sixty members, its sole material possessions were its library and a few scientific instruments. It was therefore different from many other provincial scientific societies, the chief functions of which by the 1850s were running and funding imposing museums in which local specimens of various kinds were proudly displayed. As the Society was primarily a discussion club concerned with general science, it avoided until the early 1870s the formal bureaucracy of publishing annual reports which were so important in other societies and, if minutes of its committee were kept, they have not survived for the mid-Victorian period.

As a publishing society it was effective, with both *Proceedings* and *Transactions* appearing regularly from the early 1830s to the late 1850s. But in the 1860s it became less functional for its leading members for two reasons. Firstly, in 1861 most professors of science, who were core members of the Society, moved into the new University Museum, a science centre in which daily communication was easy: their new habitation began to challenge the Society as an important locus of meeting and discussion, whether bilateral or multilateral, arranged or accidental. Secondly, like many provincial scientific societies, the Ashmolean was adversely affected in the 1860s by the growing lure for its research-minded members of their national specialist societies. In 1867 its president, Main, deplored its lack of impact in Oxford: the specialist societies in London had siphoned papers, leading to suspension of the Ashmolean's *Proceedings*. Main thought that the Society would survive only if it encouraged review and not research papers.[3]

Phillips was elected a member of the Ashmolean Society in November 1853, before he began to teach in Oxford. With his wide scientific attainments he soon became the central figure in it. On five occasions he was its annually elected president, he often sat on its committee, and at least twice acted as auditor. Though he was well aware of the difficulties faced by the Society, he remained loyal to it not only as an officer but also as a regular and versatile speaker. His presence ensured that discussion did not flag; and for twenty years he gave almost a paper a year, in which he revealed his latest research or reviewed recent research by others. His topics were varied, embracing geology, geography, meteorology, and astronomy. Within geology, he ranged over local stratigraphy, metamorphism, and palaeontology. Except in meteorology, most of his contributions had been or were about to be given elsewhere. He remained active to the last. In October 1873 he reported on the results up to that month of the bore being sunk in the Weald. A month before his death, he discussed the important discovery of odontopteryx toliapicus, a fossil bird, from the eocene period, found in the London clay. With its webbed feet, teeth on its bill, and affinities with pterosaurians, it seemed to advocates of evolutionary theory to furnish further evidence that reptiles had evolved into birds. With characteristic fairness and openness Phillips revealed to the Society details of a discovery which, in some hands, favoured a theory which he suspected.[4]

At Oxford Phillips lost contact with the York institutions he had supported up to 1854, except for the Yorkshire Philosophical Society. This connection was in part based on Phillips' strange position at Oxford apropos accommodation. He neither owned nor, it seems, rented property there. He lived at Magdalen Bridge, apparently courtesy of Daubeny and the College, and then, in the keeper's house built and owned by the University. While resident in Oxford he remained until 1870 the tenant of St Mary's Lodge, owned by the YPS, and he made money by sub-letting it whenever possible. Phillips also had in York valued old friends, especially Kenrick and Davies, who kept him informed about the Society's difficulties in filling key posts, recruiting members, promoting its objects, and encouraging papers and attenders at the monthly meetings. Though Davies and Kenrick lamented Phillips' departure from York as the heaviest blow and greatest discouragement suffered by the YPS, he gave distant support as one of its vice-presidents from 1854 until his death and retained his affection for the city.[5]

As such Phillips chaired one Council meeting in 1857, he attended another in 1861, he responded to Kenrick's plea by giving a paper in 1857 on Roman remains at Filey, and he was persuaded by Davies and Kenrick to give a public lecture in 1866 on the physical geography of Yorkshire. In the same year he donated a cast of the head of the celebrated Oxford dodo, gave copies of at least two of his publications, and in 1857 sent two pounds for the successful Yorkshire Museum Extension Fund which paid for three new rear galleries. At important moments he gave telling advice. He was concerned to preserve the repose of St Mary's Lodge and of the YPS's gardens. In 1858 he offered useful hints about putting its geological collection in good order so that visitors from the imminent meeting of the British Association in Leeds would be impressed. The previous year he supported the purchase of an ichthyosaurus from Whitby for £110 because he thought the YPS's chief aim was still to illustrate the geology of Yorkshire.[6]

In 1857 Phillips and Kenrick persuaded the YPS to rescind its decision to name and shame its keeper, Charlesworth, a friend of Phillips. Their intervention was occasioned by a meeting in York in September to promote a fund for the dependants of those killed in the Indian mutiny. Charlesworth became known as the sepoys' friend because in an impromptu speech he not only opposed retaliatory vengeance but also pointed out that an intelligent sepoy would say that the mutineers had copied the methods of British conquerors and had learned from Christian scriptures, containing examples of atrocities committed in God's name, given out by Christian missionaries in India. Charlesworth soon realised that he had been an embarrassment to the YPS, pledged discretion in the future, and offered to resign. In response the Society's Council resolved that in order to continue as keeper he should agree that he would not 'either in public or in private, use any expressions tending to disparage the authority of the holy scriptures or calculated to injure the Society in the opinion of the Christian public'. It was also decided that this agreement and the Council resolutions were to be circulated to all members of the YPS. Within a week, as a result of pleas by letters from Kenrick and Phillips, the Council decided that it was sufficient to record in its minutes that Charlesworth accepted the agreement. Some members of the YPS, whom Kenrick regarded as zealous, maintained their campaign to dismiss Charlesworth, so he resigned his keepership in January 1858.[7]

In the early 1870s Phillips became anxious about the future of the YPS. He was worried that it lacked funds to buy specimens which illustrated the natural history and antiquities of Yorkshire. In late 1870 he surrendered his lease of St Mary's Lodge to the YPS as part of a scheme he devised: the YPS would let it on the open market, charge far more rent than Phillips had paid, and from its increased rental revenue set aside up to £30 pa as a purchase fund for specimens. This scheme began to operate in 1871 but Phillips thought the fund too small so in 1872 he intimated to the YPS his intention to bring it up to £500 from his own pocket, provided the Society would not spend anything from the fund until it had reached £500. It seems that the YPS, though grateful for this promised largesse, remained committed to the 1871 arrangements which still operated when Phillips died. As a result of these negotiations, the YPS asked him late in 1871 for advice about how to promote science in it, especially as some members, who opposed the pursuit of scientific aims, wished it to spend spare income on fêtes and flower shows. Before composing his reply, which was printed and circulated for the benefit of the YPS Council, Phillips visited the Yorkshire Museum in 1872. In a wide-ranging review, his key recommendation was the appointment of a well-paid permanent keeper, of high intellectual culture and energy, who would be the driving force of the Society and its Museum. The YPS valued his suggestions but could not afford more than £100 pa as the salary of its keeper, a dilemma which led to its resorting to unsatisfactory temporary appointments from 1868 for ten years.[8]

12.2 The British Association

In his Oxford years Phillips' involvement in the British Association had two phases. Until 1862 he was its assistant general secretary and then served as one of two general secretaries until his resignation in 1863. Subsequently he remained a leading figure in the Association, being its president in 1865, an office-holder and regular performer in sections A and C, and a promoter of some important research projects in geology and astronomy. Overall he failed to attend only one of the twenty meetings held in his Oxford period. In 1871 he did not go to Edinburgh where he was missed: he was feeling his age, he was saddened by the recent death of Harcourt with whom he had worked so closely in the BAAS, and he was sending to press his big book on the geology of Oxford.[9]

In 1860 Phillips began to think of withdrawing as assistant secretary. Next year in May he informed the Association's Council that he had been assisted since 1860 by George Griffith, an *élève* who was fit to be his successor in summer 1861. Unfortunately for Phillips, Walker's illness led to his resignation as general secretary in 1861. With Phillips' support Hopkins replaced Walker that year and Phillips himself suggested that for the sake of preserving continuity he would continue as assistant secretary until 1862 when he intended to resign. Once again illness, this time of Hopkins, upset Phillips' plan. He was prevailed upon by the Association to serve as a general secretary with Hopkins for one year only until 1863. Also in 1862 Griffith was appointed assistant secretary on Phillips' recommendation. In 1863 Francis Galton replaced Phillips as a general secretary, an appointment supported by Phillips who thought that one general secretary should reside in London. Though Phillips relinquished paid office later than he wished and never coveted the unpaid general secretaryship, even for one year, he had the satisfaction of placing a protégé in the post of assistant secretary.[10]

Until 1862 Phillips continued as the general factotum of the Association and editor of its annual reports, on a salary of £300 pa. His work and deportment as assistant secretary were still highly valued. James Nasmyth regarded him as the Atlas of the Association. In 1856 Pengelly marvelled at Phillips' ability to explain administrative details in a humorous speech which 'kept lords and commons in a roar'. Next year when the Association met in Dublin, under the presidency of Lloyd, an old friend of Phillips, an honorary doctorate was conferred on him by Trinity College. In 1861 and 1865 Murchison went so far as to claim that Phillips' hard work, heartiest goodwill, and kindest manners had made him the beloved mainstay of the BAAS since 1831. When Phillips finally resigned in 1863 the *Anthropological review* deplored the loss of its most energetic and able manager whose wide popularity was well earned because he was disinterested and impartial. It is clear, however, that at the 1862 meeting held at Cambridge he was not as efficient as usual, perhaps because of the death of his sister that spring and his wish to resign. The *Athenaeum*, usually a friend of the Association, issued a scathing attack on the incompetence of the local committee and of the general arrangements.[11]

Phillips' responsibilities as assistant secretary remained unchanged. Some were made matters of routine but were still time-consuming. Though he employed a printed standard letter of acknowledgement which he annotated or completed, his letter-writing was incessant. Some of it involved thought, experience, and judgement, as when he advised various presidents about their addresses, evening lectures, and dates of meetings. Each year he prepared for the annual assembly by spending two to three weeks at or near the place of meeting. His aim was always to combine the unwritten usages of the Association with local peculiarities so that he gave to the locals the impression of friendly co-operation. During the meeting he was very busy administratively because he sat on the Council, the General Committee, and the Committee of Recommendations, and often spoke at general gatherings of the Association.[12]

In these years he was neither a policy-maker nor a promoter of lobbies of government; but he did contribute to the report of 1855 of the Parliamentary Committee of the Association about measures which could be adopted by government or parliament to improve the position of science or its cultivators. Phillips wanted all undergraduates to know some physical science; he thought that medals should be bestowed by government only for exceptional published work and as rewards for public service; and he favoured the establishment of a board of science as a responsible, authoritative, and knowledgeable adviser of government, which would circulate widely its annual report to Parliament. In 1857 the Association and the Royal Society combined in sending to Palmerston, the prime minister, resolutions about improving the position of science and its cultivators, including the creation of a board of science. Phillips was not only a member of the appropriate Royal Society committee but he summarised, bound, and indexed the replies to a circular issued to the Association's members. His efforts were in vain: for decades governments preferred to continue using the Royal Society as their chief adviser on scientific matters. Another venture was far more successful. After Humboldt died in May 1859, Phillips sent out in October on behalf of the BAAS a circular about the Humboldt testimonial, the subscription forms for which were to be returned to him. He issued a second circular in February 1860 for what had become the Humboldt Foundation for physical science and travel, to which he gave a guinea. By summer 1861 £8,000 had been raised.[13]

Much of Phillips' editorial work on the annual reports was routine and time-consuming but on occasion required discretion. For instance, after Sedgwick's outburst against Murchison at the 1853 meeting apropos the classification of lower palaeozoic rocks, Phillips persuaded Sedgwick that the printed version in the *1853 report* should not contain personal attacks. With Sedgwick's reluctant permission, Phillips used his gelding knife to excise Sedgwick's complaints against Murchison. Subsequently Sedgwick attacked Murchison vehemently at the meetings held in 1854, 1856, and 1861; but Phillips ensured that Sedgwick's 1853 paper was the last to be published as more than a short abstract in the Association's reports. For Phillips and the Association, proper deportment in science mattered. As editor Phillips was also largely responsible for the still valuable *Index* of 1864 to the annual reports of the meetings held up to 1860. Aided by a BAAS grant of £100 which paid for assistance from Griffith and William Askham, it had author, subject, and place indexes for the commissioned reports and the transactions of the sections.[14]

From 1854 to 1862 inclusive Phillips was named as a member of various co-operative ventures, to most of which he contributed nothing. He and Ramsay never produced the columnar section of British strata for which they were given £15 in 1854. He found himself marginalised in the magnetic survey of Britain in 1857. He left the donkey work of committees about provincial museums (1854) and zoological nomenclature (1860) to Henslow and Jardine respectively. As secretary of a committee on the moon's surface (1860), he revealed to section A in 1861 his own views about its physical history but was replaced as secretary by Birt in 1862.[15] As a solo performer he was far more prominent. In 1858 he gave an evening lecture in Leeds Town Hall on the ironstones of Yorkshire. His report in 1856 on cleavage in rocks helped to establish the mechanical explanation of it. Though he gave to the statistics section a paper on the Association's research grants (1857) and two to section A (self-registering thermometer, 1856; moon's history, 1861), it was in section C that he was most visible, with two contributions on the geology of the Thames Valley (1860, 1861), two on ironstones and ores (1857, 1858), and two on igneous geology (dyke swarms on the Isle of Arran, 1855; contact and regional metamorphism as effects of the earth's internal heat, 1858). In the discussions in section C he continued to express reservations about the glacial theory as advocated by Ramsay and to maintain that several geological systems were arbitrary because they were based on a limited area and ignored the general question of what amount of difference in the totality of past life constituted a system.[16]

From 1863 Phillips attended the meetings unburdened by administrative chores and for the first time in thirty years had total freedom to contribute as he wished. In 1865 he served as president of the Association, a high station which generated honorary doctorates for him next year at Oxford and Cambridge. It also made him an ex officio though not prominent member of the Association's Council. At the last meeting he attended, at Bradford in 1873, he was a vice-president of the Association in recognition of his contributions to the geology and scientific life of Yorkshire. He was president of section C twice (1864, 1873) and on four occasions a vice-president (1863, 1866, 1870, 1872). In 1872 he was also a vice-president of section A, a double honour rare in the annals of the Association. He remained a reliable evening lecturer: in 1869 he expatiated about Vesuvius, using diagrams, apparatus, and a large working model of a geyser.[17]

Phillips' presidential address, delivered in the famous Town Hall, Birmingham, was characteristic in several ways (Fig 12.1). He avoided current controversies about scientific instruction, the advancement of science, and Lubbock's recently published *Prehistoric times*, omissions for which he was criticised: in his view discussion of such topics could easily degenerate into acrimonious controversy which degraded science and scientists. By conviction and temperament he was not a politician or statesman of science dedicated to improving its public position via regular speaking, writing, or lobbying. He played no role in the genesis of the weekly periodical *Nature*, first published in 1869. Unusually for a geologist he orated knowledgeably in his address about the importance of exact measurement in physical science and he evaluated recent research on the nebular theory, spectral analysis, and the dimensions of the solar system. He emphasised the importance of chemical geology as revealed by Bischof, Delesse, and Daubrée. Using the example of BAAS-sponsored research on iron, he told his audience in Birmingham that science was a friend of art and a guide to industry. In a justificatory retrospect of the Association's achievements he claimed that, in focussing on new secure facts, new laws of phenomena, and better instrumentation, it had practiced Baconian induction and had helped to raise slowly but securely 'the steady columns of physical truth'. He urged that scientists should

MEETING OF THE BRITISH ASSOCIATION AT BIRMINGHAM: PROFESSOR PHILLIPS, THE PRESIDENT, DELIVERING THE INAUGURAL ADDRESS IN THE TOWNHALL.—SEE PAGE 21.

**Fig. 12.1   Phillips orating in 1865 as president of the British Association to
a packed audience in Birmingham Town Hall, a popular venue for
music festivals and political meetings.**

avoid hastiness in adopting extreme opinions or in fearing final results. He indicated
his opposition to evolution by natural selection and to the antiquity of man not
least because, like several of his predecessors and successors, he saw in the natural
world endless examples of contrivance which had been planned and created by a
benevolent God.[18]

As president, Phillips found himself involved in a bitter controversy between the
Ethnological Society and the Anthropological Society of London, formed in 1863
by secession from the Ethnological Society to promote a physical and polygenist
approach to the science of man. The Ethnologicals were recognised by the BAAS

via its section E which from 1851 had been devoted to geography and ethnology; the Anthropologicals merely gave papers to sections E and D (zoology, botany, and physiology). With a rapidly growing membership, the Anthropologicals, led by James Hunt, were outspoken and ambitious. At the 1863 BAAS Hunt gave an infamous paper on the physical and mental character of the negro which defended his subjection and even slavery. Next year Hunt tried unsuccessfully to gain formal recognition for anthropology in section E. At the General Committee in 1865 Hunt proposed that the Association should establish a new section for anthropology. In a placatory move which was passed almost unanimously, Phillips proposed that Hunt's proposal be referred to the Committee of Recommendations to avoid a hasty or faulty decision. This Committee advised that anthropology be granted neither its own section nor a sub-section but become a department of section D, to be renamed biology in 1866, leaving ethnology undisturbed in section E. Resentful of the triumph of the ethnologists and of Murchison, who wished his geography section to be free from contamination by anthropology, the anthropologists saw Phillips as a pliant mouthpiece of the Council. Though the Anthropologicals were inconsistent in their accounts of Phillips' behaviour as president, they did accuse him of irregular jobbery and sectarianism in science. For his part he thought it sensible for a contentious proposal about any new section to be referred to the Committee of Recommendations. Given the bitter conflicts between the ethnologists and anthropologists until the late 1860s, when Huxley secured peace between them and ethnology disappeared from section E, it was impossible for Phillips to be all things to all men so the irenic president was hurt by what he regarded as baseless accusations made by Hunt and his cronies.[19]

As a performer in sections A and C Phillips gave on average a paper a year. To the former he spoke about astronomy (the moon, 1863; the sun, 1864) and meteorology (Lake District rainfall, 1868; aneroid barometer, 1872). To the latter he expatiated on glaciation (drift beds at Mundesley, Norfolk, 1863; erratics, 1864; striation, 1865); on geological time, as indicated by the recent research of Morlot on the conical mounds of Lake Geneva (1864) and by the spirals of ammonites (1873); on the famous St Acheul deposits in relation to the antiquity of man (1863); on the formation of valleys (1864); and on the alleged floatation power of ammonites (1873). As president of section C in 1864 he publicised recent work of which he approved, such as the computations of Thomson and Haughton on the age of the earth. In his last official utterance at the BAAS, when president of the section in 1873, he pronounced on some favourite themes such as the earth's central heat, glaciation theory, and proper scientific method which he believed should establish the laws of phenomena before those of causation. He also signalled that, having been a firm opponent of Darwinian evolution, he was moving towards an agnostic position.[20]

As president of section C at Bath Phillips was informed by the contending parties, Charles Moore and James Knipe, about an unusual and double breach of gentlemanly behaviour by geologists. Moore, the driving force of the Bath Museum, was a local secretary at the 1864 meeting at which he shone when he gave a paper in which he highlighted his important discovery in England of rhaetic beds between the lias and trias formations. In contrast, Knipe was mainly an accomplished producer of geological maps of Britain aimed particularly at members of the Association. During his performance Moore passed around some fossil specimens, a few of which disappeared. Suspecting that Knipe had appropriated them, Moore procured a plain-clothes detective to accompany him to Knipe's lodgings. Unwarranted and uninvited, they entered and found Knipe in bed. The specimens were discovered in Knipe's possession and the detective searched his lodgings for more. Moore was outraged by Knipe's theft and as punishment wanted him banned from the Association. Knipe, who admitted he was not faultless but claimed he was not criminal, was incensed by Moore's invasion of his privacy. As each party was not guiltless and their complaints reached Phillips only after he had left Bath, the bizarre incident did not become a cause célèbre.[21]

From 1864 until his death Phillips was a member of several research committees of the Association, never less than three and no fewer than six in 1866 and 1867. Some were inactive, on others he was a passenger, but on four he was active. On the lunar committee he gave essential support and telling technical advice to Birt, its secretary.[22] But most of Phillips' efforts went into three geological projects paid for by the Association, the explorations of Kent's Cavern, Torquay, and the Settle Caves, Yorkshire, and the sub-Wealden bore. From 1864 to 1880 the Association spent around £1,800 on the research on Kent's Cavern carried out by Pengelly whose reports on his results were highly popular at the BAAS's meetings. The first instalment (£100) of his grant, all told the largest in geology in the Association's history, was procured for him in 1864 by Lyell as president and by Phillips as president of section C, who had visited Torquay with Daubeny in January 1864 and met Pengelly there. Phillips was subsequently so impressed by Pengelly's vast voluntary labour that, as president of section C in 1873, he suggested that a testimonial, including cash, be presented to Pengelly. With Lee as treasurer of the fund, £600 was quickly raised, Burdett-Coutts giving £50 and Phillips £10. In March 1874 Phillips made a special journey to Torquay to inspect the excavation and to present the testimonial and money to Pengelly, who was delighted to have Phillips as presenter because he averred that he owed more of his geological knowledge to Phillips than anyone. The Kent's Cavern excavation was important because it provided new information about the association of early human artefacts and extinct and living animals, and therefore illuminated debate about the antiquity of man.[23]

Phillips greatly encouraged the research on the Victoria Cave at Settle in Yorkshire not least because it was carried out by two of his favourite pupils. Launched in 1869 by Dawkins, who was soon joined by Tiddeman, the project was funded by subscription and also by the BAAS with a grant of £450 all told from 1872 to 1878. Using the techniques of excavation developed by Pengelly and trying to take account of glacial deposits in the Cave, Dawkins and Tiddeman concluded from the layering of mammal remains and artefacts that it was probably a hyena den which had been inhabited by stone-age hunters and subsequently by Romano-Celtic people. Phillips was so interested in this research by his two pupils that he visited the Cave in September 1873.[24] Thus the ageing Phillips used his influence in the Association to gain financial support for important cave explorations at Settle and Torquay partly because they were carried out by his protégés and partly because they promised to provide new information about early humans.

One of Phillips' final acts was to support the Sub-Wealden Exploration Fund which had been launched in 1872 by Henry Willett who was advised by William Topley of the Geological Survey. They wished to end interminable debate about what was under the Weald formation, to reach palaeozoic strata, if existent, within 2,000 feet of the surface, to discover whether carboniferous strata extended across the English Channel from the continent, to celebrate the visit of the BAAS to Brighton in 1872, but not to search for coal. Phillips gave £10 to the Fund, offered advice, and was a member and once chairman of a committee for reference and scientific purposes which guided the project for which £2,000 had been subscribed by summer 1873. The bore, begun in autumn 1872, was sunk at Netherfield near Battle, Sussex, a propitious site because the Ashburnham beds, stratigraphically the lowest of the strata of the Weald, occurred there at about 600 feet above sea level on the ridge of an anticline. It was soon claimed that the Ashburnham beds were identical with the well-known Purbeck beds. At the same time subterranean temperature measurements at various depths, which Phillips advocated, were made using his maximum thermometer in the version made by Negretti and Zambra. At the 1873 BAAS Phillips, as president of section C, spoke strongly in favour of the project and procured for it the first instalment (£25) of the £150 which the Association gave it over two years. By September Phillips rejoiced publicly that the fossil evidence, which he examined, indicated that the bore, carried out until November by Bosworth, had gone through the Portland beds and reached Kimmeridge clay at 300 feet depth. Phillips' optimism, shared by Prestwich, Ramsay, and Bristow, that much older palaeozoic rocks would be easily reached was posthumously shown to be misplaced: at 1,900 feet depth the bore was still in Kimmeridge clay, an upper jurassic and a middle mesozoic formation. Even so, Phillips' interest in the project and his exemplary behaviour qua geologist were so greatly appreciated by Willett that in 1875 he presented to the Geological Society a portrait in oils of Phillips painted about 1860.[25]

Behind the scenes Phillips was consulted by the Association's leaders. In 1868 he persuaded Grove that next year the Association should visit a new centre, Exeter, in preference to Plymouth, in order to excite science there and to encourage other unvisited places to host a meeting. As a general secretary, Francis Galton consulted Phillips about the presidency and sectional officers for the Dundee meeting of 1867, at which Forbes gave a welcoming speech which drew on what Phillips had told him about the early Association. A subsequent secretary, Douglas Galton, consulted Phillips about how vice-presidents were appointed and wanted Phillips as a vice-president to support James Joule as president at Bradford in 1873. Presidents turned to Phillips for help with their speeches. In 1872 William Carpenter sought advice on the structure of his address. The previous year Sir William Thomson, desperate about his speech, 'a gigantic black cloud', drew heavily on material derived from Phillips about the roles in the Association of Harcourt and Herschel, both of whom had died in spring 1871. Thomson was thus enabled to stress that Brewster initiated it but Harcourt was its architect and lawgiver. As an expert on its finances Phillips was asked as late as 1874 by its treasurer, William Spottiswoode, to audit its accounts. After the deaths of Murchison and Yates in 1871, Phillips was the last survivor of those responsible for the success of the 1831 meeting where he was Harcourt's clerk of works. Faithful and evergreen, Phillips served the Association for almost forty-three years from its inception. At the 1874 meeting Tyndall, the president, ignored Phillips' services but Hull, as president of section C, did not. He expressed the general and deep regret at the irreparable loss the Association had suffered and paid tribute to Phillips as an imitable paragon, intellectually, morally, and socially.[26]

12.3 The Geological Society

Phillips was elected for the first time to the Council of the Geological Society in February 1853 before his appointment as Buckland's deputy at Oxford; but it was only when he was settled in Oxford that its proximity to London via a railway journey of about one and a half hours permitted him to become a frequent office-holder, including the presidency, which he and many geologists regarded as the headship of English geology. From 1853 to 1862 he was never off the Council, except for one year (1857-8). For three years he was an ordinary Councillor (1853-4, 1856-7, 1860-1), for three a vice-president (1854-6, 1861-2), and for two president (1858-60) (Fig. 12.2). From 1862 he never held office, a withdrawal occasioned by his diminished energy for regular travel to and from London after the death of his sister that year, his teaching load in term time, and his non-stop responsibilities as keeper of the new University Museum which became fully functional in 1861. Only on rare occasions after 1862 did he try to use his authority: he successfully promoted the award of the Wollaston medal in 1865 to Davidson, but failed to secure it next year for Thomas Wright who had to wait until 1878 for it.[27]

Once Phillips had settled in Oxford, Murchison sounded him out about becoming a secretary of the Society but he declined. Yet his value to it was such that twice in 1856 he was asked to be president but rejected the invitations. The first occasion occurred in February when Sharpe was elected, Phillips having begged to be let off for two years. The second took place in June after Sharpe's unexpected death on 31 May when he was thrown from his horse. Though the Council unanimously wanted Phillips to assume the geological episcopacy and Murchison, Portlock, and Ramsay implored him to do so, Portlock was elected as consulate president until Phillips was ready. In 1858 Phillips at last accepted but with qualms: unlike London gentlemen geologists he was neither a resident in London nor a free agent unencumbered with duties. In 1872 he could have joined that select group of Greenough, Murchison, Buckland, Lyell, Horner, and William John Hamilton, who served as president more than once. The outgoing president, Prestwich, and Ramsay, who had just become director general of the Geological Survey, told him he was popular, eminent, an Oxford professor, and had the desired weight; and they played on his sense of responsibility by revealing that the next president might be a 'semi-geologist and without general grasp'. But Phillips felt he was too old and not well enough to accept, leaving the way open for the eighth duke of Argyll to be elected.[28]

Phillips' contributions to the Geological Society were mostly made when he was president. Of the seven papers he gave in his Oxford period, five were delivered as president. Most of the seven were devoted to the geology of north-east Yorkshire and Oxford, with a strong emphasis on sections.[29] In his two presidential addresses given in 1859 and 1860 in February, Phillips followed the more enterprising of his predecessors who had dilated on important themes suggested by recent research instead of merely summarising papers recently given to the Society. In his first address Phillips deliberately avoided palaeontology in order to concentrate on the constitution of rocks and especially the phenomena explicable by his favourite theories of the earth's central heat and cooling. He concluded with reflections on metamorphism as revealed in the Lake District. The second address focussed on the state of palaeontology prefaced by a strong statement that in spite of its data being incomplete they were sufficient as the basis of generalisations about past life on the earth. Phillips gave particular attention to zones of life, palaeontological periods, and the succession of life. To this palaeontological conspectus he added a coda on geological time and its conversion to historical time, which was not long planned but provoked by Darwin's views on geological time given just three months previously in his *Origin*.[30]

Fig. 12.2   Phillips photographed in 1860 when he was exceptionally busy and looking a little tired.

In both addresses Phillips avoided glaciology, a topic which by 1859 was discussed in public with considerable personal animosity by Forbes and by Tyndall and his ally Huxley. Phillips was well qualified to discuss glacial physics: in 1857 and 1858 he had refereed for the Royal Society of London a paper by Tyndall and Huxley on the structure and motion of glaciers (1857) and one by Tyndall on physical properties of ice (1858). He reported that they provided valuable new facts which were compatible with a true general theory, which Forbes had developed but not completed; but Phillips also felt that the language of Tyndall and Huxley was unnecessarily acrimonious and lacked scientific dignity. Privately Phillips revealed to Whewell his desideratum that all the phenomena should be looked at mechanically in order to discover what occurs and happens in ice, without indulging in controversy about theoretical notions such as its viscosity, plasticity, and regelation (the power of broken ice in freezing water to weld itself). In his addresses Phillips could have adumbrated such views or given an overview of the veined structure of glaciers which he had promised in 1856 to report on for the BAAS. Even though he was generally keen to promote physical geology, he was silent about glacial physics as president of the Geological Society because he had no wish to give the combatants the opportunity of dragging him into personal controversy, which he found distasteful and demeaning. In March 1859 he welcomed the omission in Forbes' *Glaciana* of the more controversial parts of his research. Though Forbes was one his oldest friends, Phillips was not involved in the intrigues of autumn 1859 when Tyndall, Huxley, and their front-man Edward Frankland, conspired to deny Forbes the Copley medal of the Royal Society which William Thomson, Murchison, Whewell, and Lyell had tried to secure for him.[31]

Two important initiatives were launched in 1858 during Phillips' presidency. The first was the Society's committee which directed but did not contribute money for a systematic exploration by Pengelly of the bone cave at Brixham in Devon, which had been discovered in January. Though Phillips sat ex officio on the committee and recognised the importance of research on this particular cave, he left it to Falconer and Prestwich, whom he rightly regarded as experts, to direct Pengelly on behalf of the Society.[32] The second project was the co-operative revision of the second edition of Greenough's geological map of England and Wales, paid for by part of the bequest made to the Society by Greenough who had died in 1855. Phillips thought the enterprise important, wanted to do justice to his uncle, and had the appropriate expertise, so he not only sat on the appropriate committee but he was one of a quartet of individuals who contributed most to the collaborative effort. Aided by Tate and Binney, Phillips revised the whole of the north of England. When the third edition of Greenough's map appeared in 1865, the full title made amends to Smith because it acknowledged that Greenough's work was based on Smith's map of 1815.[33.] Like his predecessors Phillips shelved a third project, one dear to Horner, because it was too expensive. Horner wished to transform the Society's museum by appointing a permanent salaried curator who would be so well paid that the membership subscription would have to be raised.[34]

Again like his predecessors as president, Phillips tried to organise lively meetings and to promote good fellowship at the anniversary dinner held each February and at the Society's Club. Founded in 1824 and limited to forty members, the Club was a gathering at which from ten to twenty prominent geologists dined, drank, and talked in a tavern for a couple of hours before the Society's evening meetings. The Club encouraged the exchange of gossip, information, and schemes. It helped to ensure decent attendance and informed discussion at the Society's meetings; and, though not a ginger group, it probably contributed to the way in which the Society was governed because some of its members sat on the Society's Council.[35]

The big disappointment of Phillips' presidency was that he failed to lure back to the Society no less than Sedgwick, who had withdrawn from it totally in 1854 in protest against what he regarded as its unjustified and unjustifiable censorship of his views about palaeozoic nomerclature, the relation between his Cambrian system and Murchison's Silurian system, and Murchison's geological imperialism. As president Phillips publicly praised Sedgwick's genius in 1859 and called on him to complete his research on the Lake District and submit it to the Society. Phillips did not reveal that in January he had persuaded his Council to name Sedgwick as vice-president unanimously and had thrice failed to persuade Sedgwick to accept the post. Phillips felt cruelly mortified by Sedgwick's refusal which he equated with a sad loss of his own influence as president. Sedgwick also refused Phillips' invitation to attend his second anniversary dinner.[36]

As an old and close friend of both Sedgwick and Murchison, Phillips could not avoid the Cambrian-Silurian controversy, the existence of which he deplored. In the mid-1850s he told each of them that personal attacks and a controversial tone were neither necessary nor desirable in their publications. He preferred 'the mildest and most friendly form of controversial statement'. Phillips was also involved in the controversy via the Geological Society for whom he refereed Sedgwick's paper of 1852 on the classification of older palaeozoic rocks. Phillips advised that Sedgwick expunge his violent language. Sedgwick complied but, as a result of interference by Hopkins, the president, the paper was printed in unchanged form. The Council of the Society then took the step, unique in Victorian science, of cancelling the last fifteen pages of Sedgwick's article. It soon rescinded this decision but subsequently forbad him to present to the Society personally controversial matter about lower palaeozoic strata. These judgements paved the way for Sedgwick's withdrawal.[37]

Worried by the bitterness shown by Sedgwick in his anti-Murchisonian paper at the 1853 BAAS, Phillips tried unsuccessfully that year to induce the Geological Society to set up a committee to deliberate on Cambrian-Silurian nomenclature. But he could not dodge the question in his revised *Guide* (1854) and *Manual* (1855). Impressed by Sedgwick's research on the sandstone of May Hill, Phillips felt by mid-1853 that there was a case for seeing a break at the base of Murchison's upper Silurian group of palaeozoic life and that the lower Silurian of Murchison was synonymous in time with the upper Cambrian of Sedgwick. Phillips continued to oppose any dogmatic use of the term 'system' because this sort of classification was, in his view, often provisional, premature, and dependent on the district investigated and the varying languages adopted by different palaeontologists. With these caveats, he was the first to resuscitate in print the compromise term 'Cambro-Silurian' which Sedgwick had coined in 1846 and then dropped. In his *Guide* of 1854 Phillips used the phrase Cambro-Silurian group as preferable to Murchison's lower Silurian strata, his aim being to represent the alliances of nature by terms which would preserve the memory of a long friendship in science. Next year in his *Manual* Phillips applied the term Siluro-Cambrian to strata of the USA but not to those of Britain. Yet on the basis of his favourite method of comparing the numbers of fossil species and genera common to adjacent geological strata, he proposed that the boundary between the upper Cambrian and lower Silurian groups occurred between the Llandeilo and Arenig formations (Fig.12.3). He thus gave to the Cambrian more formations than Jukes, the other leading advocate of the Cambrian-Silurian compromise, who in 1857 put

	1	2	3	4	5	6	7	8	9
SEDGWICK 1855	Silurian			Upper Cambrian	Middle Cambrian				Lower Camb.
MURCHISON 1859	Upper Silurian			Lower Silurian (Primordial Silurian)					Camb.
GEOLOGICAL SURVEY 1866	Upper Silurian			Lower Silurian					Camb.
JUKES 1857	Upper Silurian			Cambro-Silurian					Camb.
PHILLIPS 1855	Upper Silurian			Cambro-Silurian			Cambrian		
LYELL 1865	Upper Silurian		Middle Sil.	Lower Silurian			Cambrian		
LAPWORTH 1879	Silurian			Ordovician			Cambrian		
	1	2	3	4	5	6	7	8	9
Principal Formations	Ludlow	Wenlock	Upper Llandovery = May Hill / Lower Llandovery	Bala = Caradoc Sandstone	Llandeilo	Arenig	Tremadoc	Lingula Flags	Longmynd

Fig. 12.3 Classifications of British Lower palaeozoic rocks, 1855-79, adapted from Secord, 1986a, p. 287.

the boundary between the Cambro-Silurian and the Cambrian as occurring between the Lingula flags and the Longmynd rocks. Indeed, in 1855 Phillips allocated more formations to the Cambrian than Murchison, Lyell, and the Survey subsequently gave in the next ten years. Furthermore, Phillips provided in 1855 what Sedgwick had hitherto failed to give, ie the palaeontological basis for the Cambrian group; and it was more secure than the fossil evidence which Sedgwick offered later that year in his *Synopsis of the classification of British palaeozoic rocks* because Sedgwick relied on nothing more than Frederick McCoy's biased lists of fossils. Phillips' use of the mediating term Cambro-Silurian as equivalent to Murchison's lower Silurian was regarded by Sedgwick as a cowardly half-measure; but it claimed that formations which previously had been widely assumed to be Silurian had a hybrid Cambro-Silurian character. Thus in 1854 and 1855 Phillips did his best to preserve the notion of a Cambrian group of rocks and he reversed in part what Sedgwick regarded as Murchison's encroachment into the Cambrian. Posthumously Phillips' three-fold classification of lower palaeozoic rocks into Silurian, Cambro-Silurian, and Cambrian helped to pave the way for Charles Lapworth's suggestion of 1879 that, while retaining the triple division, the Cambro-Silurian strata should be called Ordovician, a solution which was gradually adopted and is still accepted.[38]

For the rest of his life Phillips remained sympathetic to the Cambrian, whether considered as a group of rocks or as a geological period. He thought that Sedgwick's task had been more difficult than Murchison's: the former had worked upwards on the lowest and least understood rocks, whereas the latter descended from upper deposits which were rich in fossils and clear in their succession and character. Because Phillips deplored the acrimonious controversy between Sedgwick and Murchison, he continued in his writings 'to employ the combination of Siluro-Cambrian, or Cambro-Silurian' to commemorate their 'gigantic labours'. Privately he leaned towards Sedgwick. Phillips twice revised the speech of Sabine who as president of the Royal Society presented its Copley medal to Sedgwick in 1863: Sabine made it clear that in north Wales, the physical structure of which had been solved first by Sedgwick, the Cambrian was equivalent to Murchison's lower Silurian. Phillips changed his lecture day to attend the Society's anniversary meeting and dinner, at which he proposed Sabine's health and paid tribute to Sedgwick. Ten years later when Geikie applied to Phillips for letters apropos his biography of Murchison, Phillips refused to send him those from Murchison and Sedgwick written after their unhappy division of opinion and particularly those which related to Murchison's *Siluria* after it became aggressive. Only after Sedgwick's death and Phillips' signed tribute to him in *Nature* did Phillips send to Geikie all but the last of Murchison's letters.[39]

12.4 The Royal Society

In his Oxford period Phillips was not a core fellow of the Royal Society of London. He served twice as a councillor, on each occasion only for a year. As soon as it was known that he would be moving to Oxford, he was elected to the Council for 1853-4, the last year of the presidency of the Earl of Rosse, who knew Phillips well. Then, in spite of the wish of the next president, Lord Wrottesley, who admired Phillips' work for the BAAS, that he be on the Council, he demurred: he preferred to serve metropolitan science by sitting frequently on the Council of the Geological Society from 1853 to 1862. When his friend Sabine became president of the Royal Society in 1861, Phillips continued to resist but eventually succumbed, serving as councillor 1867-8 and acting as a vice-president that year, having been named by Sabine. Phillips responded to these indicators of esteem by providing for the Society obituaries of Buckland, Conybeare, and Harcourt, by advising Wrottesley on 'knotty points', and by guiding Sabine informally on meteorological matters in the 1860s.[40]

From 1856 to 1870 the Royal Society was important for Phillips because no fewer than eleven of his papers on astronomy were delivered to it, with ten appearing in its *Proceedings* and one in *Philosophical Transactions*. For a geological-astronomer like Phillips, the Royal Society was an outlet preferable to the Astronomical Society of which he was not a member. It was also appropriate because his private observatory was financially supported by the Royal Society. As a performer he gave in the 1860s far more papers on astronomy to the Royal Society than on geology to the Geological Society. Not surprisingly he was used by the Royal Society from the late 1850s as a referee of papers by leading scientists on important research in physical science as well as on physical geology and palaeontology. He reported favourably on the solar physics of De la Rue, Stewart, and Loewy carried out with the photoheliograph at Kew Observatory; and he commended the account by Robinson and Thomas Grubb of the famous large reflecting telescope at Melbourne. Though he disliked the polemical tone of Tyndall and Huxley on the physical properties of glaciers, he regarded their results as important. He approved Haughton's results about geological jointing but suspected his mechanical explanation. In commending Ramsay on the relative duration of certain geological periods, he stressed that a little *was* known about their absolute duration. In 1871 as a third referee of Owen on thylacoleo carnifax, an Australian fossil mammal, Phillips reported that Owen was biased, had suppressed objections, had indulged in *ad hominem* sarcasm about the living and the dead, and his paper could not be published as it stood. With Phillips' permission, his report was sent to Owen to try to end his improper behaviour.[41]

Phillips made occasional contributions to discussions in the Royal Society about promoting physical science in the colonies. In 1859 Sir William Denison, governor of New South Wales, consulted the Society about how to promote geology, botany, and geology in the colonies. In its advice to the Colonial Office, the Society drew on Phillips' view that terrestrial physics (meteorology, magnetism, and physical geography) as well as natural history should be nurtured, though his notion of local directors in each colony working under a scientific head in England came to naught. If in 1859 Phillips advocated one of his pet scientific interests, in 1866 he tried to promote his favourite instrument maker, Cooke, by arguing in favour of sending to India expensive physical equipment, including a big achromatic telescope to be made by Cooke. Phillips had the temerity to pit himself against no less than Sir John Herschel, a great authority on colonial astronomy, who decisively squashed the proposal.[42]

In the cases of medals and fellowships Phillips was not a frequent agitator but on occasion exerted himself in supporting what he regarded as deserving cases. He took the initiative of urging successfully that Davidson be awarded a Royal medal in 1870 for his magisterial research on fossil brachiopods. For fellowships he supported his favourite pupil, Dawkins, and worthy provincials not well known in London such as Sanders, Lowe, and Townshend Hall. Phillips could hardly avoid backing Francis Buckland, son of William. At the end of his life Phillips took up the cause of Seeley after the death of Sedgwick, a key signer of his proposal. In 1865 Phillips was involved in the curious case of William K. Parker, whose research on foraminifera he admired and used. As an impecunious father of eight children and a doctor in Pimlico, London, Parker could not afford the total FRS fee of £50, made up of £10 for admission and £40 for composition in lieu of £4 for the annual subscription. A group of his supporters, led by William Carpenter and including Phillips, who gave £7, generously raised the full sum.[43] The episode illustrated Phillips' belief that lack of money should not inhibit the career of a proved talent and his continuing commitment to good mutual feeling among geologists.

Notes

1. Sanders to Phillips, 2 Nov, 16 Dec 1867, Kenrick to Phillips, 19 Feb 1868, all PP.
2. See section 9.7.
3. *Proceedings of Ashmolean Society*, 1866-8, vol. 3, 11 Feb 1867. There is no standard history of the Society and its records are incomplete.

4. Hutchins, 1994, p. 216; *Proceedings of Ashmolean Society*, 1853-8, 1866-8, vol. 3; *Report of the Ashmolean Society 1872-7*; Mss of papers and proceedings of Ashmolean Society, 1853-82, Bodleian Library, Dep. C. 657; J. Parker to Phillips, 8 June 1872, PP.

5. Kenrick to Phillips, 7, 18 Feb, 3 May, 9 Sept 1855, 16 Nov 1857, 20 March, 10 June 1864, 13 Jan 1873; Davies to Phillips, 30 Nov 1854, 7, 14 Feb 1857, all PP.

6. YPS Council minutes, 6, 12, April 1857, 7 June 1858, 18 March, 30 Sept 1861; *YPS 1857 report*, pp. 7, 23, 29, 32; *1866 report*, pp. 13, 38; *1869 report*, p. 34; Phillips to Harcourt, 12 April 1858, HP.

7. Pyrah, 1988, pp. 71-7; YPS Council minutes, 5 Oct (q), 11 Oct 1857, 16 Jan 1858.

8. *YPS 1871 report*, pp. 9-10; *1872 report*, p. 9; YPS Council minutes, 3 Oct 1870, 15 Feb, 4 Nov 1872 (reproduces Phillips to Noble, 16 Oct 1872), 6 Jan 1873; Noble to Phillips, 29 Nov 1871, 17 Oct 1872, Kenrick to Phillips, 13 Jan 1873, PP.

9. Milne-Home to Phillips, 3 July, Lee to Phillips, 1 Nov 1871, PP.

10. BAAS Council minutes, 16 Nov 1860, 3 May, 4 Sept 1861, 1 Oct 1862, 5 June 1863; Phillips to Sorby, 17 Oct 1861, Sorby P; *BAAS 1861 report*, pp. xxxi-xxxii; *1862 report*, p. xxxi; *1863 report*, p. xxxi.

11. Nasmyth to Phillips 18 Sept 1855, PP; Pengelly, 1897, p. 58; *BAAS 1861 report*, p. 105; *1865 report*, p. 48; *Anthropological review*, 1863, vol. 1, p. 464; *Athenaeum*, 1862, ii, p. 463.

12. BAAS Council minutes, 3 May 1861, for Phillips' description of his post; Phillips to Pengelly, 21 July 1856, annotated letter, Torquay Museum; Harrowby to Phillips, 28 Aug 1854; Sabine to Phillips, 26 June 1857, PP; Phillips to Owen, 4 Aug, 28, 31 Dec 1858, OP; *British Association, Leeds ...1858. Report of local committee*, Moxon, Leeds, pp. 13, 15; Dudley Ryder, second Earl of Harrowby (1798-1882), *DNB*, president 1854; Owen, president 1858.

13. BAAS Council minutes, 12 Sept 1855, 16 Jan 1857; *BAAS 1855 report*, pp. xlviii-lx; *1857 report*, pp. xx-xxix, xxxviii; Howarth, 1931, pp. 221-4; Hall, 1984, pp. 178-9; Bodleian Library, BAAS Dep 5, Miscellaneous Papers, ff. 218-52 on Humboldt testimonial and foundation; Murchison to Phillips, 19 Sept 1859, PP; Phillips to Sabine, 11, 16 Dec 1859, Sa P.

14. Secord, 1986a, pp. 254-5, 263-4; BAAS Council minutes, 16 Nov 1860, 1 Oct 1862; BAAS, 1864, unpaged notice.

15. *BAAS 1854 report*, p. xlvi; on magnetic survey, see section 10.7; *1860 report*, p. xlvii; Phillips, 1861b.

16. *1858 report*, p. xlviii; Phillips, 1856a; *1857 report*, p. 167; Phillips, 1856c; Phillips, 1861b; *1860 report*, p. 90; Phillips, 1861a; *1857 report*, p. 89; Phillips, 1858e, 1855e, 1858d; *Athenaeum*, 1854, p. 1243, 1856, p. 1030.

17. *Scientific opinion*, 1869, vol. 2, p. 310.

18. Phillips, 1865b, pp. lii (q), criticised in *Quarterly journal of science*, 1865, vol. 2, pp. 727-30; Lubbock, 1865.

19. *Anthropological review*, 1864, vol. 2, pp. 294-5; 1865, vol. 3. pp. 354-71; 1866, vol. 4, pp. vi-xiii; Stocking, 1987, pp. 248-54; Richards, 1989; *Athenaeum*, 1863, ii, pp. 341-2, 685; Phillips to Galton, 7 Nov 1865, Galton Papers; James Hunt (1833-69), *DNB*, first president Anthropological Society 1863-7 and first editor of *Anthropological review.*

20. Phillips, 1863d, 1864c; 1868d, 1872b; 1863c, 1864b, 1865c; 1864e, 1873c; 1863f; 1873d; 1864f, 1873b; Charles Adolphe Morlot (1820-67), of Lausanne, expert on lake dwellings.

21. Knipe to Phillips, 24 Sept 1864, Moore to Phillips, 25 Sept 1864, PP; Phillips to Moore, 30 Sept 1864, Moore P; Charles Moore (1815-81), *DNB*; James Anthony Knipe (b 1804). For details about the BAAS 1864 meeting and Knipe I am indebted to Hugh Torrens.

22. See section 10.6.

23. Pengelly, 1897, pp. 152, 159, 235-7; Pengelly to Phillips, 9 Oct, 21 Nov 1865, PP; Phillips, 1873b, p. 72; Lee document, 14 Oct 1873, PP; *BAAS 1874 report*, p. 1.

24. Dawkins, 1870b; Dawkins and Tiddeman, 1872; Tiddeman, 1874.

25. Documents pertaining to Sub-Wealden exploration, PP, box 101, f. 10; Phillips, 1873b and e; Topley, 1872; Willett and Topley 1873, 1874; Willett 1878; Henry Willett supported the Brighton Museum and was a local secretary of the BAAS in 1872; William Topley (1841-94), *DNB*, published in 1875 his classic memoir on the Weald.

26. Grove to Phillips, 25 Dec 1867, PP; Phillips to Grove, 11 Feb 1868, Gr P; F. Galton to Phillips, 8 Jan 1866, Forbes to Phillips, 19 Oct 1866, D.S. Galton to Phillips, 2 Aug 1872, Carpenter to Phillips, 3 June 1872, Thomson to Phillips, 4 July 1871 (q), Spottiswoode to Phillips, 16 Jan 1874, all PP; Phillips to Thomson, 11 July 1871 (two letters), KP; *BAAS 1871 report*, pp. lxxxiv-lxxxv; Hull, 1874, pp. 67-8; Douglas Strutt Galton (1822-99), *DNB*, general secretary BAAS, 1871-95; James Prescott Joule (1818-99), *DNB*, was too ill to be president in 1873; William Spottiswoode (1825-83), *DNB*, treasurer of BAAS 1862-74.

27. Phillips to De la Beche, 25 Sept 1846, Horner to De la Beche, 20 Sept 1846, DLB P; Davidson to Phillips, 26 Jan 1865, Wright to Phillips, 27 June 1865; PP.

28. Phillips to Murchison, 19 May 1854, MP; Salter to Phillips, 2 June 1856, Murchison to Phillips, 4, 7 June 1856, Portlock to Phillips 4 June 1856, Ramsay to Phillips, 6 June 1856; Portlock to Phillips, 23 Sept 1857, 12 Jan 1858; Prestwich to Phillips, Ramsay to Phillips (q), both 27 Dec 1871, all PP; William John Hamilton (1805-67), *DNB*; George Douglas Campbell, eighth duke of Argyll (1823-1900), *DNB*, was president of the Geological Society 1872-4 whilst secretary of state for India 1868-74.

29. Phillips, 1858a and b; 1859c, 1860c and d, 1865i, 1868c.

30. Phillips, 1859a, 1860a.

31. Phillips to Royal Society, 16 March 1857, 29 March 1858, RSL archives, RR, 3, 267, 269; Tyndall and Huxley, 1857; Tyndall, 1858; Phillips to Whewell, 17 April 1859, WP; Phillips to Forbes, 19 March 1859, FP about Forbes, 1859; Phillips, 1856a, p. 392; Cunningham, 1990, pp. 255-79; Rowlinson, 1971; Russell, 1996, pp. 426-8; Edward Frankland (1825-99), *DNB*; Tyndall and Huxley united again from 1864 in the X-Club, an influential cabal.

32. Woodward, 1907, pp. 208-9; Van Riper, 1993, pp. 74-100; Phillips, 1859a, pp. xl-xli.

33. Woodward, 1907, p. 208; Phillips to ? 8 Aug 1859, American Philosophical Society; *QJGS*, 1865, vol. 21, pp. lv-lvi; Hamilton to Phillips, 7 Feb 1865, Mylne to Phillips, 15 Feb 1865, PP; the other three chief contributors were Murchison, Prestwich, and Godwin-Austen.

34. Horner to Phillips, 24 Dec 1858, Mylne to Phillips, 1 Jan 1859, T.R. Jones to Phillips, 10 June 1859, all PP; Woodward, 1907, pp. 245-6, stressed that from the 1850s the library became more important than the museum.

35. Woodward, 1907, pp. 65, 97, 133, 200, 204; Geikie, 1875, vol. 1, p. 96; Rudwick, 1985, p. 21; Phillips to ?, 26 Oct 1858, Welcome Medical Library, London.

36. Secord, 1986a, pp. 231-3; Phillips, 1859a, p. xxxi; Phillips to Sedgwick, 20, 24, 30 Jan 1859, Se P; Sedgwick to Phillips, 11 Feb [1860], 26 Nov 1860, PP.

37. Phillips to Murchison, 29 Oct 1854, EUL, Gen 523/4; Phillips to Sedgwick, 20 Feb 1856, Se P (q).

38. Forbes to Ramsay, Sat 28 [Nov] 1853, RP; Phillips to Sedgwick, 8 Nov 1854, 11 March, 20 Aug 1855, 3 Jan, 20 Feb 1856, Se P; Jukes to Sedgwick, 21 April 1854, Se P; Phillips, 1854b, p. 121; Phillips, 1855c, pp. vi, 103-5; Sedgwick and McCoy, 1855; Sedgwick to Jukes, 20 Nov 1855, Se P; Secord, 1986a, pp. 189, 243, 255, 286-311; Frederick McCoy (1823-99), *DNB*, assisted Sedgwick 1846-54; Charles Lapworth (1842-1920), *DNB*.

39. Phillips, 1873a; Phillips, 1860b, pp. vii (q), 25; Sabine to Phillips, 12 Nov 1863, in Burkhardt and Smith, 1999, vol. 11, pp. 668-9; Sabine, 1864, pp. 31-5; Clark and Hughes, 1890, vol. 2, pp. 397-8; Phillips to Sedgwick, 18 Jan 1873, Se P; Phillips to Geikie, 18 Jan 1873, 6 Jan 1874, EUL, Gen 525.

40. Wrottesley to Phillips, 10 April 1855 (q), 21 Oct 1856, Sabine to Phillips, 17 Dec 1867, PP; *Proceedings of Royal Society*, 1868, vol. 16. p. 209; Phillips to Sabine, 27 Oct 1863, SaP; Phillips, 1857b, 1859b, 1872a; for meteorological guidance, see section 10.7.

41. For Phillips' astronomy, see section 10.6; Phillips' referee's reports, RSL, RR. 6.101; RR. 6.239; RR.3, 267, 269; RR.5, 110; RR. 7, 336; Phillips to Stokes, 4, 17 March 1871, Stokes to Owen, St P, NB7, Phillips to Stokes, 17 March 1871, private, St P, P335; the refereed papers were De la Rue, Stewart, and Loewy, 1869; Robinson and Grubb, 1869; Tyndall and Huxley, 1857; Tyndall, 1858; Haughton, 1864; Ramsay, 1874; and Owen, 1871. Thomas Grubb (1800-78), *DNB*, constructed reflecting telescopes, the largest being that at Melbourne with a mirror of four feet diameter.
42. Hall, 1984, p. 170; Phillips memo, 12 May 1859, RSL, MM, 14.28; Phillips' memo 17 Nov 1866, Phillips to Sabine, 30 Nov 1866, St P, RS 547, 550; Sabine to Harcourt, 29 Nov 1866, HP; Herschel to Sabine, 11 Aug, 1866, Sabine to Herschel, 3, 8 Dec 1866, in Crowe, 1998, pp. 630, 634; William Thomas Denison (1804-71), *DNB*.
43. On Davidson, Prestwich to Phillips, 24 May 1869, Huxley to Phillips 28 April 1870, Davidson to Phillips, 26 April, 9 Nov 1870, PP; Phillips to Dawkins, 25 Oct 1864, Wellcome Medical Library autographs; Sanders of Bristol to Phillips, 12 June 1864; Lowe of Nottingham to Phillips, 28 Dec 1865, 18, 20 Jan 1866, 16 Jan 1867; Hall of Barnstaple to Phillips, 12 Feb 1869; Buckland to Phillips, 28 Feb 1866, all PP; Phillips to White and Sharpey, 5 March 1874, RSL, MC, 10.80, on Seeley; Carpenter to Phillips, 22, 29 May 1865, PP; Townshend Monckton Hall (1845-99); Edward Joseph Lowe (1825-1900). FRS elections were announced in the 1860s in June. Buckland and Hall were never elected FRS, whereas Seeley was elected in 1879.

Chapter 13

Evolution, the Earth, Man, and God

From 1854 Phillips was a public figure as the teacher of geology in the University of Oxford, and a president of both the Geological Society and the British Association. As such he overcame his dislike of acrimonious controversy and pronounced, albeit reluctantly, on some of the most important issues which exercised the intelligentsia. Via addresses, books, and reviews, he contributed to the debates which raged in the 1850s and 1860s about evolution of species, Darwinian natural selection, design in nature, the succession of life on the earth, the age of the earth, the antiquity of man, Lyellian uniformitarianism, the interpretation of scripture, and the relation between science and religion. In this chapter I discuss his reactions to *Vestiges of the natural history of creation* (1844) and Darwin's *Origin of species* (1859); his own views about *Life on the earth, its origin and succession* (1860); his evaluations of *Essays and reviews* (1860) and Lyell's *Antiquity of man* (1863); and his sentiments from 1864 about the issues these works raised. In so doing I try to show that it was Phillips' wide range as a geologist, from palaeontology to geochronology, that made him one of the most formidable opponents of Darwinian evolution.

Some of his views were clearly related to his Christian providentialism in a mutually reinforcing way. After the death of his sister in 1862, Phillips' Christianity was not weakened but strengthened. For him there was, it seems, no crisis of faith or anxiety about it. Avoiding allegiance to any party in the Church of England, he was an unsectarian, irenic, and incarnationist Christian, who eschewed dogma, conflict, and the beliefs that humankind was fallen and life a probation. This sort of Christianity informed his notions of proper behaviour in scientific discussion, for example, in his open-minded and unaggressive opposition to Darwinian evolution. In the concluding section of this chapter I try to show that Phillips' version of Christianity was more than a device for surviving in the highly charged religious atmosphere of Oxford. It infused his social and intellectual relations there with colleagues and friends, most of whom were Anglican clergymen. In the 1860s it even enabled him to guide no less than Edward Pusey, the leader of the high church ritualists, on matters geological and theological.

As professor of geology in the 1860s in what has sometimes been regarded as an inflexible high church university, Phillips' teaching and publications were in no way constrained by his Christian beliefs or by his University. For him there was no conflict between his science and religion or between Genesis and geology. Unlike Powell in the 1850s, Phillips did not isolate himself at Oxford. Indeed, his career there qualifies the older interpretations that in the 1860s the controversy between science and religion took fire and that scientific naturalism drove out a providential view of nature. As Brooke, Cantor, Kenny, and Turner have stressed, the terms 'science', 'secularity', and 'religion' can be dangerously distorting when applied to Victorian savants. I have therefore used them cautiously, as useful linguistic approximations, to explore the intellectual and social dimensions of Phillips' deepest convictions as a Christian philosopher.[1]

13.1 Vestiges of creation

In 1844 Churchill published *Vestiges of the natural history of creation*, a best-selling, wide-ranging, and sensational evolutionary epic which made it the most controversial scientific book of its time. In response to much criticism, the anonymous author, Robert Chambers, brought out next year *Explanations*, an anonymous sequel to *Vestiges*. In both works he argued that his so-called law of development applied not only to astronomy and geology but to all existing organisms, which had evolved by natural means analogous to gestation. In his account of the creation of new species by transmutation, Chambers suggested that human beings had evolved from apes and he offered a materialistic interpretation of the highest aims, capacities, and feelings of humans. As he himself said, he had annulled the cherished distinctions between the spiritual and the material and between the moral and the physical. Such views were publicly denounced in 1845 by two leading divines at Cambridge, both of whom were friends of Phillips. In *Indications of the Creator* Whewell rejected the notions that life had grown out of dead matter, higher animals out of lower ones, and humans out of brutes. Sedgwick was so outraged by *Vestiges* that he made his debut as a reviewer in an invective of eighty-five pages published in the *Edinburgh*. Five years later, in the fifth edition of his *Discourse on the studies of the University of Cambridge*, Sedgwick attacked with deliberate vituperation the notion of transmutation of species and denounced deists, pantheists, sceptics, materialists, Catholics, tractarians and idolaters.[2]

Phillips made no immediate public response, vehement or gentle, to *Vestiges*, because he did not see it as a threat. Privately he thought the book replete with bold assumptions rashly poised on a slight basis; and he judged that Sedgwick's *Discourse*, described as luminous, had answered *Vestiges* effectively. In his inaugural lecture at Oxford Phillips showed that he retained his long-held suspicion of evolution. He claimed that life was not indefinitely variable, that species were constant within narrow limits, that present life was not derived from earlier types, and that humans were geologically recent and headed the system of nature. He also stressed that in the past there had been many successive systems of life and renewals of it, of which only the most recent resembled present-day forms. He offered as an illustration the way in which land plants and some reptiles appeared for the first time in the upper palaeozoic period. He also emphasised that sometimes there was a connection between biological and geological change, giving as an example the end of the palaeozoic area, an epoch of extraordinary violence which folded, fractured, and broke up the seabed and which corresponded to a distinct change in the system of life.[3]

It is not widely realised that in *Vestiges* and *Explanations* Chambers used Phillips' research in four ways. Firstly, in claiming that the earth had a history Chambers adduced Phillips' views on the different ages of different mountain ranges and of the violent disturbances of strata, some of which as Chambers said involved large scale extinction of species. Secondly, in a discussion of the earliest forms of life and in buttressing his notion of development as analogous to foetal changes, Chambers twice alluded to Phillips' detection of minute fossil fish in the Aymestry limestone of the Silurian system. This discovery, of what Chambers regarded as the oldest vertebrate remains, was for him clear evidence of 'laws presiding over the development of animated tribes on the face of the earth, and that of the individual in embryo'. Thirdly, when Chambers attacked Sedgwick for rash palaeontological generalisation, from study of merely a small region, he cited Phillips as a high authority on the way in which the variety of physical conditions, under which the detached Silurian deposits of England were formed, produced '*a corresponding diversity in the traces of organic life in each situation*'. Finally, in rebutting Sedgwick's claim that there had been breaks in the chain of past life, Chambers twice cited Phillips' view that there was 'no break in the vast chain of organic development till we reach the existing order of things'. Chambers was honest enough to add that Phillips' notion of development did not involve previous types of life being related as parent and offspring.[4] Presumably Phillips did not relish the way in which the author of *Vestiges* had deftly appropriated his views in order to support the notion of transmutation and to attack Sedgwick's critique of *Vestiges*.

13.2　The origin of species

Until 1859 the contacts between Phillips and Darwin, who had lived in rural Kent from 1842, were infrequent. Though they knew each other from the late 1830s via the Geological Society, they rarely met because they were geographically separated and belonged to different coteries. But there were two bursts of correspondence between them, both launched by Darwin who wished to serve his own interests. In 1848 he was desperate for data to support his theory that erratic boulders in Britain had been transported not by diluvial action, as Hopkins had recently argued, but by coast ice while the land was subsiding. Darwin therefore pumped Phillips relentlessly for information about the way in which some erratic boulders had moved upwards from their place of origin to that of rest. In his published paper Darwin not only cited two of Phillips' publications on the matter but on Phillips' 'high authority', privately communicated, ruled out the possibility that the boulders were derived from higher strata now entirely denuded. In 1856 Darwin wrote out of the blue to help Phillips with his forthcoming report to the BAAS on cleavage in rocks. Phillips was happy to include the references to South American cleavage which Darwin supplied and to accept a present of a book. It is clear from these exchanges and from Darwin's notebooks that he respected and exploited Phillips as a geologist.[5]

When Phillips became president of the Geological Society in 1858 their relation changed. In January 1859 Phillips informed Darwin that he had been awarded its Wollaston medal. Though Darwin was grateful for Phillips' extremely kind manner of communication, he was too ill to attend the anniversary meeting to receive it. As president Phillips dilated on Darwin's valuable geological researches, including the distribution of boulders, and had the piquant pleasure of presenting the medal to Lyell who accepted it on behalf of Darwin. Later that year Phillips was one of about eighty people to whom Darwin sent a complimentary copy of his *Origin*. Darwin also defended his book in two letters to Phillips, whom he regarded as an important judge who was likely 'to fulminate awful anathemas' against it. Though Phillips disagreed with Darwin's main conclusions, he replied that he would cautiously reflect and perhaps say nothing about them in his presidential address in February 1860. For a short time Darwin hoped that Phillips was shaken and had 'the fear of Oxford before his eyes'. Simultaneously, Darwin was irritated by Phillips' calm judiciousness: like Huxley, Darwin felt that Phillips would go to that part of Hell which accommodated, according to Dante, 'those who are neither on God's side nor on that of the Devil's'.[6]

In his presidential address Phillips took positions that were no surprise to Darwin. In the *Origin* Darwin had devoted a chapter to the imperfection of the geological record and another to the geological succession of organic beings. In response Phillips made it clear that for him the palaeontological record of marine deposits was reasonably complete and it showed no evidence of evolution having occurred. Moreover Phillips averred that evolution from common ancestors was a desperate view: for him the structures and adaptations of living things were various, special, 'determinate', and examples of contrivance which led him to conclude that there was 'a plan of creation'. Though he was not explicit in this address about his belief in a providential God, he had ended his previous address with the statement that the permanence of the laws of nature resulted from the 'constant will of the great Maker, with whom is no variableness, neither shadow of turning, the same yesterday, today, and for ever'.[7]

Darwin was well aware that for evolution to have been produced very slowly by natural selection, incomprehensively vast periods of past time were required. He thought that the rate of denudation of strata offered the best evidence of this inconceivably huge lapse of time. He gave just one crucial quantitative example, that of the denudation by marine action of the chalk dome of the Weald, which he estimated had taken 300 million years. This conclusion was attacked in late December in an anonymous review of the *Origin* in the *Saturday review*. The reviewer claimed that Darwin's figure was far too high because he had ignored undercutting of cliffs by marine action, a point that Darwin immediately accepted: in the second edition of the *Origin* published in January 1860 the figure was reduced to 150 million years. The reviewer made other points which Phillips was to make the next year but did not mention other agents of denudation.[8]

In his presidential address of February 1860 to the Geological Society, Phillips cut up Darwin by exposing his 'abuse of arithmetic' in his 'geological calculus'. For Phillips the figure of 300 million years was inconceivable for three reasons. Firstly, Darwin assumed that the only agent of denudation was marine action, thus ignoring the powerful effects of the atmosphere and rivers. Secondly, Darwin's assumption of one inch per century as the rate of marine erosion sat uncomfortably with the maximum rate of erosion on the Yorkshire coast of eighty inches per year. Finally, Darwin ignored the effect of undercutting on the rate of erosion of cliffs. Phillips' own estimate for the denudation of the Weald by all three agents was no more than a million years. This semi-quantitative conclusion deprived Darwin of what he thought was a telling indication of the 'incomprehensively vast' age of the earth as postulated by Lyell. Lyell heard Phillips' address and immediately wrote about it to Darwin. It is inconceivable that Lyell did not grasp and communicate to Darwin the importance of Phillips' critique, about which Darwin was very curious.[9]

3.3   Life on the earth

In May 1860 Phillips delivered a stronger attack on evolution by natural selection when he gave the second annual Rede lecture at the University of Cambridge. In expanded form it was published as *Life on the earth* in November that year. The main themes of the succession and origin of past life had intrigued Philips for several years. In 1854 his lectures on ancient life in the sea, fresh water, and on land, had concluded with a review of the succession of life on the earth. His 1855 *Manual* was mainly a palaeontological illustration of that succession. In thanking Murchison in 1857 for a copy of his *Siluria*, Phillips said it had suggested to him ' a coup d'essai on the origin of life, a beautiful and grand theme'. Late in 1859 he was asked by the reverend Latimer Neville, vice-chancellor at Cambridge, to deliver the Rede lecture for 1860. It seems that Whewell, backed by Sedgwick, persuaded Neville to approach Phillips. Provoked by the contents of Darwin's *Origin*, Phillips accepted and chose as his theme the origin and succession of life on the earth.[10]

The terms of the Robert Rede lectures, endowed in 1524 by him, were reorganised in 1858 so that just one lecture per year in term time was to be delivered by a man of eminence in science or literature, appointed annually by the vice-chancellor. The first lecturer was Owen who in May 1859 spoke about the classification and distribution of mammals. He repeated his view, first given in 1857, that the human brain had unique structural features; and in his peroration expatiated on the uniqueness of humans as the supreme work of the Maker. These views were underpinned by Owen's belief that the diversity of life, past and present, was the result of the embodiment of a divine idea in successive, diverse, and intricately adapted forms of life. From early 1860 Owen became the fiercest and most implacable of the opponents of Darwinian evolution.[11]

When Sedgwick heard in April 1860 that Phillips was to give the Rede lecture, he was delighted. He and Whewell thought Phillips the very man, not least because there was no fear that Phillips would 'fly off into Darwinian or any transcendental nonsense'. Sedgwick liked Phillips' subject: it was just what he would have chosen for a substantial rebuttal of the *Origin*, but aged seventy-five he was too tired and ill to give a repeat performance of his extensive published diatribes against *Vestiges*. He confined himself to just two pages of unsigned reviews of the *Origin* in the *Spectator*, leaving a full evaluation to Phillips. Yet he summoned up enough energy to mount with Phillips a joint public attack in Cambridge on Darwinian evolution in May 1860. At the Cambridge Philosophical Society on 7 May Sedgwick discussed the succession of living things as revealed by palaeontology, and concluded yet again that the fossil record was fatal to the notion of transmutation of species. He also reviewed certain theories which professed to account for the origin of new species. Characteristically he noted the sins of Darwin who had attacked man's uniqueness, rejected final causes, and given a toehold to Lucretian materialism.[12]

Sedgwick's decided rejection of Darwinian evolution pleased Phillips who took two hours to deliver his Rede lecture on 15 May. In dilating on nature as a divine idea and on successive forms of life as creations of a benevolent Almighty Hand and Power, Phillips emphasised that the historical record as revealed by palaeontology gave no evidence for the existence of intermediate species or for major groups of readily fossiled species having common ancestors. Thus within eight days the two professors of geology at Oxford and Cambridge made clear their opposition to Darwinian evolution. Their joint campaign was noticed by Joseph Hooker, assistant director of the Royal Botanic Garden, Kew, and an early convert to Darwinism. Having heard that Phillips' Rede lecture was 'mere twaddle', Hooker saw both Phillips and Sedgwick as being 'beside the mark' because they were not naturalists and had no understanding of natural history; and he commended Henslow's defence in his botany lectures of Darwin's *Origin* against the attacks of Phillips and Sedgwick. For his part Sedgwick was delighted with Phillips' lecture and urged him to strike while the iron was hot by publishing an extended version of it.[13]

Before delivering his lecture, Phillips had been approached by Macmillan, a commercial publisher new to him, to do what Sedgwick wanted. Based in Cambridge, Alexander Macmillan and his lady companions attended Phillips' lecture, which they liked. Macmillan was impressed by Phillips' command of his material as revealed to a university audience, but feared that the book, provisionally entitled 'The origin and succession of life on the earth', might not be widely intelligible and popular. He jollied Phillips along, decided on a print run of 1,500 copies selling at 6/6 each, and publicised the book in a way that was new to Phillips. Presentation copies bound in morocco leather were sent to Prince Albert, Burdett Coutts, and Neville. Standard copies winged their way to opponents, supporters, and friends, including Darwin, Hooker, Sedgwick, Whewell, Ramsay, Hopkins, Ansted, Lyell, Murchison, Morris, Salter, Owen, Seeley, Huxley, Henslow, and S.P. Woodward. Macmillan also sent review copies to no fewer than twenty-eight journals and newspapers. Published in mid-November 1860 *Life on the earth* made such a capital start that by early January Macmillan was thinking of a second edition of 1,250 to 1,500 copies.[14]

A leading theme of *Life* was that the natural world was 'the expression of a DIVINE IDEA', ie that nature was the result of an intention or plan which involved forethought, foreknowledge, and pre-arrangement by a Great Maker, Author, and Designer. Contrary to the opinion of his favourite pupil, Dawkins, Phillips was not pandering to theological sectarians when he expressed such sentiments. After the publication of *Life*, he confided to Sedgwick 'that all that was and is and is to come stands as known and appointed before the Master of Nature and the Maker of Man'. Phillips was happy to see 'the Divine Plan' in the visible works of nature. Without that reassurance he felt that 'we should indeed drift idly on the stream of time, and be reduced to the "general form of life" of which the moderns speak with such filial respect'. His notion of the divine idea was similar to the views expressed by Owen in his presidential address of 1858 to the British Association, which Phillips heard. Owen alluded to the axiom of the continuous operation of creative and divine power or of the ordained becoming of living things. Two years later in his *Palaeontology* Owen reiterated this axiom of ordained becoming and made it quite clear that he believed in a benevolent, pervasive, and non-mechanical great First Cause, which possessed an active and anticipating intelligence. Phillips was also familiar with Owen's well-known notion of the archetypal idea by which he meant that in the case of vertebrates their structural affinities were manifestations of a pre-existent archetypal idea planned by the divine mind who also foreknew its modifications. This Platonic and Christian approach of Owen was very similar to that of Phillips who told his Oxford students in 1858 that his own notion of the divine idea was equivalent to Empedocles' concept of the intelligible universe as interpreted by neo-Platonist philosophers. For Phillips it followed from his concept of the divine idea that any system or classification of nature was not a photograph but a rude and misty sketch of phenomena, imperfectly seen, understood, and recorded by human beings. But the inadequate sketch could be improved by studying nature in a reverent and devout way, using the senses endowed by the Almighty, to establish a vast number of special facts as the basis of diverse and comprehensive laws of nature, the operation of which indicated divine design and sustenance.[15]

Phillips' notion of creation of species was like that of Owen who believed that its causes were not discoverable by experiment or observation and therefore the process or event was inscrutable. For Phillips creation was a word which indicated the first appearance of something; in the case of species its proximate cause was a mystery because it could not be derived from experience or by analogy. But its ultimate cause was an act of God's power, transcending all human thought and experience, which provided for the appearance of new forms and structures of life at definite times and in certain places. Thus, like his old friend Harcourt, Phillips believed that there had been many separate acts of creation of new species; and that these acts, of unknown and unknowable cause, were systematic because each species seemed to have one place and one time of origin. For Phillips and many palaeontologists at the time, the frequent appearance of new structures at definite times in the past was adequately explained by separate creations, even though they were aware that the notion of creation was a convenient word for their ignorance of the laws by which species originated.[16]

For a Darwinian such as Alfred Wallace the ideas of special adaptation of species and of intelligent foresight became unintelligible if creation were regarded as the unknown manner in which species came into existence. But for Phillips and Owen these related ideas were anything but unintelligible. For over thirty years Phillips had been impressed by the pervasiveness of special adaptations in living things, whether resident in land, water, or air, not only currently but in all geological periods. Every species, it seemed, was adapted during its existence to its mode of life in a way that to Phillips was inexplicable unless one assumed 'a wise coordination of the different parts of the structure to answer an appointed end'. Phillips used as illustrations of contrivance the familiar examples of the peculiarities of the woodpecker and chameleon, both of which he himself had studied, and the less familiar ones of the adaptations of the human eye, far superior to any optical instrument, and of the trilobites' eyes made up of many reticulated lenses. Such examples of intricate and successful adaptation, plus the whole fossil record, indicated to Phillips, and many of his contemporaries, that nature past and present was the work of divine design, thought, and wisdom. It was inconceivable to him that the co-ordinations of past and present life had originated in chance variations and were 'the fortunate offspring of a few atoms of matter, warmed by the *anima mundi*, a spark of electricity, or an accidental ray of sunshine'.[17]

Phillips' review of the fossil record drew on his long experience as a palaeontologist. He was not alone in his conviction that Darwin had greatly over-rated the incompleteness of the fossil record. In Phillips' view the record was sufficiently complete for shell-bearing marine creatures to be a sample which could be used to discover whether any species had slowly evolved into others. It showed that there was only limited change in each group of animals and plants and not the 'infinitely numerous transitional links' between species, which Darwin had postulated. Moreover discontinuity characterised the fossil record. Phillips stressed that certain organisms such as belemnites occurred only in certain geological formations: they suddenly appeared and disappeared. He was impressed by the way in which in the least disturbed strata new genera and families with new types of structure appeared frequently and suddenly at definite times, a fact which Darwin had acknowledged as a grave difficulty. Phillips noted that the earliest known mollusc, the brachiopod, occurred in all the geological systems of strata and was still living; yet it showed no transitional forms as it should if Darwin were right. Phillips' most telling argument was that the earliest known forms of life, those of the lower palaeozoic period, ie the Cambrian, were highly complicated and well differentiated organisms; and the total absence of simpler ancestors could not be explained away by assuming that strata and life had existed in pre-palaeozoic times, but the latter had been totally destroyed by metamorphism, as Darwin had suggested. Phillips was impressed by the fact that in the thick lowest Cambrian rocks there were no signs of life even though these deposits were generally undisturbed and therefore favourable to fossilisation, and these rocks were unaffected by metamorphism. The absence in the fossil record of earlier progenitors of the oldest known marine organisms was for Phillips and many others a conclusive objection to Darwinian evolution.[18] With characteristic visual inventiveness Phillips enshrined this point in a diagram which represented the fluctuating diversity of life in the three great periods of the earth's history. It showed that life began in the early palaeozoic area and overall became more diverse over time (Fig. 13.1). It should be noted that even today the Cambrian is the oldest geological system sufficiently rich in fossils for them to be used for correlating strata.

Phillips' book contained a surprising calculation, ie what Burchfield has called 'the first important quantitative determination of the earth's age based exclusively on geological data'. Phillips gave only a page to the time taken for the Weald to be denuded, which he guesstimated at no more than three million years, because he was after the bigger game of showing that the age of the earth was less than Darwin's first estimate of 300 million years for the denudation of the Weald. Thus Phillips' geochronology was a new weapon he devised to rebut Darwinian evolution. From the late 1830s Phillips had viewed the age of the earth as a magnificent but insoluble problem. In his *Manual* of 1855 he reviewed no fewer than nine possible ways of establishing geological chronology, all of which used stratified deposits and not one of which was viable. The approach he adopted in 1860 was one he had thought of in the late 1830s. Knowing the mean or extreme rate of production of stratified deposits at the present day and their total thickness, he calculated the time taken for their deposition as about 96 million years. He was aware that he had made several questionable assumptions and that his calculation would be totally vitiated if the cooling of the earth, which he accepted as a fact, had deranged the steady rate of sedimentation. He therefore tried to take account of this derangement by making a second calculation which gave a result of between 38 and 64 million years. Though his conclusions about the age of the earth were semi-quantitative, Phillips had for a second time in 1860 effectively denied Darwin the vast amount of time he needed for his two postulates of slow evolution and of huge gaps in the fossil record.[19]

Darwin reacted to Phillips' *Life* in two ways. Within a week of receiving a copy in mid-November 1860, Darwin decided to strike out altogether his 'confounded Wealden calculation' from the third edition of the *Origin* which was published in April 1861. He knew he had burnt his own fingers severely and unpleasantly apropos this key point of geochronology. Secondly, Darwin was irritated by the style of *Life*, which Phillips published only because he felt that someone in his public positions had to reveal his reactions to the *Origin*. As a reluctant controversialist he avoided Sedgwick's vituperation, Wilberforce's facetiousness, and above all Owen's malevolence, all of which had been displayed earlier in 1860. Phillips' tone was calm and, as Asa Gray and Huxley noted, his arguments were fair and without malice even when his conclusions were dead against those of the *Origin*. Darwin found Phillips' *Life* exasperating: he preferred to be vigorously attacked than 'handled in the namby-pamby, old-woman style of the cautious Oxford professor'. Two close friends of Darwin joined him in denigrating Phillips' book. Like Darwin, Hooker thought it dull and full of 'weak, washy stilted stuff'. Lyell, who at that time was not a convert to Darwinian evolution, judged it 'fearfully retrograde' perhaps because Phillips had used Lyell's uniformitarianism to reach a geochronological conclusion

LIFE ON THE EARTH.

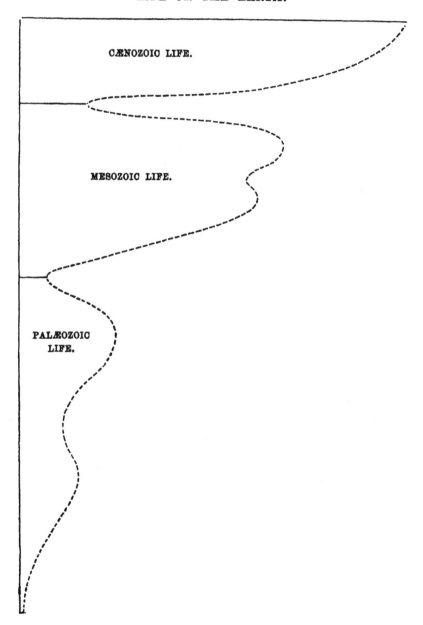

Fig. 13.1 A diagram by Phillips for Phillips, 1860b. The curve represents the
fluctuating but overall increasing variety of life during the earth's
history, which Phillips had divided in 1841 into three great eras;
and it shows his belief that life began at the start of the Cambrian
period. Phillips thought it remarkable that the diversity of life
diminished at the ends of the palaezoic and mesozoic eras and then
increased at the beginnings of their respective successors.

at odds with Lyell's geological timescale and had argued against Lyell's views about the imperfection of the geological record. Lyell was certainly displeased that Phillips had quoted at length Lyell's arguments of 1832 for his still-held belief that species exist and each was endowed when created with the features by which it is currently distinguished.[20]

The main arguments of *Life* were welcomed by Sedgwick, who recommended it warmly to his class, and Murchison because it was compatible with his own conclusion in his *Siluria* that life had a beginning. Phillips' book was widely seen as a reliable conspectus of the geological arguments against Darwinian evolution. But some anti-Darwinians, such as Sanders, S.P. Woodward, and Salter found odd faults in Phillips' palaeontology, criticised his statistical tables, drew attention to his absurdly impossible precision when he stated that the Ganges carried annually 6,368,077,440 cubic feet of sediment, and found him washy and tame on key points. For his part, Macmillan, who wanted a second edition, thought Phillips' book intimidating and formidable because of its technical language. He disliked Phillips' suggestion of a technical glossary: he wanted Phillips, who had made £50 profit, to lighten his pages by putting them in a popular shape; and he urged Phillips to 'forget the geological compeers', to remember the 'populus', and to be unorthodox in order to gain publicity from hostile bishops. By summer 1861 *Life on the earth* had sold 1,300 copies and had received the accolade of imitation from David Page's *Life of the globe*. Even so it was clear to Macmillan that Phillips was unlikely to produce a second edition considerably more popular and appealing to a wider audience, whose current taste was for death and not life on earth. In spring 1863 Phillips proposed to Macmillan a second edition of *Life* and to Longman an edition of his *Manual* as a companion volume. Macmillan accepted this publication scheme, but wanted Phillips to write a new book which would take the reader 'through the ape to the snail when looking for our whence' in response to Lyell's *Antiquity of man* and Huxley's *Man's place in nature*. Phillips demurred: though capable of reviewing Lyell's *Antiquity* at length, he was not a comparative anatomist and he disliked the feuding between Owen and Huxley in section D of the British Association in 1862 because it diminished faith in science and the reputation of scientists for calmness and candour.[21]

13.4 Genesis and geology revisited

Phillips made a significant contribution to the debate engendered by the publication of *Essays and reviews* in March 1860. The volume sold 22,000 copies in two years and it generated a huge controversy involving 400 publications by 1865. Written by seven Anglicans, six of whom were ordained, it was widely denounced as subversive and heterodox. It brought to a head a gathering crisis in the Church of England. Two of 'the seven against Christ', Rowland Williams and Henry Wilson, were synodically indicted for heresy and in 1862 were suspended from their posts for a year; but early in 1864 the privy council quashed their indictment. Outraged by this triumph of civil over ecclesiastical jurisdiction, no fewer than 11,000 Anglican clergy signed a declaration drawn up at Oxford and promoted by Wilberforce. It recorded their belief in the plenary inspiration and divine authority of scripture, which was the word of God, and in everlasting punishment. After further agitation by Wilberforce, in June 1864 the Church of England, via both houses of convocation of Canterbury, condemned the book but this measure had no legal effect.[22]

Three essays were of particular interest to Phillips. Frederick Temple, then headmaster of Rugby School and later archbishop of Canterbury, pleaded for toleration as the opposite of dogmatism, so he welcomed the way in which geology had undermined a literal reading of the first chapters of Genesis. Powell, Phillips' mathematical colleague at Oxford, developed the view he had expounded as long ago as 1838, ie that miracles were open to scientific explanation and could not be used to support Christian faith. This notorious rejection of miracles seemed to Wilberforce to be 'scarcely veiled atheism'. Powell also argued that, as knowledge and faith were different and separate, the more knowledge advanced then the more 'Christianity, as a real religion, must be viewed apart from connexion with physical things'. As examples of advancing knowledge which palpably contradicted the letter of scripture, he adduced the discoveries of geology, the antiquity of man, the development of species, and the rejection of the idea of creation. Having dissociated the spiritual from the physical, Powell professed his admiration for Darwin's masterly *Origin* with its 'grand principle of the self-evolving powers of nature'. Phillips had been close to Powell in the late 1830s but, when settled in Oxford, seems to have had only formal contact with him. Powell had lost interest in promoting science at Oxford and in University Museum agitation. In the 1850s he became ever more concerned with publishing controversial works about the relation between the scientific and religious realms.[23]

A third essay, by Charles Goodwin, a barrister and Egyptologist, on Mosaic cosmogony contained a long attack on Buckland's Bridgewater treatise for attempting to reconcile the Mosaic narrative and modern geology. In so doing Buckland, claimed Goodwin, evaded the plain meaning of the Mosaic narrative, introduced obscurity into it, and disrespectfully despoiled it of consistency and grandeur by representing it as a series of elaborate equivocations. Buckland's reconciliatory enterprise seemed to Goodwin to be painful, humiliating, shuffling, stumbling, and pitiful. For Goodwin the Mosaic narrative was 'not an authentic utterance of divine knowledge but a human utterance which it has pleased providence to use in a special way for the education of mankind' about the goodness of the Creator to all his creatures, especially humans. Though Phillips was not unsympathetic to this view, presumably he did not welcome Goodwin's attack on Buckland because as recently as 1858 he had joined Owen and Robert Brown in updating Buckland's Bridgewater treatise geologically, palaeontologically, and botanically. Phillips' contribution was mainly restricted to technical footnotes commenting on the text. He refrained from any comments about Genesis and geology and natural theology. But he had associated himself generally with Buckland's 'good old book' which, in using geology to serve natural theology, tried to show that the earth's history teemed with endless examples of design and to strengthen belief in a wise, good, and supreme creator and all-sustaining providence.[24.]

Three contributors to *Essays and reviews* were clerical Oxonians, Powell, Mark Pattison, and Benjamin Jowett, so it was not surprising that two important quasi-official responses were organised by Oxonian clerics, Wilberforce and William Thomson. Not content with denouncing *Essays and reviews* in 1861 in the *Quarterly review*, Wilberforce was the driving spirit behind *Replies to Essays and reviews* (1862). Just as in his opposition to Darwinian evolution Wilberforce was smart enough to seek the advice of a leading scientific authority, Owen, so in *Replies* he was happy to include solicited contributions from two prominent Oxford scientists, Phillips and Main, both of whom saw adaptation, design, and a designer in the natural world. The approach of Main, a clerical astronomer, showed that he still thought it worthwhile to attempt that reconciliation between scripture and science which Goodwin had deplored. Main argued that some biblical statements were beyond scientific research, some were metaphorical, but some were amenable to re-expression using the best modern science. In contrast, Phillips, the solitary lay contributor to the volume, made no attempt to reconcile geology and scripture and refrained from exegesis of the latter. Drawing on the familiar two books approach, Phillips averred that 'we do not seek our Christianity in the rocks, nor our geology in the Bible; we do not confound two independent records'. He was also very suspicious of the assumption made by Main and Powell that current science was reliable because, given the limitations of human reason and senses, it gave no more than a feeble, imperfect, and provisional indication of the Divine Idea and of the laws by which God regulates nature. Phillips directed his main points against Powell but, because of his recent death, without naming him. Forced off his fence by Powell, Phillips aligned himself with what he called the clerical geological tradition of Conybeare, Sedgwick, and Buckland in that they saw in the geological record not 'an anti-Mosaic history of the creation of man, but pre-Mosaic tables of stone, inscribed by the hand of the Divine Master, and bearing earlier terms of the one creative series, whose latest period includes the history of man'. In his conclusion Phillips was hardly triumphant about the impacts of geology: it had added to the defences of natural theology, was not hostile to deductions from scripture, and was not at all opposed to Christian faith. These were not the ringing endorsements about the beneficial effects of geology which Wilberforce wanted.[25]

He did, however, hear them from one of his contributors, Gilbert Rorison, a Scottish episcopal minister who discussed the creative week described in Genesis. Drawing considerably on Phillips' *Life* apropos the age of the earth and the nature of the fossil record, Rorison was convinced that the strata were registers of creative and supernatural divine acts and that all nature was and is foreseen and piloted by divine thought and will. One alleged positive effect of geology was revealed in *Aids to faith*, a collection of theological essays edited by William Thomson, a good friend of Phillips. In an analysis of the Mosaic record, Alexander McCaul, professor of Hebrew and divinity at King's College, London, attempted a type of reconciliation of geology and scripture which Phillips generally avoided and deplored. McCaul claimed that Moses anticipated some of the most startling of recent scientific discoveries and therefore was inspired.[26]

13.5 The age of man

From the 1830s Phillips' publications made it clear that, like many geologists, he believed in the great antiquity of the earth but the relative novity of man. This confidence was fundamentally related to his long-held and firm belief in the theory of the cooling earth. Like his friend Harcourt, Phillips was sure that it was only after a long period of cooling, which produced a moderate temperature and other suitable conditions, that man had appeared on the earth and prospered. His belief in the late date of man on geological grounds had the advantage of not disagreeing with the authority of scripture. Yet he was not dogmatic about the novity of man. In his *Manual* (1855) he considered the well-established arguments for and against the co-existence of humans and extinct mammals. He was aware that in a few caves artefacts and human bones had been found in layers that also contained the remains of extinct mammals, but suspected that this stratigraphic association did not necessarily indicate contemporaneity of the various deposits. He was also aware that nowhere in Europe had human bones been found in the unconsolidated deposits known as drift. But he expected that further research might well demonstrate the co-existence of humans with mammoths and extinct carnivores.[27]

Between 1858 and 1860 his expectation was spectacularly confirmed not just by field work at two sorts of site but by intense discussion about the results of that field work by Pengelly, Falconer, Prestwich, Evans, and Lyell. The first site was Brixham cave, meticulously excavated from 1858 by Pengelly under the superintendence of a committee of the Geological Society. Its most active members were Prestwich and Falconer, who were Britain's leading authorities on European tertiary and quaternary strata and on the fossils they contained. By summer 1859 they were convinced that at Brixham flint implements occurred in association with the fossil bones of extinct mammals which had flourished in the pleistocene, ie ice-age period; and that the tools and bones were contemporaneous because they and their immediate surroundings seemed undisturbed. The second site was the gravel of the Somme valley in which Jacques Boucher de Perthes had discovered flint hatchets and remains of the mammoth. Published in 1846 his views were ridiculed, but in spring 1859 Prestwich and Evans, who had been stimulated by Falconer's account in late 1858 of Boucher's collection, vindicated his results. More importantly they claimed that at Amiens and Abbeville the hatchets and bones of extinct large mammals were contemporaneous. The gravel in which they were found was hard, compact, and apparently undisturbed; the hatchets, the bones, and their neighbouring delicate fossil shells were neither worn nor broken so the bones had not been washed into the gravel from older deposits. The new view about the antiquity of man was cogently expressed by Lyell as president of section C of the British Association in 1859. Having seem the Somme valley sites that summer, Lyell was sure that man co-existed with mammals now extinct, and existed when Europe was different in climate and topography from its present state. He stressed that the flint instruments were of great antiquity if compared with the times of history and separated from them by 'a vast lapse of ages'. Thus man was not new and, according to Lyell, had lived on the earth for far longer than 6,000-8,000 years.[28]

In 1863 Lyell published his famous and popular synthesis of arguments and data in favour of the *Antiquity of man*. It immediately annoyed Falconer and Prestwich (and later Lubbock) because they felt that Lyell had slighted their work by not sufficiently acknowledging it. Falconer thought that Lyell wrote as if he were the Louis XIV of geology who believed that 'La geologie: c'est moi! L'ancienneté d'homme; c'est moi aussi!' Much of Lyell's book was concerned with reviewing the evidence for the geological age of man who, he claimed, had appeared at the end of the ice age but before the extinction of the mammoth. On the assumption that geological processes of the past were of the same intensity as those of the present, Lyell implied that man had existed for 100,000 years. He devoted his concluding chapters to transmutation of species. To Darwin's disappointment and anger Lyell showed himself to be a half-hearted convert to Darwinian evolution. Lyell's reluctance was based not just on the arguments he had made in public for thirty years but also on his recent conversion to the view that Unitarianism was more comforting and consolatory than Anglicanism apropos creation and divine dispensation. In his *Antiquity* he revealed he derived comfort and happiness from his unprovable faith in the immortality of the soul as well as his belief in the uniqueness of man, with his moral faculty and power of self-improvement. In his final pages Lyell supported the view of Asa Gray that Darwinian evolution and natural selection did not destroy the arguments in favour of design and a designer; and he expressed the view, so dear to Owen and Phillips, that nature was the material embodiment of pre-concerted arrangement by a divine mind.[29]

Though Phillips continued to regard Lyell as dogmatic, biased, and too hypothetical in some of his principles of geology, he respected Lyell as an illustrious geologist distinguished by his comprehensive view of his subject. In return Lyell, who still had a strongly developed sense of *amour- propre*, deplored some of Phillips' views but unlike Darwin did not merely exploit him. In 1853 Lyell welcomed Phillips' appointment at Oxford and continued to view him as a formidable expositor in print of their subject. He gave detailed and useful practical advice on toasts and toasters when Phillips was about to preside over his first anniversary meeting at the Geological Society. On occasion he invited Phillips to his home. In 1855 he wanted Phillips to join him for ten days in field work at sites where, according to Ramsay, glacial action had occurred in the Permian period.[30]

As president of the Geological Society Phillips showed excessive caution when he avoided the question of the antiquity of man in his two addresses. Yet in 1858 Portlock, his predecessor, and in 1861 Horner, his successor, did not shirk the responsibility of describing the state of play about what they regarded as a hot topic and a great geological question. Horner not only supported the high antiquity of man but, having attacked marginal notes in some current bibles that the earth was created in 4004 BC, he urged that they should be removed because they were untrue and mischievous. It was ironic that Horner cited Phillips' *Life* not only on the time of 500,000 years required to deposit certain strata but also for Phillips' tentative views that early man was post-glacial, that he was contemporaneous with the extinct hippopotamus, and that the Somme flint instruments might indicate the higher antiquity of man.[31]

In 1863 Phillips was so concerned about Lyell's *Antiquity* that in August he visited St Acheul, near Amiens, to see the gravel beds which Lyell had examined previously on four separate occasions. Phillips' conclusions, given to the British Association that year, were not entirely in Lyell's favour. It seemed to Phillips that the layers of gravel had been disturbed by river action and that the age of the flint instruments was that of the latest disturbance of gravel. Though he accepted as probable the contemporaneity of man and the rhinoceros, he thought the last gravel deposit was no more than a few thousand years old because it was post-glacial.[32]

Phillips followed this evaluation of the Somme finds with a long anonymous critique of Lyell's *Antiquity of man* published in October 1863 in the *Quarterly review*. Murchison, who welcomed Phillips' opposition to Lyell's unqualified uniformitarianism, suspected Phillips was the reviewer. So too did Hooker who recognised 'that old frump Phillips' hand'. For his part Phillips thought he handled Lyell's pet notions mercifully and he refrained from mentioning the accusations of plagiarism that had been levelled against Lyell. Not for the first time Phillips commented on Lyell's 'simple faith in dogmatic geology' and addiction to favoured hypotheses, some of which were highly conjectural. Phillips doubted Lyell's belief that the materials overlying the flints and bones in the Somme valley had accumulated gradually over an immense span of time. Like James Forbes, Phillips criticised severely Lyell's estimate of 224,000 years as the duration of the glacial period. This 'dreamy computation' was based on the one case of the present-day rate of elevation of Scandinavian land, from which Lyell had arrived at the general conclusion that continents rose by two and a half feet per century. Phillips went on to ask why in those 224,000 years there were no traces of evolution. Having stressed that the steps of the fossil record were incompatible with the smooth sheet of transmutation, Phillips nevertheless admitted that more research was needed on the imperfection of the fossil record and the definition of species.[33]

In his review Phillips argued that Christian faith, which was based on moral evidence and the nature of man, did not need the support of geology, so Christians could look calmly on controversial topics such as evolution and the age of man. He made it clear that research on these matters should be pursued because man enjoyed the glorious destiny of 'never ending advancement' by diligently exercising God-given faculties in studying God's works, which being a form of worship was more valuable than winning arguments. Guided by these beliefs, Phillips came to two conclusions. Firstly, he accepted provisionally that man co-existed with mammoths, though he was still worried about the possible conflation of times of origin and deposition of materials. Secondly, he rejected as very questionable Lyell's computation of the immense antiquity of man. It was methodologically unsound; and the expected remains of such long-enduring men were not abundant but restricted to just flints. Phillips' tentative computation of the age of man, revealed privately but merely hinted at publicly, was around 10,000 years. This figure did not make him a convinced member of the Prestwich camp which settled on a figure of about 20,000 years as middle ground between the estimates of 100,000+ years of the Lyellians and of the 6,000 years of the traditionalists. Phillips' estimate, which put him between the Prestwichians and the traditionalists, preserved the novity of man on the geological timescale and was not incompatible with the authority of scripture.[34]

13.6 Last testimonies

For the last ten years of his life Phillips did not change appreciably his views on Darwinian evolution of species, the antiquity of man, design in nature, the age of the earth, and its long-term cooling, all of which he discussed in a cautious, unaggressive, and undogmatic way before the British Association, particularly as its general president and as president of its section C. Apropos what he called Darwin's 'elegant treatise', his position remained that there were insurmountable difficulties in accepting that forms of life were indefinitely variable with time and circumstance. His own view was that, though transmutation might be possible at the species level, it was not applicable to genera, families, orders, and classes, either in the past or the present. He still believed that the current state of knowledge of the fossil record gave no instance of supra-species evolution; that plants and animals had occurred in 'natural groups' which had common characters; and that each group had a limited distribution in space and time. But in the early 1870s he began to doubt his previous view that the fossil record was totally opposed to evolution by natural selection. After all, in the late 1860s he himself had helped Huxley to provide an important instance of a bird-like reptile, which he construed as a missing link between reptiles and birds, and then explained by postulating that these separate classes of animals had evolved from common ancestors. This experience persuaded Phillips that more research might fill gaps in the fossil record, so in 1873 he urged that the alleged imperfection of the geological record should be tested and remedied by collating results from different regions about the life periods of species, genera, and families. Thus at the end of his life Phillips was not complacent about fixity of species: for him Darwinian evolution had become a more open question and the arguments against it somewhat less decisive.[35]

Phillips' belief in the relative novity of man was reinforced in 1864 on his visit to Switzerland where he met Morlot who took him to see the flattened delta cones produced at the foot of a precipice by the torrent, La Tinière, on the Lake of Geneva near Villeneuve. One cone was composed of arched bands, about twenty feet thick, which contained respectively Roman, bronze, and stone age artefacts. Using literary evidence about the date of the Roman remains and making assumptions about the rate of deposition of layers, Morlot had concluded that the stone age occurred 6,000-7,000 years ago. At La Tinière Phillips examined another cone, estimated by Morlot to be 10,000 years old, and extracted from its lowest band, containing pottery and the bones of animals, a human cranial bone which he tried unsuccessfully to date. Morlot regarded Phillips as *the* appreciator of the cones but failed to persuade him to return to excavate them more systematically. For his part, in his 1863 review Phillips had welcomed Morlot's approximate estimates and then next year at the British Association publicised Morlot's latest conclusions about the ages of the cones. He also promoted Morlot's views in Lockyer's short-lived journal, the *Reader*.[36]

As president of the Association in 1865 Phillips re-affirmed his view that on the geological timescale the most ancient man was new. But he also called for further research on cave and river gravel deposits to establish whether humans and quadrupeds now extinct lived at the same time or whether the relics of different times had been gathered into one locality by natural processes. By the early 1870s he thought he had the answer with respect to the implements found in the gravel of the Thames valley: early man appeared after it had been deposited in the post-glacial period by fresh water. This result harmonised with Phillips' view that the earth's history and man's appearance on it were providentially arranged. After the former had undergone many changes it had become adapted to man's physical, intellectual, and moral nature, at which time 'it pleased the Giver of all good to place man upon the earth and bid him look up to heaven'. This rejection of an evolutionary ancestry for man was underpinned by Phillips' belief in man's uniqueness, especially the immortality of the soul which for him could not have been produced by natural selection. It was also buttressed by his continuing belief in a comprehensive plan of nature devised by a perpetual governor, the maker and preserver of nature. For him the forces of nature were selected by its ruler, and matter itself was a visible expression of divine purpose. Thus Phillips gave short shrift to materialism, atheism, pantheism, and to chance as enthroned by Darwin in his theory of natural selection.[37]

In the 1860s Phillips made common cause with William Thomson, the famous physicist, in criticising Darwin and his followers for their prodigal use of Lyell's vast timescale. Moreover Thomson and Phillips produced their own estimates of an alternative scale of geological time which was so restricted that evolution by slowly acting natural selection was impossible. They had met regularly before 1860 at meetings of the British Association. From 1855 at the latest Thomson knew that by the late 1830s Phillips had not only espoused the theory of the cooling of the earth but had also himself measured the rise in temperature with depth in its crust. In 1855 Thomson cited as reliable Phillips' figure of 1 °F for 45-60 feet. Seeing Phillips as an uncommon geologist who welcomed the assistance which physics could give to geology, Thomson consulted him in June 1861 about whether he or geologists in general subscribed to Darwin's 'prodigious durations for geological epochs'.

Thomson told Phillips that his preliminary calculations gave the age of the sun as 20 million years and that of the earth at the widest stretch as 200 to 1,000 million years. In his prompt reply Phillips informed Thomson that Darwin's computations were absurd and that Phillips' own calculations, given in his *Life on earth*, gave 96 million years as the age of stratified rocks.[38]

In 1862 Thomson published his estimate of the age of the sun's heat as falling between 100 and 500 million years. The following year, in the full version of his paper on the gradual cooling of the earth from its original molten state, he estimated the age of the earth's crust as 98 million years on the basis of certain assumptions. If they did not hold then the range became 20 to 400 million years. Thomson's method was not surprising to Phillips because over twenty years before Phillips had suggested in print that the age of the earth's crust could be determined in principle from the rate of long-term refrigeration of the globe. Thus by 1863 Phillips had used geological data and arguments to arrive at 96 million years as the age of the earth's sedimentary rocks, while Thomson used physical data and principles to generate 98 million years as the age of the earth's crust. These geological and physical conclusions were so strikingly concordant that together they made Lyell's steady-state picture of the earth's history even more implausible and they deprived Darwinian evolution of the huge timescale it needed.[39]

Thomson said nothing in public about Phillips' estimate of the age of the earth until in 1869 he castigated 'orthodox geologists' for ignoring it. Privately he was glad that Phillips was on his side, accepted Phillips' corrections, and rejoiced that they were in remarkable agreement about the age of the earth, in opposition to what Thomson called 'the Hutton, Lyell and Huxley banking and discount company (unlimited)'. He could have added Hooker who was also unconvinced by Thomson's physical estimates. In contrast Phillips was the first leading British geologist to welcome Thomson's thermodynamic approach. As president of section C of the British Association in 1864, Phillips recommended the computations of Thomson and Haughton who had estimated the age of the earth as 98 million and 2,300 million years respectively. Simultaneously, Lyell, as president of the Association, reiterated his opposition to the cooling earth theory. Next year Phillips, as president of the BAAS, again spoke favourably about Thomson's work on the earth's age. No wonder that Thomson regarded Phillips as a true geologist, because he welcomed physical instruments, approaches, and conclusions just at the time when Thomson was failing to persuade Huxley in public debate that physics could aid geology. Indeed, to the end of his life Phillips retained his enthusiasm for physical geology in general and for Thomson's thermodynamics in particular. As president of section C of the BAAS in 1873 Phillips dilated on the consequences of the cooling earth considered as a fact; and he took care to cite Thomson's view that the radiant heat of the sun was ineluctably decreasing and therefore not capable of arresting the gradual refrigeration of the earth.[40]

Phillips also shared with Thomson a strong belief in a designed universe. In his presidential address to the British Association in 1871 Thomson stressed that the argument for design had been badly served by those devotees of William Paley who had succumbed to the frivolities of naïve teleology. Thomson believed in an ever-acting creator and ruler because he saw strong proofs of intelligent and benevolent design all round him. Like Stokes, BAAS president in 1869, Phillips practised the fearless inductive investigation of secondary causes and their effects, without forgetting the great first cause and ignoring the pervasive and wonderful proofs of design in nature. To the end of his life Phillips saw natural phenomena as emblems of God and science as a reverent pursuit and a form of worship of the Creator. His theistic view of nature gave him the comfort and security which atheism, pantheism, materialism, and agnosticism could not provide. Though the Church of England was racked in the 1860s by several crises concerning faith and authority, Phillips' belief in a natural theological view of nature remained undisturbed.[41]

Apropos the relation between geology and scripture, Phillips maintained the position that he had held since the 1830s. He thought it pointless and injurious to try to reconcile or integrate the independent readings of the word of God in science and revelation, though he hoped that ultimately these readings would be in accord. For him there was no conflict between Genesis and geology and apparently no anxiety about their relation. It seems that he regarded Genesis as a narrative which used imagery, concepts, and information available at the time and place of its composition to convey the great spiritual truths of a creating and superintending God and his relation to man. Accordingly, it was futile to try to adjust readings of Genesis so that they were not at odds with or even fitted the recent conclusions of geology; and it was equally futile to try to find in Genesis anticipations or vindications of such conclusions.[42]

In the late 1860s Phillips had some success in promoting his views via one of the most important commentaries on the Bible published in the Victorian era. In 1864 John Denison, speaker of the House of Commons, was so worried by the damage being done to the Anglican church by *Essays and reviews* and by Colenso's attack on the historical veracity of the Pentateuch, that he suggested the publication of an informed and up-to-date commentary on the Bible for the laity. He consulted the archbishop of York, who backed the project strongly. Published in ten volumes from 1871 and edited by Frederic Cook, it was known as the *Speaker's Commentary*. Genesis was allocated to Edward Browne, bishop of Ely, who in 1863 had attacked Colenso, while admitting that there were apparent inconsistencies between Genesis and modern geology and that Moses was not a teacher of geology. Probably on the advice of the archbishop, Browne consulted Phillips several times from 1868 about the six days of creation, the Noachian deluge and the antiquity of man. In his introduction and commentary on Genesis, Browne was generally happy to adopt Phillips' responses without naming him: Genesis was not a revelation of scientific truths and Lyell's figure of 100,000 years for the antiquity of man was far too high. But on the Flood Phillips failed to persuade Browne who took the view that there had been an extensive local submergence in 'a signal departure from the common course of nature and providence', which taught a great moral lesson.[43]

13.7 Irenic Christianity

The death of Phillips's sister in 1862 was the great grief of his adult life. It strengthened his views about the Christian life, increased his concern with the immortality of the soul, and sharpened his belief that death was merely a terrestrial separation. Many private poems he wrote in these years contrasted the sadness, woe, and shortness of human life with God's redeeming power, the return of the God-like aspects of humans to God after physical death, and heavenly rest. In one poem of 1867, entitled 'No sect in heaven,' he revealed his vision of heavenly beatitude in which Christians were indistinguishable because they lacked the accoutrements, sacred objects, and behaviour, which he regretted were encouraged by terrestrial sects. He deplored sectarianism within Christianity because it divided Christians, promoted dogmatism, and at its worst encouraged bigotry, authoritarianism, aggression, and conflict. His ideal was that of the Christian irenarch, promoting peace between sects and recognising that their members were trying by their own lights to be Christians. In the intellectual sphere his irenicism was associated with advocating and implementing cautious induction, avoiding dogmatic conjectures, and conducting scientific discussion in a non-pugilistic way. Socially it involved discouraging feuding, avoiding schadenfreude, treating colleagues politely, and generally being gentle and considerate. Politically it meant avoiding extremes and being a convinced but unobtrusive Gladstonian liberal, especially when Gladstone was a Peelite member of Parliament for the University from 1847 to 1865. No doubt Phillips welcomed Gladstone's sense of providential mission, his respect for old usages, and his Peelite view that change was welcome if it was the only alternative to incompetence or injustice. In 1865 Phillips showed his support for Gladstone, who lost his University seat in July 1865, by giving a pound for his election expenses and the same sum for his memorial fund.[44]

It seems from the testimony of those who knew Phillips in his Oxford period that, remarkably, he lived up to his irenic ideals. Without ever being judged as an insufferable paragon, he was widely regarded as uncompetitive, humble, cheerful, and reverent. He grieved at the unseemly contests provoked by odium theologicum and odium scientificum; and he deplored 'the wasted time of intolerant controversy'. Above all he was seen by friends and critics as genial. Hull had no reason to exaggerate when he declared that at the British Association Phillips' 'genial face and lucid words brought sunshine wherever he appeared, and threw light on every topic he handled'. It is tempting to see Phillips' irenicism as being rooted in an incarnationist approach to the Christian life. Though he did not write about this question per se and, like Acland, did not obtrude his Christianity on others, the notion of incarnationism throws useful light on Phillips' behaviour. Incarnationists, who flourished for around thirty years from 1850, rejected the evangelical obsessions with sin, guilt, spiritual danger, mortification, retribution, and eternal damnation. They saw Christ as divine and a perfect human whose example was to be followed via piety, reverence, faith, rectitude, and brotherly love, in an attempt to create a sort of heaven on earth. They hoped that in the afterlife a loving and gracious God would provide beatitude. Some of them, such as Lord Morpeth, stretched the notion of brotherly love to include the unobtrusive performance of public duty, clubable friendship, the encouragement of social cohesion, and even kind looks. Two of Phillips' obituarists, Willett and Hull, averred that the best way of mourning his death was to copy his purity of life. Willett went so far as to claim that Phillips represented the '*highest style of man*' devoted to 'intellectual pursuits in an active and earnest life, tempered by considerate words and gentle actions'.[45]

As an exponent of irenic and incarnationist Christianity Phillips had easy working relations with many colleagues and friends of several sectarian persuasions both within and without the Church of England. In York his oldest friends were Kenrick (a Unitarian divine), Harcourt (an Anglican clergyman), and Allis (a Quaker). In Cambridge his oldest and closest friend was Sedgwick, whose evangelical tendencies and denunciatory style he avoided. From the mid-1860s Phillips taught and advised Cookson, the conservative master of Peterhouse and a vice-chancellor of the University. Phillips was so impressed by the reverend William Selwyn, professor of divinity, that he subscribed a guinea to Selwyn's Cambridge Young Men's Christian Society. No doubt Phillips deplored Huxley's lecture to it in March 1870 when he claimed that science, often vilified by Christians, was neither Christian nor unChristian but extra Christian.[46]

Even in Oxford, a cockpit of religious controversy, Phillips found it possible to have productive relations with Anglicans of different sectarian persuasions, whether they were scientific colleagues, university officials, or college heads. In the 1850s he aided the scientific research of Walker, an evangelical cleric, and Johnson, a lay tractarian, in experimental philosophy and astronomy respectively. In the 1860s and early 1870s he assisted two accomplished clerical astronomers: he collaborated with Main on astronomical research and worked successfully with Pritchard to promote astronomy in the University; but he did not share their desire, revealed in sermons, to reconcile science and religion.[47] Phillips enjoyed straightforward relations with important University officials, all clergymen, including six successive vice-chancellors, two of whom (Cotton and Jeune) were evangelicals. Phillips was a good friend of Henry Coxe, a cleric who as head of the Bodleian Library excluded party spirit from it and ruled it with tactful courtesy, much as Phillips did at the University Museum.[48] Phillips was also friendly with several heads of colleges, again all clergy, who were not vice-chancellors during his Oxford period. He was on good terms with Cradock, Plumptre, Griffiths, Charles Williams, John Matthias Wilson, and especially William Thomson who until 1860 was a broad churchman. When Thomson, appointed archbishop of York in 1862, delivered his inaugural address in 1866 as president of the Yorkshire Philosophical Society, he praised his 'intimate friend' Phillips as a true philosopher who combined patient research, 'unaffected goodness and piety', and 'the reasonable caution which is so needful a counterpoise to the pride and over-confidence of successful research'. For his part Thomson was Phillipsian in some of his characteristic attitudes. He was an ecumenical bishop, always generous to nonconformists. He believed in the indivisibility of truth, the harmony between science and religion as two lines of research into truth, nature as God's creation, the arguments from scientific laws to design and from design to a designer, and in science as requiring induction, deduction, and imagination. As archbishop he spoke at the BAAS and local scientific societies and was well informed about current science. On occasion he castigated the Church of England for not doing enough to understand recent developments in science.[49]

Phillips counted as friends several local clergymen, some of whom were nationally known and others destined to be so. He gave a large donation to the new school promoted by Nicholas Moody, rector of St Clement's Oxford, who also wanted Phillips to give a geological talk at a tea party for his congregation on the wonders we tread on and God's wisdom. Phillips was on dining terms with Bishop Wilberforce, William Kay, a well-known biblical scholar, and two leading high churchmen, Burgon and Henry Mansel. Phillips presumably met Mansel at Magdalen College where he was reader in theology from 1855; and he may well have sympathised with the views expressed in Mansel's Bampton lectures of 1858 on the limits of religious thought in which he applied Kantian notions about the limits of human knowledge to argue that only revelation provided a secure basis for religion and the Bible's chief use was to guide conduct. After Mansel became dean of St Paul's Cathedral in 1868, Phillips still knew him well enough to lobby him on behalf of Wiltshire, the clergyman who ran the Palaeontographical Society. At Oxford Phillips met Adam Farrar, whose interest in science was reflected in books on science in theology and the history of free thought and maintained when he moved in 1864 to Durham to be professor of divinity. Another Oxford friend was William Jones, who moved to York in 1863 to be the archbishop's right hand man. A moderate evangelical, he liked a little high church ritual for its colour and became bishop of St David's Cathedral in 1874.[50]

Finally, Phillips contributed to the publications of three prominent local clergymen, Wilberforce, Pattison, and Pusey, a trio whose members were not always *ad idem*. In 1862 Wilberforce was happy to have Phillips' name with his own on the title page of *Replies to essays and reviews*. For his part in 1873 Phillips lamented the death of the eloquent bishop. In 1869 Pattison published an edition of Alexander Pope's *Essay on man* and was grateful to Phillips for corrections to the introduction. One of Phillips' professorial colleagues was Pusey, professor of Hebrew and canon of Christ Church, Oxford. Within the Oxford or tractarian movement of the 1830s he was the leader of the high church ritualists who were known from 1838 as Puseyites. After Newman's secession to the Roman Church in 1845, Pusey and his followers had some success in reviving catholic doctrine and observances in the Church of England. In the 1850s he appeared to Tuckwell as a 'devout Casaubon, unconscious of contemporary trivialities, aloof in patristic reverie and spiritual pathology'. In 1870 Pusey succeeded in squashing a proposal to confer an honorary degree on Huxley, whom he regarded as a black heretic. Yet Pusey was not an opposer of science in general. At Oxford in the 1850s he supported the new Museum because it was devoted to science conceived as reading the book of God's works. During the next decade he turned for advice to Phillips who became not only a geological but also theological guide.[51]

In 1862 John Colenso, bishop of Natal, claimed in his *Critical examination of the Pentateuch* that the first five books of the old testament, traditionally ascribed to Moses, were forgeries so they were neither literally accurate nor inspired. He was soon deposed and excommunicated for heresy by Robert Gray, bishop of Cape Town, but after his case had become a cause célèbre, the Privy Council decided in 1865 that Gray had no power to try Colenso. Meanwhile in 1863 the bench of bishops in England had condemned Colenso. In 1869 it took firm action when on its authority a new bishop of Natal was consecrated by Gray to replace Colenso. From 1869 until Colenso's death in 1883, there were two rival bishops sitting in Natal.[52]

Pusey supported Gray but was so worried by Colenso's interpretation of the Pentateuch and his rejection of a universal flood that in 1863 he consulted Phillips twice, first about the physical evidence for a universal flood and then about Colenso's statement, derived from Lyell, that in the Auvergne the volcanic hills were formed ages before the Noachian deluge and their loose covering had never been disturbed. In the only reply that survives, Phillips expressed the Lyellian view that these hills were not recent but mainly of eocene age and doubted whether there was any special evidence for a flood. He went on to tell Pusey that there were two sorts of lava in Auvergne, which Daubeny (whom Pusey knew) had called antediluvian and postdiluvian because they dated from before and after the erosion of the present valleys. Phillips stressed that in Buckland's time these dates were known as pre- and post- Noachian but that in the 1860s such terms meant before and after the valleys were formed. Without complicating matters by alluding to Scrope's and Lyell's denial of the classification of these volcanoes into ante and postdiluvian, and their assumption of a vast scale of geological time, Phillips told Pusey clearly that there had not been a Noachian flood in that part of France.[53]

At the Church Congress held in October 1865 Pusey led a discussion about the spirit in which science should be applied to the study of the Bible. His intended irenicon, in which he argued that faith and science were not antagonistic because each had its distinctive features, was deeply indebted to Phillips, twice called an eminent geologist, for several key points. Pusey claimed that theology should be fearless and calm apropos the results of science, that it was mischievous to try to square God's words with imperfectly understood facts, that it was thus premature of theologians to welcome Buckland's theory of a universal deluge for which currently there was no evidence, and that the grounds for the antiquity of man were precarious. Pusey not only cited Phillips' exposure of Lyell's abuse of arithmetic in *Antiquity of man* but also reproduced from Phillips' 1860 presidential address to the Geological Society three paragraphs on the computation of geological time.[54]

Early in 1866 Pusey preached a sermon in Oxford on the miracles of prayer in relation to natural phenomena. It was a contribution to a long-running dispute, fuelled from the early 1860s by Tyndall who claimed that prayer could not be physically efficacious, that praying indicated ignorance of science, and that official decreeing of special days for prayer at a time of national crisis was a gross clerical abuse. The immediate context of the sermon was the cattle plague of 1865 and 1866. Phillips heard a preliminary version of it but felt that Pusey had erred in asserting that prayer was physically effective apropos phenomena, such as meteorological ones, which were not yet subject to scientific laws. In a long letter to Pusey, Phillips argued that the fixed laws of science applied, actually or potentially, to all material phenomena so he tried to demonstrate at length that the certain occurrence of sunrise and the apparently uncertain occurrence of a rain shower were equally subject to the operation of scientific laws, because the Divine Architect had created a system of nature composed of a general plan and special adjustments. If prayer were efficacious then fresh adjustments, which could not happen in the usual sequence of cause and effect, occurred as a consequence of human intercession. Phillips' answer to what he regarded as a perplexing problem was that such fresh adjustments were simultaneously 'examples of unfathomed law and proofs of foreseeing condescension [by God] to particular supplications of His creatures'. Pusey replied that he yearned to see no evidence of fixed laws in just those matters about which people should pray and that God could be thought of as being like a chemist who when asked exerted his will to perform reactions with entirely regular and predictable results. But he also accepted Phillips' argument that meteorological phenomena were potentially as lawlike as those of astronomy and, without mentioning Phillips by name, modified the sermon accordingly.[55] Phillips therefore enabled Pusey to produce not only an informed response to Tyndall but also, as he told Phillips, to avoid pitting theology against the possible results of science in the future. Thus Phillips, who suspected Puseyism as a sect, changed the views of its leader. Like Acland, Phillips helped Pusey to keep an open mind about science and to avoid judging it solely from the aggressive and irreverent claims of scientists such as Tyndall and Huxley.

Notes

1. For an older view Chadwick, 1970, p.3; for a revisionist view and criticisms of the conflict thesis, Brooke and Cantor, 1998; Brooke, 2001b; Cantor, 1991; Cantor and Kenny, 2001; Turner, 1993.

2. Chambers, 1844b; Secord, 2000, pp. 222-47; Whewell, 1845; Sedgwick, 1845, 1850.
3. Phillips, 1860b, pp. 188-91; Phillips, inaugural lecture notes, PP, box 112, f. 1.
4. Chambers, 1994, Vestiges, pp. 60, 63, 74, 93, 202 (q); Explanations, pp. 39 (q), 92 (q), 164.
5. Darwin to Phillips, 14 Feb, 7 and 12 March, 7/14 April 1848 in Burkhardt and Smith, 1988, vol. 4, pp. 113-14, 121-3, 131-2; Darwin to Phillips, 18 and 28 Jan 1856 in Burkhardt and Smith, 1990, vol. 6, pp. 22-3, 33; Hopkins, 1848, pp. 80, 94, 98 and Darwin, 1848, pp. 315-17 both used Phillips' data about erratics being carried upwards; Phillips, 1856a, pp. 375-6.
6. *QJGS*, 1859, vol. 15, pp. xxiii-xxiv; Darwin to Phillips, 21 Jan, 8 Feb, 11 and 26 Nov 1859 (q), Darwin to Huxley, 25 Nov 1859, Darwin to Lyell, 2 Dec 1859 (qs)in Burkhardt and Smith, 1991, vol. 7, pp. 237-8, 245-6, 371-2, 403, 399-400. 409-10.
7. Phillips, 1860a, pp. xxxii, xxxvii (qs); Phillips, 1859a, p. lxi.
8. Darwin, 1859, pp. 293-7; *Saturday review*, 24 Dec 1859, pp. 775-6;
9. Phillips, 1860a, pp. li-lii (qs); Darwin, 1859, p. 293 (q); Darwin to Lyell, 18/19 Feb 1860, in Burkhardt and Smith, 1993, vol. 8, pp. 92-4.
10. Phillips, printed synopsis of twelve lectures on the fossil remains of plants and animals, Easter and Trinity terms, 1854, PP, box 112, f. 2; Phillips to Murchison, 6 March 1857, MP; Sedgwick to Phillips, 15 April 1860, PP; reverend Latimer Neville (1827-1904) vice-chancellor 1859-61, master of Magdalene College 1853-1904, to whom Phillips, 1860b, was dedicated. The other Rede lecturers of the 1860s were Robert Willis, Sabine, Ansted, Airy, Tyndall, William Thomson, Ruskin, Max Müller and Huggins.
11. Owen, 1859; for a sympathetic account of Owen's approach, see Rudwick, 1976a, pp. 207-14; Robert Rede (d 1519), *DNB*.
12. Sedgwick to Phillips, 15 (q), 23 April 1860, PP; Browne, 2002, pp. 117-18; Clark and Hughes, 1890, vol. 2, pp. 361-2, 603.
13. Phillips, 1860b, pp. 203-4; Phillips, notes for Rede lecture, PP, box 103, f. 9; Hooker to Henslow, June 1860 (q), Hooker to Harvey, 1860, in Huxley, 1918, vol. 1, pp. 514-15, 519-20; Sedgwick to Phillips, 19 May 1860, PP; Joseph Dalton Hooker (1817-1911), *DSB*, director of Kew Garden 1865-85.
14. Phillips to Sedgwick, 1 May 1860, Se P; Macmillan to Phillips, 30 April, 16, 21 May, 24 Sept, 16, 17, 19 Nov 1860, 5, 16 Jan 1861, list of recipients, PP, mainly box 47; Alexander Macmillan (1818-96) was publisher to Cambridge University 1860-80
15. Phillips, 1860b, p. 3 (q) 4-5, 45; Dawkins, annotations on Phillips, 1860b, p. 45, Dawkins papers, Buxton Museum; Phillips to Sedgwick, 22 Nov, 24 Dec 1860 (qs), Se P; Owen 1858, p. li; Owen, 1860a, pp. 3, 407, 413-14; Phillips, lecture notes, hilary term 1858, PP, box 112, f. 9; neo-Platonists regularly interpreted Empedocles' cosmology as equivalent to their distinction between the intelligible and perceptible universes.
16. Owen, 1860a, pp. 397-8; Phillips, 1860b, pp. 24, 46, 187, 213; Harcourt, 1860, p. 176.
17. Wallace to Darwin, 30 Nov 1861, in Burkhardt and Smith, 1994, vol. 9, pp. 356-8; Phillips, 1860b, pp. 9-10, 31-45 (42 q), 217 q; independently of Darwin, Alfred Russell Wallace (1823-1913), *DSB*, adumbrated the theory of evolution by natural selection.
18. Phillips, 1860b, pp. 83-4, 98-9, 113-14, 119, 171-3, 206-7, 212-15; Darwin, 1859, p. 315 q; Rudwick, 1976a, pp. 228-39.
19. Burchfield, 1974, p. 59; Phillips, 1855c, p. 618; Phillips, 1838a, pp. 293-4; Phillips, 1860b, pp. 122-37.
20. Darwin to Lyell, 20 (q), 25 Nov 1860, Darwin to Hooker, 17 Dec 1860 (q), in Burkhardt and Smith, 1993, vol. 8, pp. 479-82, 494-6, 531-3; Owen, 1860b; Gray, 1861; Huxley to Phillips, 28 Nov 1860, PP; Darwin to Hooker, 15 Jan 1861, Darwin to Horner, 20 March 1861 (q), in Burkhardt and Smith, 1994, vol. 9. pp. 8-9, 62-3, 162-3; Lyell to Ticknor, 29 Nov 1860, in Lyell, 1881, vol. 2, pp. 340-2; Phillips, 1860b, pp. 185-7.
21. Macmillan to Phillips 16, 21 Nov, 4 Dec 1860, 5 Jan (q), 4 March (q), 12, 29 June 1861, 23 March 1863 (q); Murchison to Phillips, 19 Nov 1860; Sanders to Phillips, 1 March 1861, all PP; Salter to Sedgwick, 5 Jan 1861, Se P; Phillips, 1860b, p. 126; Page, 1861; on

section D in 1862, Rupke, 1994, pp. 296-8, Phillips to Acland, 15 Oct 1862, 24 Feb 1863, Acland papers, Bodleian Library, d. 200; the copy in PP, box 103, f. 8 of Phillips, 1860b, annotated by him in 1861 for a second edition and re-titled *Life and death on the earth*, shows that he made little attempt to make it more popular; the proposal for a companion *Manual* was never implemented; David Page (1814-79), *DNB*, author of geological primers.

22. Ellis, 1980; Altholz, 1994; Meacham, 1970, pp. 217-32, 246-51; Rowland Williams (1817-70), *DNB*; Henry Bristow Wilson (1803-88). *DNB*.

23. Temple, 1860, pp. 43, 47; Powell, 1860, pp. 128, 139 (qs); Wilberforce, 1861, p. 252; for Powell's theological preoccupations in the 1850s see Corsi, 1988, pp. 194-224; Frederick Temple (1821-1902), *DNB*.

24. Goodwin, 1860, pp. 230, 249-253 (253q); Buckland, 1858; Phillips to ?, 30 Sept 1857, Bu P (q); Charles Wycliffe Goodwin (1817-78), *DNB*.

25. Meacham, 1970, pp. 217-32; Goulburn et al., 1862, esp pp. 501-13 (Main), 514-16 (Phillips, 1862b, solicited by Cotton); Mark Pattison (1813-84), *DNB*, rector of Lincoln College, 1861-84; Benjamin Jowett (1817-93), *DNB*, professor of greek 1855-93.

26. Rorison, 1862, pp. 322-3; Thomson, 1861; Thomson, 1919, pp. 109-28; McCaul, 1861; Gilbert Rorison (1821-61); Alexander McCaul (1799-1863), *DNB*.

27. Harcourt, 1860, p. 185; Phillips, 1860b, p. 47; Phillips, 1855c, pp. 435-8.

28. Grayson, 1983, pp. 168-98; Van Riper, 1993, pp. 74-116 (115 q).

29. Bynum, 1984; Porter, 1982a, p. 29 (q); Lyell, 1863, pp. 111, 144-6, 498, 505-6; Lyell, 1881, vol. 2, pp. 320-2; Bartholomew, 1973.

30. Lyell to Phillips, 21 Sept 1853, 8 Feb, 8 June, 25 July 1855, 16, 18 Feb 1858, PP.

31. Portlock, 1858, pp. clxi-clxii; Horner, 1861, pp. lx-lxx.

32. Phillips, St. Acheul, Aug 1863, PP, notebook 81; Phillips, 1863f.

33. Phillips, 1863f, pp. 407-8 (qs), 411-12; Murchison to Harcourt, 1 Nov 1863, HP; Hooker to Darwin, 9 March 1864, in Burkhardt and Smith, 2001, vol. 12, pp. 64-5; Phillips to Murray, 12 Oct 1863, Mu P; Forbes, 1863, esp p. 291.

34. Phillips, 1863g, pp. 415(q), 416; Phillips to Murray, 12 Oct 1863; Van Riper, 1993, pp. 144-83.

35. Phillips, 1865b, pp. lxii-lxiii (qs); Huxley, 1870; Phillips, 1873b, p. 74

36. Phillips, 1864e; Morlot to Phillips, 20 Sept 1864, PP, box 101, f. 9; Lockyer to Phillips, 19 Dec 1864, Morlot to Phillips, 10 Feb 1865, PP; Phillips, 1863g, pp. 381-2.

37. Phillips, 1865b, p. lix (q), lxi; Phillips, 1871a, pp. 463-92; Phillips, Nebular hypothesis, [1860s], PP, box 111, f. 9.

38. Thomson, 1855; Thomson to Phillips, 7 June 1861, PP; Phillips to Thomson, 12 June 1861, KP; for Thomson on the cooling and age of the earth, see Smith and Wise, 1989, pp. 526-33, 552-611.

39. Thomson, 1862, 1863; Phillips, 1838a, p. 294; Rudwick, 1976a, pp. 257-60.

40. Thomson 1871a, read 1869; Thomson to Phillips, 22 Dec 1868, 1 June 1869 (q), PP; Phillips, 1864d, 1865b, p. liv; Lyell, 1864, p. lxix; Phillips, 1873b, p. 72.

41. Thomson, 1871b, p. cv; Stokes, 1869, p. civ; on Stokes as public religious scientist, see Wilson, 1984.

42. Phillips, notes on Genesis [1860s], PP, box 114, f, 2.

43. Thomson, 1919, pp. 233-43; Browne, 1863, pp. 80-1; Browne, 1871, pp. 30, 36, 59-63, 74-8 (78 q); Browne to Phillips, 25 Nov 1868, EUL Gen 784/1, 2, 7, 23 Dec 1868, 9 June 1869, PP; John Evelyn Denison (1800-73), *DNB*, speaker 1857-72; Frederic Charles Cook (1810-89), *DNB*; Edward Harold Browne (1811-91), *DNB*, bishop of Ely 1864-73 and Winchester 1873-90; John William Colenso (1814-83), *DNB*.

44. Phillips, Poems 1856-73, University of London Library, MS 517; Phillips, no sect in heaven [c 1867], PP, box 102, f. 6; for Phillips' Gladstonian liberalism, Bernard to Phillips, 24 April, Palmer to Phillips 24 April, Wayte to Phillips, 4 Aug 1865, PP; for irenic behaviour in science, see Cantor, 2001, esp pp. 340-2.

45. Hull, 1874, p. 68; Willett, 1878, 7th report, June 1874, p. 8 (q), 10th report, March 1875, p. 8 (q); Hilton, 1988, 1994.
46. Cookson to Phillips, 16 May 1866, 29 April, 7 June 1867, 17 Oct 1871, Selwyn to Phillips, 13 May 1870, all PP; Huxley, 1918, vol. 1, p. 328; William Selwyn (1806-75), *DNB*.
47. Walker to Phillips, 2 Jan 1856, 13 Sept 1857, Johnson to Phillips, 28 Dec 1858, Main to Phillips, 2, 9 Nov 1865, 29 April 1867, Pritchard to Phillips, 14 Feb 1870, all PP; for co-operation with Pritchard see section 10.6; for Pritchard's reconciliation of science and religion, see Pritchard, 1897, pp. 84-8, 90-4, 113, 116, 176, 187, 190-2, 204-12.
48. The vice-chancellors were: Cotton, 1852-6; David Williams, 1856-8; Jeune, 1858-62; John Prideaux Lightfoot (1803-87), 1862-6, rector of Exeter College 1854-87; Leighton, 1866-70; and Liddell 1870-4. Lightfoot to Phillips, 22 Sept 1865, 4 Jan 1866, Coxe to Phillips, nd, all PP; Henry Octavius Coxe (1811-81), *DNB*, Bodleian librarian 1860-81.
49. Cradock to Phillips, 10 Dec 1867, Plumptre to Phillips, 20 April 1867, Griffiths to Phillips 3 Aug 1866, 21 April 1868, Charles Williams to Phillips, 5 Dec 1869, 3 Jan, 21 Aug, Wilson to Phillips, nd. [*c* 1870], Thomson to Phillips, 5 Nov 1863, all PP; Thomson, 1866, p. 4 (q); Bullock, 1891, pp. 15, 17, 19, 27-8; Kirk Smith, 1958 pp. 139-41; John Matthias Wilson (1813-81), *DNB*, professor of moral philosophy 1846-74, president Corpus Christi College 1872-81.
50. Moody to Phillips, 10, 19 Oct; Wilberforce to Phillips, 1 Jan 1866; Kay to Phillips, 15 Jan 1866; C.A. Mansel to Phillips, 16 Jan 1865; H.L. Mansel to Phillips, 20 Jan 1866, 6 July 1871; Wiltshire to Phillips, 4 July 1871; Farrar to Phillips, 24 Jan 1869; Davies to Phillips, 26 March 1866; Jones to Phillips, 16 July 1867, all PP; Nicholas James Moody (1811-58); William Kay (1820-86), *DNB*, fellow of Lincoln College 1840-67; on Henry Longueville Mansel (1820-71), *DNB*, as begetter of agnosticism see Lightman, 1987, pp. 32-90; Adam Storey Farrar (1826-1905), *DNB*, fellow of Queen's College 1852-63, tutor at Wadham College 1855-64; William Basil Tickell Jones (1822-97), *DNB*, fellow of University College 1851-7, lecturer in classics and modern history 1858-65.
51. Goulburn et al, 1862; Phillips to M. Harcourt, 21 July 1873, HP; Pattison to Phillips, 1 Nov 1869, PP; Pattison, 1869; Tuckwell, 1900, p. 150, was alluding to Isaac Casaubon (1559-1614), a classical scholar; Browne, 2002, pp. 337-9; Atlay, 1903, p. 348.
52. Hinchliff, 1964, esp pp. 85-114; Robert Gray (1809-72), *DNB*, bishop of Cape Town 1847-72.
53. Colenso, 1862, pp. vii-viii; Pusey to Phillips, 12 Nov [1863], 23 Nov [1863], PP; Phillips to Pusey, 24 Nov 1863, Pusey papers, Pusey House, Oxford; for Daubeny on Auvergne, see Rudwick, 1974.
54. Liddon, 1898, vol. 4, pp. 77-81; Pusey, 1865; Pusey to Phillips, 7 Feb [1866], PP.
55. Pusey, 1866, pp. 11-13, 31-2; Turner, 1974; Phillips to Pusey, 30 Jan 1866, Pusey to Phillips, nd [Feb 1866], 7 Feb [1866], all PP. ·

Conclusion

Phillips was not only the chief pupil of William Smith, who began to be recognised in the late 1820s as the father of English geology. He was also a destitute orphan who was brought up, protected, and employed by his uncle for about seventeen years. Long after Smith's death in 1839, Phillips remained loyal to Smith and through publications tried to keep his name alive. In his *Life on the earth* (1860), Phillips claimed that Smith's discovery of the successive stages of the stratification of Britain 'gave the basis of a true palaeontology, and a true history of the succession of life on the globe'. Phillips' last big work, his *Geology of Oxford* (1871), praised Smith's 1815 map, his subsequent county maps, his still useful knowledge of the fossils of the fuller's earth rock, and for taking the first step towards using zones of ammonites to demarcate liassic deposits. Phillips' posthumous *Geology of the Yorkshire coast* (1875) was dedicated to the memory of Smith by his still affectionate nephew and grateful pupil who had begun the work under Smith's auspices.[1] As Smith's executor, Phillips was responsible for Smith's destitute and insane widow and presumably paid for her to be looked after in a lunatic asylum at York in her last years. Phillips was less interested in erecting memorials to Smith. He did not accelerate the production of a bust of Smith by Matthew Noble in 1849 for the Northampton church where Smith was buried. In 1872 Phillips offered to donate a marble bust of Smith to the Bath Museum, as a result of hearing that Samuel Sharp had presented a cast of the Northampton bust to the Geological Society and intended to do the same for the Bath Museum in which William Mitchell was keen to erect a memorial to Smith.[2]

Though Phillips learnt geologically from Smith, his uncle's career as a mineral surveyor indicated what his nephew should avoid. Smith was often in difficult financial circumstances, being forced to sell his precious collection of fossils in the 1810s, suffering imprisonment for bankruptcy in 1819, and becoming dependent on charity in the late 1820s. His circumstances denied him the opportunity for regular study and exacerbated his tendency to procrastinate apropos publication. Recognition of his achievements came to him belatedly partly because he left it to others to diffuse them in print because he was happier with oral exposition. As a lecturer he was engaging but rambling. In sharp contrast Phillips constantly sought financial security by taking a variety of paid jobs connected with science, and especially geology, until finally he attained it at Oxford. Through his paid activities he acquired opportunities for regular study and research. As a writer he worked, finished, and published, gaining a high reputation in the 1830s as a geological researcher and expositor. As a lecturer who was fluent and structured he was in such demand that in the 1830s he worked himself to exhaustion. Many of his career moves clearly involved gaining financial security, especially from 1829 when he and his sister set up home together. His final move also brought him the social security of high status at Oxford. In one other respect Smith may have provided a negative model for Phillips. No observer knew better than he the difficulties of Smith's marriage to a wife who was mentally unstable. Phillips saw painfully that marriage was not necessarily perpetual domestic bliss; and his uncle's torment may well have made him so cautious about marriage that it seems he never seriously contemplated it, even though in the late 1820s he had several unmarried women friends in Yorkshire.

In a letter of 1848 to his friend Ramsay Phillips modestly said that his own research on the geology of Yorkshire, palaeozoic fossils, and the Malvern area was a development, coloured by his own contribution, of views taught him as a boy by Smith in London. He remembered that they discussed topics such as synclines, anticlines, denudation, and the distribution of life in relation to the nature of sediments, the age of deposit, and geographical position, on all of which Smith remained silent in print. As Torrens and Knell have recently stressed, it is difficult to reconstruct from the published and manuscript sources Smith's confused but changing views about geological theory.[3] Even so, in broad terms Phillips' debts to Smith are clear. Through his uncle he learned how to identify strata in the field from their order of superposition, their lithological features, and their characteristic fossils; he was taught the use of surveying instruments and used them in constructing both traverse and columnar sections. Under his uncle he absorbed the field craft of observing and inferring the boundaries between strata and learned how to colour topographical maps geologically, sometimes on a large scale. From his uncle he derived in his teens an interest in tables showing the distribution of fossils between strata. Subsequently he developed a statistical palaeontology concerned with characteristic combinations of life, which he used to characterise geological systems and to periodise the history of the earth.

By necessity Smith had a narrow range as a geologist. He was happy to study the stratigraphy of just England. He had little interest in igneous geology and was suspicious of the aids which zoology, botany, chemistry, and physics might give to geology. In contrast, Phillips went out of his way in the late 1820s to extend his geological experience. He visited the Continent and Scotland where he studied structural geology (mountain building and folding), glacial phenomena, and igneous effects as well as stratigraphy. Especially from the early 1830s Phillips tried to exploit sciences collateral to geology, the most significant example being his interest in physical geology, particularly his obsession with the theory of the cooling earth which became central to his directionalist view of the earth's history and man's late appearance on it. In any event during Smith's lifetime Phillips learned from others besides his uncle. Via correspondence, meetings, reading, and field work, he derived benefit in his twenties from other British geologists, such as Conybeare, and from a few continental ones, such as Adolphe Brongniart and Cuvier, of whom he became a life-long devotee. From 1830 Phillips developed several of his characteristic geological concerns in opposition to Lyell's *Principles of geology*. At this time he constructed his philosophy of geology as a positive inductive science, free from the tyranny of hypotheses but not excluding the operations of the mind; and he was also Baconian in his belief in the value of co-operation in science. In all these ways Phillips was not a disciple of Smith.

There is one debt of Phillips to Smith which is often ignored. Smith often exhibited 'a considerable facility in mechanical inventions'.[4] Phillips' skill with measuring instruments for surveying was certainly derived from his uncle who encouraged him to invent new devices such as his lithographic press. Trained by Smith as an engineer, Phillips subsequently became unique among British geologists as a designer and maker of several scientific instruments which he used not only in geology but in terrestrial physics, especially meteorology, magnetism, and electricity. From the early 1850s Phillips' virtuosity in surveying, mapping, and instrumentation was deployed in astronomical research on the moon, Mars, and sun spots, in which he used his skills as a draughtsman and pioneer astronomical photographer to try to record features possibly analogous to those on earth. By the early 1860s Phillips had designed and built his own private observatory, housing an instrument made to his specification by Cooke whose early career was encouraged by Phillips. After the death of his sister in 1862, astronomy may well have provided silent consolation and a feeling of communion with her.

Smith's views on science and religion remain obscure but Phillips' were clear. From the early 1830s he regarded it as futile and dangerous to Christianity to try to reconcile the results of geology with readings of Genesis. As he wrote in the final edition of his *Guide* (1864), 'geology rests on no traditional fable, no theological dictum'.[5] But he had long believed in the certainty and constancy of laws appointed by the designer, maker, and governor of nature. Phillips saw nature as theocentric and the reading of the book of nature as a reverential, even sacred, act. From the late 1830s his Anglicanism was devout but non-sectarian and irenic. Particularly in his Oxford period it affected his behaviour in scientific institutions and controversies: in the former he was genial; in the latter undogmatic and unaggressive. It had not always been so. In the 1830s, when he was establishing his reputation and under considerable stress, he gave on occasion insufficient acknowledgement to the research of others.

Smith's wife give him no support, companionship, or intellectual help in their troubled marriage. In stark contrast, the bachelor Phillips received these abundantly for thirty-three years from his unmarried younger sister. Except in the late 1830s when Phillips was on the edge of exhaustion and breakdown, their relation was symbiotic. In exchange for protection, Anne Phillips promoted his career in vital ways. In York and Oxford she ran their residences, in which she acted as an accomplished host to his guests, whether they were ordinary Yorkists, distinguished foreigners, geological chums, visiting BAAS stars, or University dignitaries. During Phillips' short-lived professorship in Dublin, she resided with him at Kingstown. When he worked for the Survey in Pembrokeshire and Malvern, she organised and stayed in the temporary accommodation which he used as base stations for his geological forays. On occasion she accompanied him on geological and lecturing trips and to meetings of the BAAS, the leaders of which respected and liked her. Particularly in Phillips' Oxford period she was an acceptable guest with him at dinners and took trouble to have good relations with the families of leading Oxonians. She was a capable woman who in his absence read his rain gauges and, having been taught by her brother, she coloured geologically maps for his books. In his Survey period she helped to sort and catalogue the fossil and mineral spoils. She was not unique at the time in contributing to the geological success of a male relative. Mary Lyell was a tamed wife, sitting silently at the dinner table while Charles held forth at length. Charlotte Murchison gave intellectual purpose and subsequently considerable wealth to her husband. Emma Darwin's dowry made Darwin a rich leisured gentleman in a secure domestic environment, while Mary Buckland polished her husband's prose.[6] But none of these women was identified with a venerated field object named after her. Whereas Anne Phillips' brother was commemorated eponymously via a maximum thermometer, a lunar crater, and two areas of Mars, she secured eponymous recognition in the form of Miss Phillips' conglomerate, specimens of which were eagerly sought for decades by geologists visiting the Malvern Hills because it was central to any interpretation of their history and structure.

Phillips' career thus illuminates the nature of the Smithian inheritance; but it also brings into sharp relief the difficulties of making a career in science in nineteenth-century Britain. As an orphan who lacked independent means, Phillips was peculiarly aware that he had to make ends meet financially. He constantly aimed to be well paid for contributing to science pursued as a vocation under his own direction. This ideal career was not always realisable so he undertook a wide range of paid activities in order to secure a livelihood. Over a period of almost thirty years he was a provincial museum keeper, a public lecturer, an encyclopaedist, a textbook writer, an employee of the Geological Survey, a professor at King's College, London, and Trinity College, Dublin, a government commissioner, an occasional consultant, the key administrator of the British Association, a writer of popular books about Yorkshire, and a reviewer for the weekly, the *Athenaeum*.[7] In some years he pursued four or five of these activities in order to try to secure occupational and financial security. Several of his roles and posts, initially attractive, turned out to have such drawbacks that he relinquished them or resigned. Public lecturing in metropolis and province led him to the edge of a nervous breakdown in the late 1830s. The post of professor, so superficially alluring, had disadvantages. Qua professor at KCL and TCD Phillips decided to be no more than a visiting lecturer; and, having considered chairs at UCL and the Royal Dublin Society, he rejected them because their rewards and prospects were so uncertain. Phillips' pre-Oxonian life shows dramatically the difficulties and instabilities of a career in science at that time.

Important shifts in his career were not regular, planned, or expected but were initiated by sheer chance and contingencies totally outside his control and often accomplished by personal patronage. Being orphaned brought him into close contact with his uncle. Harcourt's invitation to Smith to lecture in York put Phillips in touch with the YPS and gave him a job and a powerful patron. The way in which Brewster fixed on York as the first place of meeting of the British Association led to Phillips' becoming its key administrator. Finally, Strickland's unexpected death permitted Daubeny and Sedgwick to help to bring Phillips to Oxford, thus rescuing him from relative provincial obscurity, enabling him at last to make all his talents tell fully, concluding the Odyssean phase of his career, and ending his financial insecurity.

Phillips' Oxford career qualifies the tenacious assumption that as late as the 1870s Oxford shunned science as inferior, useless, and dehumanising. The opposite was the case. As keeper of the University Museum Phillips presided over a university science centre of unrivalled size and scope in Britain and he promoted the establishment of the Clarendon Laboratory and the University Observatory. The creation of such facilities shows that for about twenty years the University supported science strongly. Indeed, by the mid-1870s it seemed that Oxford, not Cambridge, was poised to become Britain's premier university for science.

The issues of professionalisation and secularisation are also illuminated by Phillips' career at Oxford. In a celebrated article Frank Turner argued that there was a connection in mid-Victorian Britain between professionalisation and scientific naturalism: advocates of the latter, such as Huxley and other members of the X-club, aggressively promoted the former. Recently, Turner has stressed that at this time various groups attempted to wrest or to protect cultural dominance for their particular professions and philosophical outlooks. Furthermore, in a critique of the secularisation thesis, he has found it wanting because various eminent Victorians saw the relation between the secular and religious as symbiotic not hostile.[8] Turner's claims may be usefully reformulated: other groups besides the propounders of naturalism promoted professionalisation. One was that coterie of Scottish natural philosophers and engineers, identified by Crosbie Smith as the north British physicists. In the mid-Victorian period these distinguished professional scientists, led by William Thomson, transformed natural philosophy into energy physics and initiated experimental physics in laboratories. Because they pursued their science within a Christian framework, they were at bitter odds with the proponents of scientific naturalism, especially Huxley and Tyndall.[9] Another group of professional scientists, more pacific but possessing equally strong Christian commitments, was clustered around the Museum and Phillips at Oxford. While deploring anticlericalism, scepticism, materialism, and atheism, they were responsible for the impressive growth of facilities for science and some recognition of it in examinations, both of which are viewable as aspects of professionalisation. These two important cases confirm the continuing importance in the mid-Victorian period of eminent lay scientists who simultaneously saw no conflict between their science and Christianity, believed in natural theology, and promoted the professionalisation of science.

Notes

1. Phillips, 1860b, p. 187; Phillips, 1871a, pp. 3-4, 7, 166; Phillips, 1875, p. iii.
2. Baker to Phillips, 17 Oct 1849; Sharp to Phillips, 4 Dec 1871; Winwood to Phillips, 14 Dec 1871, 23 Jan, 12 Feb 1872; Mitchell to Phillips, 23 Jan, 6 Feb 1872, all PP; Sheppard, 1917, pp. 214-20; Matthew Noble (1816-76); Samuel Sharp (1814-82), *DNB*; William Stephen Mitchell (1840-92) projected a biography of Smith in the 1870s.
3. Phillips to Ramsay, 22 Oct 1848, RP; Torrens, 2001, pp. 79-81; Knell, 2000, pp. 20-7.
4. Phillips, 1844a, p. 129.
5. Phillips, 1864a, p. 6.
6. Phillips to Murray, 23 April 1847, 26 March 1849, Mu P; Desmond and Moore, 1992, pp. 271, 278; Kölbl-Ebert, 1997a and b.
7. Most of Phillips' reviews in the *Athenaeum*, nine in 1841 and four in 1842, were short notices; but in the longer ones he promoted such favourite themes as the importance of physics in geology, of the internal heat of the globe, and of local research which ideally generated data and a spirit of enquiry. I am indebted to Hugh Torrens for information about these reviews, derived from an annotated copy in the Library, City University, London.
8. Turner, 1993, pp. 3-37, 171-200.
9. Lightman, 2001; Smith, 1998.

Appendix 1
Lecture courses given by Phillips to YPS

Date	Topic	Number of lectures
1825, Feb	Geology	5
1826, Feb, March	Geology and fossils	8
1827, Feb, March	Invertebrate zoology	6
1829, July	Aquatic animals of York	6
1830, May	Natural history of quadrupeds	3
1831, March, April	Geology	6
1832, Jan	Geology of Yorkshire	4
1832, April, May	Phenomena of atmosphere	4
1833, Oct, Nov	Natural History of birds, reptiles, and whales	4
1834, March	Inferior aquatic animals	4
1839-40, winter	Zoological and geological contents of Yorkshire Museum	6
1842, March	Geology	4
1853, April	Physical geography of Yorkshire	4

Appendix 2
Lecture courses given by Phillips outside York

Date	Topic	Number of lectures	Host/venue
1825 June	Geology and fossils	6	Wakefield
1825 Dec, 1826 Jan	Geology and fossils	10	Leeds PLS
1826 April, May	Geology and fossils	10	Hull LPS
1827 Feb, March	Invertebrate zoology	6	Leeds PLS
1827 June	Invertebrate zoology	6	Royal Manchester Institution
1827 Nov, 1828 Feb, Sept, Oct	Invertebrate zoology	9	Sheffield LPS
1828 May	Invertebrate zoology	9	Hull LPS
1830 Jan	Aquatic animals of Yorkshire	6	Leeds LPS
1831 May, June	Geology	12	University College London
1831 Aug	Natural history of animals	6	Royal Manchester Institution
1831 Oct	Geology	6	Halifax LPS
1832 July	Geology	6	Royal Manchester Institution
1832 Nov, Dec	Geology	8	Hull LPS
1834 April-June	Geology	8	King's College London
1834 Oct, Nov	Geology (+ 5 field trips)	30	Newcastle LPS
1834 Dec, 1835 Jan	Inferior aquatic animals	6	Royal Manchester Institution

1835 May, June	Geology, esp heat of earth	8	King's College London
1836 June, July	Geology	6	Royal Manchester Institution
1836 June, July	Geology	6	Manchester Athenaeum
1836 June, July	Geology	6	Manchester Mechanics' Institution
1836 June, July	Geology	6	Liverpool Royal Institution
1837 May	Fossils	6	King's College London
1837 June, July	Geology	8	Birmingham Philosophical Institution
1838 Feb-June	Geology	12	Royal Institution, London
1838 April, May	Geology	6	London Institution
1838 April, May	Geology	6 (x 2)	Bristol Institution
1838 May, June	Geology	8	King's College, London
1838 June	Geology	8	Chester
1838 July	Geology	8	Royal Manchester Institution
1838 July	Geology	8	Manchester Athenaeum
1838 July	Geology	8	Manchester Mechanics' Institution
1838 Oct	Geology	6	Sheffield LPS
1839 Feb, March	Geology, practical	12	King's College London
1839 Feb, March	Heat of earth	6	Royal Institution, London
1839 March, April	Geology	6	London Institution
1840 Oct, Nov	Fossils	6	Royal Manchester Institution
1840 Oct, Nov	Fossils	6	Manchester Athenaeum

1840 Oct, Nov	Fossils	6	Liverpool Mechanics' Institute
1842 March	Geology	6	Royal Manchester Institution
1844 Jan-March	Geology	10	Royal Institution, London
1849 December	Geology	6	Royal Manchester Institution
1849 December	Geology	6	Manchester Athenaeum
1853 Jan-March	Geology	9	Royal Institution, London
1853 April	Physical geography of Yorkshire	4	Leeds Mechanics' Institute
1853 Dec – 1854 Jan	Geology	6	London Institution
1857 Jan-March	Geology	10	Royal Institution London
1860 Oct	Life on the earth	4	Royal Manchester Institution

Bibliography

Acland, H.W. and Ruskin, J. (1893), *The Oxford Museum*, Allen, London and Orpington.

Addyman, P. (1981), 'Archaeology in York, 1831-1981', in Feinstein, C.H. (ed.), *York 1831-1981. 150 years of scientific endeavour and social change*, Sessions, York, pp. 53-87.

Agassiz, J.L.R. (1833-43), *Researches sur les poissons fossiles*, 5 vols, Petitpierre, Neuchâtel.

Alborn, T.L. (1986), 'The York Unitarians: theology, science, and the tensions of modern liberalism', unpublished BA thesis, Harvard University.

Allen, D.E. (1978), *The naturalist in Britain. A social history*, Penguin, Harmondsworth.

_____ (1987), 'The natural history society in Britain through the years', *Archives of natural history*, vol. 14, pp. 243-59.

Altholz, J.L. (1994), *Anatomy of a controversy: the debate over essays and reviews 1860-1864*, Scolar Press, Aldershot.

Andrews, J.H. (1975), *A paper landscape: the Ordnance Survey in nineteenth-century Ireland*, Clarendon Press, Oxford.

Anon (1831), review of Phillips, 1829b, *Philosophical magazine*, vol. 9, pp. 342-54, 430-41.

Anon (1851), *The illustrated exhibitor ... of the Great Exhibition of the industry of all nations*, Cassell, London.

Anon (1860), review of Phillips, 1860b, *The critic*, vol. 21, pp. 671-2.

Anon (1865a), 'Professor John Phillips', *The leisure hour*, pp. 557-60.

Anon (1865b), 'British Association for the Advancement of Science. Meeting at Birmingham, September, 1865. The President's address', *Quarterly journal of science*, vol. 2, pp. 727-30.

Anon (1870), 'Eminent living geologists. John Phillips', *Geological magazine*, vol. 7, pp. 301-6.

Anon (1874), 'John Phillips', *Nature*, vol. 9, pp. 510-11.

Archaeological Institute (1848), *Memoirs illustrative of the history and antiquities of the county and city of York, communicated to the annual meeting of the Archaeological Institute of Great Britain and Ireland, held at York, July 1846, with a general report of the proceedings of the meeting, and catalogue of the museum formed on that occasion*, Archaeological Institute, London.

d'Archiac, E. and Verneuil, E. de (1842), 'On the fossils of the older deposits in the Rhenish provinces; preceded by a general survey of the fauna of the Palaeozic rocks, and followed by a tabular list of the organic remains of the Devonian system in Europe', *TGSL*, 1842, vol. 6, pp. 303-410.

Armstrong, A. (1974), *Stability and change in an English county town. A social study of York, 1801-51*, Cambridge University Press, Cambridge.

Atlay, J.B. (1903), *Sir Henry Wentworth Acland ... A memoir*, Smith, Elder, London.

Bailey, E. (1952), *Geological Survey of Great Britain*, Murby, London.

Baines, H. (1840), *The flora of Yorkshire*, Longman, London.

Bakewell, R. (1833), *An introduction to geology*, 4th ed, Longman, London.

Barrande, J. (1852-1911), *Système Silurian du centre de la Bohême*, 8 vols, Barrande, Prague.

Bartholomew, M.J. (1973), 'Lyell and evolution: an account of Lyell's response to the prospect of an evolutionary ancestry for man', *British journal for the history of science*, vol. 6, pp. 261-303.

_____ (1979), 'The singularity of Lyell', *History of science*, vol. 17, pp. 276-93.

Bell, C.R. (1969), 'The Swimming Bath on the Manor Shore 1837-1923', *YPS report*, pp. 33-42.

Bellot, H.H. (1929), *University College London 1826-1926*, University of London Press, London.

Berman, M. (1978), *Social change and scientific organization. The Royal Institution, 1799-1844*, Heinemann, London.

Berry, H.F. (1915), *A history of the Royal Dublin Society*, Longman, London.

Bigsby, J.J. (1868), *Thesaurus Siluricus. The flora and fauna of the Silurian period*, Van Voorst, London.

Binney, E.W. (1868-75), *Observations on the structure of fossil plants found in the carboniferous strata*, Palaeontographical Society, London.

Bischof, C.G.C. (1854-9), *Elements of chemical and physical geology ... translated ... by Benjamin H. Paul and J. Drummond*, 3 vols, Cavendish Society, London.

Black, A. and Black, C. (1845), *Picturesque guide to the English Lakes, including an essay on the geology of the district by John Phillips*, 2nd edn, Black, Edinburgh; 13th edn, 1865; 21st edn, 1888.

Blackwell, J.K. (1850), 'Report to the secretary of state for the home department on the ventilation of mines and collieries', *Parliamentary papers*, vol. 23, pp. 443-73.

Blainville, H.M.D. de (1827), *Mémoire sur les bélemnites, considérées zoologiquement et géologiquement*, Levrault, Paris.

Blau, E. (1982), *Ruskinian gothic: the architecture of Deane and Woodward, 1845-1861*, Princeton University Press, Princeton.

Blomfield, C.J. (1831), *The duty of combining religious instruction with intellectual culture. A sermon preached in the chapel of King's College London at the opening of the institution on the 8th of October 1831*, Fellowes, London.

Bonney, T.G. (1919), *Annals of the Philosphical Club of the Royal Society*, Macmillan, London.

Boylan, P. (1981), 'A new revision of the pleistocene mammalian fauna of Kirkdale Cave, Yorkshire', *PYGS*, vol. 43, pp. 253-80.

Brander, G. and Solander, D.C. (1766), *Fossilia Hantoniensia collecta, et in Musaeo Britannico deposita*, London.

Brewster, D. (1826), 'Results of the thermometrical observations made at Leith Fort every hour of the day and night, during the whole of the years 1824 and 1825', *Edinburgh journal of science*, vol. 5, pp. 18-32.

_____ (1827), 'Notice respecting the hourly meteorological observations proposed by the Royal Society of Edinburgh, to be made twice every year, on the 17th July and 15th January', *Edinburgh journal of science*, vol. 6, pp. 144-9.

_____ (1830), 'Decline of science in England', *Quarterly review*, vol. 43, pp. 305-42.

British Association for the Advancement of Science (1864), *Index to reports and transactions from 1831 to 1860 inclusive*, Murray, London.

Brock, W.H. (ed.) (1967), *The atomic debates. Brodie and the rejection of the atomic theory*, Leicester University Press, Leicester.

_____ (1997), *Justus von Liebig: the chemical gatekeeper*, Cambridge University Press, Cambridge.

Brongniart, Adolphe T. (1825), 'Observations sur les végétaux fossiles renfermés dans les grès de Hoer en Scandinavie', *Annales des sciences naturelles*, vol. 4, pp. 200-24.

_____ (1828a), *Histoire des végétaux fossiles, ou recherches botaniques et géologiques sur les végétaux renfermés dans les diverses couches du globe. Volume 1*, Dufour and D'Ocagne, Paris and Amsterdam.

_____ (1828b), *Prodrome d'une histoire des végétaux fossiles*, Levrault, Paris and Strasbourg.

Brongniart, Alexandre (1829), *Tableau des terrains qui composent l' écorce du globe; ou, essai sur la structure de la partie connue de la terre*, Levrault, Paris.

Brooke, J.H. (1979), 'The natural theology of the geologists: some theological strata', in Jordanova, L. and Porter, R.S. (eds), *Images of the earth: essays in the history of the environmental sciences*, British Society for History of Science, Chalfont St Giles, pp. 39-64.

_____ (1989), 'Scientific thought and its meaning for religion: the impact of French science on British natural theology, 1827-1859', *Revue de synthèse*, vol. 110, pp. 33-59.

_____ (1991), 'Indications of a creator: Whewell as apologist and priest', in Fisch, M. and Schaffer, S., *William Whewell: a composite portrait*, Clarendon Press, Oxford, pp. 149-73.

_____ (1999), 'The history of science and religion: some evangelical dimensions', in Livingstone, D.N., Hart, D.G., and Noll, M.A. (eds), *Evangelicals and science in historical perspective*, Oxford University Press, New York and Oxford, pp. 17-40.

_____ (2001a), 'The Wilberforce-Huxley debate: why did it happen?', *Science and Christian belief*, vol. 13, pp. 127-41.

_____ (2001b), 'Religious belief and the content of the sciences', in Brooke, J.H., Osler, M.J., and Van der Meer, J.M. (eds), *Science in theistic contexts. Cognitive dimensions. Osiris*, vol. 16, pp. 3-28.

_____ and Cantor, G.N. (1998), *Reconstructing nature. The engagement of science and religion*, Clark, Edinburgh.

Browne, E.H. (1863), *The Pentateuch and the elohistic psalms in reply to Bishop Colenso*, Parker, London.

_____ (1871), 'General introduction, introduction and commentary on Genesis', in Cook, F.C. (ed.), *The Holy Bible. Volume 1. Part 1. Genesis to Exodus*, Murray, London, pp. 1-236.

Browne, J. (2002), *Charles Darwin. The power of place. Volume 2 of a biography*, Cape, London.

Buckland, F.T. (ed.) (1858), *Geology and mineralogy considered with reference to natural theology. By the late William Buckland. A new edition, with additions by professor Owen; professor Phillips; Mr Robert Brown. And memoir of the author*, Routledge, London.

Buckland, W. (1818), 'Order of superposition of strata in the British Islands', in Phillips, W., *A selection of facts from the best authorities arranged so as to form an outline of the geology of England and Wales*, Phillips, London.

⸻ (1822), 'Account of an assemblage of fossil teeth and bones of elephant, rhinoceros, hippopotamus, bear, tiger, and hyena, and sixteen other animals; discovered in a cave at Kirkdale, Yorkshire in the year 1821: with a comparative view of five similar caverns in various parts of England, and others on the continent', *Philosophical transactions*, vol. 112, pp. 171-236.

⸻ (1823), *Reliquiae diluvianae; or observations on the organic remains contained in caves, fissures, and diluvial gravel, and on other geological phenomena, attesting the action of an universal deluge*, Murray, London.

⸻ (1836), *Geology and mineralogy considered with reference to natural theology*, Pickering, London.

⸻ (1840), 'Presidential address', *PGS*, 1842, vol. 3, pp. 210-67.

Bud, R.F. (1974), 'The Royal Manchester Institution', in Cardwell, D.S.L. (ed.), *Artisan to graduate. Essays to commemorate the foundation in 1824 of the Manchester Mechanics' Institution*, Manchester University Press, Manchester, pp. 119-133.

Bullock, C. (1891), *The people's archbishop: the late most reverend William Thomson, archbishop of York*, Home Words, London.

Bunbury, F. (ed.) (1890-3), *Memorials of Sir C.J.F. Bunbury, Bart.*, 9 vols, privately printed, Mildenhall (copy in Cambridge University Library).

Burchfield, J.D. (1974), *Lord Kelvin and the age of the earth*, Macmillan, London.

Burkhardt, F.H. and Smith, S. (eds) (1986), *The correspondence of Charles Darwin, Volume 2. 1837-1843*, Cambridge University Press, Cambridge; 1988, vol. 4, 1847-1850; 1990, vol. 6, 1856-1857; 1991, vol. 7, 1858-1859; 1993, vol. 8, 1860; 1994, vol. 9, 1861; 1999, vol. 11, 1863; 2001, vol. 12, 1864.

Burton, J.M.C. (1988), 'The history of the Meteorological Office to 1905', PhD thesis, Open University.

Butcher, N.E. (1983), 'The advent of colour-printed geological maps in Britain', *Proceedings of the Royal Institution of Great Britain*, vol. 55, pp. 149-61.

Bynum, W.F. (1984), 'Charles Lyell's *Antiquity of man* and its critics', *Journal of the history of biology*, vol. 17, pp. 153-87.

Cantor, G.N. (1991), *Michael Faraday: Sandemanian and scientist*, Macmillan, London.

⸻ (2001), 'Quaker responses to Darwinism', in Brooke, J.H., Osler, M.J., and Van der Meer, J.M. (eds), *Science in theistic contexts. Cognitive dimensions. Osiris*, vol. 16, pp. 321-42.

⸻ and Kenny, C. (2001), 'Barbour's fourfold way: problems with his taxonomy of science-religion relationships', *Zygon*, vol. 36, pp. 765-81.

Carrington, R.C. (1863), *Observations of the spots on the sun from November 9, 1853, to March 24, 1861, made at Redhill*, Williams and Norgate, London.

Casella, L.P. (1860), *An illustrated and descriptive catalogue of philosophical, meteorological, mathematical, surveying, optical and photographic instruments, manufactured by L.P. Casella*, Lane, London.

Chadwick, O. (1970), *The Victorian church. Part II*, Black, London.

Challinor, J. (1963), 'Some correspondence of Thomas Webster, geologist – IV', *Annals of science*, vol. 19, pp. 285-97.

Chambers, R. (1844a), 'The scientific meeting at York', *Chambers' Edinburgh journal*, vol. 2, pp. 321-4.

_____ (1844b), *Vestiges of the natural history of creation*, Churchill, London.

_____ (1994), *Vestiges of the natural history of creation and other evolutionary writings*, ed. by J.A. Secord, University of Chicago Press, Chicago and London.

Chapman, A. (1998), *The Victorian amateur astronomer. Independent astronomical research in Britain, 1820-1920*, Wiley, Chichester.

Charlesworth, E. (1835), 'Observations on the crag formation and its organic remains; with a view to establish a division of the tertiary strata overlying the London clay in Suffolk', *Philosophical magazine*, vol. 7, pp. 81-94.

_____ (1836), 'On the crag of Suffolk, and on the fallacies connected with the method now usually employed for ascertaining the relative age of tertiary deposits', *Philosophical magazine*, vol. 8, pp. 529-38.

_____ (1837), ' Observations on the crag, and on the fallacies involved in the present system of classification of tertiary deposits', *Philosophical magazine*, vol. 10, pp. 1-9.

Chester, G.J. (1881), *Notes on the present and future of the archaeological collections of the University of Oxford*, Shrimpton, Oxford.

Church, R.A. (1986), *The history of the British coal industry. Volume 3. 1830-1913: Victorian pre-eminence*, Clarendon Press, Oxford.

Clark, E.K. (1924), *The history of 100 years of life of the Leeds Philosophical and Literary Society*, Jowett and Sowry, Leeds.

Clark, J.W. and Hughes, T.M. (1890), *The life and letters of the reverend Adam Sedgwick*, 2 vols, Cambridge University Press, Cambridge.

Cleevely, R.J. (1974), 'A provisional bibliography of natural history works by the Sowerby family', *Journal of the Society for the Bibliography of Natural History*, vol. 6, pp. 482-559.

Clerke, A.M. (1887), *A popular history of astronomy during the nineteenth century. Second edition*, Black, Edinburgh.

Cockburn, W. (1838), *A letter to professor Buckland, concerning the origin of the world*, Hatchard, London.

_____ (1840), *The creation of the world. Addressed to R.I. Murchison, Esq. and dedicated to the Geological Society*, Hatchard, London.

_____ (1844), *A sermon on the evils of education without a religious basis*, Whittaker, London.

_____ (1845), *The Bible defended against the British Association: being the substance of a paper read in the geological section, at York, on the 27th of September, 1844. Fifth edition. To which is added a correspondence between the Dean and some members of the Association*, Whittaker, London and York.

Colby, T.F. (1837), *Ordnance Survey of the county of Londonderry. Volume the first. Memoir of the city and north western liberties of Londonderry. Parish of Templemore*, Hodges and Smith, Dublin.

Colenso, J.W. (1862), *The Pentateuch and book of Joshua critically examined. Part 1*, Longman, London.

Collinge, W.E. (1925), 'John Phillips', *YPS 1824 report*, pp. 37-46.

Colvin, H. (1983), *Unbuilt Oxford*, Yale University Press, New Haven and London.

Conybeare, W.D. (1830-1), 'On Mr Lyell's Principles of Geology' and 'An examination of those phaenomena of geology, which seem to bear most directly on theoretical speculations', *Philosophical magazine*, vol. 8, pp. 215-19, 359-62, 401-6; vol. 9, pp. 19-23, 111-17, 188-97, 258-70.

_____ (1833) 'Report on the progress, actual state and ulterior prospects of geological science', *BAAS 1831-2 report*, pp. 365-414.

_____ (1834), 'Reply to 'Layman' on geology', *Christian observer*, vol. 34, pp. 306-9.

_____ and Phillips, W. (1822), *Outlines of the geology of England and Wales, with an introductory compendium of the general principles of that science, and comparative views of the structure of foreign countries*, Phillips, London.

Corsi, P. (1988), *Science and religion: Baden Powell and the Anglican debate 1800-1860*, Cambridge University Press, Cambridge.

Cotta, C.B. Von (1866a), *Die geologie der gegenwart dargestellt und beleuchtet*, Weber, Leipzig.

_____ (1866b), *Rocks classified and described. A treatise on lithology. An English edition by Philip Henry Lawrence*, Longman, London.

Cox, L.R. (1942), 'New light on William Smith and his work', *PYGS*, 1942-8, vol. 25, pp. 1-99.

Crowe, M.J. (ed.) (1998), *A calendar of the correspondence of Sir John Herschel*, Cambridge University Press, Cambridge.

Cunningham, F.F. (1990), *James David Forbes. Pioneer Scottish glaciologist*, Scottish Academic Press, Edinburgh.

Curthoys, M.C. (1997), 'The examination system' in Brock, M.G. and Curthoys, M.C. (eds), *The History of the University of Oxford. Volume VI. Nineteenth-century Oxford, Part 1*, Clarendon Press, Oxford, pp. 339-74.

Curwen, E.C. (ed.) (1940), *The journal of Gideon Mantell surgeon and geologist*, Oxford University Press, London.

Cuvier, G. (1812, 1825). *Recherches sur les ossemens fossiles de quadrupès, où l'on rétablit les caractères de plusiers espèces d'animaux que les révolutions du globe paroissent avoir détruites*, Deterville, Paris, 4 vols, (3rd ed, 1825, Dufour and d'Ocagne, Paris, 5 vols).

_____ and Brongniart, A. (1808), 'Essai sur la géographie minéralogique des environs de Paris', *Annales du Muséum d'Histoire Naturelle*, vol. 11, pp. 293-326.

_____ (1810), 'Memoir on the mineral geography of the environs of Paris', *Philosophical magazine*, vol. 35, pp. 36-58.

_____ (1811), 'Essai sur la géographie minéralogique des environs de Paris', *Mémoires de la classe des sciences mathématiques et physiques de l'Institut Impérial de France Année 1810*, pp. 1-278.

Dalton, J. (1793), *Meteorological observations and essays*, Ostell, London.

Darwin, C.R. (1839), *Journal of researches into the geology and natural history of the various countries visited by H.M.S. Beagle*, Colburn, London.

_____ (1848), 'On the transportal of erratic boulders from a lower to a higher level', *QJGS*, vol. 4, pp. 315-23.

_____ (1849), 'Geology' in Herschel, J.F.W. (ed.), *A manual of scientific enquiry; prepared for the use of Her Majesty's navy: and adapted for travellers in general*, Murray, London, pp. 156-95.

_____ (1859), *On the origin of species by means of natural selection or the preservation of favoured races in the struggle for life*, Murray, London. Reprinted 1968, Penguin, Harmondsworth.

Daubeny, C.G.B. (1826), *A description of active and extinct volcanoes; with remarks on their origin, their chemical phenomena, and the character of their products, as determined by the condition of the earth during the period of their formation*, Phillips, London (2nd ed, 1848).

——— (1853), *Can physical science obtain a home in an English university?* in *Miscellanies: Being a collection of memoirs and essays on scientific and literary subjects published at various times*, Parker, Oxford, 1867, vol. 2, pp. 41-59.

——— (1856), 'Presidential address', *BAAS 1856 report*, pp. xlviii-lxxiii.

——— (1860), 'Remarks on the final causes of the sexuality of plants, with particular reference to Mr Darwin's work, "on the origin of species by natural selection"', *BAAS 1860 report*, pp. 109-10.

Daubrée, G.A. (1859), *Études et expériences synthétique sur le métamorphisme et sur la formation des roches cristallines*, Dunod, Paris.

Davidoff, L. and Hall, C. (1987), *Family fortunes. Men and women of the English middle class, 1780-1850*, Hutchinson, London.

Davidson, T. (1851-86), *British fossil brachiopoda*, Palaeontographical Society, London.

Davies, G.L. Herries (1969), 'The University of Dublin and two pioneers of English geology: William Smith and John Phillips', *Hermathena*, no. 109, pp. 24-36.

——— (1983), *Sheets of many colours: the mapping of Ireland's rocks, 1750-1890*, Royal Dublin Society, Dublin.

——— (1995), *North from the Hook. 150 years of the Geological Survey of Ireland*, Geological Survey of Ireland, Dublin.

——— and Mollan, R.C. (eds) (1980), *Richard Griffith 1784-1878*, Royal Dublin Society, Dublin.

Davies, K.C. and Hull, J. (1976), *The zoological collections of the Oxford University Museum. A historical review and general account, with comprehensive donor index to the year 1975*, Oxford University Museum, Oxford.

Davies, R. (1843), *Extracts from the municipal records of the City of York*, Nichols, London.

Davis, A.G. (1943), 'The triumvirate: a chapter in the heroic age of geology', *Proceedings of the Croydon natural history and scientific society*, vol. 11, pp. 123-46.

Davis, J.W. (1889), *History of the Yorkshire Geological and Polytechnic Society, 1837-1887. With biographical notices of some of its members*, YGS, Halifax. This volume is *PYGS*, vol. 10.

Dawkins, W.B. (1870a), 'Geological theory in Britain', *Edinburgh review*, vol. 131, pp. 39-64.

——— (1870b), 'Exploration of the Victoria Cave, Settle', *BAAS 1870 report*, p. 148-9.

——— (1874a), *Testimonials in favour of W. Boyd Dawkins, a candidate for the professorship of geology at Oxford*, np.

——— (1874b), *Cave hunting, researches on the evidence of caves respecting the early inhabitants of Europe*, Macmillan, London.

——— and Sandford, W.A. (1866-72), *A monograph of the British pleistocene mammalia, Volume 1. British pleistocene felidae*, Palaeontographical Society, London.

——— and Tiddeman, R.H. (1872), 'Report on the Victoria Cave, explored by the Settle Cave Exploration Committee', *BAAS 1872 report*, pp. 178-80.

De la Beche, H.T. (1822), 'On the geology of the coast of France and of the inland country adjoining; from Fecampe … to St. Vaast', *TGSL*, vol. 1, pp. 73-89.

_____ (1830), 'Notes on the geographical distribution of organic remains contained in the oolitic series of the great London and Paris basin, and in the same series of the south of France', *Philosophical magazine*, vol. 7, pp. 81-95.

_____ (1831), *A geological manual*, Truettel and Würtz, London.

_____ (1834), *Researches in theoretical geology*, Knight, London.

_____ (1835), *How to observe. Geology*, Knight, London.

_____ (1839), *Report on the geology of Cornwall, Devon and West Somerset. Published by order of the Lords Commissioners of Her Majesty's Treasury. Printed for Her Majesty's Stationery Office*, Longman, London.

_____ (1846), 'On the formation of the rocks of south Wales and south western England' in *Memoirs of the Geological Survey of Great Britain*, Longman, London, vol. 1, pp. 1-296.

De la Rue, W. (1859), 'On the present state of celestial photography in England', *BAAS 1859 report*, pp. 130-53.

_____ Stewart, B., and Loewy, B. (1869), 'Researches on solar physics. Heliographical positions and areas of sun spots observed with the Kew photoheliograph during the years 1862 and 1863', *Philosophical transactions*, vol. 159, pp. 1-110.

Dean, D.R. (1999), *Gideon Mantell and the discovery of dinosaurs*, Cambridge University Press, Cambridge.

Delesse, A.E.O.J. (1858), *Études sur le métamorphisme des roches*, Dalmont and Dunod, Paris.

Desmond, A. (1982), *Archetypes and ancestors: palaeontology in Victorian London, 1850-1875*, Blond and Briggs, London.

_____ (1984), 'Robert E. Grant: the social predicament of a pre-Darwinian transmutationist', *Journal of the history of biology*, vol. 17, pp. 189-223.

_____ (1989), *The politics of evolution. Morphology, medicine, and reform in radical London*, Chicago University Press, Chicago and London.

_____ (1994), *Huxley: the devil's disciple*, Joseph, London.

_____ and Moore, J. (1992), *Darwin*, Penguin, Harmondsworth.

Douglas, J.A. and Edmonds, J.M. (1950), 'John Phillips' geological maps of the British isles', *Annals of science*, vol. 6, pp. 361-75.

Drew, J. (1855), *Practical meteorology. Second edition edited by Frederic Drew*, Van Voorst, London.

Duncan, P.B. (1836), *A catalogue of the Ashmolean Museum, descriptive of the zoological specimens, antiquities, coins, and miscellaneous curiosities*, Ashmolean Museum, Oxford.

Earwaker, J.P. (1870-1), 'Natural science at Oxford', *Nature*, vol. 3, pp. 170-1.

Eastmead, W. (1824), *Historia rievallensis: containing the history of Kirkby Moorside … To which is prefixed a dissertation on the animal remains, and other curious phenomena, in the recently discovered cave at Kirkdale*, Baldwin, Cradock and Joy, London.

Edmonds, J.M. (1975a), 'The geological lecture-courses given in Yorkshire by William Smith and John Phillips', *PYGS*, vol. 40, pp. 373-412.

_____ (1975b), 'The first geological lecture course at the University of London, 1831', *Annals of science*, vol. 32, pp. 257-75.

_____ (1982), 'The first apprenticed geologist', *Wiltshire archaeological and natural history magazine*, vol. 76, pp. 141-54.

Ellis, I. (1980), *Seven against Christ: a study of 'Essays and reviews'*, Brill, Leiden.

Ellis, J. and Solander, D.C. (1786), *The natural history of many curious and uncommon zoophytes*, White, London.

Engel, A.J. (1983), *From clergyman to don. The rise of the academic profession in nineteenth-century Oxford*, Clarendon Press, Oxford.

Evans, Joan (1956), *A history of the Society of Antiquaries*, Society of Antiquaries, London.

Evans, John (1875), 'Presidential address', *QJGS*, vol. 31, pp. xxxvii-xliii.

Everett, J.D. (1869), '[Second] report of the committee for the purpose of investigating the rate of increase of underground temperature downwards in various localities of dry land and under water', *BAAS 1869 report*, pp. 176-89.

_____ (1870), 'Third report ...', *BAAS 1870 report*, pp. 29-37.

_____ (1871), 'Fourth report ...', *BAAS 1871 report*, pp. 14-25.

_____ (1872) 'Fifth report ...', *BAAS 1872 report*, pp. 128-34.

Eyles, J.M. (1967), 'William Smith: the sale of his geological collection to the British Museum', *Annals of science*, vol. 23, pp. 177-212.

_____ (1969a), 'William Smith: some aspects of his life and work', in Schneer, C.J. (ed), *Toward a history of geology*, MIT Press, Cambridge, Mass., and London, pp. 142-58.

_____ (1969b), 'William Smith (1769-1839): a bibliography of his published writings, maps and geological sections, printed and lithographed', *Journal of the Society for the Bibliography of Natural History*, vol. 5, pp. 87-109.

_____ (1971), 'William Smith (1769-1839) und die suche nach kohle in Grossbritannien', *Geologie*, vol. 20, pp. 710-14.

_____ (1975), 'William Smith', *DSB*, vol. 11, pp. 486-92.

_____ (1985), 'William Smith, Sir Joseph Banks and the French geologists', in Wheeler, A.C. and Price, J.H. (eds), *From Linnaeus to Darwin commentaries on the history of biology and geology*, Society for the History of Natural History, London, pp. 37-50.

Eyles, V.A. (1937), 'John MacCulloch, FRS, and his geological map: an account of the first geological survey of Scotland', *Annals of science*, vol. 2, pp. 114-29.

_____ (1950), 'The first national geological survey', *Geological magazine*, vol. 87, pp. 373-82.

Farey, J. (1806), 'On the stratification of England', *Philosophical magazine*, vol. 25, pp. 44-9.

_____ (1810), 'Geological remarks and queries on Messrs Cuvier and Brongniart's memoir on the mineral geography of the environs of Paris', *Philosophical magazine*, vol. 35, pp. 113-39.

_____ (1811), *General view of the agriculture and minerals of Derbyshire. Volume 1*, Macmillan, London.

_____ (1815), 'Observations on the priority of Mr Smith's investigations of the strata of England; on the very unhandsome conduct of certain persons in detracting from his merit therein; and the endeavours of others to supplant him in the sale of his maps', *Philosophical magazine*, vol. 45, pp. 333-44.

_____ (1818a), 'Mr Smith's geological claims stated', *Philosophical magazine*, vol. 51, pp. 173-80.

_____ (1818b), 'On the very correct notions concerning the structure of the earth, entertained by the Rev. John Michell, as early as the year 1760; and the great neglect which his publication of the same has received from later writers on geology; and regarding the treatment of Mr Smith, by certain persons', *Philosophical magazine*, vol. 52, pp. 183-6.

_____ (1819a), 'On the importance of knowing and accurately discriminating fossil-shells, as the means of identifying particular beds of the strata, in which they are enclosed', *Philosophical magazine*, vol. 53, 112-32.

_____ (1819b), 'Free remarks on the geological work of Mr Greenough', *Philosophical magazine*, vol. 54, pp. 127-32.

_____ (1820), 'Free remarks on Mr Greenough's geological map, lately published under the direction of the Geological Society of London', *Philosophical magazine*, vol. 55, pp. 379-83.

Feinstein, C.H. (ed.) (1981), *York 1831-1981: 150 years of scientific endeavour and social change*, Sessions, York.

Fitton, W.H. (1817a), 'Transactions of the Geological Society. Vol. II', *Edinburgh review*, vol. 28, pp. 174-92.

_____ (1817b), 'Transactions of the Geological Society. Vol. III', *Edinburgh review*, vol. 29, pp. 70-94.

_____ (1818), 'Geology of England', *Edinburgh review*, vol. 29, pp. 310-37.

_____ (1824a), 'Geology of the deluge', *Edinburgh review*, vol. 39, pp. 196-234.

_____ (1824b), 'Inquiries respecting the geological relations of the beds between the chalk and the Purbeck limestone in the south-east of England', *Annals of philosophy*, vol. 8. pp. 365-83.

_____ (1829), 'Presidential address', *PGS*, 1834, vol. 1, pp. 112-34.

_____ (1832, 1833), 'Notes on the history of English geology', *Philosophical magazine and journal of science*, vol. 1, pp. 147-60, 268-75, 442-50; vol. 2, pp. 37-57.

_____ (1847), 'A stratigraphical account of the section from Atherfield to Rocken End, on the south-west coast of the Isle of Wight', *TGS*, vol. 3, pp. 289-328.

Flett, J.S. (1937), *The first hundred years of the Geological Survey of Great Britain*, His Majesty's Stationery Office, London.

Forbes, E. (1854), 'Presidential address', *QJGS*, vol. 10, pp. xxii-lxxxi.

_____ and Hanley, S. (1853), *A history of British mollusca and their shells*, Van Voorst, London.

Forbes, J.D. (1832a), 'On the horary oscillations of the barometer near Edinburgh, deduced from 4,410 observations', *Edinburgh journal of science*, vol. 6, pp. 261-86.

_____ (1832b), 'Report upon the recent progress and present state of meteorology', *BAAS 1832 report*, pp. 196-258.

_____ (1843), *Travels through the Alps of Savoy and other parts of the Pennine chain, with observations on the phenomena of glaciers*, Black, Edinburgh.

_____ (1859), *Occasional papers on the theory of glaciers now first collected and chronologically arranged. With a prefatory note on the recent progress and present aspect of the theory*, Black, Edinburgh.

_____ (1863), 'Lyell on the antiquity of man', *Edinburgh review*, vol. 118, pp. 254-302.

Ford, J. (1877), *Memorials of John Ford*, ed. S. Thompson., Harris and Sessions, London and York.

Forgan, S. (1986), 'Context, image and function: a preliminary enquiry into the architecture of scientific societies', *British journal for the history of science*, vol. 19, pp. 89-113.

_____ (1989), 'The architecture of science and the idea of a university', *Studies in history and philosophy of science*, vol. 20, pp. 405-34.

Fox, R. (1997), 'The University Museum and Oxford science 1850-1880', in Brock, M.G. and Curthoys, M.C. (eds), *The History of the University of Oxford. Volume VI. Nineteenth-century Oxford, part 1*, Clarendon Press, Oxford, pp. 641-91.

Fox-Strangways, C.E. (1892), *The jurassic rocks of Britain. Volumes 1 and 2. Yorkshire*, HMSO, London.

Fraser, D. (ed.) (1980), *A history of modern Leeds*, Manchester University Press, Manchester.

Galloway, R.L. (1882), *A history of coal mining in Great Britain*, Macmillan, London.

Galton, F. (1874), *English men of science: their nature and nurture*, Macmillan, London.

Garnham, T. (1992), *Oxford Museum. Deane and Woodward*, Phaidon, London.

Geikie, A. (1872), 'Phillips' *Geology of Oxford*', *The Academy*, vol. 3. pp. 212-3.

_____ (1875), *Life of Sir Roderick Impey Murchison based on his journals and letters with notices of his scientific contemporaries and a sketch of the rise and growth of palaeozoic geology in Britain*, Murray, London.

_____ (1895), *Memoir of Sir A.C. Ramsay*, Macmillan, London.

_____ (1904), 'Aveline', *QJGS*, vol. 60, pp. lxvi-lxviii.

_____ (1917), *Annals of the Royal Society Club*, Macmillan, London.

George, E.S. (1837), 'On the Yorkshire coalfield', *Transactions of the Philosophical and Literary Society of Leeds*, vol. 1, pp. 135-72.

Gerstner, P. (1994), *Henry Darwin Rogers, 1808-1866: American geologist*, University of Alabama Press, Tuscaloosa.

Gifford, M.J. (ed.) (1932), *Pages from the diary of an Oxford lady 1843-1862*, Blackwell, Oxford.

Gilbertson, W. (1837), *Catalogue of fossils from the mountain limestone of Great Britain*, Preston.

Goldie, G. (1841), *A short account of the Yorkshire Museum*, Sotheran, York.

Gooday, G.J.N. (1989), 'Precision measurement and the genesis of physics teaching laboratories in Victorian Britain', PhD thesis, University of Kent.

Goodwin, C.W. (1860), 'On the Mosaic cosmogony', in Temple, F. et al., *Essays and reviews*, Parker, London; 9th edn, 1861, Longman, London, pp. 207-53.

Gordon, E.O. (1894), *The life and correspondence of William Buckland*, Murray, London.

Gordon, M.M. (1869), *The home life of Sir David Brewster*, Edmonston and Douglas, Edinburgh.

Goulburn, E.M. et al. (1862), *Replies to 'Essays and reviews'... with a preface by the Lord Bishop of Oxford; and letters from the Radcliffe observer and the reader in geology in the University of Oxford*, Parker, Oxford and London.

Grainge, W. (1864), *The geology of Harrogate: to which are appended statements and opinions, never before published of Dr William Smith; professors Phillips, Daniel [sic], and Turner; William West, of Leeds; Thomas Sopwith, professor Johnstone [sic] of Durham; John Buddle; and others; on the origin of the medicinal waters of Harrogate*, Ackrill, Harrogate.

_____ (1871), *The history and topography of Harrogate and the forest of Knaresborough*, Smith, London.

Gray, A. (1861), 'Phillips' *Life on earth*', *American journal of science and arts*, vol. 31, pp. 444-9.

Grayson, D.K. (1983), *The establishment of human antiquity*, Academic Press, New York and London.

Great Exhibition (1851a), *Official descriptive and illustrated catalogue*, 4 vols, Spicer, London.

_____ (1851b), *Official catalogue of the Great Exhibition of the works of industry of all nations, 1851 by authority of the Royal Commission*, Spicer and Clowes, London.

Green, A.H. et al. (1869), *The geology of the carboniferous limestone, Yoredale rocks, and millstone grit of north Derbyshire and the adjoining parts of Yorkshire*, Longman, London.

Greene, M.T. (1982), *Geology in the nineteenth century. Changing views of a changing world*, Cornell University Press, Ithaca and London.

Greenough, G.B. (1819), *A critical examination of the first principles of geology*, Longman, London.

_____ (1820), *A geological map of England and Wales*, Longman, London.

Greenwell, W. and Rolleston, G. (1877), *British barrows. A record of the examination of sepulchral mounds in various parts of England*, Clarendon Press, Oxford.

Guest, I. (1991), *Dr John Radcliffe and his Trust*, Radcliffe Trust, London.

Gunther, R.T. (1925), *Early science in Oxford. Volume 3*, private, Oxford.

Hall, M.B. (1984). *All scientists now. The Royal Society in the nineteenth century*, Cambridge University Press, Cambridge.

Hall, V. (1987), *A history of the Yorkshire Agricultural Society 1837-1987*, Batsford, London.

Harcourt, E.W. (1891, 1905), *The Harcourt papers*, vols. 13 and 14, private, Oxford.

Harcourt, W.V. (1826), 'An account of the strata north of the Humber near Cave', *Annals of philosophy*, vol. 11, pp. 435-9.

_____ (1829), 'On a discovery of fossil bones in a marl pit near North Cliff', *Philosophical magazine*, vol. 6, pp. 225-32.

_____ (1830), 'Further examination of the deposit of fossil bones at North Cliff in the County of York', *Philosophical magazine*, vol. 7, pp. 1-9.

_____ (1839), 'Presidential address', *BAAS 1839 report*, pp. 3-22.

_____ (1860), 'On the effects of long-continued heat, illustrative of geological phenomena', *BAAS 1860 report*, pp. 175-92.

Harrington, J.B. (1883), *Life of Sir William E. Logan*, Sampson Low, London.

Harris, A. (1968), 'The Ingleton coalfield', *Industrial archaeology*, vol. 5, pp. 313-26.

Harte, N. and North, J. (1978), *The world of University College, London, 1828-1978*, University College, London.

Haughton, S. (1856), 'On slaty cleavage and the distortion of fossils', *Philosophical magazine*, vol. 12, pp. 409-21.

_____ (1864), 'On the joint systems of Ireland and Cornwall and their mechanical origin', *Philosophical transactions*, vol. 154, pp. 393-412.

Hays, J.N. (1974), 'Science in the City; the London Institution, 1819-1840', *British journal for history of science*, vol. 7, pp. 146-62.

_____ (1981), 'The rise and fall of Dionysius Lardner', *Annals of science*, vol. 38, pp. 527-42.

Healey, E. (1978), *Lady unknown: the life of Angela Burdett-Coutts*, Sidgwick and Jackson, London.

Hearnshaw, F.J.C. (1929), *The centenary history of King's College, London, 1828-1928*, Harrap, London.

Hemingway, J.E (1946), 'Martin Simpson, geologist and curator', in Browne, H.B., *Chapters of Whitby history. The story of Whitby Literary and Philosophical Society and of Whitby Museum*, Brown, Hull and London, pp. 93-105.

Hendrickson, W.B. (1961), 'Nineteenth-century state geological surveys: early government support of science, *Isis*, vol. 52, pp. 357-71.

Herschel, J.F.W. (1833), *A treatise on astronomy*, Longman, London.

_____ (1861), *Meteorology*, Black, Edinburgh.

Higham, N. (1963), *A very scientific gentleman. The major achievements of Henry Clifton Sorby*, Pergamon Press, Oxford and London.

Hilton, B. (1988), *The age of atonement: the influence of evangelicalism on social and economic thought, 1795-1865*, Clarendon Press, Oxford.

Hilton, B. (1994), 'Whiggery, religion, and social reform: the case of Lord Morpeth', *Historical journal*, vol. 37, pp. 829-59.

Hinchliff, P. (1964), *John William Colenso Bishop of Natal*, Nelson, London and Edinburgh.

Hinsley, F.B. (1969-70), 'The development of coal mine ventilation in Great Britain up to the end of the nineteenth century', *Transactions of the Newcomen Society*, vol. 42, pp. 25-39.

Holmes, T.V. and Sherborn, C.D. (1891), *Geologists' Association. A record of excursions made between 1860 and 1890*, Stanford, London.

Hopkins, W. (1834), 'Remarks on Farey's account of the stratification of the limestone district of Derbyshire', *Philosophical magazine*, vol. 5, pp. 121-31.

_____ (1835), 'Researches in physical geology', *Transactions of the Cambridge Philosophical Society*, vol. 6, pp. 1-84.

_____ (1848), 'On the elevation and denudation of the district of the Lakes of Cumberland and Westmorland', *QJGS*, vol. 4, pp. 70-98.

Horner, L. (1811), 'On the mineralogy of the Malvern Hills', *TGSL*, vol. 1, pp. 281-321.

_____ (1861), 'Presidential address', *QJGS*, vol. 17, pp. xxiv-lxxii.

Howarth, O.J.R. (1931), *The British Association for the Advancement of Science: a retrospect 1831-1931*, BAAS, London.

Hull, E. (1862), 'On iso-diametric lines, as a means of representing the distribution of sedimentary clay and sandy strata, as distinguished from calcareous strata, with special reference to the carboniferous rocks of Britain', *QJGS*, vol. 18, pp. 127-146.

_____ (1874), 'Presidential address to section C', *BAAS 1874 report*, pp. 67-73.

Hunt, R. (1866), 'John Phillips', in Walford, E. (ed), *Portraits of men of eminence in literature, science, and art, with biographical memoirs*, Bennett, London, vol. 5, pp. 35-42.

_____ (1874), 'Prof. John Phillips', *Athenaeum*, 1, pp. 597-8.

Hunton, L. (1837), 'Remarks on a section of the upper lias and marlstone of Yorkshire, showing the limited vertical range of the species of ammonites and other testacea, with their value as geological tests', *TGSL*, vol. 5, pp. 215-22.

Hutchins, R. (1990), 'Magdalen's astronomy observatory', *Magdalen College record*, pp. 44-51.

———— (1994), 'John Phillips, 'geologist-astronomer', and the origins of the Oxford University observatory, 1853-1875', *History of universities*, vol. 13, pp. 193-249.

———— (1995a), 'Charles Daubeny (1795-1867): the bicentenary of Magdalen's first modern scientist', *Magdalen College record*, pp. 81-92.

———— (1995b), 'Charles Daubeny 1795-1867: Oxford's professor of chemistry and botany', typescript, Magdalen College Library.

Hutton, J. (1795), *Theory of the earth, with proofs and illustrations*, Cadell and Davies, Edinburgh.

Huxley, L. (1918), *Life and letters of Sir Joseph Dalton Hooker*, Murray, London.

Huxley, T.H. (1855), review of Phillips' *Manual*, *Westminster review*, vol. 64, pp. 565-9.

———— (1862), 'Secretarial address', *QJGS*, vol. 18, pp. xl-liv.

———— (1864), *British fossils. On the structure of the belemnitidae*, HMSO, London [a Geological Survey monograph].

———— (1868), 'On the animals which are most nearly intermediate between birds and reptiles', *Annals and magazine of natural history*, vol. 2, pp. 66-75.

———— (1870), 'Further evidence of the affinity between the dinosaurian reptiles and birds', *QJGS*, vol. 26, pp. 12-31.

James, F. (ed.) (1993), *The correspondence of Michael Faraday. Volume 2. 1832-December 1840*, Institution of Electrical Engineers, London.

———— (ed.) (1996), *The correspondence of Michael Faraday. Volume 3. 1841-December 1848*, Institution of Electrical Engineers, London.

———— (ed.) (1999), *The correspondence of Michael Faraday. Volume 4. January 1849-October 1855*, Institution of Electrical Engineers, London.

———— and Ray, M. (1999), 'Science in the pits: Michael Faraday, Charles Lyell and the Home Office enquiry into the explosion at Haswell colliery, county Durham, in 1844', *History and technology*, vol. 15, pp. 213-31.

Jankovic, V. (1998), 'Ideological crests versus empirical troughs: John Herschel's and William Radcliffe Birt's research on atmospheric waves, 1843-50', *British journal for history of science*, vol. 31, pp. 21-40.

———— (2000), *Reading the skies. A cultural history of English weather*, Manchester University Press, Manchester.

Jardine, W. (1858), *Memoirs of Hugh Edwin Strickland*, Van Voorst, London.

Jelly, H. (1832), 'The rev. B. Richardson', *Bath and Bristol magazine*, vol. 1, pp. 303-307.

Johnston, J.F.W. (1832), 'Account of the scientific meeting at York', *Edinburgh journal of science*, vol. 6, pp. 1-32.

Johnstone, J.V.B. (1840), 'On the application of geology to agriculture', *Journal of the Royal Agricultural Society*', vol. 1, pp. 271-5.

Jones, T.R. (1850), *Entomostraca of the cretaceous formation of England*, Palaeontographical Society, London.

———— (1883), 'The Geologists' Association: its origin and progress', *Proceedings of the Geologists' Association*, vol. 7, pp. 1-57.

Judd, J.W. (1868), 'On the Speeton clay', *QJGS*, vol. 24, pp. 218-50.

_____ (1870), 'Additional observations on the neocomian strata of Yorkshire, and Lincolnshire, with notes on their relations to the beds of the same age throughout northern Europe', *QJGS*, vol. 26, pp. 326-58.

Jukes, J.B. (1857), *The student's manual of geology*, Black, Edinburgh.

_____ (1872), *The student's manual of geology*, 3rd ed, Black, Edinburgh.

Kargon, R.H. (1977), *Science in Victorian Manchester. Enterprise and expertise*, Johns Hopkins University Press, Baltimore and London.

Keller, F. (1866), *The lake dwellings of Switzerland and other parts of Europe ... translated and arranged by John Edward Lee*, Longman, London.

Kendall, P.F. and Wroot, H.E. (1924), *Geology of Yorkshire. An illustration of the evolution of northern England*, Hollinek, Vienna.

Kenrick, J. (1834), 'On the alleged Greek traditions of the deluge', *Philosophical magazine*, vol. 5, pp. 25-33.

_____ (1835), *The union of religion with intellectual culture. A sermon delivered at the presbyterian chapel, Eustace Street, Dublin, August 16, 1835; being the Sunday after the meeting of the British Association for the Advancement of Science*, Fellowes, London.

_____ (1846), *An essay on primaeval history*, Fellowes, London.

_____ (1860), *A biographical memoir of the late rev. Charles Wellbeloved*, Whitfield, London.

Kidd, J. (1815), *A geological essay on the imperfect evidence in support of a theory of the earth, deducible either from its general structure or from the changes produced on its surface by the operation of existing causes*, Oxford University Press, Oxford.

Kirk Smith, H. (1958), *William Thomson Archbishop of York. His life and times, 1819-90*, SPCK, London.

Klein, J.T. (1778), *Naturalis dispositio echinodermatum. Edita et aucta a N.G. Leske*. Lipsiae.

Knell, S.J. (2000), *The culture of English geology, 1815-1851: a science revealed through its collecting*, Ashgate, Aldershot.

Knight, C. (1864-5), *Passages of a working life during half a century: with a prelude of early reminiscences*, Bradbury and Evans, London.

Kohlstedt, S.G. (1976), *The formation of the American scientific community: the American Association for the Advancement of Science 1848-60*, University of Illinois Press, Urbana.

Kölbl-Ebert, M. (1997a), 'Mary Buckland (née Morland) 1797-1857', *Earth sciences history*, vol. 16, pp. 33-8.

_____ (1997b), 'Charlotte Murchison (née Hugonin) 1788-1869', *Earth sciences history*, vol. 16, pp. 39-43.

Koninck, L.G. de (1842-4), *Description des animaux fossiles qui se trouvent dans le terrain carbonifère de Belgique*, Dessain, Liège.

Lamarck, J.B. (1802-9), 'Mémoires sur les fossiles des environs de Paris, comprenant la détermination des espèces qui appartiennent aux animaux marins sons vertèbres ...', *Annales du Muséum d'Histoire Naturelle*, vols. 1-14 [39 short papers].

Lankester, E.R. (1868-70), *The fishes of the old red sandstone of Britain. Part 1. The cephalaspidae*, Palaeontographical Society, London.

Laudan, R. (1976), 'William Smith. Stratigraphy without palaeontology', *Centaurus*, vol. 20, pp. 210-26.

_____ (1987), *From mineralogy to geology. The foundations of a science, 1650-1830*. University of Chicago Press, Chicago.

Launay, L. de (1940), *Une grande famille de savants. Les Brongniart*, Rapilly, Paris.

Layton, D. (1973), *Science for the people. The origins of the school science curriculum in England*, Allen and Unwin, London.

Leckenby, J. (1859), 'Note on the Speeton clay of Yorkshire', *Geologist*, vol. 2, pp. 9-11.

_____ (1864), 'On the sandstones and shales of the oolites of Scarborough, with descriptions of some new species of fossil plants', *QJGS*, vol. 20, pp. 74-82.

Lee, J.E. (1881), *Note-book of an amateur geologist*, Longman Green, London.

Lester, J. and Bowler, P. (1995), *E. Ray Lankester and the making of modern British biology*, British Society for the History of Science, Stanford in the Vale.

Lewis, C.L.E. and Knell, S.J. (eds) (2001), *The age of the earth: from 4004 BC to AD 2002*, Geological Society, London.

Lhwyd, E. (1699), *Lithophylacii Britannici ichnographia*, London.

Liddon, H.P. (1898), *Life of Edward Bouverie Pusey. Vol. IV (1860-1882)*, Longman, London.

Lightman, B. (1987), *The origins of agnosticism: Victorian unbelief and the limits of knowledge*, Johns Hopkins University Press, Baltimore.

_____ (2001), 'Victorian sciences and religions: discordant harmonies', in Brooke, J.H, Osler, M.J, and Van der Meer, J.M. (eds), *Science in theistic contexts. Cognitive dimensions. Osiris,* vol. 16, pp. 343-66

Linnaeus, C. (1758-9), *Systemae naturae*, 10th edn., Salvii, Holmiae.

Liscombe, R.W. (1980), *William Wilkins 1778-1839*, Cambridge University Press, Cambridge.

Lloyd, H. (1835), 'Observations on the direction and intensity of the terrestrial magnetic force in Ireland', *BAAS 1835 report*, pp. 117-62.

Lobley, J.L. (1868), *Mount Vesuvius: a description, history, and geological account of the volcano, with a notice of the recent eruption*, Stanford, London.

Lonsdale, W. (1832), 'On the oolitic district of Bath', *TGSL*, 1832, vol. 3, pp. 241-76.

Lubbock, J. (1865), *Prehistoric times, as illustrated by ancient remains, and the manners and customs of modern savages*, Williams and Norgate, London and Edinburgh.

Lycett, J. (1872), *A monograph of the British fossil trigoniae*, Palaeontographical Society, London.

Lyell, C. (1830, 1832, 1833), *Principles of geology, being an attempt to explain the former changes in the earth's surface, by reference to causes now in operation*, 3 vols, Murray, London.

_____ (1837), 'Presidential address', *PGS*, 1838, vol. 2, pp. 479-523.

_____ (1838), *Elements of geology*, Murray, London.

_____ (1855), *Elements of geology*, 5th edn, Murray, London.

_____ (1863), *The geological evidences of the antiquity of man with remarks on theories of the origin of species by variation*, Murray, London.

_____ (1864), 'Presidential address', *BAAS 1864 report*, pp. lx-lxxv.

_____ (1868), *Principles of geology*, 10th edn, Murray, London.

_____ (1872), *Principles of geology*, 11th edn, Murray, London.

Lyell, K.M. (1881), *Life, letters and journals of Sir Charles Lyell*, Murray, London.

McCartney, P.J. (1977), *Henry De la Beche: observations on an observer*, Friends of the National Museum of Wales, Cardiff.

McCaul, A. (1861), 'The Mosaic record of creation', in Thomson, W. (ed.), *Aids to faith; a series of theological essays*, Murray, London, pp. 189-236.

McConnell, A. (1990), 'Features of portable, travelling and mountain barometers', *Antique collecting*, vol. 24, pp. 5-9.

_____ (1992), *Instrument makers to the world. A history of Cooke, Troughton and Simms*, Sessions, York.

MacDonagh, O.O.G.M. (1967), 'Coal mines regulation: the first decade', in Robson, R. (ed.), *Ideas and institutions of Victorian Britain: essays in honour of George Kitson Clark*, Bell, London, pp. 58-86.

McDowell, R.B. and Webb, D.A. (1982), *Trinity College Dublin 1592-1952. An academic history*, Cambridge University Press, Cambridge.

MacGregor, A.G. (1997), 'The Ashmolean Museum', in Brock, M.G. and Curthoys, M.C. (eds), *The history of the University of Oxford. Volume VI. Nineteenth-century Oxford, Part 1*, Clarendon Press, Oxford, pp. 598-610.

_____ and Headon, A. (2000), 'Re-inventing the Ashmolean. Natural history and natural theology at Oxford in the 1820s to 1850s', *Archives of natural history*, vol. 27, pp. 369-406.

MacLeod, R.M. (1970), 'Science and the civil list, 1824-1914', *Technology and society*, vol. 6, pp. 47-55.

_____ (1971), 'The Royal Society and the government grant: notes on the administration of scientific research, 1849-1914', *Historical journal*, vol. 14, pp. 325-58.

_____ (1983), 'Whigs and savants: reflections on the reform movement in the Royal Society, 1830-48', in Inkster, I. and Morrell, J. (eds), *Metropolis and province: science in British culture 1780-1850*, Hutchinson, London, pp. 55-90.

McMillan, N.F. and Greenwood, E.F. (1972), 'The Beans of Scarborough: a family of naturalists', *Journal of the Society for the Bibliography of Natural History*, vol. 6, pp. 152-61.

McOuat, G. (1996), 'Species, rules and meaning: the politics of language and the ends of definitions in 19th century natural history', *Studies in history and philosophy of science*, vol. 27, pp. 473-519.

Macray, W.D. (1894-1911), *A register of the members of St Mary Magdalen College, Oxford*, Frowde, London.

Mallet, R. (1848), 'On the dynamics of earthquakes; being an attempt to reduce their observed phenomena to the known laws of wave motion in solids and fluids', *Transactions of the Irish Academy*, vol. 21, pp. 50-106.

_____ (1861), 'Report of the experiments made at Holyhead to ascertain the transit-velocity of waves, analogous to earthquake waves, through the local rock formations', *BAAS 1861 report*, pp. 201-36.

_____ (1862a), 'Proposed measurement of the temperatures of active volcanic foci to the greatest attainable depth, and of the temperature, state of saturation, and velocity of the steam and vapours evolved', *BAAS 1862 report*, pp. 33-4.

_____ (1862b), *Great Neapolitan earthquake of 1857. The first principles of observational seismology ...*, Chapman and Hall, London.

_____ and Mallet, J.W. (1858), *The earthquake catalogue of the British Association*, Taylor and Francis, London.

Mantell, G.A. (1827), *Illustrations of the geology of Sussex: containing a general view of the geological relations of the south-eastern part of England; with figures and descriptions of the fossils of Tilgate forest*, Relfe, London.

_____ (1846), *Thoughts on animalcules; or, a glimpse of the invisible world revealed by the microscope*, Murray, London.

Martin, W. (1809), *Petrificata Derbiensia; or figures and descriptions of petrifactions collected in Derbyshire*, Lyon, Wigan.

Martineau, J. (1878), *In memoriam John Kenrick*, private, London.

Meacham, S. (1970), *Lord bishop, The life of Samuel Wilberforce, 1805-1873*, Harvard University Press, Cambridge, Mass..

Meadows, A.J. (1972), *Science and controversy. A biography of Sir Norman Lockyer*, Macmillan, London.

Meenan, J. and Clarke, D. (eds) (1981), *RDS: the Royal Dublin Society 1731-1981*, Gill and Macmillan, Dublin.

Melmore, S. (1942), 'Letters in the possession of the Yorkshire Philosophical Society', *The north western naturalist*, vol. 17, pp. 317-32.

_____ (1943), 'Letters in the possession of the Yorkshire Philosophical Society', *North western naturalist*, vol. 18, pp. 21-9, 149-60.

Middleton, W.E.K. (1964), *The history of the barometer*, Johns Hopkins University Press, Baltimore.

_____ (1965), *A history of the theories of rain and other forms of precipitation*, Oldbourne, London.

_____ (1966), *A history of the thermometer and its use in meteorology*, Johns Hopkins University Press, Baltimore.

Miller, J.S. (1821), *A natural history of the crinoidea*, Frost, Bristol.

Milne-Edwards, H. and Haime, J. (1850-5), *British fossil corals*, Palaeontographical Society, London.

Mitchell, W.S. (1872), 'Notes on early geologists connected with the neighbourhood of Bath', *Proceedings of the Bath Natural History and Antiquarian Field Club*, vol. 2, pp. 303-42.

Mole, J. (1788), *Elements of algebra*, Robinson, London.

Morrell, J.B. (1983), 'Economic and ornamental geology: the Geological and Polytechnic Society of the West Riding of Yorkshire, 1837-53' in Inkster, I. and Morrell, J. (eds), *Metropolis and province: science in British culture 1780-1850*, Hutchinson, London, pp. 231-56.

_____ (1988a), 'Science and government: John Phillips (1800-74) and the early Ordnance Geological Survey of Britain', in Rupke, N.A. (ed), *Science, politics and the public good. Essays in honour of Margaret Gowing*, Macmillan, London, pp. 7-35.

_____ (1988b), 'The early Yorkshire Geological and Polytechnic Society: a reconsideration', *Annals of Science*, vol. 45, pp. 153-67.

_____ (1989), 'The legacy of William Smith: the case of John Phillips in the 1820s', *Archives of natural history*, vol. 16, pp. 319-35.

_____ (1994), 'Perpetual excitement. The heroic age of British geology', *Geological curator*, vol. 5, pp. 311-17.

_____ (2001), 'Genesis and geochronology: the case of John Phillips (1800-1874)', in Lewis, C.L.E. and Knell, S.J. (eds.), *The age of the earth: from 4004 BC to AD 2002*, Geological Society, London, pp. 85-90.

_____ and Thackray, A.W. (1981), *Gentlemen of science: early years of the British Association for the Advancement of Science*, Clarendon Press, Oxford.

_____ (1984), *Gentlemen of science: early correspondence of the British Association for the Advancement of Science*, Royal Historical Society, London.

Morris, J. (1854), *A catalogue of British fossils*, 2nd edn, the author, London.

Morris, J. and Lycett, J. (1850-4), *Monograph of the mollusca from the great oolite, chiefly from Minchinhampton and the coast of Yorkshire*, Palaeontographical Society, London.

Morton, J. (1712), *The natural history of Northamptonshire*, Knaplock and Wilkin, London.

Morton, V. (1987), *Oxford rebels. The life and friends of Nevil Story Maskelyne 1823-1911, pioneer Oxford scientist, photographer and politician*, Sutton, Gloucester.

Murchison, R.I. (1827), 'On the coal-field of Brora in Sutherlandshire, and some other stratified deposits in the north of Scotland' *TGSL*, vol. 2, pp. 293-326.

_____ (1839), *The Silurian system, founded on geological researches in the counties of Salop, Hereford, Radnor, Montgomery, Caermarthen, Brecon, Pembroke, Monmouth, Gloucester, Worcester, and Stafford; with descriptions of the coal-fields and overlying formations*, Murray, London.

_____ (1841), 'First sketch of some of the principal results of a second geological survey of Russia', *Philosophical magazine*, vol. 19, pp. 417-22.

_____ (1842), 'Presidential address', *PGS*, 1842, vol. 3, pp. 637-87.

_____ (1843a), 'Presidential address', *PGS*, 1846, vol. 4, pp. 65-151.

_____ (1843b), 'The "Permian system" as applied to Germany, with collateral observations on similar deposits in other countries', *BAAS 1843 report*, pp. 52-4.

_____ (1854), *Siluria. The history of the oldest known rocks containing organic remains, with a brief sketch of the distribution of gold over the earth*, Murray, London.

_____ (1867), *Siluria. A history of the oldest rocks in the British Isles and other countries; with sketches of the origin and distribution of native gold, the general succession of geological formations, and changes of the earth's surface.* 4th edn, Murray, London.

Nasmyth, A. (1841), *Three memoirs on the development and structure of the teeth and epithelium*, Churchill, London.

Negretti, E.A.L. and Zambra, J.W. (1864), *A treatise on meteorological instruments: explanatory of their scientific principles, method of construction, and practical utility*, Negretti and Zambra, London.

Neison, E. (1876), *The moon and the condition and configuration of its surface*, Longman, London.

Newton, C. (1847), *Map of British and Roman Yorkshire, prepared under the direction of the central committee of the Archaeological Institute for the annual meeting at York in 1846*, Archaeological Institute, London.

_____ (1855), 'On the British and Roman antiquities of Yorkshire and on a map of Roman Yorkshire' [letter of Newton to Phillips, 13 March 1847, read April 1847], *Proceedings of the YPS*, vol. 1, pp. 29-33.

Nichol, J.P. (1837), *Views of the architecture of the heavens*, Tait, Edinburgh.

North, F.J. (1934), 'Further chapters in the history of geology in south Wales; Sir H.T. De la Beche and the Geological Survey', *Transactions of the Cardiff Naturalists' Society*, vol. 67, pp. 31-103.

O'Dwyer, F. (1997), *The architecture of Deane and Woodward*, Cork University Press, Cork.

Oldroyd, D.R. (1990), *The Highlands controversy. Constructing geological knowledge through fieldwork in nineteenth-century Britain*, Chicago University Press, Chicago.

_____ (1992), 'The Archaean controversy in Britain: Part II – the Malverns and Shropshire', *Annals of science*, vol. 49, pp. 401-60.

Oppel, C.A. (1863), *Ueber jurassiche cephalopoden*, Ebner and Seubert, Stuttgart.

Orange, A.D. (1973), *Philosophers and provincials; the Yorkshire Philosophical Society from 1822 to 1844*, YPS, York.

Orbigny, A.D. d' (1842-9), *Paléontologie française ... Terrains jurassiques*, Masson, Paris.

Otley, J. (1823), *A concise description of the English Lakes, and adjacent mountains, with general directions to tourists; and observations on the mineralogy and geology of the district*, Otley, Keswick.

Outram, D. (1984), *Georges Cuvier: vocation, science and authority in post-revolutionary France*, Manchester University Press, Manchester.

Ovenell, R.F. (1986), *The Ashmolean Museum 1683-1894*, Clarendon Press, Oxford.

Owen, R. (1851-89), *Mesozoic fossil reptiles*, Palaeontographical Society, London.

_____ (1858), 'Presidential address', *BAAS 1858 report*, pp. xlix-cx.

_____ (1859), *On the classification and geographical distribution of the mammalia, being the lecture on Sir Robert Rede's Foundation. Delivered before the University of Cambridge ...May 10, 1859. To which is added an appendix 'on the gorilla' and 'on the extinction and transmutation of species'*, Parker, London.

_____ (1860a), *Palaeontology or a systematic summary of extinct animals and their geological relations*, Black, Edinburgh.

_____ (1860b), 'Darwin on the origin of species', *Edinburgh review*, vol. 111, pp. 487-532.

_____ (1871), 'On the fossil mammals of Australia – Part 4. Dentition and mandible of thylacoleo carnifex, with remarks on the arguments for its herbivority', *Philosophical transactions*, vol. 161, pp. 213-66.

Page, D. (1861), *The past and present life of the globe. Being a sketch in outline of the world's life system*, Blackwood, Edinburgh and London.

Palmieri, L. (1873), *The eruption of Vesuvius in 1872, with notes and an introductory sketch of the present state of knowledge of terrestrial vulcanity, the cosmical nature and relations of volcanoes and earthquakes by Robert Mallet*, Asher, London.

Parker, James (1858), *The Ashmolean Museum and the Ashmole collection of antiquities*, np.

Parker, John H. (1870),*The Ashmolean Museum: its history, present state, and prospects. A lecture delivered to the Oxford Architectural and Historical Society, November 2, 1870*, Parker, Oxford.

Parkinson, J. (1804, 1808, 1811), *Organic remains of a former world*, 3 vols, Robson, London.

_____ (1811), 'Observations on some of the strata in the neighbourhood of London, and on the fossil remains contained in them', *TGSL*, vol. 1, pp. 324-54.

Pattison, M. (ed.) (1869), *Alexander Pope. Essay on man*, Clarendon Press, Oxford.

Peacock, A.J. (1981), 'Adult education in York, 1800-1947', *York history*, no. 5, pp. 225-62.

Peacock, R.A. (1866), *On steam as the motive power in earthquakes and volcanoes and on cavities in the earth's crust*, Le Lievre, Jersey.

Peckham, M. (1951), 'Dr Lardner's Cabinet Cyclopaedia', *Papers of the Bibliographical Society of America*, vol. 45, pp. 37-58.

Pengelly, H. (1897), *A memoir of William Pengelly, of Torquay, geologist with a selection from his correspondence. With a summary of his scientific work by the reverend professor Bonney*, Murray, London.

Pengelly, W. and Heer, O. (1862), 'The lignites and clays of Bovey Tracey', *Philosophical transactions of the Royal Society*, vol. 152, pp. 1019-86.

Penn, G. (1822), *A comparative estimate of the mineral and Mosaical geologies*, Ogle and Duncan, London.

Phillips, J. (1827), 'On the direction of the diluvial currents in Yorkshire', *Philosophical magazine*, vol. 2, pp. 138-41.

_____ (1828a), 'Remarks on the geology of the north side of the Vale of Pickering', *Philosophical magazine*, vol. 3, pp. 243-9.

_____ (1828b), 'Geological observations made in the neighbourhood of Ferrybridge, in the years 1826-1828', *Philosophical magazine*, vol. 4, pp. 401-9.

_____ (1829a), 'Prospectus for *A description of the strata of the Yorkshire coast* ...', *Philosophical magazine*, vol. 5, pp. 74-5.

_____ (1829b), *Illustrations of the geology of Yorkshire; or, a description of the strata and organic remains of the Yorkshire coast: accompanied by a geological map, sections, and plates of the fossil plants and animals*, Wilson, York.

_____ (1829c), 'On a group of slate rocks ranging e.s.e. between the rivers Lune and Wharfe, from near Kirkby Lonsdale to near Malham; and on the attendant phenomena', *TGSL*, vol. 3, pp. 1-20.

_____ (1830), 'On the geology of Havre', *Philosophical magazine*, vol. 7, pp. 195-8.

_____ (1831), 'On some effects of the atmosphere in wasting the surfaces of buildings and rocks', *PGSL*, 1834, vol. 1, pp. 323-4.

_____ (1832a), 'On the lower or ganister coal series of Yorkshire', *Philosophical magazine*, vol. 1, pp. 349-53.

_____ (1832b), 'Description of a new self-registering maximum thermometer', *BAAS 1832 report*, pp. 574-5.

_____ (1833a), 'On a modification of the electrophorus of Volta', *Philosophical magazine*, vol. 2, pp. 363-5.

_____ (1833b), 'Report of experiments on the quantities of rain falling at different elevations above the surface of the ground at York, undertaken at the request of the Association', *BAAS 1833 report*, pp. 401-412, 472.

_____ (1834a), *A guide to geology*, Longman, London.

_____ (1834b), 'Further report of the result of twelve months' experiments on the quantities of rain falling at different elevations above the surface of the ground at York, undertaken at the request of the Association', *BAAS 1834 report*, pp. 560-3.

_____ (1834c), 'On the ancient and partly buried forests of Holderness', *Philosophical magazine*, vol. 4, pp. 282-8.

_____ (1834d), 'On subterranean temperature observed at a depth of 500 yards at Monk-Wearmouth', *Philosophical magazine*, vol. 5. pp. 446-51.

_____ (1835a), *Illustrations of the geology of Yorkshire; or a description of the strata and organic remains: accompanied by a geological map, sections, and plates of the fossil plants and animals. Part I. The Yorkshire coast*, 2nd ed., Murray, London.

_____ (1835b), 'Third report of experiments on the quantities of rain falling at different elevations above the surface of the ground at York', *BAAS 1835 report*, pp. 171-9.

_____ [and Daubeny, C.G.B.] (1835c), 'Geology' in *Encyclopaedia metropolitana*, 1845, vol. 6, pp. 529-808.

_____ (1835d), 'Notice of a newly discovered tertiary deposit on the coast of Yorkshire', *Philosophical magazine*, vol. 7, p. 486.

_____ (1836a), *Illustrations of the geology of Yorkshire; or a description of the strata and organic remains: accompanied by a geological map, sections, and diagrams, and figures of fossils. Part II. The mountain limestone district*, Murray, London.

_____ (1836b), *A guide to geology*, 3rd edn, Longman, London.

_____ (1837), *A report on the probability of the occurrence of coal and other minerals in the vicinity of Lancaster. Addressed to the Lancaster Mining Company*, Barwick, Lancaster.

_____ (1837, 1839), *A treatise on geology*, 2 vols, Longman, London.

_____ (1838a), *A treatise on geology, forming the article under that head in the seventh edition of the Encyclopaedia Britannica*, Black, Edinburgh.

_____ (1838b), *An index geological map of the British Isles*, Weale, London.

_____ (1838c), 'Geology', *Penny Cyclopaedia*, vol. xi, pp. 127-51.

_____ (1838d), 'Supplementary note, on the geological evidence of former conditions of organized life, and its unbroken succession', in Powell, B., *The connexion of natural and divine truth; or the study of the inductive philosophy considered as subservient to theology*, Parker, London, pp. 309-13.

_____ (1839a), 'Biographical notice of William Smith', *Magazine of natural history*, vol. 3, pp. 213-20.

_____ (1839b), 'Remarks on a note in Prof. Sedgwick and Mr. Murchison's communication in the last number', *Philosophical magazine*, vol. 14, pp. 353-4.

_____ (1840a), 'New experimental researches on rain', *BAAS 1840 report*, pp. 45-7.

_____ (1840b), 'Organic remains', *Penny cyclopaedia*, vol. 16, pp. 487-91.

_____ (1840c), 'Palaeozoic series', *Penny cyclopaedia*, vol. 17, pp. 153-4.

_____ (1841a), *Figures and descriptions of the palaeozoic fossils of Cornwall, Devon and West Somerset; observed in the course of the Ordnance Geological Survey of that district. Published by order of the Lords Commissioners of Her Majesty's Treasury*, Longman, London.

_____ (1841b), 'Further researches on rain at York', *BAAS 1841 report*, pp. 30-2.

_____ (1841c), 'Saliferous system', *Penny cyclopaedia*, vol. 20, pp. 354-5.

_____ (1841d), 'On the temperature of the air in York Minster', *BAAS 1841 report*, pp. 29-30.

_____ (1842a), 'On the occurrence of shells and corals in a conglomerate bed, adherent to the face of the trap rocks of the Malvern hills, and full of rounded and angular fragments of those rocks', *Philosophical magazine*, vol. 21, pp. 288-93.

_____ (1842b), 'On the microscopic structure of coal', *BAAS 1842 report*, pp. 47-8.

_____ (1843a), 'Notice of the Ordnance Geological Museum', *BAAS 1843 report*, p. 61,

_____ (1843b), 'On certain movements in the parts of stratified rocks', *BAAS 1843 report*, pp. 60-1.

_____ (1843c), 'On the occurrence of trilobites and agnosti in the lowest shales of the palaeozoic series, on the flanks of the Malvern Hills', *Philosophical magazine*, vol. 22, pp. 384-5.

_____ (1844a), *Memoirs of William Smith*, Murray, London.

_____ (1844b), 'On the curves of annual temperature at York', *BAAS 1844 report*, p. 21.

_____ (1844c), 'On the quantity of rain received in gauges at unequal elevations upon the ground', *BAAS 1844 report*, p. 21.

_____ (1844d), 'On simultaneous barometric registration in the north of England', *BAAS 1844 report*, p. 21.

_____ (1846a), *Introductory address delivered at the opening of the new hall of the York Institute of popular science and literature. March 31, 1846*, Coultas, York.

_____ (1846b), 'On anemometry', *BAAS 1846 report*, pp. 340-7.

_____ (1848a), *The Malvern Hills compared with the palaeozoic districts of Abberley, Woolhope, May Hill, Tortworth, and Usk. [With Salter, J.W.,] Palaeontological appendix. Memoirs of the Geological Survey of Great Britain, and of the Museum of Practical Geology in London. Volume II, Part I*, Longman, London.

_____ (1848b), 'On the remains of microscopic animals in the rocks of Yorkshire', *Proceedings of the Yorkshire Geological Society*, vol. 2, pp. 274-85.

_____ (1848c), 'Observations on the process of petrification', *Proceedings of the Yorkshire Geological Society*, vol. 2, pp. 305-8.

_____ (1848d), 'Notice of further progress in anemometrical researches', *BAAS 1848 report*, p. 97.

_____ (1848e), 'Ethnographical note on the vicinity of Charnwood Forest', *BAAS 1848 report*, pp. 99-100.

_____ (1849a), 'Contributions to anemometry – the thermanemometer', *BAAS 1849 report*, pp. 28-9.

_____ (1849b), 'On tumuli in Yorkshire', *BAAS 1849 report*, p. 86.

_____ (1850a), 'On isoclinal magnetic lines in Yorkshire', *BAAS 1850 report*, p. 14.

_____ (1850b), *Report to the secretary of state for the home department on the ventilation of mines and collieries, Parliamentary papers*, vol. 23, pp. 475-569.

_____ (1850c), *Geological map of the British Isles and adjacent coast of France*, Society for Promoting Christian Knowledge, London.

_____ (1851a), 'Experiments made at York (lat. 53° 58' N) on the deviation of the plane of vibration of a pendulum from the meridional and other vertical planes', *Philosophical magazine*, vol. 2, series 4, pp. 150-4.

_____ (1851b), 'On the structure of the crag', *BAAS 1851 report*, pp. 67-8.

_____ (1853a), *The rivers, mountains, and sea coast of Yorkshire. With essays on the climate, scenery, and ancient inhabitants of the county*, Murray, London; 1855, 2nd edn.

_____ (1853b), 'On the dispersion of erratic rocks at higher levels than their parent rock in Yorkshire', *BAAS 1853 report*, p. 54.

_____ (1853c), 'On magnetic phenomena in Yorkshire', *BAAS 1853 report*, pp. 6-7.

_____ (1853d), 'On photographs of the moon', *BAAS 1853 report*, pp. 14-18.

_____ (1853e), 'Report on the physical character of the moon's surface as compared with that of the earth', *BAAS 1853 report*, pp. 84-7.

_____ (1853f), 'Notes on a living specimen of priapulus caudatus dredged off the coast of Scarborough', *BAAS 1853 report*, p. 70.

_____ (1853g), 'On some of the relations of archaeology to physical geography in the north of England', *Archaeological journal*, vol. 10, pp. 179-86.

_____ (1853h), *Railway excursions from York, Leeds, and Hull. Part I. Excursions from York. Part II. Excursions from Hull. Part III. Excursions from Leeds*, Goddard and Lancaster, Hull; 1854, 2nd edn; 1855, 3rd edn retitled *Excursions in Yorkshire by the North Eastern Railway*, Sampson, York.

_____ (1853i), *A map of the principal features of the geology of Yorkshire*, Phillips, York; 1862, 2nd edn.

_____ (1854a), 'Geological sketches around Ingleborough', *Notices of the proceedings of the meetings of the Royal Institution, with abstracts of the discourses delivered at the evening meetings*, vol. 1, pp. 278-80.

_____ (1854b), *A guide to geology*, 4th ed, Longman, London.

_____ (1854c), 'Second report on the physical character of the moon's surface', *BAAS 1854 report*, pp. 415-17.

_____ (1854d), 'On micrometrical and photographic drawings of the lunar surface', *BAAS 1854 report*, pp. 25-6.

_____ (1854e), 'On the temperature of the interior of York Minster', *Proceedings of the Ashmolean Society*, no. 32, pp. 37-40.

_____ (1855a), 'Thoughts on ancient metallurgy and mining among the Brigantes and in some other parts of Britain suggested by a page of Pliny's Natural History' [read March 1848], *Proceedings of the YPS*, vol. 1, pp. 77-92.

_____ (1855b), 'The neighbourhood of Oxford and its geology', in Sandars, T.C. (ed.), *Oxford essays contributed by members of the University*, Parker, London, vol. 1, pp. 192-212.

_____ (1855c), *Manual of geology: practical and theoretical*, Griffin, London and Glasgow.

_____ (1855d), 'On temperature of sea, on the east coast of England and coasts of Ireland', *Proceedings of the Ashmolean Society*, no. 33, pp. 70-2.

_____ (1855e), 'Remarks on certain trap dykes in Arran', *BAAS 1855 report*, p. 94.

_____ (1856a), 'Report on cleavage and foliation in rocks, and on the theoretical explanations of these phenomena', *BAAS 1856 report*, pp. 369-96.

_____ (1856b), 'The meteor of January 7th, 1856', *Proceedings of the Ashmolean Society*, no. 34, pp. 101-2.

_____ (1856c), 'On a new method of making maximum self-registering thermometers', *BAAS 1856 report*, p. 41.

_____ (1857a), 'Notes on the drawing of Copernicus presented to the Royal Society by P.A. Secchi', *Proceedings of the Royal Society of London*, vol. 8, pp. 73-5.

_____ (1857b), 'Buckland', *Proceedings of the Royal Society of London*, vol. 8, pp. 264-8.

_____ (1858a), 'On some comparative sections in the oolitic and ironstone series of Yorkshire', *QJGS*, vol. 14, pp. 84-98.

_____ (1858b), 'On the estuary sands in the upper part of Shotover Hill', *QJGS*, vol. 14, pp. 236-41.

_____ (1858c), 'On the Malvern Hills', [13 March 1857], *Notices of the proceedings of the meetings of the Royal Institution*, vol. 2, pp. 385-93.

_____ (1858d), 'Notice of some phenomena at the junction of the granite and schistose rocks in west Cumberland', *BAAS 1858 report*, p. 106.

_____ (1858e), 'On the haematite ores of north Lancashire and west Cumberland', *BAAS 1858 report*, pp. 106-7.

_____ (1859a), 'Presidential address', *QJGS*, vol. 15, pp. xxv-lxi.

_____ (1859b), 'Thoughts on ancient metallurgy and mining among the Brigantes and in some other parts of Britain suggested by a page of Pliny's Natural History', *Archaeological journal*, vol. 16, pp. 7-21.

_____ (1859c), 'On a fossil fruit found in the upper part of the Wealden deposits in Swanage Bay, Isle of Purbeck', *QJGS*, vol. 15, pp. 46-9.

_____ (1859d), 'Conybeare, *Proceedings of the Royal Society of London*, vol.9, pp 50-2.

_____ (1860a) 'Presidential address', *QJGS*, vol. 16, pp. xxvii-lv.

_____ (1860b), *Life on the earth. Its origin and succession*, Macmillan, Cambridge and London.

_____ (1860c), 'On some sections of the strata near Oxford. No. 1. The great oolite in the valley of the Cherwell', *QJGS*, vol. 16, pp. 115-19.

_____ (1860d), 'On some sections of the strata near Oxford. No. II. Sections south of Oxford (Culham)', *QJGS*, vol. 16, pp. 307-11.

_____ (1861a), 'Notice of post-glacial gravels of the valley of the Thames', *BAAS 1861 report*, p. 129.

_____ (1861b), 'Contributions to a report on the physical aspect of the moon', *BAAS 1861 report*, pp. 180-1.

_____ (1862a), *Geological map of the British Isles, and adjacent coast of France*, Society for Promoting Christian Knowledge, London.

_____ (1862b), 'Letter to Cotton' in Goulburn, E.M. et al, *Replies to 'Essays and reviews'* ... , Parker, Oxford and London, pp. 514-16.

_____ (1863a), 'Suggestions for the attainment of a systematic representation of the physical aspect of the moon', *Proceedings of the Royal Society*, vol. 12, pp. 31-7.

_____ (1863b), *Notices of rocks and fossils in the University Museum, Oxford*, Parker, Oxford.

_____ (1863c), 'On the drift beds of Mundesley, Norfolk', *BAAS 1863 report*, p. 85.

_____ (1863d), 'Researches on the moon', *BAAS 1863 report*, pp. 9-10.

_____ (1863e), 'On the telescopic appearance of the planet Mars', *Proceedings of the Royal Society*, vol. 12, pp. 431-7.

_____ (1863f), 'On the deposit of the gravel, sand, and loam with flint instruments at St Acheul', *BAAS 1863 report*, pp. 85-6.

_____ (1863g), 'Antiquity of man', *Quarterly review*, vol. 114, pp. 368-417.

_____ (1864a), *A guide to geology*, 5th ed, Longman, London.

_____ (1864b), 'On the distribution of granite blocks from Wasdale Craig', *BAAS 1864 report*, pp. 65-6.

_____ (1864c), 'Notice of the physical aspect of the sun', *BAAS 1864 report*, p. 7.

_____ (1864d), 'Presidential address to section C', *BAAS 1864 report*, pp. 45-9.

_____ (1864e), 'On the measure of geological time by natural chronometers', *BAAS 1864 report*, pp. 64-5.

_____ (1864f), 'On the formation of valleys near Kirkby Lonsdale', *BAAS 1864 report*, pp. 63-4.

_____ (1865a), 'Oxford fossils. No. 1', *Geological magazine*, vol. 2, pp. 292-3.

_____ (1865b), 'Presidential address', *BAAS 1865 report*, pp. li-lxvii.

_____ (1865c), 'On glacial striation', *BAAS 1865 report*, pp. 71-2.

_____ (1865d), 'Further observations on the planet Mars', *Proceedings of the Royal Society*, vol. 14, pp. 42-6.

_____ (1865e), 'The planet Mars', *Quarterly journal of science*, vol. 2, pp. 369-81.

_____ (1865f), 'Notices of the physical aspect of the sun', *Proceedings of the Royal Society*, vol. 14, pp. 46-52.

_____ (1865g), 'Notices of the surface of the sun', *Proceedings of the Royal Society*, vol. 14, pp. 476-9.

_____ (1865h), 'Notice of a spot on the sun, observed at intervals during one rotation', *Proceedings of the Royal Society*, vol. 14, pp. 479-80.

_____ (1865i), 'Note on the geology of Harrogate', *QJGS*, vol. 21, pp. 232-4.

_____ (1865-70), *A monograph of British belemnitidae: jurassic*, Palaeontographical Society, London.

_____ (1866a), 'Meteoric shower', *Athenaeum*, vol. ii, pp. 644-5.

_____ (1866b), 'Pluie de météores, le 13-14 Novembre 1866', *Les Mondes*, vol. 12, pp. 495-6.

_____ (1867a), 'Notice of a zone of spots on the sun', *Proceedings of the Royal Society*, vol. 15, pp. 63-70.

_____ (1867b), 'Observations of temperature during two eclipses of the sun', *Proceedings of the Royal Society*, vol. 15, pp. 421-3.

_____ (1868a), 'Notices of some parts of the surface of the moon', *Philosophical transactions*, vol. 158, pp. 333-45.

_____ (1868b), 'Obituary notice of Dr Daubeny', *Proceedings of the Ashmolean Society*, vol. 5, pp. 8-22.

_____ (1868c), 'Notice of the Hessle drift, as it appeared in sections above forty years since', *QJGS*, vol. 24, pp. 250-5.

_____ (1868d), 'On the quantity of rain measured in the Lake District', *BAAS 1868 report*, pp. 472-3.

_____ (1869a), *Vesuvius*, Clarendon Press, Oxford.

_____ (1869b), 'Earthquakes', *Quarterly review*, vol. 126, pp. 80-121.

_____ (1870), 'Ceteosaurus', *Athenaeum*, vol. i, p. 454.

_____ (1871a), *Geology of Oxford and the valley of the Thames*, Clarendon Press, Oxford.

_____ (1871b), 'Geology', in Herschel, J.F.W. original editor, 4th edn superintended by R. Main, *A manual of scientific enquiry*, Murray, London, pp. 248-75.

_____ (1871c), 'Observations of the eclipse at Oxford, December 22, 1870', *Proceedings of the Royal Society*, vol. 19, pp. 290-1.

_____ (1872a), 'William Vernon Harcourt', *Proceedings of the Royal Society of London*, vol. 20, pp. xiii-xvii.

_____ (1872b), 'On the temperature correction of an aneroid', *BAAS 1872 report*, pp. 61-2.

_____ (1873a), 'Sedgwick', *Nature*, vol. 7, pp. 257-9.

_____ (1873b), 'Presidential address, section C', *BAAS 1873 report*, pp. 70-5.

_____ (1873c), 'On the ammonitic septa in relation to geological time', *BAAS 1873 report*, p. 86.

_____ (1873d), 'On the ammonitic spiral in reference to the power of floatation attributed to the animal', *BAAS 1873 report*, pp. 85-6.

_____ (1873e), 'The Sub-Wealden exploration – important discovery', *Geological magazine*, vol. 10, p. 527.

_____ (1875), *Illustrations of the geology of Yorkshire; or, a description of the strata and organic remains. Part 1. The Yorkshire coast, third edition, edited by R. Etheridge*, Murray, London.

_____ (1885), *Manual of geology theoretical and practical, edited by Etheridge, R. and Seeley H.G.*, 2 vols, Griffin, London.

Pickering, J. (1997), 'William J. Burchell's South African mammal collection, 1810-1815, *Archives of natural history*, vol. 24, pp. 311-26.

_____ (1998), 'William John Burchell's travels in Brazil, 1825-1830, with details of the surviving mammal and bird collections', *Archives of natural history*, vol. 25, pp. 237-65.

Pickstone, J.V. (1988), 'Science in France', *History of science*, vol. 26, pp. 201-11.

_____ (1994), 'Museological science? The place of the analytical/comparative in nineteenth-century science, technology and medicine', *History of science*, vol. 32, pp. 111-38.

_____ (1995), 'Past and present knowledges in the practice of the history of science', *History of science*, vol. 33, pp. 203-24.

_____ (2000), *Ways of knowing. A new history of science, technology and medicine*, Manchester University Press, Manchester.

Playfair, J. (1802), *Illustrations of the Huttonian theory of the earth*, Creech, Cadell, and Davies, Edinburgh.

Plot, R. (1677), *The natural history of Oxfordshire*, The Theatre, Oxford.

Porter, R.S. (1977), *The making of geology. Earth science in Britain 1660-1815*, Cambridge University Press, Cambridge.

_____ (1982a), 'Charles Lyell: the public and private faces of science', *Janus*, vol. 69, pp. 29-50.

_____ (1982b), 'The Natural Sciences Tripos and the "Cambridge school of geology", 1850-1914', *History of universities*, vol. 2, pp. 193-216.

Portlock, J.E. (1843), *Report on the geology of the county of Londonderry, and of parts of Tyrone and Fermanagh*, Milliken, Dublin.

_____ (1858), 'Presidential address', *QJGS*, vol. 14, pp. lxxix-clxiii.

Powell, B. (1838), *The connexion of natural and divine truth; or the study of the inductive philosophy considered as subservient to theology*, Parker, London.

_____ (1860), 'On the study of the evidences of Christianity' in Temple, F. et al., *Essays and reviews*, Parker, London; 9th edn, 1861, Longman, London, pp. 94-144.

Prestwich, G.A. (1899), *Life and letters of Sir Joseph Prestwich*, Blackwood, Edinburgh and London.

Prestwich, J. (1840), 'On the geology of Coalbrook Dale', *TGSL*, vol. 5, pp. 413-95.

_____ (1851), *A geological enquiry respecting water-bearing strata of the country around London*, Van Voorst, London.

_____ (1875), *The past and future of geology. An inaugural lecture given on January 29, 1875*, Macmillan, London.

Pritchard, A. (1897), *Charles Pritchard. Memoirs of his life. With an account of his theological work by the Lord Bishop of Worcester and of his astronomical work by professor H.H. Turner*, Seeley, London.

Pritchard, C. (1868-9), 'Cooke', *Monthly notices of the Royal Astronomical Society*, vol. 29, pp. 130-5.

Procter, W. (1855), 'Report of the proceedings of the Yorkshire Antiquarian Club, in the excavation of barrows from the year 1849', *Proceedings of the YPS*, vol. 1, pp. 176-89.

Proctor, R.A. (1869), *Chart of Mars: from 27 drawings by Mr Dawes*.

Pusey, E.B. (1865), 'The spirit in which the researches of learning and science should be applied to the study of the Bible', in *Authorized report of the proceedings of the Church Congress held at Norwich ... October 3rd, 4th, and 5th, 1865*, Cundall and Miller, Norwich, 1866, pp. 181-90, 320.

_____ (1866), *The miracles of prayer. A sermon preached before the University in the cathedral church of Christ, in Oxford, on Septuagesima Sunday, 1866*, printed in *Ten sermons preached before the University of Oxford between 1864-1879*, Parker, Oxford, 1880.

Pyrah, B.J. (1988), *The history of the Yorkshire Museum and its geological collections*, Sessions, York.

Quenstedt, F.A. (1858), *Der Jura*, Laupp, Tübingen.

Ramm, H. (1971), 'The Yorkshire Philosophical Society and archaeology, 1822-55', *YPS 1971 report*, pp. 66-73.

Ramsay, A.C. (1855), 'On the occurrence of angular, subangular, polished, and striated fragments and boulders in the Permian breccia of Shropshire, Worcestershire, &c; and on the probable existence of glaciers and ice-bergs in the Permean period', *QJGS*, vol. 11, pp. 185-205.

_____ (1860), *The old glaciers of Switzerland and north Wales*, Longman, London.

_____ (1862), 'On the glacial origin of certain lakes in Switzerland, the Black Forest, Great Britain, Sweden, north America, and elsewhere', *QJGS*, vol. 18, pp. 185-204.

_____ (1874), 'On the comparative value of certain geological ages (or groups of formations) considered as items of geological time', *Proceedings of the Royal Society*, vol. 22, pp. 145-8, 334-43.

_____ (1894), *The physical geology and geography of Great Britain*, 6th edn, Stanford, London.

Richards, E. (1989), 'Huxley and woman's place in science: the "woman question" and the control of Victorian anthropology', in J.R. Moore (ed.), *History, humanity and evolution. Essays for John C. Greene*, Cambridge University Press, Cambridge, pp. 253-84.

Robinson T.R. and Grubb, T. (1869), 'Description of the great Melbourne telescope', *Philosophical transactions*, vol. 159, pp. 127-62.

Rogers, E. (1896), *Life and letters of William Barton Rogers edited by his wife*, Houghton Mifflin, Boston and New York.

Rolleston, G. (1875), 'On the people of the long barrow period', *Journal of the Anthropological Institute*, vol. 5, pp. 120-73.

_____ (1884), *Scientific papers and addresses*, Clarendon Press, Oxford.

_____ and Smith, H.J.S. (1874), 'In memoriam John Phillips', *Proceedings of the Ashmolean Society*, pp. 2-11.

Rorison, G. (1862), 'The creative week', in Goulburn, E.M. et al., *Replies to 'Essays and reviews'* ..., Parker, Oxford and London, pp. 277-345.

Rowell, G.A. (1870), *A list of donations to the antiquarian and ethnological collections in the Ashmolean from the year 1836 to the end of the year 1868*, Ashmolean Museum, Oxford.

Rowlinson, J.S. (1971), 'The theory of glaciers', *Notes and records of the Royal Society of London*, vol. 26, pp. 189-204.

Royal Commission on historical monuments (1975), *An inventory of the historical monuments in the city of York. Volume IV. Outside the city walls east of the Ouse*, HMSO, London.

Rudwick, M.J.S. (1963), 'The foundation of the Geological Society of London: its scheme for cooperative research and its struggle for independence', *British journal for the history of science*, vol. 1, pp. 325-55.

_____ (1969), 'The glacial theory', *History of Science*, vol. 8, pp. 136-57.

_____ (1974), 'Poulett Scrope on the volcanoes of Auvergne: Lyellian time and political economy', *British journal for the history of science*, vol. 7, pp. 205-42.

_____ (1975a), 'Charles Lyell, F.R.S. (1797-1875) and his London lectures on geology, 1832-33', *Notes and records of the Royal Society of London*, vol. 29, pp. 231-63.

_____ (1975b), 'Caricature as a source for the history of science: De la Beche's anti-Lyellian sketches of 1831', *Isis*, vol. 66, pp. 534-60.

_____ (1976a), *The meaning of fossils. Episodes in the history of palaeontology*, 2nd edn., Science History Publications, New York.

_____ (1976b), 'The emergence of a visual language for geological science, 1760-1840', *History of science*, vol. 14, pp 149-95.

_____ (1978), 'Charles Lyell's dream of a statistical palaeontology', *Palaeontology*, vol. 21, pp. 225-44.

_____ (1985), *The great Devonian controversy. The shaping of scientific knowledge among gentlemanly specialists*, Chicago University Press, Chicago and London.

_____ (ed.) (1990-1), 'Introduction' to facsimile reprint of Lyell, C., *Principles of geology*, 1st edn, University of Chicago Press, Chicago and London, vol. 1, pp. vii-lviii.

_____ (1996), 'Cuvier and Brongniart, William Smith, and the reconstruction of geohistory', *Earth Sciences History*, vol. 15, pp. 25-36.

_____ (1997), *Georges Cuvier, fossil bones, and geological catastrophes. New translations and interpretations of the primary texts*, University of Chicago Press, Chicago and London.

Rupke, N.A. (1983), *The great chain of history. William Buckland and the English school of geology (1814-1849)*, Clarendon Press, Oxford.

_____ (1994), *Richard Owen: Victorian naturalist*, Yale University Press, New Haven and London.

_____ (1997), 'Oxford's scientific awakening and the role of geology', in Brock, M.G. and Curthoys, M.C. (eds), *The history of the University of Oxford. Volume VI. Nineteenth-century Oxford, Part I*, Clarendon Press, Oxford, pp. 543-62.

Ruskin, J. (1856), *Modern painters. Volume 4*, reproduced in Cook, E.T. and Wedderburn, A. (eds), *Complete works of John Ruskin*, Allen, London, 1903, vol. 6.

Russell, C.A. (1996), *Edward Frankland. Chemistry, controversy and conspiracy in Victorian England*, Cambridge University Press, Cambridge.

Sabine, E. (1838), 'A memoir on the magnetic isoclinal and isodynamic lines in the British Isles, from observations by professors Humphrey Lloyd and John Phillips, Robert Were Fox, Captain James Clark Ross, and Major Edward Sabine', *BAAS 1838 report*, pp. 49-196.

_____ (1861), 'Report on the repetition of the magnetic survey of England, made at the request of the General Committee of the British Association', *BAAS 1861 report*, pp. 250-79.

_____ (1864), 'Presidential address', *Proceedings of the Royal Society of London*, vol. 13, pp. 22-39.

Salter, J.W. (1864-83), *British trilobites*, Palaeontographical Society, London.

Schaffer, S. (1989), 'The nebular hypothesis and the science of progress', in, Moore, J.R. (ed.), *History, humanity and evolution. Essays for John C. Greene*, Cambridge University Press, Cambridge, pp. 131-64.

_____ (1998a), 'On astronomical drawing', in Jones, C.A. and Galison, P. (eds), *Picturing science, producing art*, Routledge, London, pp. 441-74.

_____ (1998b), 'The leviathan of Parsonstown: literary technology and scientific representation', in Lenoir, T. (ed.), *Inscribing science. Scientific texts and the materiality of communication*, Stanford University Press, Stanford, pp. 182-222.

Scott, R.H. (1887), *Elementary meteorology*, Kegan Paul, London.

Scott, R.H. and Galloway, W. (1872), 'On the connexion between explosions in collieries and weather', *Proceedings of the Royal Society*, vol. 20, pp. 292-305.

Scrope, G.J.P. (1862), *Volcanoes. The character of their phenomena, their share in the structure and composition of the surface of the globe, and their relation to its internal forces. With a descriptive catalogue of all known volcanoes and volcanic formations*, Longman, London.

_____ (1869), 'Phillips' Vesuvius', *Geological magazine*, vol. 6, pp. 122-5.

Secord, J.A. (1982), 'King of Siluria: Roderick Murchison and the imperial theme in nineteenth-century British geology', *Victorian studies*, vol. 25, pp. 413-42.

_____ (1985), 'John W. Salter: the rise and fall of a Victorian palaeontological career', in Wheeler, A. and Price, J.H. (eds), *From Linnaeus to Darwin: commentaries on the history of biology and geology*, Society for the History of Natural History, London, pp. 61-75.

_____ (1986a), *Controversy in Victorian geology: the Cambrian-Silurian dispute*, Princeton University Press, Princeton.

_____ (1986b), 'The Geological Survey of Great Britain as a research school, 1839-1855', *History of science*, vol. 24, pp. 223-75.

_____ (ed.) (1997), 'Introduction' to Lyell, C., *Principles of geology*, 1st edn, Penguin, Harmondsworth, pp. ix-xliii.

_____ (2000), *Victorian sensation. The extraordinary publication, reception, and secret authorship of Vestiges of the natural history of creation*, University of Chicago Press, Chicago and London.

Sedgwick, A. (1826), 'On the classification of the strata which appear on the Yorkshire coast', *Annals of philosophy*, vol. 11, pp. 339-62.

_____ (1829), 'On the geological relations and internal structure of the magnesian limestone and the lower portions of the new red sandstone series in their range through Nottinghamshire, Derbyshire, Yorkshire and Durham to the southern extremity of Northumberland', *TGSL*, vol. 3, pp. 37-124.

_____ (1830), 'Presidential address', *PGS*, 1834, vol. 1, pp. 187-212.

_____ (1831), 'Presidential address', *PGS*, 1834, vol. 1, pp. 270-316.

_____ (1833), *A discourse on the studies of the University*, Parker, London.

_____ (1835a), 'Remarks on the structure of large mineral masses, and especially on the chemical changes produced in the aggregation of stratified rocks during different periods after their deposition', *TGSL*, vol. 3, pp. 461-86.

_____ (1835b), 'Description of a series of longitudinal and transverse sections through a portion of the carboniferous chain between Penigent and Kirkby Stephen', *TGSL* (1836), vol. 4, pp. 69-102.

_____ (1838), 'A synopsis of the English series of stratified rocks inferior to the old red sandstone', *PGS*, vol. 2, pp. 675-85.

_____ (1843), 'Outline of geological structure of north Wales', *PGS*, 1846, vol. 4, pp. 212-24.

_____ (1845), 'Natural history of creation', *Edinburgh review*, vol. 82, pp. 1-85.

_____ (1850), *A discourse on the studies of the University of Cambridge*, 5th edn, Parker, London.

_____ and McCoy, F. (1855), *A synopsis of the classification of the British palaeozoic rocks, with a systematic description of the British palaeozoic fossils in the Geological Museum of the University of Cambridge*, Parker, London.

_____ and Murchison, R.I. (1839a), 'Classification of the older stratified rocks of Devonshire and Cornwall', *Philosophical magazine*, vol. 14, pp. 241-60.

_____ (1839b), 'Supplementary remarks on the "Devonian" system of rocks', *Philosophical magazine*, vol. 14, pp. 354-8.

Shairp, J.C., Tait, P.G., and Adams-Reilly, A. (1873), *Life and letters of James David Forbes*, Macmillan, London.

Sharpe, D. (1847), 'On slaty cleavage', *QJGS*, vol. 3, pp. 74-105.

_____ (1849), 'On slaty cleavage', *QJGS*, vol. 5, pp. 111-29.

Sheppard, T. (1917), 'William Smith: his maps and memoirs', *PYGS*, vol. 19, pp. 75-253.

_____ (1922), 'Martin Simpson and his geological memoirs', *PYGS*, vol. 19, pp. 298-315.

_____ (1934), 'John Phillips', *PYGS*, vol. 22, pp. 153-87.

Shortland, M. (1994), 'Darkness visible: underground culture in the golden age of geology', *History of science*, vol. 32, pp. 1-61.

Simcock, A.V. (1984), *The Ashmolean Museum and Oxford Science 1683-1983*, Museum of the History of Science, Oxford.

Simson, R. (ed.) (1756), *The elements of Euclid, viz. the first six books, together with the eleventh and twelfth*, Foulis, Glasgow.

Smiles, S. (1859), *Self-help with illustrations of conduct and perseverance*, Murray, London.

_____ (1874), *Life of a Scottish naturalist, Thomas Edward*, Murray, London.

_____ (1878), *Robert Dick baker, of Thurso geologist and botanist*, Murray, London.

Smith, A.Z. (1986), *A history of the entomological collections in the University Museum Oxford with lists of archives and collections*, Clarendon Press, Oxford.

Smith, C.W. (1985), 'Geologists and mathematicians: the rise of physical geology', in Harman, P.M. (ed.), *Wranglers and physicists. Studies on Cambridge physics in the nineteenth century*, Manchester University Press, Manchester, pp. 49-83.

_____ (1998), *The science of energy: a cultural history of energy physics in Victorian Britain,* Athlone, London.

Smith, C.W. and Wise, M.N. (1989), *Energy and empire: a biographical study of Lord Kelvin*, Cambridge University Press, Cambridge, 1989.

Smith, J.P. (1839), *On the relation between the holy scriptures and some parts of geological science*, Jackson and Walford, London.

Smith, W. (1806), *Observations on the utility, form and management of water meadows, and the draining and irrigating of peat bogs*, Bacon, Norwich.

_____ (1815a), *A delineation of the strata of England and Wales, with part of Scotland*, Cary, London.

_____ (1815b), *A memoir to the map and delineation of the strata of England and Wales, with part of Scotland*, Cary, London.

_____ (1816-19), *Strata identified by organized fossils, containing prints on coloured paper of the most characteristic specimens in each stratum*, Arding and Sowerby, London.

_____ (1817a), *Stratigraphical system of organised fossils, with reference to the specimens of the original geological collection in the British Museum: explaining their state of preservation and their use in identifying the British strata*, Williams, London.

_____ (1817b), *Geological section from London to Snowdon, showing the varieties of the strata, and the correct altitudes of the hills*, Cary, London.

_____ (1818a), *Geology of England: Mr. Wm. Smith's claims*, Smith, London, reproduced in Sheppard, T. (1917), 'William Smith: his maps and memoirs', *PYGS*, vol. 19, pp. 213-18, original in Scarborough Public Library.

_____ (1818b), *Report on the state of Monmouth,* Heath, Monmouth.

_____ (1818-19), One report (1818) and three plans (1818-9) for the proposed Aire and Don canal, and Went branch, details in Eyles, J.M. (1969), 'William Smith (1769-1839): a bibliography …', *Journal of the society for the bibliography of natural history*, vol. 5, p. 101.

_____ (1827), 'On retaining water in rocks for summer use', *Philosophical magazine*, vol. 1, pp. 415-17.

Sorby, H.C. (1853), 'On the origin of slaty cleavage', *Edinburgh new philosophical journal*, vol. 55, pp. 137-50.

_____ (1858), 'On the microscopical structure of crystals, indicating the origin of minerals and rocks', *QJGS*, vol. 14, pp. 453-500.

_____ (1897), *Fifty years of scientific research*, Independent Press, Sheffield.

Sowerby, J. (1812, 1818, 1821, 1823), *The mineral conchology of Great Britain*, 4 vols, Meredith, London.

Spearman, T.D. (1981), 'Humphrey Lloyd, 1800-1881', *Hermathena*, pp. 37-52.

Spring-Rice, T. (1836), 'The Budget: speech of the Right Honourable Thomas Spring Rice … in the House of Commons on Friday, May 6, 1836', in *Mirror of Parliament*, London.

Stafford, R.A. (1989), *Scientist of empire. Sir Roderick Murchison, scientific exploration and Victorian imperialism*, Cambridge University Press, Cambridge.

Stephen, L. (ed.) (1902), *Letters of John Richard Green*, Macmillan, London.

Stewart, B. (1868), 'An account of certain experiments, on aneroid barometers, made at Kew Observatory', *Proceedings of the Royal Society*, vol. 16, pp. 472-80.

Stocking, G.W. (1987), *Victorian anthropology,* Collier Macmillan, London.

Stokes, G.G. (1869), ' Presidential address', *BAAS 1869 report*, pp.lxxxix-cv.

Stone, L. (1975), 'The size and composition of the Oxford student body 1580-1910', in Stone, L. (ed.), *The University in society. Volume 1*, Princeton University Press, Princeton, pp. 3-110.

Strickland, H.E. (1851), 'On the elevatory forces which raised the Malvern Hills', *Philosophical magazine*, vol. 2, pp. 358-65.

_____ (1852), *On geology in relation to the studies of the University of Oxford*, Vincent and Bell, Oxford.

Sydenham. P.H. (1979), *Measuring instruments: tools of knowledge and control*, Peregrinus, Stevenage.

Symons, G.J. (1864), 'On the fall of rain in the British Isles during the years 1862 and 1863', *BAAS 1864 report*, pp. 367-407.

_____ (1867), *Rain: how, when, where, why it is measured*, Stanford, London.

Tate, R. and Blake, J.F. (1876), *The Yorkshire lias*, Van Voorst, London.

Taylor, F.S. (1952), 'The teaching of science at Oxford in the nineteenth century', *Annals of science*, vol. 8, pp. 82-112.

Temple, F. et al. (1860), *Essays and reviews*, Parker, London: 9th edn, 1861, Longman, is cited.

Temple, F. (1860), 'The education of the world,' in Temple et al. 1860, pp. 1-49.

Thackeray, A.D. (1972), *The Radcliffe Observatory 1772-1972*, Radcliffe Trust, London.

Theakston, S.W. (1841), *Guide to Scarborough*, Theakston, Scarborough.

Thompson, W.H. (1904), 'John Phillips, the geologist', *Gentleman's magazine*, vol. 296, pp. 554-63.

Thomson, E.H. (1919), *The life and letters of William Thomson Archbishop of York*, Lane, London.

Thomson, W. (1819-90) (ed.) (1861), *Aids to faith; a series of theological essays*, Murray, London.

_____ (1819-90), (1866), *The inaugural address of the archbishop of York, president of the [Yorkshire Philosophical] Society ... 17 January 1866*, Murray, London.

_____ (1824-1907), (1855), 'On the use of observations of terrestrial temperature for the investigation of absolute dates in geology', *BAAS 1855 report*, pp. 18-19.

_____ (1824-1907), (1862), 'On the age of the sun's heat', *Macmillan's magazine*, vol. 5, pp. 288-93.

_____ (1824-1907), (1863), 'On the secular cooling of the earth', *Philosophical magazine*, vol. 25, pp. 1-14.

_____ (1824-1907), (1871a), 'Of geological dynamics', *Transactions of the Glasgow Geological Society*, vol. 3, pp. 215-40.

_____ (1824-1907), (1871b), 'Presidential address', *BAAS 1871 report*, pp. lxxxiv-cv.

Thorp, W. (1841), 'On the agricultural geology of part of the Wolds district of Yorkshire and of the oolite in the neighbourhood of north and south Cave', *Transactions of the Yorkshire Agricultural Society*, vol. 4, pp. 37-139.

Thurnam, J. (1849), 'Description of an ancient tumular cemetery, probably of the Anglo-Saxon period, at Lamel Hill, near York', *Archaeological journal*, vol. 6, pp. 27-39, 123-36.

_____ (1855), 'Description of an ancient tumular cemetery, probably of the Anglo-Saxon period, at Lamel Hill, near York', [read June 1848], *Proceedings of the PYS*, vol. 1, pp. 98-105.

Tiddeman, R.H. (1874), 'Second report of the committee for the exploration of the Settle Caves (Victoria Cave)', *BAAS 1874 report*, pp. 133-8.

_____ (1890), 'Physical history of the carboniferous rocks in upper Airedale', *PYGS*, vol. 11, pp. 482-92.

Timbs, J. (1866), 'John Phillips', *The year-book of facts in science and art*, Lockwood, London, pp. 3-9.

Topham, J.R. (1992), 'Science and popular education in the 1830s: the role of the Bridgewater Treatises', *British journal for the history of science*, vol. 25, pp. 397-430.

_____ (1998), 'Beyond the "Common Context": the production and reading of the *Bridgewater Treatises*', *Isis*, vol. 89, pp. 233-62.

_____ (2000), 'Scientific publishing and the reading of science in nineteenth-century Britain: a historiographical survey and guide to sources', *Studies in history and philosophy of science*, vol. 31, pp. 559-612.

Topley, W. (1872), 'On the sub-Wealden exploration', *BAAS 1872 report*, pp. 122-3.

Torrens, H.S. (1975), 'The Bath geological collections', *Geological Curators Group Newsletter*, vol. 1, pp. 87-124.

_____ (1978), 'The Sherborne School Museum and the early collections and publications of the Dorset Natural History and Antiquarian Field Club', *Proceedings of the Dorset Natural History and Archaeological Society*, vol. 98, pp. 32-42.

_____ (1990), 'The scientific ancestry and historiography of *The Silurian System*', *Journal of the Geological Society*, vol. 147, pp. 657-62.

_____ (1992a), 'William Smith – the truth', *Geoscientist*, vol. 2. p. 30.

_____ (1992b), 'When did the dinosaur get its name?' *New Scientist*, 4 April, pp. 40-2.

_____ (1994), 'Patronage and problems: Banks and the earth sciences', in Banks, R.E.R. (ed.), *Sir Joseph Banks: a global perspective*, Royal Botanic Gardens, Kew, pp. 49-75.

_____ (1997), 'Politics and palaeontology: Richard Owen and the invention of dinosaurs', in Farlow, J.O. and Brett-Surman, M.K. (eds), *The complete dinosaur*, Indiana University Press, Indianapolis, pp. 173-91.

_____ (1998), 'Coal hunting at Bexhill 1805-1811: how the science of stratigraphy was ignored', *Sussex archaeological collections*, vol. 136, pp. 177-91.

_____ (1999), 'William Edmond Logan's geological apprenticeship in Britain 1831-1842', *Geoscience Canada*, vol. 26, pp. 97-110.

_____ (2001), 'Timeless order: William Smith (1769-1839) and the search for raw materials 1800-1820', in Lewis, C.L.E. and Knell, S.J. (eds), *The age of the earth: from 4004 BC to AD 2002*, Geological Society of London, London, pp. 61-83.

_____ (2003), 'Introduction' to reprint of Phillips, 1844a, Royal Literary and Scientific Institution, Bath, pp. xi-xxxviii.

_____ and Ford, T.D. (1989), 'John Farey (1766-1826): an unrecognised polymath', in Farey, J. *General view of the agriculture and minerals of Derbyshire, 1811*, vol. 1, (reprinted by Peak District Mines Historical Society, Matlock, 1989), pp. 1-28.

_____ and Getty, T.A. (1984), 'Louis Hunton (1814-1838) – English pioneer in ammonite biostratigraphy', *Earth sciences history*, vol. 3, pp. 58-68.

_____ and Winston, J.E. (2002), 'Eliza Catherine Jelly (28th September 1829 – 3rd November 1914): pioneer female bryozoologist', *Annals of bryozoology*, pp. 299-325.

Townsend, J. (1813, 1815), *The character of Moses established for veracity as an historian: recording events from the creation to the deluge*, 2 vols, Longman, London.

Trimmer, J. (1845), 'Review of Phillips' Memoirs of Smith', *North British review*, vol. 4. pp. 96-125.

Tuckwell, W. (1865), *Practical remarks on the teaching of physical science in schools. With letters from Charles Daubeny; Henry Acland; John Phillips*, Rivington, London.

_____ (1900), *Reminiscences of Oxford*, Cassell, London.

Turner, F. M. (1974), 'Rainfall, plagues, and the Prince of Wales: a chapter in the conflict of science and religion', *Journal of British studies*, vol. 13, pp. 46-65.

_____ (1993), 'The religious and the secular in Victorian Britain', in Turner, F.M., *Contesting cultural authority. Essays in Victorian intellectual life*, Cambridge University Press, Cambridge, pp. 3-37.

Turner, G. L'E. (1983), *Nineteenth-century scientific instruments*, Sotheby, London.

Tweedale, G. (1991), 'Geology and industrial consultancy: Sir William Boyd Dawkins (1837-1929) and the Kent coalfield', *British journal for the history of science*, vol. 24, pp. 435-52.

Twyman, M. (1967), 'The lithographic hand press 1796-1850', *Journal of the Printing Historical Society*, no. 3, pp. 3-50.

_____ (1970), *Lithography 1800-1850. The techniques of drawing on stone in England and France and their application in works of topography*, Oxford University Press, London.

_____ (1972), 'Lithographic stone and the printing trade in the nineteenth century', *Journal of the Printing Historical Society*, no. 8, pp. 1-41.

_____ (ed.), (1976), *Henry Bankes' treatise on lithography. Reprinted from the 1813 and 1816 editions*, Printing Historical Society, London.

Tyndall, J. (1856a), 'Comparative view of the cleavage of crystals and slate rocks', *Philosophical magazine*, vol. 12, pp. 35-48.

_____ (1856b), 'Observations on the theory of the origin of slaty cleavage by H.C. Sorby', *Philosophical magazine*, vol. 12, pp. 129-35.

_____ (1858), 'On some physical properties of ice', *Philosophical transactions*, vol. 148, pp. 211-30.

_____ and Huxley, T.H. (1857), 'On the structure and motion of glaciers', *Philosophical transactions*, vol. 147, pp. 327-46.

University of Oxford (1872), *Notice of the board of studies for the natural science school of the University of Oxford*, in Gu P, 65.

Van Riper, A.B. (1993), *Men among the mammoths. Victorian science and the discovery of human prehistory*, University of Chicago Press, Chicago and London.

Vernon, H.M. and Vernon, K.D. (1909), *A history of the Oxford Museum*, Clarendon Press, Oxford.

Vincent, E.A. (1994), *Geology and mineralogy at Oxford 1860-1986. History and reminiscence*, Department of Earth Sciences, Oxford.

Voltz, P.L. (1830), *Observations sur les bélemnites*, Levrault, Paris.

Ward, J.C. (1877), 'Jonathan Otley, the geologist and guide', *Transactions of the Cumberland Association for the Advancement of Literature and Science*, vol. 2, pp. 125-69.

Walcott, J. (1779), *Descriptions and figures of petrifications found in the quarries, gravel pits, etc. near Bath*, Hazard, Bath.

Warner, R. (1801), *The history of Bath*, Cruttwell, Bath.

Way, A. (1859), 'Enumeration of blocks or pigs of lead and tin, relics of roman metallurgy, discovered in Great Britain', *Archaeological journal*, vol. 16, pp. 22-40.

Webster, T. (1814), 'On the freshwater formations of the Isle of Wight, with some observations on the strata over the chalk in the south-east part of England', *TGSL*, vol. 2, pp. 161-254.

_____ (1825), 'Reply to Dr Fitton's paper ... respecting the geological relations of the beds between the chalk and the Purbeck limestone in the south-east of England', *Annals of philosophy*, vol. 9, pp. 33-50.

Wellbeloved, C. (1828), *The large extent of the subjects of knowledge, a motive to diffidence and humility. An address delivered to the members of the York Mechanics' Institute, on Thursday, the 27th of March, 1828*, Hargrove, York.

_____ (1829), *Account of the ancient and present state of the Abbey of St Mary, York, and of the discoveries made in recent excavations conducted by the Yorkshire Philosophical Society*, Society of Antiquaries, London.

_____ (1842), *Eburacum, or York under the Romans*, Sunter and Sotheran, York.

Werner, A.G. (1809), *New theory of the formation of veins*, Constable, Edinburgh.

Whewell, W. (1845), *Indications of the Creator. Extracts, bearing upon theology, from the history and the philosophy of the inductive sciences*, Parker, London.

_____ and Henslow, J.S. (eds) (1833), *Lithographed signatures of the members of the British Association for the Advancement of Science, who met at Cambridge, June MDCCCXXXIII*, Cambridge University Press, Cambridge.

Wilberforce, S. (1861), 'Review of Essays and reviews', *Quarterly review*, vol. 109, pp. 248-305.

Willett, H. (1878), *The record of the Sub-Wealden exploration*, Smith, Brighton.

_____ and Topley, W. (1873, 1874), 'First and second reports of the Sub-Wealden Exploration Committee', *BAAS 1873 and 1874 reports*, pp. 490-5, 21-2.

Williamson, W.C. (1837), 'On the distribution of fossil remains on the Yorkshire coast, from the lower lias to the Bath oolite inclusive', *TGSL*, vol. 5, pp. 223-42.

_____ (1884), 'Biographical notices of eminent geologists: III. John Williamson', *PYGS*, vol. 8, pp. 295-313.

_____ (1896), *Reminiscences of a Yorkshire naturalist*, Redway, London.

Wilson, D.B. (1984), 'A physicist's alternative to materialism: the religious thought of George Gabriel Stokes', *Victorian studies*, vol. 28, pp. 69-96.

Wilson, G. and Geikie, A. (1861), *Memoir of Edward Forbes*, Macmillan, Cambridge and London.

Wilson, L.G. (1972), *Charles Lyell. The years to 1841: the revolution in geology*, Yale University Press, New Haven and London.

_____ (1998), *Lyell in America: transatlantic geology, 1841-1853*, Johns Hopkins University Press, Baltimore.

Winch, N.J. (1821), 'Observations on the eastern part of Yorkshire', *TGSL*, vol. 5, pp. 545-57.

Wood, H.H. (1865), *On the theory of development, and the antiquity of man: a letter to J. Phillips*, Rivington, London.

_____ (1867), 'Lake dwellings of Switzerland', *Contemporary review*, vol. 4, pp. 380-94.

Wood, S.V. elder (1848-82), *The crag mollusca*, Palaeontographical Society, London.

Wood, S. V. junior and Rome, J.L. (1868), 'On the glacial and postglacial structure of Lincolnshire and south-east Yorkshire', *QJGS*, vol. 24, 146-84.

Woodward, H.B. (1907), *The history of the Geological Society of London*, Geological Society, London.

Woodward, J. (1728, 1729), *An attempt towards a natural history of the fossils of England in a catalogue*, 2 vols, Fayram, London.

Wright, T. (1857-80), *The British fossil echinodermata of the oolitic formations*, Palaeontographical Society, London.

_____ (1860), 'On the subdivisions of the inferior oolite in the south of England, compared with the equivalent beds of that formation on the Yorkshire coast', *QJGS*, vol. 16, pp. 1-48.

_____ (1864-82), *A monograph on the British fossil echinodermata from the cretaceous formations*, Palaeontographical Society, London.

Wylie, W.M. (1852), *Fairford graves. A record of researches in an Anglo-Saxon burial place in Gloucestershire*, Parker, Oxford.

Yanni, C. (1999), *Nature's museums. Victorian science and the architecture of display*, Athlone Press, London.

Yeo, R. (1985), 'An idol of the market place: Baconianism in nineteenth-century Britain', *History of science*, vol. 23, pp. 251-98.

_____ (1991), 'Reading encyclopaedias: science and the organization of knowledge in British dictionaries of arts and sciences, 1730-1850', *Isis*, vol. 82, pp. 24-49.

_____ (1993), *Defining science: William Whewell, natural knowledge, and public debate in early Victorian England*, Cambridge University Press, Cambridge.

Yorkshire Philosophical Society (1824), *Objects and laws of the Yorkshire Philosophical Society: with the annual report for 1823*, Alexander, York.

Young, G. (1817), *A history of Whitby, and Streoneshalh; with a statistical survey of the vicinity to the distance of twenty five miles*, Clark and Medd, Whitby.

_____ (1838), *Scriptural geology; or an essay on the high antiquity ascribed to the organic remains imbedded in stratified rocks*, Simpkin and Marshall, London.

_____ and Bird, J. (1822, 1828), *A geological survey of the Yorkshire coast: describing the strata and fossils occurring between the Humber and the Tees, from the German Ocean to the Plain of York*, Clark, Whitby (1828 edn, Kirby, Whitby).

Zittel, K.A. von (1901), *History of geology and palaeontology to the end of the nineteenth century. Translated by Ogilvie-Gordon, M.M.*, Scott, London.

Index

Aberdeen, Earl of 216, 234

Acland, H.W. 1, 247, 250, 261, 297, 307, 310-12, 315-17, 319-21, 323, 367, 370

Adaptation 54, 80, 146, 335, 355

Adare, Lord 199-200

Agassiz, J.L.R. 102, 109, 152, 213, 217-18, 233, 279

Aire-Don canal 22-5

Airy, G.B. 75, 79, 219, 297, 302

Akerman, J.Y. 309

Albert, Prince 126, 231, 354

Alderson, J. 80

Allis, T. 102, 367

American Association for the Advancement of Science 216-17

American Philosophical Society 4

Anglicanism, liberal 133-5, 149

Ansted, D.T. 136, 225, 354

Anthropology 335-6

Apjohn, J. 197-8

Arago, D.F. 121

Archaeological Institute 229-30, 291

Archaeology 106, 209-10, 223, 228-30, 232-3, 262-3, 290-1, 308-9, 331, 337

Archiac, É.J.A.D. de St-Simon d' 172

Argyll, Duke of 339

Ashmolean Society 4, 247, 255, 299, 329-30

Astronomy 75, 209-10, 217, 220-2, 249, 273, 291-8, 334, 336, 344-5

Athenaeum 4, 126, 219, 281, 281, 296, 332, 378

Athenaeum club 172

Atkinson, James 41-2, 75

Atkinson, John 31, 48, 56

Audubon, J.J. 59

Aurora borealis 79, 105, 209, 219

Austen, R.A.C. 164-5

Auvergne 81-3, 127, 243, 278, 280, 369

Aveline W. T., 184, 186-7, 260

Babbage, C. 88, 90, 104, 119, 137

Bache, A.D. 123

Backhouse, T. 50, 53

Bacon, F. 73

Baconian induction 42, 55, 63-4, 78, 106, 111, 116, 122, 147, 299, 334

Baily, F. 137

Baily, W.H. 185, 194

Baines, E. 31

Baines, H. 53, 73, 106, 208

Baker, A.E. 231

Baker, G. 231

Bakewell, R. 29, 139

Bankes, H. 25-6

Banks, J. 13, 17, 20, 27-9, 34

Baring, F.T. 162,165

Barker, E. 56

Barrande, J. 257

Barry, C. elder 162-3

Barry, C. junior 297-8

Barry, E.M. 310

Bathurst, W.H. 57

Beale, D. 268

Bean, W. 50, 62

Beaumont, J.B.A.L.L.É de 114, 145, 191, 193

Beckwith, S. 208

Beer, W. 221, 292

Belcher, H. 117

Belemnites 83, 146, 219, 275-8

Bell, G.C. 259

Bell, T. 135-6

Bernard, M. 1

Bessel, F.W. 215

Bielsbeck excavation 76-7, 107

Bigsby, J.J. 257

Bilton, W. 127

Binney, E.W. 257, 341

Bird, J. 25, 43, 62, 65

Birt, W.R. 220, 292, 294, 334, 337

Bischof, C.G.C. 257, 334

Black, A. 142, 232

Blackwell, J. 202, 225, 227-8

Blainville, H.M.D. de 197, 219, 276

Blake, J.F. 287

Blau, E. 312

Blomfield, C.J. 133-5

425

Milton Keynes UK
Ingram Content Group UK Ltd.
UKHW031138141024
449569UK00024B/1234